Chirality in Industry II

CHIRALITY IN INDUSTRY II

Developments in the Commercial Manufacture and Applications of Optically Active Compounds

Edited by

A. N. COLLINS

ZENECA Specialities, Manchester, UK

G. N. SHELDRAKE

The Queen's University of Belfast, UK

J. CROSBY

ZENECA Pharmaceuticals, Macclesfield, UK

JOHN WILEY & SONS

Chichester • New York • Weinheim • Brisbane • Singapore • Toronto

Baffins Lane, Chichester,
West Sussex PO19 1UD, England

National 01243 779777
International (+44) 1243 779777

e-mail (for orders and customer service enquiries): cs-books@wiley.co.uk
Visit our Home Page on http://www.wiley.co.uk
or http://www.wiley.com

Other Wiley Editorial Offices

John Wiley & Sons, Inc., 605 Third Avenue,
New York, NY 10158-0012, USA

VCH Verlagsgesellschaft mbH,
Pappelallee 3, D-69469 Weinheim, Germany

Jacaranda Wiley Ltd, 33 Park Road Milton,
Queensland 4064, Australia

John Wiley & Sons (Asia) Pte Ltd, 2 Clementi Loop #02–01,
Jin Xing Distripark, Singapore 0512

John Wiley & Sons (Canada) Ltd, 22 Worcester Road,
Rexdale, Ontario M9W 1L1, Canada

Library of Congress Cataloguing-in-Publication Data

Chirality in industry : the commercial manufacture and applications of optically active compounds / edited by A. N. Collins, G. N. Sheldrake, and J. Crosby.
p. cm.
Includes bibliographical references and index.
ISBN 0 471 93595 6
1. Chirality. 2. Enantiomers—Separation. 3. Enantiomers–Biotechnology. I. Collins. A. N. (Andrew N.) II. Sheldrake, G. N. III. Crosby, J.
QP517.C57C47 1997
660′.6—dc20 92–16000
CIP

British Library Cataloguing in Publication Data

A catalogue record for this book is available from the British Library

ISBN 0 471 96680 0

Typeset in 10/12pt Times by Thomson Press (India) Ltd, New Delhi
Printed and bound in Great Britain by Biddles Ltd, Guildford, Surrey
This book is printed on acid-free paper responsibly manufactured from sustainable forestation, for which at least two trees are planted for each one used for paper production.

Contents

List of Contributors vii

Foreword xi

Preface xiii

Acknowledgements xv

1 **Introduction** 1
J. Crosby

2 **Chiral Drugs: Regulatory Aspects** 11
J. J. Blumenstein

3 **Production Methods for Chiral Non-steroidal Anti-inflammatory Profen Drugs** 19
G. P. Stahly and R. M. Starrett

4 **Synthesis of Enantiomerically Pure Nucleosides** 41
B. L. Bray, M. D. Goodyear, J. J. Partridge and D. J. Tapolczay

PHYSICAL METHODS AND CLASSICAL RESOLUTION

5 **Rational Design in Resolutions** 81
A. Bruggink

6 **Resolution Versus Stereoselective Synthesis in Drug Development: Some Case Histories** 99
B. A. Astleford and L. O. Weigel

7 **Crystal Science Techniques in the Manufacture of Chiral Compounds** 119
W. M. L. Wood

8 **Membrane Separations in the Production of Optically Pure Compounds** 157
J. T. F. Keurentjes and F. J. M. Voermans

BIOLOGICAL METHODS AND CHIRAL POOL SYNTHESES

9 **Four Case Studies in the Development of Biotransformation-based Processes** 183
R. McCague and S. J. C. Taylor

10 (*S*)-2-Chloropropanoic Acid: Developments in its Industrial Manufacture 207
S. C. Taylor

11 Development of a Full-scale Process for a Chiral Pyrrolidine 225
H.-J. Federsel

12 Development of a Multi-stage Chemical and Biological Process for an Optically Active Intermediate for an Anti-glaucoma Drug 245
A. J. Blacker and R. A. Holt

13 Synthesis and Applications of Chiral Liquid Crystals 263
D. Pauluth and A. E. F. Wächtler

14 Bio-transformation in the Production of L-Carnitine 287
Th. P. Zimmerman, K. T. Robins, J. Werlen and F. W. J. M. M. Hoeks

ASYMMETRIC SYNTHESES BY CHEMICAL METHODS

15 Asymmetric Hydrocyanation of Vinylarenes 309
A. L. Casalnuovo and T. V. Rajanbabu

16 Enantioselective Protonation in Fragrance Synthesis 335
C. Fehr

17 Thiamphenicol: a Manufacturing Process Involving a Double Inversion of Stereochemistry 353
L. Coppi, C. Giordano, A. Longoni and S. Panossian

18 Sharpless Asymmetric Epoxidation: Scale-up and Industrial Production 363
W. P. Shum and M. J. Cannarsa

19 Asymmetric Synthesis of Sulphoxides: Two Case Studies 381
P. Pitchen

20 Asymmetric Reduction of Prochiral Ketones 391
J.-C. Caille, M. Bulliard and B. Laboue

Index 403

List of Contributors

B. A. Astleford
Central Process R&D, Lilley Research Laboratories, Lilley Corporate Center 4813, Indianapolis, IN46285, USA

A. J. Blacker
ZENECA Process Technology Department, Huddersfield Works, P.O. Box A38, Leeds Road, Huddersfield, UK

J. J. Blumenstein
Janssen Research Foundation, Technical Regulatory Affairs, Titusville, NJ 08560, USA

B. L. Bray
Glaxo Wellcome Chemical Development Division, Girolami Research Centre, Research Triangle Park, NC, USA

A. Bruggink
Chemferm, P.O. Box 3933, 4800 DX Breda, The Netherlands

M. Bulliard
SIPSY, Route de Beaucouz, B.P. 79, 49242 Avrillé, France

J.-C. Caille
SIPSY, Route de Beaucouz, B.P., 79, 49242 Avrillé, France

M. J. Cannarsa
ARCO Chemical Co., 3801 West Chester Pike, Newtown Square, PA 19073, USA

A. L. Casalnuovo
DuPont Agricultural Products, Stine-Haskell Research Center, S 300/124, Newark, DE 19714, USA

L. Coppi
Zambon Group SpA, Via Lillo del Duca 10, 20091 Bresso (Milan), Italy

J. Crosby
ZENECA Pharmaceuticals, Macclesfield, Cheshire SK10 2NA, UK

H.-J. Federsel
Process Chemistry I, Astra Production Chemicals AB, S-151 85 Sodertalije, Sweden

C. Fehr
Firmenich SA, Route des Jerne 1, CH-1211, Geneva 8, Swetzerland

C. Giordano
Zambon Group SpA, Via Lillo del Duca 10, 20091 Bresso (Milan), Italy

M. D. Goodyear
Glaxo Wellcome Chemical Development Division, Medicines Research Centre, Stevenage, Hertfordshire, UK

F. W. J. M. M. Hoeks
Department of Biotechnology, R&D, LONZA AG, CH 3930 Visp, Switzerland

R. A. Holt
ZENECA LifeScience Molecules, P.O. Box 2, Billingham, Cleveland, UK

J. T. F. Keurentjes
Akzo, International Research Laboratories, Velperweg 76, Postbus 9300, 6800 SB, Arnhem, The Netherlands

B. Laboue
SIPSY, Route de Beaucouz, B.P. 79, 49242 Avrillé, France

A. Longoni
Zambon Group SpA, Via Lillo del Duca 10, 20091 Bresso (Milan) Italy

R. McCague
Chiroscience Ltd, Cambridge Science Park, Milton Road, Cambridge CB4 4WE, UK

S. Panossian
Zambon Group SpA, Via Lillo del Duca 10, 20091 Bresso (Milan), Italy

J. J. Partridge
Glaxo Wellcome Chemical Development Division, Girolami Research Centre, Research Triangle Park, NC, USA

D. Pauluth
Merck Germany/BDH, 64271 Darmstadt, Germany

P. Pitchen
Rhône-Poulenc Rorer Inc., 500 Arcola Road, P.O. Box 1200, Collegeville, PA 19426-0107, USA

T. V. Rajanbabu
Department of Chemistry, Ohio State University, Columbus, OH 43210, USA

K. T. Robins
Department of Biotechnology, R&D, LONZA AG, CH 3930 Visp, Switzerland

W. P. Shum
ARCO Chemical Co., 3801 West Chester Pike, Newtown Square, PA 19073, USA

G. P. Stahly
Albemarle Corporation, 451 Florida Blvd, Baton Rouge, LA 70801, USA

R. M. Starrett
Albemarle Corporation, 451 Florida Blvd, Baton Rouge, LA 70801, USA

D. J. Tapolczay
Glaxo Wellcome Chemical Development Division, Medicines Research Centre, Stevenage, Hertfordshire, UK

S. C. Taylor
ZENECA Life Science Molecules, P.O. Box 2, Billingham, Cleveland, UK

S. J. C. Taylor
Chiroscience Ltd, Cambridge Science Park, Milton Road, Cambridge CB4 4WE, UK

F. J. M. Voermans
Akzo, International Research Laboratories, Velperweg 76, Postbus 9300, 6800 SB, Arnhem, The Netherlands

A. E. F. Wächtler
E. Merck Germany/BDH, 64271 Darmstadt, Germany

L. O. Weigel
Central Process R&D, Lilley Research Laboratories, Lilley Corporate Center 4813, Indianapolis, IN46285, USA

J. Werlen
Department of Biotechnology, R&D, LONZA AG, CH 3930 Visp, Switzerland

W. M. L. Wood
ZENECA Huddersfield Works, P.O. Box A38, Leeds Road, Huddersfield, West Yorkshire, UK

Th. P. Zimmermann
Department of Biotechnology, R&D, LONZA AG, CH 3930 Visp, Switzerland

Foreword

There is an increasing willingness of industrial organic chemists to publish the results of their process research and development work, much of which is high calibre, and this has been reflected in the number of books and symposia devoted to these topics. This can only be of great benefit to the industry, since we can all learn from each other by examining the details of successes and failures, particularly in conversion of laboratory methods to fully scaled up manufacturing processes. This edition of *Chirality in Industry II* is therefore most welcome.

The editors are to be congratulated, first of all on the successful Volume 1 of this series (now available in paperback—this must reflect its popularity!) but now also on persuading so many industrial chemists to write up their results on methods of making optically active compounds, on large scale as well as in the laboratory. *Chirality in Industry II* continues the high standard set in Volume 1, and also contains new case studies covering the diversity of methodologies—resolutions, asymmetric syntheses, biological methodologies—required to make chiral molecules.

I have enjoyed reading the manuscript prior to publication and I know readers will benefit from reading this text. I sincerely hope the editors will be planning *Chirality in Industry III* in the near future.

Trevor Laird
Scientific Update
(Editor OPRD)

Preface

The topic of enantioselective synthesis/manufacture continues to be served by a stream of new books[1] and symposia;[2] many of the latter are specifically targeted at industrial aspects of chirality and include individual company symposia. The appearance of the first volumes on stereoselective synthesis in the Houben-Weyl series[3] is particularly timely. Also, we await with keen anticipation the appearance in early 1997 of the new American Chemical Society – Royal Society of Chemistry journal *Organic Process Research and Development*, expecting it to be a rich source of examples of the successful scale-up of methods for making single enantiomer products.

In introducing the second volume of *Chirality in Industry*, we note that once again all the authors are from industry or have strong industrial connections. Whilst it is hoped that Volume II will stand on its own, the intention is that Volumes I and II taken together should present a balanced, up-to-date and comprehensive picture of the technologies required for the production of optically active molecules on the multi-kilogram to high tonnage scales, as well as illustrating the breadth of application of the products of these technologies: pharmaceuticals, agrochemicals, electronics, food, flavours and fragrances. We are gratified to learn that Volume I has found use as a source of case histories for graduate courses and we hope this volume will prove even more useful in that respect.

The present volume is divided into sections on Physical Methods and Classical Resolutions, Biological Methods and Chiral Pool Syntheses and Asymmetric Syntheses. However, these are preceded by a chapter dealing with regulatory issues for chiral drugs and by then by two chapters on chiral drug development which illustrate how widely the industrial practitioner must cast the net in pursuit of practical production methods. Since enantioselective drug manufacture is probably the major driving force for the development of new technologies, regulatory attitudes (Chapter 2) clearly play an important role in dictating the pace and direction of development. Chapter 3 provides an overview of methods for making single enantiomers of one important class of drugs (profens) whilst Chapter 4 presents a more detailed account of the choices made between routes during the development of three specific drug candidates in another very important pharmaceutical class. Here three case histories of the development of processes for anti-viral products are described that embrace enzymatic resolutions, classical resolutions, selective crystallisations of desired diastereoisomers, asymmetric synthesis and the use of membranes.

As with Volume I, we hope that this volume will prove of particular interest to readers who are professionally involved in the scale-up of methods for the production of optically active materials, but also we hope that students and researchers involved in a more academic pursuit of optical activity will benefit from some of the facets of 'large-scale' thinking. An economic solution is most likely to be a simple, elegant solution.

REFERENCES AND NOTES

1. (a) Sheldon, R. A., *Chirotechnology: Industrial Synthesis of Optically Active Compounds*, Marcel Dekker, New York, 1993; (b) Rahman, A., and Shah, Z., *Stereoselective Synthesis in Organic Chemistry*, Springer, New York, 1993; (c) Jannes, G., and Dubois, V. (Eds), *Chiral Reactions in Heterogeneous Catalysis*, Plenum Press, New York, 1995; *Proceedings of ChiCat, The First European Symposium on Chiral Reactions in Heterogeneous Catalysis*, Brussels, 25–26 October 1993; (d) Seyden-Penne, J., *Chiral Auxiliaries and Ligands in Asymmetric Synthesis*, Wiley–Interscience, New York, 1995; (e) Ojima, I. (Ed.), *Catalytic Asymmetric Synthesis*, VCH, New York, 1993; (f) Noyori, R., *Asymmetric Catalysis in Organic Synthesis*, Wiley, New York, 1994; (g) Eliel, E. L., Wilen, S. H., and Mander, L. N., *Stereochemistry of Organic Compounds*, Wiley–Interscience, New York, 1994; (h) Nogradi, M., *Stereoselective Synthesis: A Practical Approach*, 2nd edn, VCH, Weinheim, 1995; (i) Aitken, R. A., and Kilenyi, S. N. (Eds), *Asymmetric Synthesis*, Blackie, Glasgow, London, 1992; (j) Červinka, O., *Enantioselective Reactions in Organic Chemistry*, Ellis Horwood, London, 1995; (k) Procter, G., *Asymmetric Synthesis*, Oxford Science Publications, Oxford, 1996.
2. The annual Chiral Europe and Chiral USA series of meetings organised by Spring Innovations Ltd, Stockport, UK, have become established features of the 'chiral' calendar. These are supplemented by various fine chemicals conferences and trade shows, such as ChemSpec, which attract many offerings from the chiral manufacturing sector. There are also the International Symposia on Chiral Discrimination, the eighth of which is scheduled to take place in Edinburgh in mid 1996.
3. Helmchen, G., Hoffmann, R. W., Mulzer, J., and Schaumann, E. (Eds), *Stereoselective Synthesis. Houben-Weyl, Methods of Organic Chemistry*, Vols E21a onwward Georg Thieme Verlag, Stuttgart, 1995

March, 1996

A. N. C.
G. N. S.
J. C.

Acknowledgements

We would like to express our thanks to all of the contributors for their patient cooperation in the preparation of this volume, to ZENECA Specialties and the Queen's University of Belfast for practical assistance, and finally to our families for agreeing to go through the process a second time.

1 Introduction

J. CROSBY
ZENECA Pharmaceuticals, Macclesfield, UK

1.1 Chiral Pool . . . 3
1.2 Chiral Auxiliaries . . . 3
1.3 Asymmetric Catalysis . . . 3
1.4 Classical Resolution and Crystallisation Methods . . . 4
1.5 Biological Methods . . . 5
1.6 Targets . . . 6
1.7 Enabling Science and Technology . . . 6
1.7.1 Crystallisation . . . 6
1.7.2 Membranes/Separation Technologies . . . 7
1.7.3 Supercritical Fluid Technology . . . 8
1.8 Conclusions and the Future . . . 8
1.9 References and Notes . . . 9

In surveying developments since Volume I was published in 1992, it is clear that the needs of the pharmaceutical industry continue to be a major driving force for development of new and cost-effective methods for manufacturing optically active materials and, inevitably, much of the content of this volume is pharmaceutically oriented. Speed to market remains of paramount importance and can affect the selection of technology, particularly for pharmaceuticals and agrochemicals, and the pressure to achieve rapid commercialisation is even greater than it was four years ago.

The introduction of racemic drugs is becoming increasingly unattractive owing to policy changes made by regulatory agencies. As a result, the practical preparation of non-racemic drugs is a critical issue in the pharmaceutical industry. In 1990, 25% of the 'synthetically' derived drugs on the market were chiral, but only 3% were marketed as pure enantiomers[1] (Figure 1).

The picture is changing very rapidly. Although accurate and up-to-date statistics are difficult to obtain and open to different interpretations, it has been predicted that 75% of man-made pharmaceuticals will be single enantiomers by the year 2000.[2] The corresponding figure for agrochemicals has been estimated at about 20%.[3] The proportion of sales accounted for by resolved chiral agrochemical products is still relatively modest,[4] e. g. *ca.* US$0.7bn for pyrethroid insecticides in 1991[5] out of a $7.7bn insecticide market.[6] The market for enantiopure

Chirality in Industry II. Edited by A. N. Collins, G. N. Sheldrake and J. Crosby

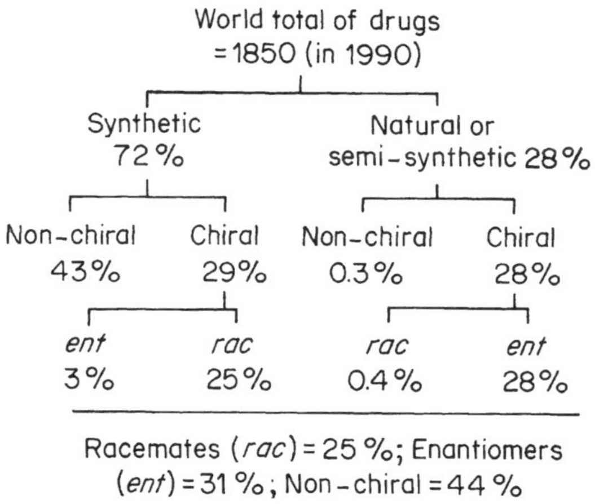

Figure 1 Distribution of single-enantiomer, chiral and non-chiral drugs

final, formulated pharmaceutical products could be \$60–90bn by the year 2000,[7] although the value of resolved chiral intermediates going into these will be much smaller, perhaps \$1–2bn;[8] this is nonetheless a big opportunity for the chiral manufacturing sector to address. In a recent report,[9] it was noted that the world-wide market for single enantiomer forms of chiral drugs surged in 1994. The market for dosage forms reached \$45.2bn.

The current move to single enantiomer drugs received a push in May 1992 when the FDA issued a policy statement. Although not an edict, and stating that it might be entirely appropriate to develop a chiral drug as a racemate, such official recognition of the chiral drug issue has persuaded drug firms that were not already doing so to start developing single enantiomers. Sales of enantiopure drugs are projected almost to double between 1994 and 1997.[10] This trend is being reinforced by a number of 'racemic switches' in which compounds previously marketed as racemates are relaunched as single isomers. There are only about six racemic switches on the market so far but about 20 compounds are said to be in redevelopment as single isomers, with Merck, Chiroscience and Sepracor being most active in the area. Merck is developing (*S*)-(+)-ibuprofen, Chiroscience compounds include (*S*)-ketoprofen and (−)-bupivacaine whilst Sepracor has a large number of candidates including both (*R*) and(*S*)-ketoprofen for different applications.[9] Switching must be largely a one-off phenomenon, a catching up exercise, but one with the potential to give a significant short to medium-term boost to the demand for chiral intermediates and to provide additional vehicles for technology development.

It is appropriate to include some comments on agrochemicals and on the position of agrochemicals versus pharmaceuticals. There are only two chapters specifically devoted to agrochemical products/intermediates: Chapter 10 in

Volume II, (*S*)-chloropropionic acid, and Chapter 4 in Volume I, pyrethroids. This is probably a fair reflection of the less developed application of single enantiomers in this area.[11] The values of bulk actives for agrochemicals fall roughly in the range £10s–£100s kg^{-1}; this is, again very approximately, an order of magnitude lower in value than bulk drugs and governs the level of sophistication of the chemistry/technology that can be brought to bear on their production. However, although historically pharmaceuticals have been able to tolerate higher manufacturing costs, this will not necessarily remain the case as we enter the era of healthcare reforms and a highly competitive generics market. These changes to the market structure might be expected to stimulate further development of chiral methodologies from within pharmaceuticals that will come closer to meeting the cost constraints of agrochemicals production.

To help set the scene for this volume, the following are presented as examples of developments in the main technology areas which have taken place since the publication of Volume I.

1.1 CHIRAL POOL

The 'pool' continues to expand rapidly through the addition of many new single enantiomer products and their intermediates, together with auxiliaries which have been made in response to synthesis needs.

1.2 CHIRAL AUXILIARIES

More sophisticated chiral auxiliaries are becoming available in commercial quantities, such as bornane-10,2-sultam and oxazolidinones[12] and (*S*)-4-(phenylmethyl)-2-oxazolidinone as a 'spin-off' from the manufacture of phenylalanine for the sweetener aspartame.

1.3 ASYMMETRIC CATALYSIS

The introduction and scale-up of asymmetric processes continue apace and the practical application of large-scale asymmetric synthesis is certainly not in doubt. The Takasago process for (−)-menthol and related chiral terpenes has, since its inception in 1982, passed the 20 000 tonne production milestone, at an impressive average consumption of only 6 g of the chiral BINAP ligand per tonne of product.[13] Elsewhere, ARCO Chemical Company and SIPSY have developed the Sharpless asymmetric epoxidation (SAE) reaction on an industrial scale.[14] SIPSY has run this process safely at 4000 litre scale to produce multi-tonne quantities of (*R*)- and (*S*)-glycidol (Chapter 18). Modified SAE has also been used to produce single-enantiomer sulphoxides on the multi-kilogram scale

(Chapter 19). Seprachem has been particularly active in developing and scaling up several key asymmetric technologies which include the Sharpless asymmetric dihydroxylation and the Jacobsen asymmetric epoxidation;[15] uses of the latter include the production of enantiomers of indene oxide required for HIV protease inhibitors. Another important reaction which is being scaled up is the Corey asymmetric reduction catalysed by oxazaborolidines, as described in Chapter 20.

However, to present a balanced picture, some of the earlier euphoria concerning the prospects for the large-scale use of asymmetric catalysis is now tempered by the realisation that although a reaction may be wonderfully catalytic and selective, in some instances large amounts of other, achiral, reagents may be required to effect the reaction and, overall, create more by-products and result in a higher cost than a superficially less attractive method using auxiliaries/resolving agents. In turn, the use of high molecular weight auxiliaries may be less effective than a simple resolution; hidden steps may also lie in its recovery and re-use. Another aspect which must be considered is selectivity: asymmetric catalysis which only delivers moderate *ee* (or *de*) and moderate yield may not be acceptable. For example, removal of an unwanted enantiomer and chemical purification from an 80% yield reaction with 80% *ee* might sacrifice 25% of the desired enantiomer, netting only 55% of the desired enantiomer. An efficient resolution, on the other hand, might yield an 'uncomplicated' 45% and involve much simpler processing.

Whilst rapid progress is being made in homogeneous asymmetric catalysis, progress in heterogeneous catalysis, which would have obvious processing advantages, lags behind. It has been pointed out[16] that the *ee*s and the range of reaction types obtained by other methods may not be easy to match by heterogeneous catalysis. The field is still largely restricted to hydrogenation and it might be asked whether the way forward here is not to produce heterogeneous catalysts *per se* but rather to focus on methods for rendering proven homogeneous catalysts recoverable.

1.4 CLASSICAL RESOLUTION AND CRYSTALLISATION METHODS

Resolution methods employing selective crystallisations are still far from predictable and chirality is still mainly treated as a qualitative property (cf. Chapter 5).[17] Nonetheless, their use is widespread (cf. Chapter 6) and a recent literature survey showed that the number of patents being granted for resolution processes exceeds that for asymmetric synthesis.[18]

A process for diltiazem has been reported[19] which exploits an entrainment resolution. This is said to be very competitive and operate for more than 25 cycles for each enantiomer.

1.5 BIOLOGICAL METHODS

At one end of the spectrum, abzymes still seem a long way from providing the basis for industrial processes. Although chiral synthesis is an area where catalytic antibodies should score highly, the industrial potential is still seen as long-term and academic researchers have to resolve some fundamental problems.[20]

At the other end of the spectrum, poly(amino acid) catalysts, which might be regarded as the simplest of enzyme analogues, recently evoked the headline 'The world's cheapest chiral catalyst.'[21] These materials appear to be being considered for development products. For example, poly(L-leucine) has been employed as the chiral epoxidation catalyst in a key step of a reported practical enantioselective synthesis of SK&F 104353, a possible candidate for the treatment of bronchial asthma.[22] Although this asymmetric epoxidation was developed by Julia *et al.*[23] in the early 1980s, it has not been widely exploited. There are obvious attractions for large-scale application; poly(L-leucine) is stable and inert and could be re-used.

A recent development aimed at improving the robustness of enzymes has been the crosslinking of enzyme microcrystals by bifunctional agents such as glutaraldehyde. The resulting crosslinked enzyme crystals, CLECs®, are reported to have properties superior to both soluble and conventionally immobilised enzymes; they remain active in environments that are otherwise incompatible with enzyme function, including high temperatures and extremes of pH,[24,25] e.g. thermolysin CLECs were fully active for 18 days at 55 °C in ethyl acetate.

A large number of enzymes have become commercially available for organic synthesis. Of about 2500 identified thus far, some 300 are available in a partly purified form. Moreover, because of advances in molecular biology, fermentation and purification techniques, the costs are being reduced to industrially attracive levels.[26] In addition, recognition and use of 'non-nameplate' applications will increase;[27] this is where an enzyme successfully catalyses a reaction other than that for which it is normally employed.

Use of oxynitrilases to produce single-enantiomer cyanohydrins is a process which appears to have good potential (cf. Volume 1, Chapter 14), and has been developed to the semi-technical scale by Peboc, but there does not appear to have been any large-scale use. Another interesting biotransformation which seems to be falling short of its potential is enantioselective nitrile hydrolysis;[28] reasons here are in part related to lack of commercial sources of the enzymes. However, Lonza has recently developed a process for (*S*)-dimethylcyclopropylcarboxamide.[29]

With respect to bio-production of chiral epoxide targets, there is now more emphasis on enantioselective conversion of racemates versus the earlier monooxygenase approach. Hydrolytic enzymes may offer more promise as they are cofactor-independent and the degradation product may be recovered as optically pure diol.[30]

Searches for extremophiles, organisms obtained from hostile environments, continue[31] and will undoubtedly enrich the armoury of enzymes tolerant of harsh operating conditions.

1.6 TARGETS

New general targets include resolved sulphoxides, which are increasingly being sought in the pharmaceuticals area (cf. Chapter 19); they are not amenable to a 'chiral pool' strategy as there are no readily available chiral sulphoxides. Whether accessed by the Kagan modification of the Sharpless epoxidation or enzymically, steric distinction is usually essential for success.

1.7 ENABLING SCIENCE AND TECHNOLOGY

There are a number of enabling sciences/technologies which are important to large-scale chiral synthesis, in particular analytical science, crystal science and the use of membranes. It might be asked why neither this volume nor its predecessor has a chapter devoted to analysis, the most important enabling science. Without an ability to analyse, the basis for manufacture, control and legislation all disappear. However, there are to our knowledge no scale-dependent aspects or, at least, none which is peculiar to chiral analysis. For this reason, analytical aspects are omitted but their importance does not go unrecognised.

1.7.1 CRYSTALLISATION

Crystallisation science is an enabling science with a high degree of relevance to large-scale manufacture. An understanding of crystallisation principles and techniques is an essential competence for anyone developing large-scale processes for single-enantiomer materials and lies behind many commercially important and successful processes. It was felt appropriate to include a substantial chapter on crystallisation techniques and, although some of it (Sections 7.2 and 7.3) is of general use, it is of particular importance in the production of single-enantiomer products. Examples which may be selected from elsewhere in this volume include the case histories presented in Chapter 6 and the importance of crystal form in processes for naproxen and ibuprofen (Chapter 3). The eutectic composition is a most important parameter to be considered when planning the separation of a racemate. For instance, the eutectic composition dictates that ibuprofen must be at least 90% pure (*S*)-isomer for the crystallisation to result in optical purification. Also, in this case by operating with the sodium salt, a more favourable eutectic is obtained.

1.7.2 MEMBRANES/SEPARATION TECHNOLOGIES

Separation and membrane technologies are developing rapidly. In addition to the Wandrey/Degussa membrane reactors, developed principally for amino acid production and which use ultrafiltration membranes (Volume I, Chapter 20), there is the Bend Research system using selectively permeable membranes[32] and the Sepracor hollow-fibre system,[33] all of which have been well described. To these can now be added the development by Akzo (Chapter 8) of a 'symmetrical' countercurrent extraction/separation system and simulated moving bed technology (SMB), the use of which is being translated from very high tonnage achiral separations to chiral separations.

Chapter 8 describes a broadly applicable membrane technology for the separation of enantiomers: a symmetrical countercurrent extraction system.[34] Two liquids, at least one of which is chiral or contains a chiral adjuvant and which are wholly miscible with each other, flow countercurrently either side of a non-miscible phase, consisting of a liquid supported on a microporous membrane; hollow-fibre modules are used. Enantiomer separation by this method should permit the production of almost any optical isomer. It does not (overtly) require functional handles; clearly interactions are required and direct crystallisation remains the only truly functionality-independent method. Method development should be relatively easy; scale-up is reported to be simple and both enantiomers can be collected. A crucial aspect of the method development is the choice of a chiral selector which will enable sufficient complex formation with the enantiomers, e.g. by H-bonding or electrostatic interaction. In practice, satisfactory results can be obtained with cheap tartaric esters and commercial applications are expected in the next few years.

Although preparative high-performance liquid chromatography (LC) is used for laboratory- and pilot-scale production, chromatographic processes have generally been considered too expensive for large-scale use. However, developments in scale-up continue, with polysaccharide chiral stationary phases proving particularly useful. As an indication of current capabilities of the technology, an example has recently been described in which, using a 50 cm × 10 cm i.d. column, 4 kg of a racemate were processed in 45 h using 510 l of methanol solvent.[35]

UOP has applied the principles of countercurrent adsorption processes, in particular its Sorbex™ SMB process, widely used for the production of commodity chemicals, to the separation of optically pure pharmaceuticals.[36] In the petrochemical sector, the economics of scale reduce separation costs to a few cents/kg; for low-tonnage chiral separations costs are significantly higher owing to high depreciation charges resulting from low throughput rates, high stationary phase costs and high solvent recovery costs. SMB uses an array of, typically, eight or more chromatographic columns linked in a circle. Although solvent consumption is less than for elution chromatography, because SMB takes much more packing to fill the columns the investment is higher, with chiral column packings costing *ca* \$8000 kg^{-1}.[37] Examples of outputs range up to 25 kg per day. It has been estimated that one of the larger units could produce 10 000 kg per year of (*R*)-3-chloro-1-phenylpropanol, an intermediate for (*S*)-(+)-fluoxetine, at a cost of \$700 kg^{-1}. The cost estimate includes racemising/recycling the (*S*)-enantiomer. Since the stationary phase is a significant proportion of the cost, and there is as yet no commercial experience of its lifetime, significant reductions can be expected with future experience and improvements.[38] Projects on the scale of 100 tonnes per annum are said to be under consideration.[39]

1.7.3 SUPERCRITICAL FLUID TECHNOLOGY

Is supercritical fluid technology going to emerge as another enabling technology for single-enantiomer production? It could have roles in both enzyme-catalysed reactions and in preparative-scale chiral chromatography. The unique pressure dependence of the physical properties of supercritical fluids presents many opportunities for the control of enzyme activity, specificity and stability in biocatalytic reactors and the rates of reactions involving gaseous reactants such as oxygen should be enhanced.[40] In LC the lower viscosity should lead to greater efficiencies.

1.8 CONCLUSIONS AND THE FUTURE

Except perhaps for some areas of amino acid production, pharmaceuticals have been the major user of technologies for the large-scale production of single-enantiomer products, with agrochemicals next, but a long way behind. Single-enantiomer products are also required for flavours and fragrances, in electronics and foodstuffs, e.g. sweeteners (Volume I, Chapter 11) and carnitine (Chapter 14). Chapter 16 illustrates how the needs of the flavour and fragrance industry have stimulated some elegant asymmetric syntheses and Chapter 13 shows how chirality is introduced into liquid crystals. The latter, relatively unsung, corner of chirals manufacture, although small in tonnage and, by pharmaceuticals standards, not very complex, has arguably one of the greatest impacts on our lives through the part it can play in the information technology explosion, for example by enhancing the performance of display devices.

Within the wider field of organic chemistry, the special status of 'chiral' chemistry is sometimes questioned. It is special, witness the activity in symposia and publications devoted to the topic and noted in the Preface, but it cannot stand alone; its successful application usually depends on intelligent integration within a spectrum of supporting technologies. The bio-examples, in particular, reveal the need for a multidisciplinary approach (cf. Chapters 9, 10, 12 and 14). It is of interest to note in this context the changes taking place in the industry after the somewhat enthusiastic spawning of 'chirals' companies in the late 1980s-early 1990s which were often seeking to exploit a single technique or reaction. There has been the realisation that such skills alone are not necessarily sufficient for commercial viability; there have been some departures and some strategic alliances and licensing deals.[41] It is possible to draw parallels with the development of the modern biotechnology industry,[42] which underwent major restructuring in the late 1980s. Companies who have succeeded are those able to integrate their technological capabilities within a broader set of skills, concentrate on selected niches or team up with appropriate partners. Many end users have developed in-house capabilities, reducing the opportunities for independent players. Much of this can be seen to apply to chirals. There is evidence from publications and symposia reports that, in addition to the activities of specialist 'chirals' companies, many of the major chemical companies have active in-house programmes to develop, in particular, catalysts for asymmetric reductions; much of this activity has probably been stimulated by the desire to have proprietary systems which match and extend the capabilities of the Noyori/Takasago technology (cf. Volume I, Chapter 17).

At the research level, all necessary methods exist for making initial quantities of enantiomers for the exploration of property differentiation. There is no longer any question as to whether a desirable isomer can be made, but only if this can be effected economically. Single-isomer development on the large scale had been constrained by the availabilities of techniques, intermediates, catalysts, etc. Now the chiral infrastructure is largely in place, there is no reason why large-scale production of even moderately priced single-enantiomer products should not be contemplated.

We noted in the Epilogue to Volume I that '... the future will see an interesting race between biology, conventional chemistry and separation technology... .' Four years later, all three areas have seen major advances; separation technologies have made a particularly strong showing; abzymes have still to realise their potential; crystallisation techniques continue to be of signal importance.

1.9 REFERENCES AND NOTES

1. *Scrip*, Feb. 16 (1993)
2. *Speciality Chem.* Jan./Feb., 6 (1994).
3. Gardner, J. C., and DiCicco, R. L., *Speciality Chem.*, July/Aug., S9 (1994).
4. Silvon, M., *Chem. Mkt. Rep.*, 7 Sept., SR22 (1992).
5. Gupta, N., and Eisberg, N., *Performance Chem.*, Aug./Sept., 19 (1991).
6. Heaton, A., in *The Chemical Industry*, 2nd edn (ed. A. Heaton), Blackie, Glasgow, 1994, p. 218.
7. See, for example, Burke, M., *Chem. Ind. (London)*, 3 Jan., 10 (1994).
8. See, for example, *Performance Chem.*, Feb., 9 (1995).
9. Stinson, S. C., *Chem. Eng. News*, 9 Oct., 44 (1995).
10. Stinson, S. C., *Chem. Eng. News*, 19 Sept., 38 (1994).
11. For a recent review of the manufacture of optically active agrochemicals, see Crosby, J., *Pestic. Sci.*, **46**, 11 (1996).
12. Stinson, S. C., *Chem. Eng. News*, 28 Sept., 76 (1992).
13. Based on figures presented by S. Akutagawa at Chiral '95 USA Symposium, Boston, MA, 15–16 May 1995.
14. Bulliard, M., and Schum, W., in *Proceedings of Chiral '95 USA Symposium*, Boston, MA, 15–16 May 1995, Spring Innovations, 1995, p. 5.

15. Pettman, R., *Speciality Chem.*, July/Aug., S12 (1994).
16. Jannes, G., and Dubois, V. (Eds), *Chiral Reactions in Heterogeneous Catalysis*, Plenum Press, New York, 1995; *Proceedings of ChiCat*, the *First European Symposium on Chiral Reactions in Heterogeneous Catalysis*, Brussels, 25–26 Oct. 1993, p.1.
17. Bruggink, A., and Ariaans, G. J. A., *Pharm. Manuf. Int.*, 85 (1995).
18. Eliel E.L., Wilen S. H., and Mander L. N., *Stereochemistry of Organic Compounds*, Wiley–Interscience, New York, 1994, p. 388.
19. Villa, M., and Pozzoli, C., in *Proceedings of Chiral '95 USA Symposium*, Boston, MA, 15–16 May 1995, Spring Innovations, 1995, p. 13.
20. Johnson, J., *Chem. Ind. (London)*, 20 Feb., 128 (1995).
21. *Chem. Ind (London)*, 4 Sept., 675 (1995).
22. Flisak, J. R., Gombatz, K. J., Holmes, M. M., Jarmas, A. A., Lantos, I., Mendelson, W. L., Novack, V. J., Remich, J. J., and Snyder, L., *J. Org. Chem.*, **58,** 6247 (1993).
23. Julia, S., Masana, J., Vega, J. C., *Angew. Chem., Int. Ed. Engl.*, **19,** 929 (1980).
24. Margolin, A. L., in *Proceedings of Chiral '94 USA Symposium*, Reston, VA, 5–6 May 1994, Spring Innovations, 1994, p. 51.
25. Margolin, A., *Perf. Chem.*, April/May, 23 (1994).
26. See, for example, *Kirk–Othmer's Encyclopedia of Chemical Technology*, 4th edn., Wiley–Interscience, New York, 1993, Vol. 9, p. 672.
27. See, for example, de Zoete, M. C., van Dalen, A. C. K., van Rantwijk, F., and Sheldon, R. A., *J. Chem. Soc., Chem. Commun.*, 1831 (1993); use of lipases to effect aminolysis, a mild procedure for enantioselective synthesis of amides.
28. See, for example, Crosby, J., Moilliet, J., Parratt, J. S., and Turner, N. J., *J. Chem. Soc., Perkin Trans. 1*, 1679 (1994), and references cited therein.
29. Chassin, C., in *Proceedings of InBio'96 Symposium*, 20–21 Feb. 1996, Manchester, Spring Innovations, 1996.
30. See, for example, Weijers, C. A. G. M., and deBont, J. A. M., in *Proceedings of Chiral Europe'95 Symposium*, London, 28–29 Sept. 1995, Spring Innovations, p. 87.
31. (a) Adams, M. W. W., and Kelly, R. M., *Chem. Eng. News*, 18 Dec., 32 (1995);. (b) Govardhan, C. P., and Margolin, A. L., *Chem. Ind. (London)*, 4 Sept., 689 (1995).
32. van Eikeren, P., Brose, D. J., Muchmore, D. C., West, J. B., and Colton, R. H., in *Proceedings of Chiral '92 Symposium*, Manchester, 24–25 March 1992, Spring Innovations, 1992, p. 63.
33. See, for example, Young, J. W., in *Proceedings of the Chiral Synthesis Symposium and Workshop*, 18 April 1989, Stockport, Spring Innovations, 1989, p. 39.
34. Keurentjes, J. T. F., *PCT Int. Pat Appl.* WO 94/07814, 1994.
35. Tachibana, K., in *Proceedings of Chiral Europe '95 Symposium*, 28–29 Sept. 1995, London, Spring Innovations, 1995, p.139.
36. *Performance Chem.*, Feb., 27 (1995).
37. *Chem. Eng. News*, 9 Oct., 52 (1995).
38. Gattuso, M. J., McCulloch, B., House, D. W., and Baumann, W. M., in *Proceedings of Chiral '95 USA Symposium*, 15–16 May 1995, Boston, MA, Spring Innovations, 1995, p. 51.
39. *Chem. Eng. News*, 9 Oct., 58 (1995).
40. Russell, A. J., Beckman, E. J., and Chaudhary, A. K., *Chemtech*, March, 33 (1994).
41. Examples include Eastman/Peboc, *Chem. Eng. News*, Oct. 9, 58 (1995); ARCO/SIPSY, *Chem. Eng. News*, Oct. 9, 44 (1995); Chiroscience's acquisition of rights to DuPont's proprietary asymmetric catalysts for manufacture of pharmaceutical intermediates and actives (Press Release, 6 Nov. 1995).
42. Polastro, E., in *Chiral Reactions in Heterogeneous Catalysis* (ed. Jannes, G., and Dubois, V.), Plenum Press, New York, 1995; *Proceedings of ChiCat*, the *First European Symposium on Chiral Reactions in Heterogeneous Catalysis*, Brussels, 25–26 Oct. 1993, p. 5.

2 Chiral Drugs: Regulatory Aspects

J. J. BLUMENSTEIN
Janssen Research Foundation, Titusville, NJ, USA

2.1 Introduction 11
2.2 Racemates Versus Single Enantiomers 12
2.3 Pharmacokinetic Evaluation of Enantiomers 13
2.4 FDA Decision Tree for Optically Active Drugs 14
2.5 Differences Between FDA and CPMP Guidelines for Undesired Enantiomers 16
2.6 Drug Product Considerations 16
2.7 Verification of Starting Materials and Intermediates 17
2.8 Conclusion 17
2.9 References and Notes 17

2.1 INTRODUCTION

While the importance of drug stereochemistry has been recognised almost since the beginning of pharmacological drug therapy, regulatory consideration of stereochemistry has really only come to the fore in the past decade. Increased regulatory attention to stereoisomeric drugs is due to both technological advances in stereosensitive analytical methods (without which regulation would be impossible) and the increased commercial practicality of stereoselective synthesis. In an era of multinational drug development, the development of policies on stereochemical aspects of drug synthesis now transcends the borders of any one regulatory agency. At the moment, pharmaceutical companies must adapt their strategies to all the regulatory environments where their product will be registered, although attempts to standardise the requirements are being made by bodies such as the International Committee on Harmonisation (ICH). Recently, several regulatory bodies around the world have issued guidelines on the development of drugs with stereogenic centres.[1] The eventual working policy will probably be a conglomeration of the various existing guidelines.

The US Food and Drug Administration (FDA) first raised the topic of stereochemical regulation in its 1987 Drug Substance Guideline,[2] although little emphasis was given to the topic at that time. An internal committee was formed in 1989 to consider further the regulatory concerns related to stereoisomeric drugs and, in response, in 1990 the Pharmaceutical Manufacturers Association

Chirality in Industry II. Edited by A. N. Collins, G. N. Sheldrake and J. Crosby

(PMA) set out its position on the development of drugs with stereogenic elements in a paper published in *Pharmaceutical Technology.*[3] In 1992 a Drug Information Association conference in Paris provided an international forum to examine regulatory requirements for chiral drugs, and these were summarised in a 1993 paper published in *Drug Information Journal*.[4] While international harmonisation may still be a little while in coming, worldwide regulatory bodies have been developing individual positions. The FDA published its stance in 1992, and requested public comment on this. Although individual industries and trade associations provided comments, a revised document has not been issued. More detailed guidelines have been presented in several open forums, but a final guidance document has not become available. The Committee for Proprietary Medicinal Products (CPMP)[5] issued its position in a 1994 policy. Canadian authorities issued draft guidance for comment in 1993, but a final policy has still to be published. This chapter will focus on the FDA policy and discuss recent developments elsewhere which are related to the sections described in that document.

There are two principal scenarios in chiral drug development: the first is the *de novo* development of an optically pure chiral drug, and the second the switch from an existing racemic drug to a single stereoisomer of that drug. Most of the regulatory guidance has focused on the first scenario, although its applicability to switches will also have to be addressed because of the growth of this area.

2.2 RACEMATES VERSUS SINGLE ENANTIOMERS

The development of chiral drugs, whether racemates or single enantiomers, raises issues of acceptable control of the manufacture, adequate pharmacological and toxicological assessment, proper characterisation of the metabolism and distribution and appropriate clinical evaluation. The pharmacological activity of the individual enantiomers should be determined, whether developing a single isomer or the racemate. In cases where the pharmacological activities of both enantiomers have been examined, there have been instances (1) in which both enantiomers have similar, desirable activities, (2) where one enantiomer is active and the other lacks pharmacological activity and (3) where each of the enantiomers has a completely different activity (see Chapter 1 in Volume I and Chapter 11 in this volume for examples). There are also intermediate situations, e.g. one isomer has an attenuated activity compared with the other.

Despite the problems identified with some racemates, the hitherto common practice of developing racemates has resulted in few recognised adverse consequences. Although technological advances have, in many cases, made the manufacture of enantiomers more feasible, the development of racemates will still continue to be appropriate for many development programmes. This situation is recognised in all the recent guidelines, although with varying levels of enthusiasm by the different regulators, all of whom have expressed a preference for the development of a single isomer, (even though there are cases where racemates are both appropriate and acceptable). Often that preference is a result of the clarity that is added by only having to study the effects of a single isomer, rather than a history of problems that have resulted from the use of racemates. The decision process when choosing whether to proceed with a racemate or a single-isomer candidate is the same as that for an achiral drug. Factors to be considered include the commercial practicality of the synthesis, the ability to formulate the dosage form, pharmacokinetics and metabolism and the risk–benefit considerations that evolve from safety and efficacy studies. With chiral drugs there is the advantage that these issues may be examined for both the single isomer and the racemate; where the racemate may not be suitable for final development, the single isomer may be, and vice versa. This is recognised by all of the regulators, and the guidelines allow the company to make this decision.

2.3 PHARMACOKINETIC EVALUATION OF ENANTIOMERS

Pharmacokinetic evaluations in animals and in clinical studies that do not use a chiral assay may be misleading if the disposition of the enantiomers is different. Therefore, regulators are focusing on techniques to quantify individual stereoisomers in pharmacokinetic studies early in development. The pharmacokinetic profile of each isomer should be characterised in animals and the profile of each enantiomer obtained from Phase I clinical studies. If the pharmacokinetics of the enantiomers are demonstrated to be the same, or exist as a fixed ratio in the target population, an achiral assay or an assay which monitors only one of the enantiomers may be used subsequently. If a racemate is being developed and the pharmacokinetic profiles of the isomers are different, the enantiomers should be monitored individually.

Similarly, in order to evaluate the pharmacokinetics of a single enantiomer, or a mixture of enantiomers, stereoselective assays for *in vivo* samples need to be available early in the development process. This will allow the assessment of the potential for interconversion and the absorption, distribution, metabolism and excretion (ADME) profiles of the individual enantiomers. Unless it proves particularly difficult, the main pharmacological activities of the isomers should be compared *in vitro*, in animals and/or in humans. It is usually sufficient to carry out toxicity studies on the racemate. A relatively benign toxicological profile for the racemate would ordinarily support further development without separate toxicological evaluation of the individual enantiomers. If toxicity is found other than that related to the desired pharmacological effects of the drug, the individual isomers should be evaluated in the study where the toxicity was detected.

Where little difference is observed in the activity and disposition of the enantiomers, racemates may be developed. There are some situations where the development of a single enantiomer is particularly desirable, such as when one enantiomer has a toxicity that the other (pharmacologically desirable) enantiomer does not have. Conversely, when there is rapid interconversion between the isomers *in vivo*, little advantage may be gained by developing a single isomer. In general, it is more important to evaluate both enantiomers clinically when both enantiomers are pharmacologically active but differ significantly in potency, specificity or maximum effect, than when one isomer is essentially inert.

If a racemate has been marketed and the sponsor wishes to develop the single enantiomer (a so-called 'racemic switch'), an appropriate abbreviated pharmacological/toxicological evaluation could be conducted to extend the existing knowledge of the racemate to that of the single isomer. Bridging studies, including a racemate control group, would usually include the longest repeat-dose toxicity study conducted and reproductive toxicity studies in the most sensitive species. Clinical evaluation should include the determination of whether there is a significant conversion to the other isomer, and whether the pharmacokinetics of the single isomer are the same as they were for that isomer as part of the racemate.

2.4. FDA DECISION TREE FOR OPTICALLY ACTIVE DRUGS

The FDA has not publicly issued more detailed guidance for the Chemistry, Manufacturing and Controls (CMC) section of applications for new drugs. The other regulatory authorities have expressed only general considerations to be taken into account in the manufacture of chiral drugs, and current regulatory thinking on CMC issues for chiral drugs has been described at recent symposia.[6] The CMC section should contain information about the identity, quality, purity and strength of the drug substance and the drug product, in addition to consideration of the stereogenic characteristics of the drug. Applications for single enantiomer and racemic drugs should include a stereochemically specific identity test and a stereochemically selective assay for both the drug substance (the active ingredient) and the drug product (the final formulated drug). The choice of the controls should be based on the methods of manufacture and the stability characteristics of the drug substance. The approach depends on whether there is one or more than one stereogenic centre present. For compounds with a single chiral centre, inversion may be detected only through the use of stereoselective assay method. However, those with multiple stereogenic centres may be able to be monitored using a standard achiral analytical techniques, since the simultaneous inversion of both centres is unlikely, and inversion of a single chiral centre will afford a diastereoisomer which may be detected through achiral techniques. This concept is illustrated in Figure 1.

The decision tree outlined in Figure 1 breaks down the FDA's requirements for pure enantiomers and racemates. The requirements for racemates are relatively straightforward. The FDA sees the need for an identity test for racemates, whether they have single or multiple chiral centres. Although it may not be intuitively obvious that this is necessary from an analytical standpoint, the

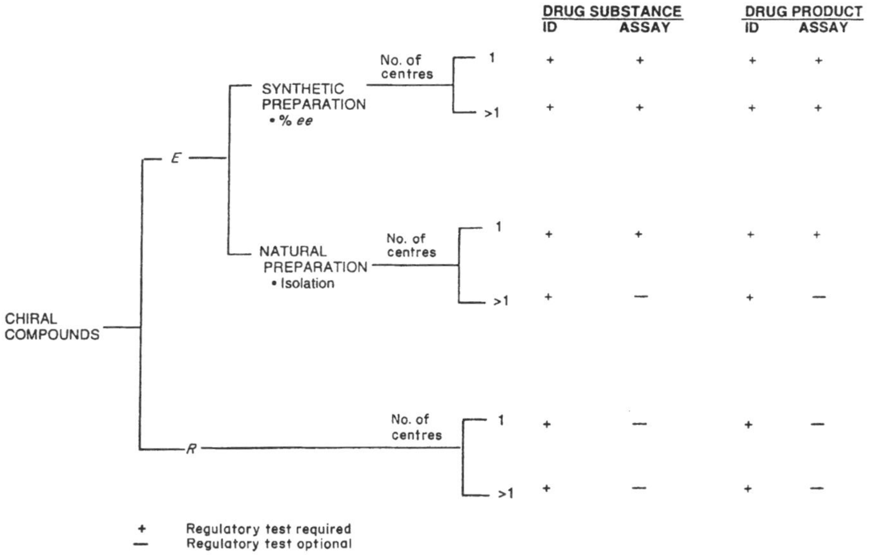

Figure 1 Enantioselective test requirements for chiral compounds. E, single enantiomer; R, racemate

regulatory stance is that the racemic nature is an identity characteristic that should be included in the specifications. While discussion on this issue has been wide-ranging, the question is often whether the substance exists as a racemic compound or a racemic mixture (conglomerate). Given that some identity tests (e.g. optical rotation) are not sensitive to subtle variations in stereochemical purity, and spontaneous resolution of racemic mixtures to afford an unanticipated, highly stereochemically enriched mixture is very unlikely, the value of this test for racemates should probably be questioned.

As may be expected, the situation with an enantiomerically pure compound is more complex. Single-enantiomer drugs may be further categorised into synthetic preparations and products isolated from natural sources. All synthetic preparations, whether having one or more chiral centres, should have both a stereospecific identity test (to determine the absolute configuration) and a stereoselective assay (to determine the enantiomeric purity). Natural products which have multiple chiral centres will require only an identification method which is stereospecific. However, natural products with only one chiral centre will require a stereoselective assay method in addition to the stereospecific identification method. As with all regulatory guidance, each case should be considered on its own merits. For example, when a stereoselective synthesis of a single isomer with multiple chiral centres is being evaluated, one must consider, *inter alia*, the number of chiral centres, whether it is a convergent or sequential synthesis and the lability of all of the chiral centres.

2.5 DIFFERENCES BETWEEN FDA AND CPMP GUIDELINES FOR UNDESIRED ENANTIOMERS

When setting specifications, there is a subtle but important difference in the way the undesired isomer is considered by the FDA and the CPMP. The CPMP specifically regards the undesired isomer as an impurity which, when coupled with recent guidelines on identification and qualification of impurities,[7] provides a clear approach to setting appropriate specification limits. The FDA's comments have suggested that diastereoisomers are regarded as impurities, but the position on the opposing enantiomer is not specified. In the absence of such guidance, it would seem advisable to follow the CPMP position and apply the same criteria to the specification limit as any other impurity.

2.6 DRUG PRODUCT CONSIDERATIONS

Stereochemical considerations for the drug *product* (the formulated drug) focus on the specifications and the stability of the drug product. The above discussion (Section 2.4) regarding the decision tree addresses most of the concerns about tests and specifications for both the drug substance and drug product. The FDA has expressed a desire (reflected in the decision tree) to mirror the controls used for the drug substance in the drug product. The CPMP focuses on validating the process to establish that there has been no impact on the stereochemical integrity of the active substance. Given the difficulty of developing validated identification methods for use with a complex pharmaceutical matrix, and the previous discussion regarding identity tests for racemates, the need for stereochemical identity tests in drug products should be re-evaluated. Except where there is a demonstrated need to monitor the level of stereochemical purity (e.g. a single enantiomer with a single chiral centre), the identity test usually provides little additional valuable information beyond the standard chemical identity tests.

The stability considerations for the drug substance and drug product parallel one another. Stressed stability studies at raised temperatures, humidities and light levels on an enantiomerically pure compound should be monitored with a stereochemically selective assay method to identify conditions which may lead to isomerisation. In most cases the products of accelerated stability studies should also be analysed with a stereoselective assay. If isomerisation does not occur, then a stereochemical assay method may not be required for long-term studies, but if isomerisation is observed, a stereochemically selective assay will usually be required. The requirements for long-term studies when isomerisation has been observed under stressed conditions,[8] but not accelerated storage conditions, are difficult to generalise, and must be dealt with on a case-by-case basis following consideration of the storage times and conditions.

2.7 VERIFICATION OF STARTING MATERIALS AND INTERMEDIATES

While the decision tree provides an overview of the analytical requirements from a stereochemical standpoint for the final specifications and test methods for a drug substance and drug product, there are many other areas in the CMC section of a registration document which require further attention from a stereochemical viewpoint. In the synthesis of the new drug substance, stereochemical considerations start with controls of the starting materials and reagents. Starting materials with a single chiral centre will require a stereoselective assay method, whereas those with multiple chiral centres may not require a stereochemical assay, since diastereoisomers will usually be detectable by achiral methods and the presence of opposing enantiomers with multiple centres is unlikely.

The issue of the suitability of controls for the stereochemical purity of intermediates is more difficult to handle generically. Stereochemically selective 'in-process' controls will be required when the procedure dictates it is necessary, e.g. to determine when a resolution step is optimal. When a given synthetic step establishes a new chiral centre to give a key intermediate, controls will usually be required to demonstrate that the functionality has been properly introduced. The assay and specifications for a key intermediate should demonstrate that the impurity levels, including the stereochemical impurities, are within established limits. Any subsequent synthetic steps which have the potential to epimerise a chiral centre will require appropriate controls to establish that the optical purity remains within established limits.

2.8 CONCLUSION

Although there is currently no harmonised guidance for the development and registration of chiral drugs, regulatory organisations have developed local guidelines. Multinational firms must interpret these local guidelines in developing their approaches to new therapeutic candidates. The local guidelines may allow companies some latitude to make their own development and registration decisions with respect to pure enantiomers and racemates; however, there is a growing consensus on the types of additional information required to support those decisions.

2.9 REFERENCES AND NOTES

1. (a) Food and Drug Administration, *FDA's Policy Statement for the Development of New Stereoisomeric Drugs*, FDA, Washington, DC, May 1, 1992; (b) CPMP Working Party on Quality of Medicinal Products, CPMP Working Party on Safety of Medicinal Products, CPMP Working Party on Efficacy of Medicinal Products, *Note for*

Guidance: Investigation of Chiral Active Substances; (c) HPB Working Group on Drug Stereochemistry, *Guidelines on the Development of Stereoisomeric Drugs*, Draft, December, 1993.

2. Center for Drugs and Biologics Food and Drug Administration Department of Health and Human Services, *Guideline for Submitting Supporting Documentation in Drug Applications for the Manufacture of Drug Substances*, (FDA, Washington, DC, February, 1987.
3. PMA Ad Hoc Committee on Racemic Mixtures, 'Comments on Enantiomerism in the Drug Development Process', *Pharm. Technol.*, May, 46 (1990).
4. Gross, M., *et al.*, Regulatory Requirements for Chiral Drugs, *Drug Inf. J.* **27**, p. 453 (1993).
5. The CPMP is a scientific committee in the EC comprised of Member States and the Commission which coordinates the registration of applications for new drugs.
6. Jamali, F., *et al.*, Stereochemical Aspects of Drug Development, Drug Information Association Workshop, Rockville, MD, 1994.
7. Food and Drug Administration, HHS. *Guideline on Impurities in New Drug Substances*, *Fed. Regist.*, **61**, 372 (1996).
8. In this context, the term 'stressed stability studies' refers to severe conditions intended to afford visible chemical and/or physical degradation. Accelerated storage conditions, on the other hand, are meant to be a short-term predictor of the behaviour over the full-length storage period.

3 Production Methods for Chiral Non-steroidal Anti-inflammatory Profen Drugs

G. P. STAHLY AND R. M. STARRETT
Albemarle Corporation, Baton Rouge, LA USA

3.1 Introduction 19
3.2 Industrial Processes for (*S*)-(+)-Naproxen 22
3.3 Industrial Processes for (*S*)-(+)-Ibuprofen 27
3.4 Other Profens and the Future 37
3.5 References 38

3.1 INTRODUCTION

Non-steroidal anti-inflammatory drugs (NSAIDs) comprise a structurally diverse group of compounds that have similar physiological effects.[1] The NSAID designation is somewhat misleading, as these drugs exhibit varying degrees of analgesic and antipyretic, as well as anti-inflammatory, activity. They are widely prescribed to relieve the inflammation encountered with diseases such as arthritis and for relief of low to moderate pain such as headache. In 1986, the 100 most widely prescribed drugs in the USA included seven NSAIDs.[2]

Most NSAIDs are organic acids, aspirin being the prototype. The profens are a subset of the class, each bearing as the common structural feature a 2-aryl-substituted propionic acid. Variation of substituents on the aryl group leads to a variety of different profens, some of which are shown in Figure 1. The most commercially significant are naproxen and ibuprofen.

The mechanism of action of NSAIDs is thought to be inhibition of prostaglandin synthesis.[3] Prostaglandins are not stored in cells but are instead synthesised and released in response to cell damage, appearing in high concentrations in inflammatory exudates. They are also associated with the development of pain resulting from injury or inflammation. The prostaglandin biosynthesis pathway is known, as is the specific step, catalysed by a cyclooxygenase enzyme, which is subject to NSAID interference. All NSAIDs inhibit cyclooxygenase activity, although not all by the same mechanism. Prostaglandin synthesis is believed to occur in all tissues, implying that the effectiveness of an NSAID is primarily influenced by its ability to reach the cyclooxygenase enzyme.

Chirality in Industry II. Edited by A. N. Collins, G. N. Sheldrake and J. Crosby

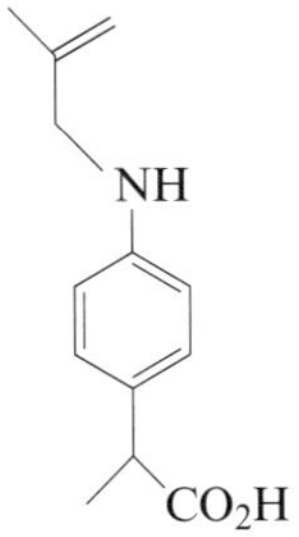

Alminoprofen

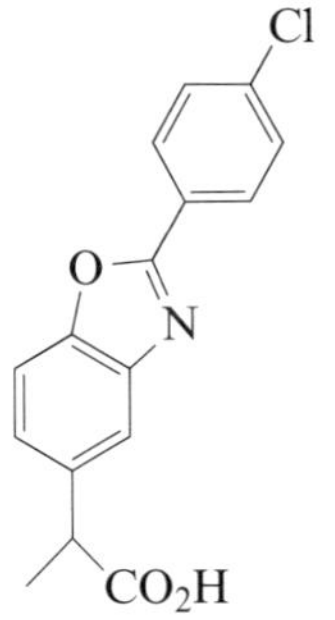

Benoxaprofen

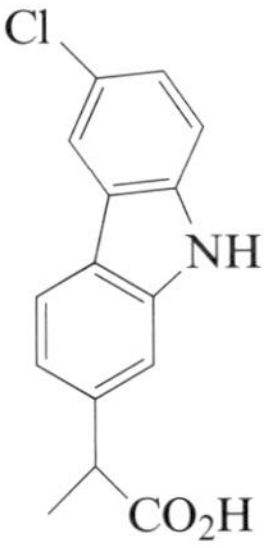

Carprofen

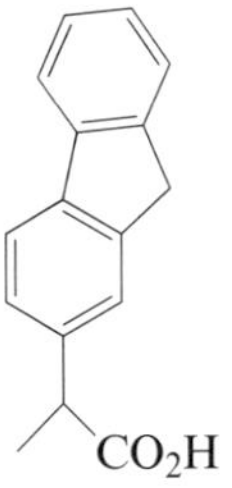

Cicloprofen

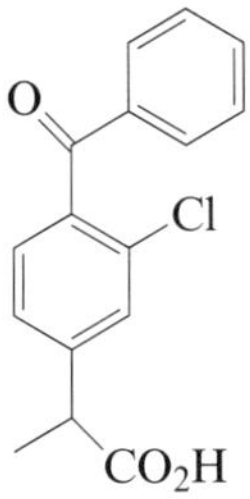

Cliprofen

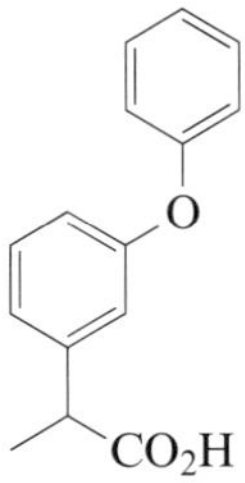

Fenoprofen

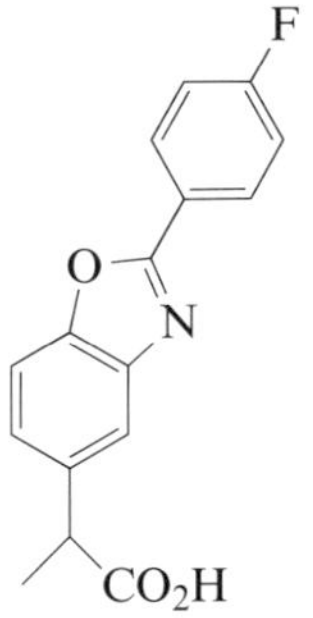

Flunoxaprofen

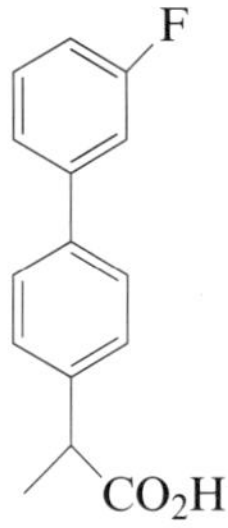

Fluprofen

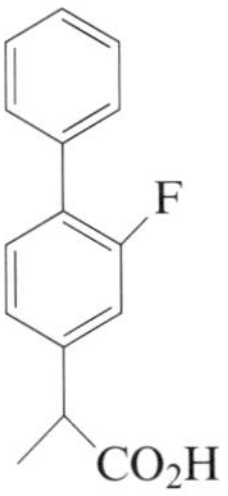

Flurbiprofen

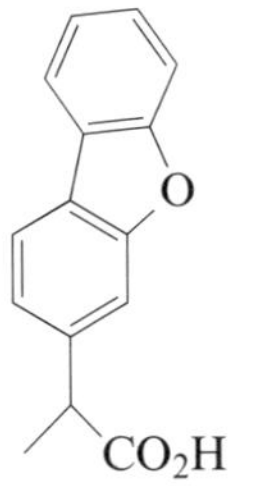

Furcloprofen

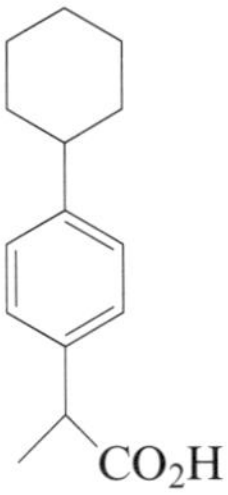

Hexaprofen

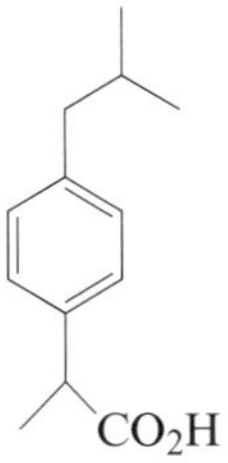

Ibuprofen

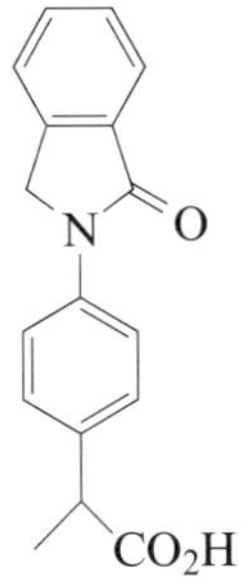

Indoprofen

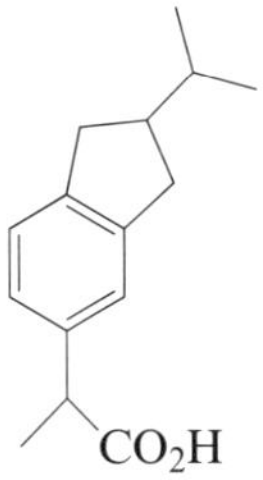

Isoprofen

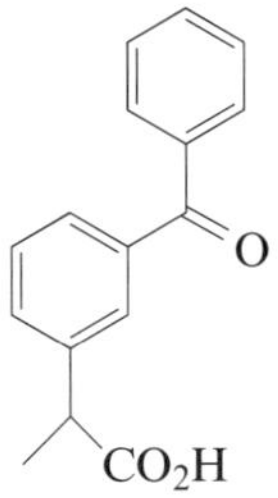

Ketoprofen

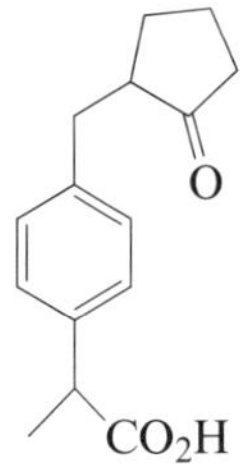

Loxoprofen

Figure 1 *(continued)*

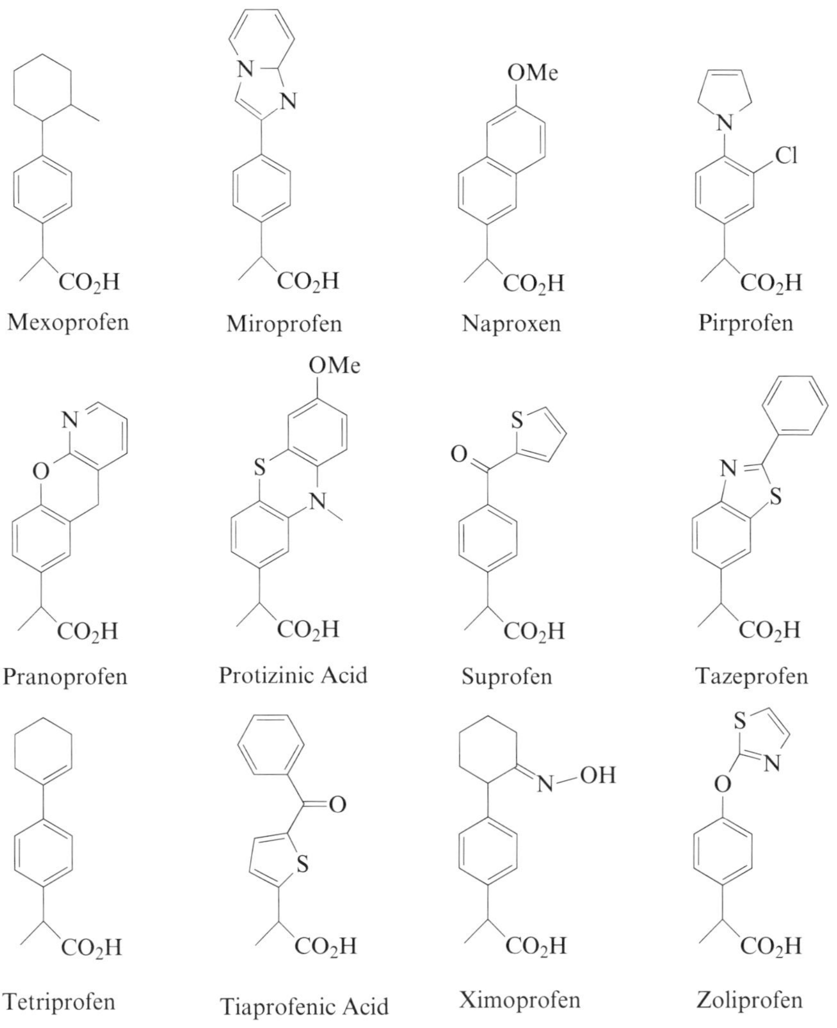

Figure 1

Thus, the rate and extent of distribution of NSAIDs in the body may be important in determining the effectiveness of a given NSAID, and the proliferation of different profen structures can probably be attributed to this fact. Delivery of the active inhibiting moiety to the cyclooxygenase enzyme is influenced by the ability of the whole molecule to reach the necessary systems. For example, although both naproxen and ibuprofen have similar analgesic and anti-inflammatory effects, naproxen has a much longer biological half-life.

All of the profen drugs contain a stereogenic centre, which is the carbon atom bearing the carboxylic acid group. In the case of naproxen, early testing clearly showed that the (*S*)-isomer was more active pharmacologically than the (*R*)-isomer, although the latter is not thought to be harmful; thus, only the (*S*)-isomer of naproxen is sold. Ibuprofen, on the other hand, was originally marketed as the racemate. Later, advantages were claimed for the administration of pure (*S*)-(+)-ibuprofen and development of products containing (*S*)-(+)-ibuprofen is under way. The development of an optically pure drug where the racemate is already in use has been called a 'racemic switch', and details of the isomer issue with relation to naproxen and ibuprofen are discussed below.

Other profen drugs have also been sold as racemic mixtures. However, the claim that (*S*)-(+)-ibuprofen may have advantages over the racemate has prompted frantic efforts in the industry to examine all profens for racemic switch potential. Some specific examples are given later in this chapter.

The commercial importance of (*S*)-(+)-naproxen, and the potential importance of (*S*)-(+)-ibuprofen and other chiral profens, have resulted in much research into chiral methodologies. Many new ways to synthesise these materials in high optical purity have appeared in the literature.[4] Frequently, new asymmetric reactions are demonstrated using intermediates leading to naproxen or ibuprofen. In this chapter, we review methods for producing chiral profen drugs from an industrial perspective. Specific discussions of naproxen and ibuprofen are followed by a more general section on other profens.

3.2 INDUSTRIAL PROCESSES FOR (*S*)-(+)-NAPROXEN

Naproxen was developed by the Syntex Corporation in the 1960s. Early screening of compounds based on naphthaleneacetic acid showed not only that naproxen is a potent anti-inflammatory agent, but also that the (*S*)-isomer is much more effective than either the (*R*)-isomer or the racemate. For example, in one *in vivo* test, (*S*)-(+)-naproxen was reported to be 28 times more effective than the (*R*)-isomer, and three times more effective than the racemate.[5] Resolved material for these studies was obtained by crystallisation of the diastereoisomeric salt pair formed from racemic naproxen and (−)-cinchonidine.

Naproxen was commercialised in the UK in 1973 and in the USA in 1977.[6] Manufacturing methods available to Syntex at this time probably did not include catalytic asymmetric synthesis, as the time necessary to develop this or other strategies (e.g. biological processes) would not have allowed a rapid introduction to the market. Consequently, patents indicate that Syntex chose to produce the racemate and develop an efficient resolution.

Scheme 1 shows a likely synthesis of the racemate.[7] Although early production may have relied on (−)-cinchonidine resolution, more efficient technology was developed subsequently and patented. This utilises *N*-alkyl-D-glucamines as resolving agents,[8] which produce, in the case of *N*-*n*-propyl-D-glucamine, (*S*)-(+)-naproxen (100% *ee*) with only two diastereoisomeric salt crystallisations.[8a]

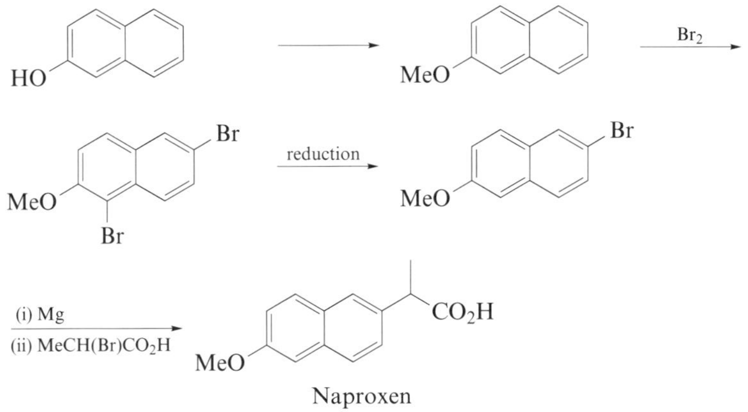

Scheme 1

The crystalline form of naproxen is important when considering process choices. Naproxen is a racemic compound (see Chapter 7 and Volume I, p. 20). This is determined by examination of a plot of melting point vs composition for a mixture of the (*R*)- and (*S*)-isomers (melting point phase diagram), as shown in Figure 2.[9] A racemic compound has a eutectic (low-melting) composition different from the racemic mixture [50% (*R*)-isomer and 50% (*S*)-isomer]. This is the case for naproxen, whose eutectic composition is about 80% (*S*)-isomer–20% (*R*)-isomer (or, since binary melting point phase diagrams of enantiomeric mixtures are symmetrical, 80% (*R*)-isomer–20% (*S*)-isomer).

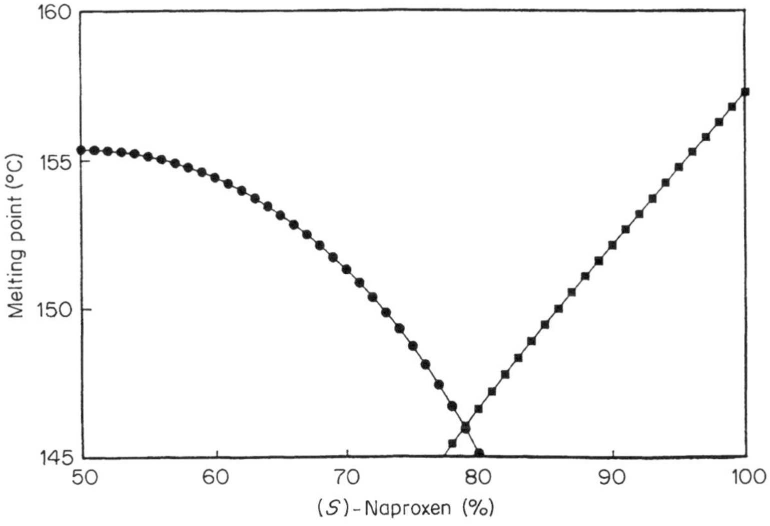

Figure 2 Calculated melting point phase diagram for naproxen

The eutectic composition is a most important feature to be considered when planning racemate separations. Crystallisation of a given isomer mixture will usually lead to a crystal crop whose composition is 'uphill' on the phase diagram curve compared with the starting composition. This is because the lowest melting (most soluble) composition remains in the mother liquor. Both melting and dissolution processes involve disruption of the crystal lattice, which means that the lowest melting composition is usually the same as the most soluble composition. Also, for a racemic compound, the eutectic composition may not be passed by crystallisation under thermodynamic conditions, so crystallisation of a racemic compound will only yield a crystal crop of increased optical purity if the starting composition lies between the eutectic composition and pure enantiomer.

In the case of naproxen, a mixture only requires about 80% of one isomer in order to allow the optical purity to be upgraded by recrystallisation of the underivatised acid. Thus, any process (resolution or synthesis) that gives >80 wt% (*S*)-(+)-naproxen can be followed by recrystallisations to yield 100 wt% (*S*)-(+)- naproxen. The commercial importance of (*S*)-(+)-naproxen made it a prime target for those seeking new methods for the synthesis of chiral molecules.

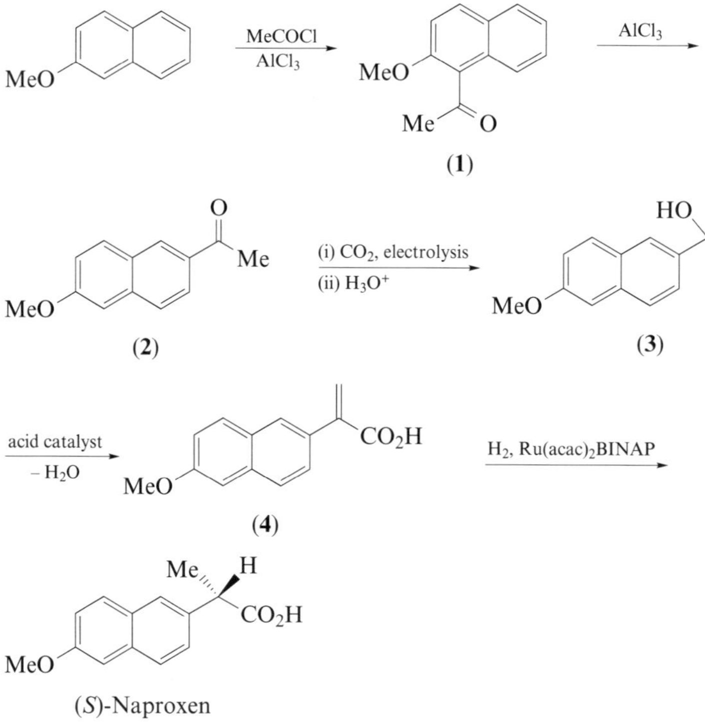

Scheme 2

Consequently, many new routes to (*S*)-(+)-naproxen were reported, the most widely publicised being a number of asymmetric syntheses.

Monsanto chemists developed a route starting with 6-acyl-2-methoxynaphthalene (**2**), as shown in Scheme 2.[9] 2-Methoxynaphthalene, which is the starting material in almost all (*S*)-(+)-naproxen syntheses, can be acylated with high regioselectivity to give **2**. It is interesting that the kinetic product of acylation, **1**, can be converted to the thermodynamically stable **2** under acylation conditions.[11] This intermediate is then electrochemically carboxylated to give hydroxy acid **3**, the dehydration of which gives the acrylic acid derivative **4**, which is the prochiral starting material for the following asymmetric reduction.This reduction is accomplished using a chiral catalyst similar to that reported by Noyori, and co-workers consisting of Ru(0) bearing a chiral BINAP ligand.[11,12] Monsanto developed a variant that was claimed to show an increased reaction rate compared with the Noyori system[14] and, using this catalyst, the acrylic acid derivative **4** is converted into (*S*)-(+)-naproxen in 100% yield and up to 98% *ee.*[10,15] This optical purity is within USP optical purity specifications,[16] but can be improved by recrystallisation if desired. The asymmetric catalyst is critical to this process and, although expensive, it is both robust and recyclable.[10] More recently, a heterogeneous chiral Ru catalyst was developed that converts **4** into (*S*)-(+)-naproxen with high enantioselectivity.[17]

DuPont researchers developed a process based on the catalytic asymmetric hydrocyanation of 6-methoxy-2-vinylnaphthalene (Scheme 3).[18] The prochiral intermediate **5** can be made most simply by a reduction–dehydration sequence starting with **2**, but also by Ni-catalysed coupling of 6-methoxy-2-bromonaphthalene with vinyl Grignard reagents.[19]

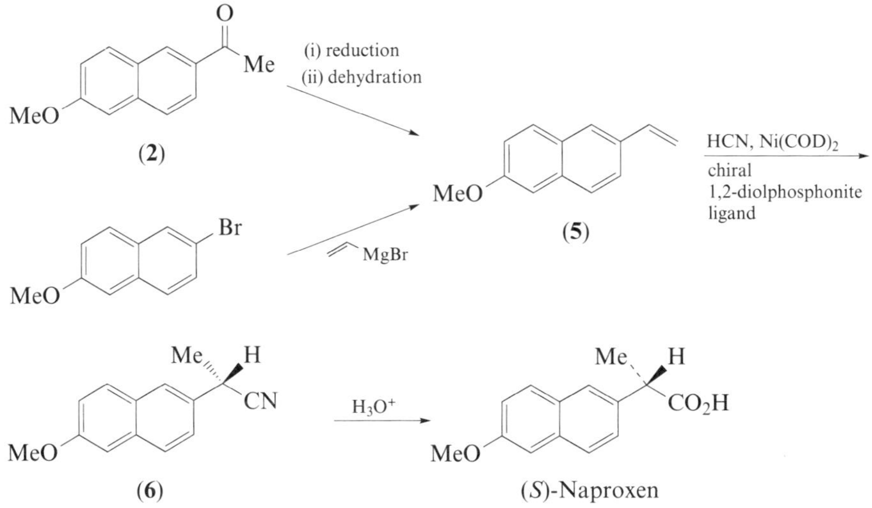

Scheme 3

Key to the asymmetric hydrocyanation process was the invention of asymmetric phosphinites, which induce enantioselectivity in the hydrocyanation through ligation of the Ni catalyst (see also Chapter 15). These phosphinites were made by elaboration of mono- and disaccharide hydroxyl groups, affording relatively inexpensive catalysts (based on Ni and sugars) which can be easily modified to optimise the desired activity. Thus, **5** was converted into **6** in >85% yield with reasonable (75–85%) *ee*. The product nitrile exhibits a favourable phase diagram, as does naproxen, and was recrystallised twice to give material in >99% *ee*. Hydrolysis then gave (*S*)-(+)-naproxen.

Another industrial process utilising asymmetric synthesis was developed by Zambon[20] (Scheme 4). An achiral version of the 1,2-aryl shift reaction was invented by Blaschim (Italy), and developed by Zambon into a manufacturing process, the key intermediate being the chiral ketal **7**. The chiral auxiliary, a tartaric acid ester, influences both the bromination and rearrangement reactions so that naproxen of 99% *ee* results. Since the auxiliary is used stoichiometrically, it must be recovered and re-used, as must a resolving agent.

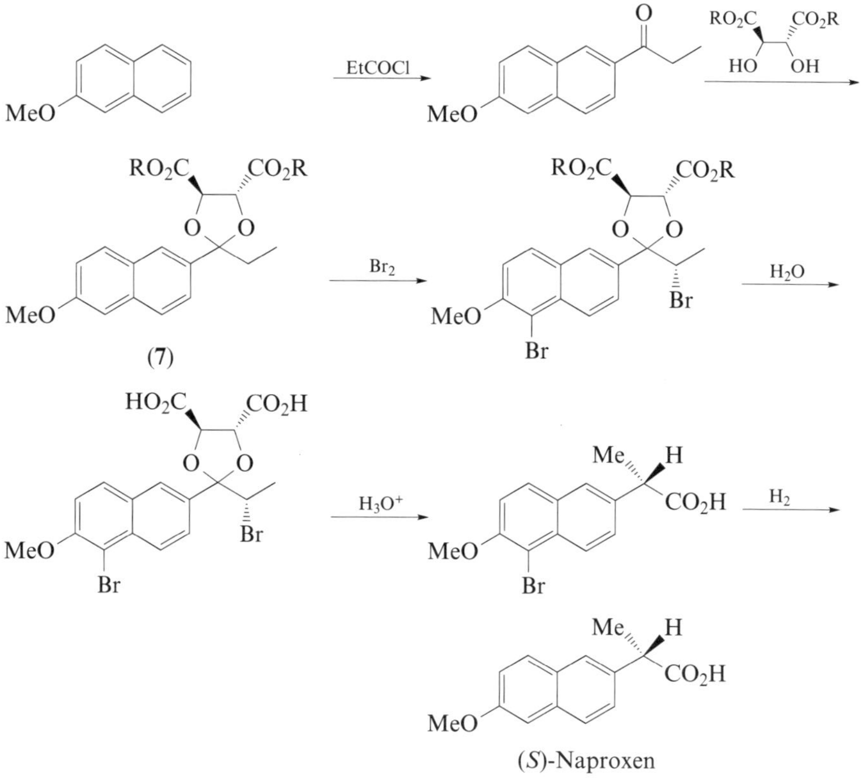

Scheme 4

At first glance, some of the newer methods, exemplified by those above, would seem preferable to a classical resolution. For example, the Monsanto asymmetric synthesis utilises selective, high-yielding reactions and produces a product of sufficient optical purity with a chiral catalyst, avoiding the recycle and racemisation steps necessary with resolution. However, other factors must also be considered. Firstly is the price obtainable for the final product, and, in particular, the price stability. It is critical to use the least expensive process available because these costs comprise an increasing percentage of the final bulk drug price. As the price of the bulk drug decreases it becomes more of a commodity chemical, and low-cost production becomes an even more important issue. Secondly, capital expenses are very important. A new process requiring a new plant may not compete economically with an old process in an existing plant, even if the new process has chemical and processing advantages. The cost of building a new plant can override these advantages. Thirdly, the importance of the requirements of governmental regulatory agencies cannot be overlooked (see also Chapter 2). In the USA the Food and Drug Administration (FDA) requires that a significant change to the process to manufacture a drug is preceded by extensive tests, possibly involving clinical trials to show that the resulting product is identical with the original one.[21]

3.3 INDUSTRIAL PROCESSES FOR (*S*)-(+)-IBUPROFEN

Ibuprofen was patented by the Boots Pure Drug Co. in the late 1960s.[22] Efficacy testing at that time revealed that (1) (*R*)- and (*S*)-isomers have similar *in vivo* potency[23], (2) only the (*S*)-isomer inhibits prostaglandin synthetase *in vitro*[23] and (3) after oral administration of the individual enantiomers, metabolites found in the urine are dextrorotatory.[24] Thus, the conclusion was that chiral inversion of the (*R*)-isomer to the active (*S*)-isomer occurs *in vivo*,[24] and an enzyme system was proposed to account for this.[25] Subsequent work led to further insights into the inversion process.[26]

All of these studies led to the production and sale of ibuprofen as a racemic mixture. Market introduction in the UK was in 1969 and in the USA in 1974. Numerous research projects looking for new racemate syntheses have been initiated, resulting in a flood of methods, many of which are applicable to a wider range of profen drugs.[27] In 1985 the US patent[22b] on ibuprofen expired, opening the generic prescription market. Even more significant was the FDA decision at that time to allow racemic ibuprofen to be sold over the counter (OTC) in the USA. These events prompted interest in commercial production of racemic ibuprofen throughout the world. Suppliers of bulk active racemate during this period included Boots in the UK and Albemarle Corporation, which was the only US producer; the original route patented by Boots is shown in Scheme 5.[28]

Illustrative of the chemistry developed more recently by others are the routes of the Nippon Petrochemicals Company shown in Scheme 6. Both routes

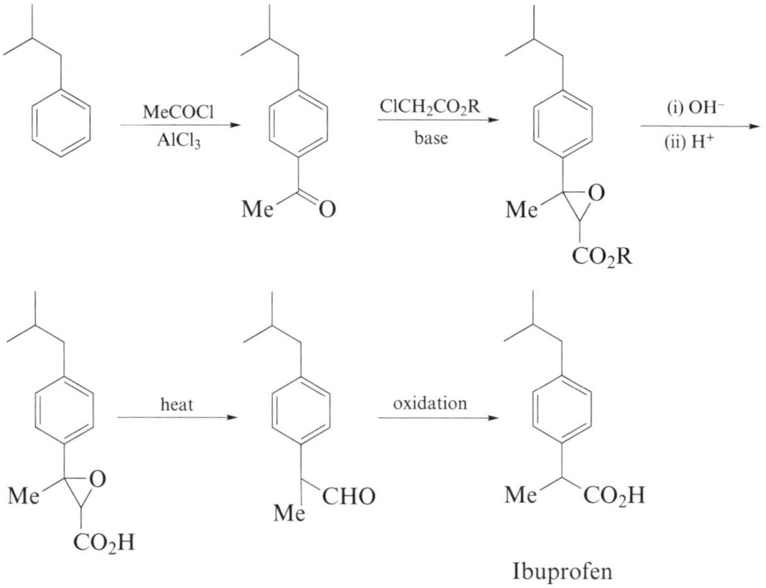

Schemes 5

produce isobutylstyrene (**8**), which is carbonylated in a final step to give ibuprofen. Route A[29] involves acid-catalysed arylation of acetaldehyde, followed by cracking of the resulting adduct to give equal amounts of isobutylstyrene (**8**) and isobutylbenzene, the latter being recycled. In route B,[30] reaction of toluene and ethylene gives perethylated toluenes, which are passed, along with isobutylbenzene, over an acidic transalkylation catalyst. The resulting mixture is fractionally distilled to remove the desired *p*-ethyl(isobutyl)benzene (present at 5–10%). Light fractions, containing lightly ethylated toluenes, are returned to the ethylation reactor and heavy fractions, containing heavily ethylated toluenes and isobutylbenzenes, are fed with fresh isobutylbenzene back to the transalkylation reactor. Dehydrogenation of *p*-ethyl(isobutyl)benzene then gives isobutylstyrene (**8**).

In the USA, a Boots–Hoechst–Celanese joint venture was established which began production in 1993,[31] and the route shown in Scheme 7 is believed to be used in that facility. This chemistry involves three steps, all catalytic, and is a 100% atom-efficient process (i.e. every atom put into the process either becomes part of the final product or is recycled).

In 1989, a US patent assigned to Analgesic Associates was issued claiming that analgesia in animals (including humans) was more rapidly attained and enhanced in effect when (*S*)-(+)-ibuprofen was used compared with the use of the racemate.[32] At about the same time that the Analgesic Associates patent application was undergoing prosecution, a European application was filed by

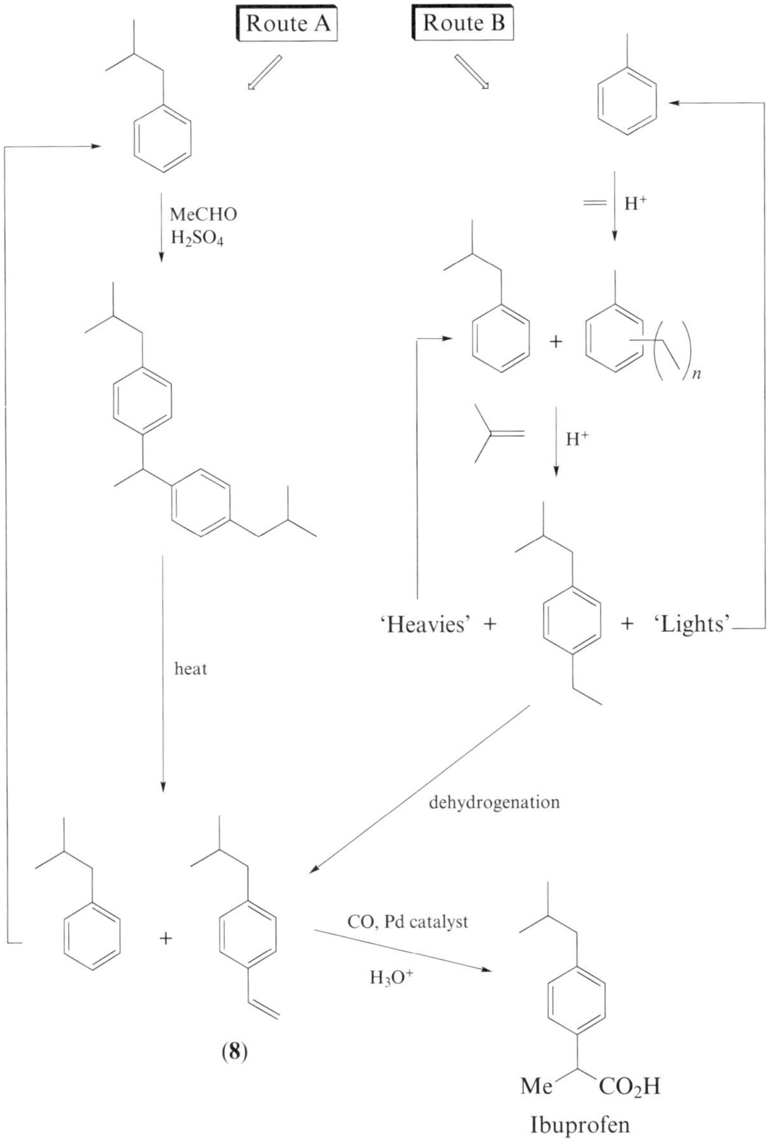

(8)

Schemes 6

Medice Chem.-Pharm. Fabrik of West Germany.[33] Ultimately, each patent was licensed to a pharmaceutical company: Sterling Drug licensed Analgesic Associate's rights, Merck & Company licensed Medice's US and Canadian rights and Boots Drug Company licensed Medice's UK rights. The claim that (*S*)-(+)-ibuprofen exhibits superior performance compared with the racemate has led to

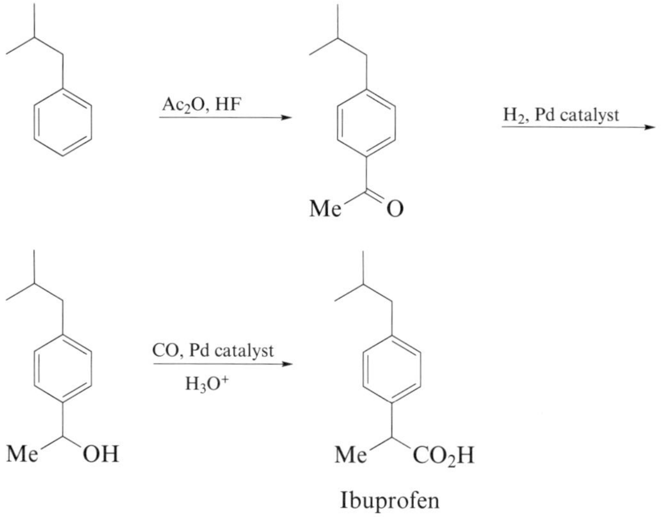

Schemes 7

recent attempts to market products containing only the (*S*)-isomer, and the sales potential of (*S*)-(+)-ibuprofen has prompted research into its synthesis.

Although naproxen was traditionally the molecule of choice for demonstrating new synthetic routes to profen drugs, ibuprofen has received more attention in the last few years, and in some cases these newly developed methods seem to be as effective for ibuprofen as naproxen. For example, Monsanto can apply their electrocarboxylation–asymmetric reduction chemistry to (*S*)-(+)-ibuprofen synthesis (Scheme 8). The hydrogenation is reported to give (*S*)-(+)-ibuprofen of 96% *ee.*[10] A commercially significant variant was reported by Albemarle researchers, who found that an effective chiral catalyst can be generated *in situ* from relatively inexpensive Ru(III) acetylacetonate.[34]

However, there can be differences in the efficiency of a given process when applied to naproxen vs ibuprofen. For example, the DuPont hydrocyanation reaction is reported to hydrocyanate isobutylstyrene (**8**) with much poorer enantioselectivity (50% *ee* product) than the corresponding reaction for (*S*)-(+)-naproxen.[18]

Union Carbide Corporation has also developed an asymmetric route to (*S*)-(+)-ibuprofen (Scheme 9). This process incorporates Carbide's Rh-catalysed hydroformylation (oxo) reaction.[35] The key to the original achiral hydroformylation method was the use of phosphine ligands to modify the Rh catalyst, giving not only the desired reactivity but also control over regioisomer distribution. Thus, olefins can be converted into aldehydes with high linear:branched ratios, the linear isomers being the products of highest value.[36] By using phosphite instead of phosphine ligands, branched rather than linear aldehydes result. In

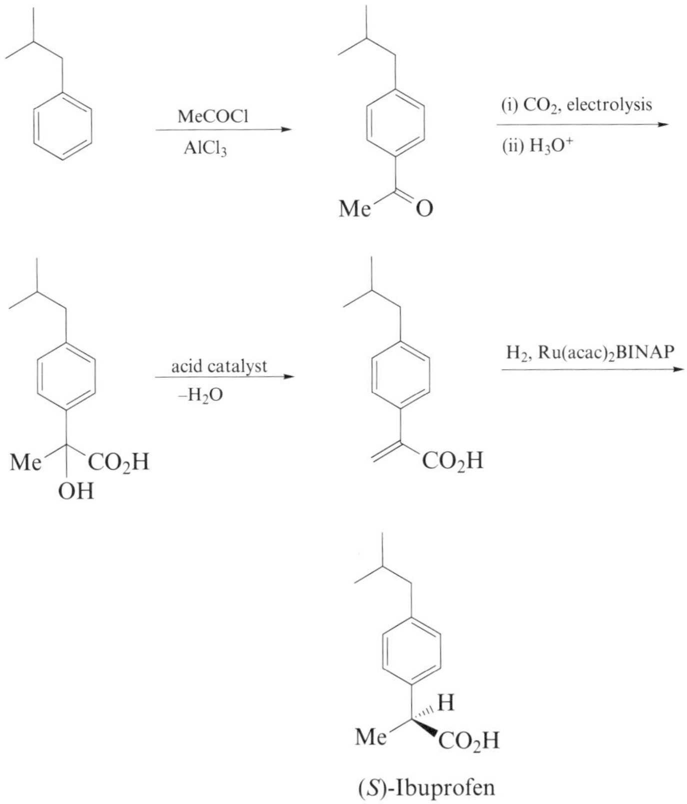

Scheme 8

addition, the use of chiral phosphites allowed enantioselective hydroformylation of isobutylstyrene (**8**).[37] The aldehyde product has been obtained in 82% *ee*, and stereospecific oxidation then produces (*S*)-(+)-ibuprofen.

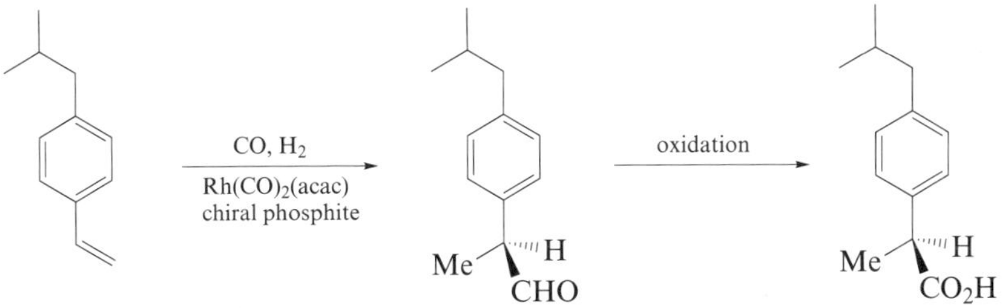

(**8**) (*S*)-Ibuprofen

Scheme 9

The prochiral precursor used by Union Carbide, isobutylstyrene (**8**), can be made in several ways. The methods of the Nippon Petrochemical Company have already been described (Scheme 6), and other routes are analogous to those used to make methoxyvinylnaphthalene (Scheme 3, compound **5**), including reduction and dehydration of isobutylacetophenone[38] and reaction of isobutylmagnesium chloride with bromostyrene (Scheme 10).[22]

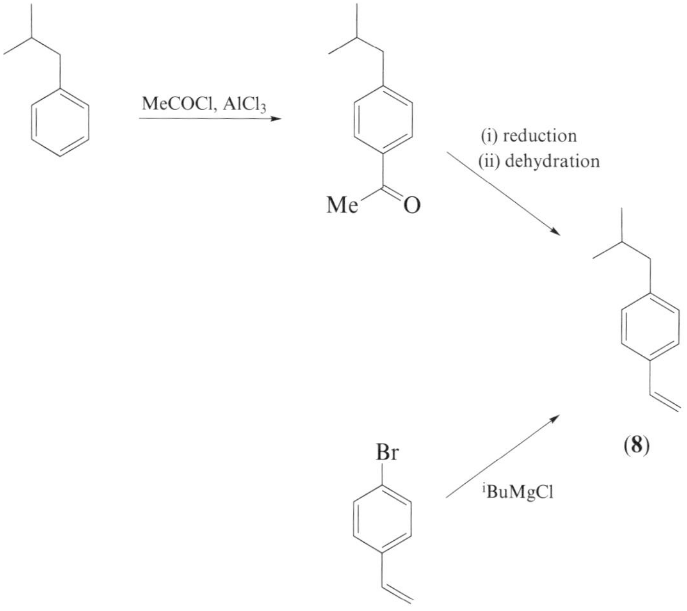

(**8**)

Scheme 10

Even more straightforward than hydroformylation would be hydrocarboxylation, in which the acid is produced directly. The achiral version of this reaction works very well on isobutylstyrene (**8**) to give racemic ibuprofen.[29] Alper and Hamel[39] reported such a reaction in 1990.

Merck researchers reported an (*S*)-(+)-ibuprofen synthesis based on the asymmetric addition of an alcohol to a ketene (Scheme 11).[40] This sequence starts with racemic ibuprofen, which is converted into the prochiral ketene. Readily available (lactic acid-derived) chiral alcohols are then used to carry out an asymmetric protonation reaction, affording diastereoisomeric esters with good diastereoselectivity. Thus, all of the racemic mixture can be converted into one enantiomer. When the starting material was ibuprofen, (*S*)-(+)-ibuprofen esters were obtained in 94–99% diastereoisomeric excess (*de*), and an ester of (*S*)-(+)-naproxen was produced in 80% *de*; as with a resolution, the chiral auxiliary must be recovered.

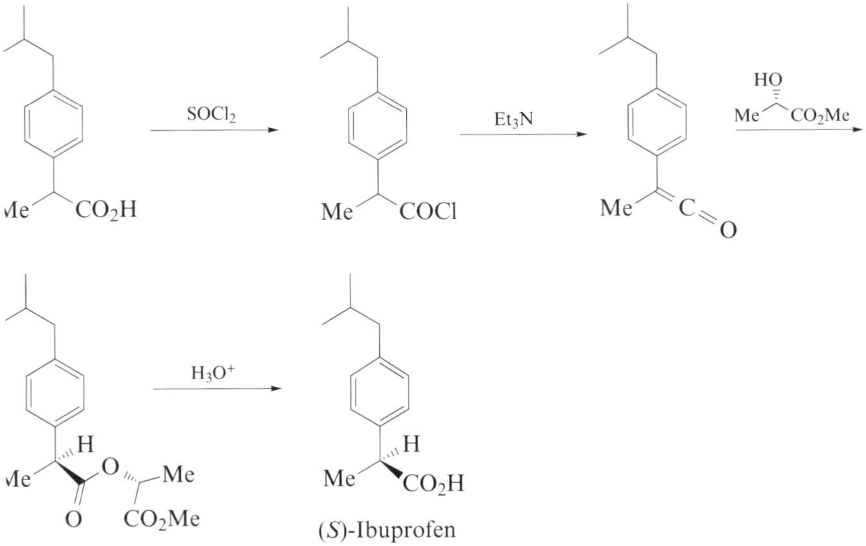

Scheme 11

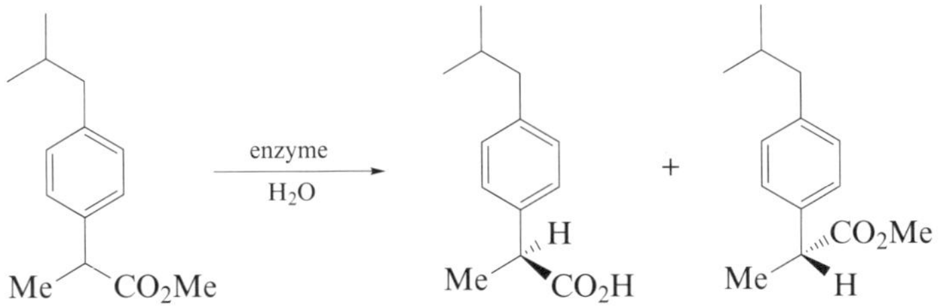

Scheme 12

Enzymatic resolution processes based on selective hydrolyses of racemic mixtures of ibuprofen esters have been investigated by several companies. A kinetic resolution of this type is shown in Scheme 12. The Wisconsin Alumni Research Foundation (WARF) made one of the earliest claims to a process of this type,[41] and Sepracor,[42] Gist-Brocades[43] and Rhône-Poulenc[44] have also developed similar chemistry. In each of these cases the processes are reported to have broad substrate specificity, in terms of both the profen moiety and the corresponding starting derivative. The methyl esters of naproxen and ibuprofen are the most common starting materials included in patent examples.

One advantage typically obtained by using enzymes is very high enantioselectivity. For example, Gist-Brocades found that addition of the methyl ester of racemic ibuprofen to growing cultures of *Pseudomonas, Mucor, Arthrobacter* or *Bacillus* species produced (*S*)-(+)-ibuprofen in 74–98% *ee*.[43] Gist–Brocades uses a growing microorganism, Sepracor uses isolated enzyme preparations and Rhône-Poulenc uses highly purified enzymes called isozymes. There are advantages and drawbacks to each of the systems, which must be evaluated case by case. The purer the enzyme, the more expensive is its production, but high purity often means high enantioselectivity. Use of a growing microorganism provides a relatively inexpensive enzyme source, but the aqueous systems necessary for fermentation may limit the amount of organic substrate that can be present.

Another important process consideration is the sense of the enantioselectivity. For example, a racemic mixture of the methyl ester of ibuprofen can afford a mixture of (*S*)-acid and (*R*)-ester or (*R*)-acid and (*S*)-ester, depending on the enzyme used. If the (*S*)-acid is the desired product, the former situation is better because the product is obtained directly and the unwanted isomer can be racemised and recycled as the methyl ester. If the (*R*)-acid were to be produced, two additional process steps would be necessary: hydrolysis of the (*S*)-ester to give product and re-esterification of the (*R*)-acid for racemisation and recycle.

The use of enzymes depends on their stabilities in non-aqueous solvents. Historically, efforts to overcome poor performance in this area include the membrane reactor system designed by Sepracor, in which the enzyme is immobilised in hollow-fibre reactors, allowing continuous kinetic resolutions.[45] Altus Biologics have developed crosslinked, crystalline enzymes that function in organic solvents, and one example, a lipase from *Candida rugosa*, hydrolyses racemic ibuprofen methyl ester to give 95% *ee* (*S*)-(+)-ibuprofen.[46] From a commercial standpoint, the enzyme is viewed simply as a catalyst, whose cost, productivity, selectivity and re-usability have to compete with those of a conventional resolving agent.

Because many companies already have the capacity to produce racemic ibuprofen, they have chosen to investigate new diastereoisomeric salt resolution processes rather than asymmetric routes. Ibuprofen can be resolved by common resolving agents such as α-methylbenzylamine[47] and lysine.[48] Designer amines, **9** and **10**, were developed in Japan specifically for the resolution of ibuprofen.[49] Crystallisation of a mixture of racemic ibuprofen and (−)-**9** from methanol–water, followed by one recrystallisation of the resulting salt, gave a salt containing (*S*)-(+)-ibuprofen with 100% *ee*.[49b]

As discussed earlier, the crystalline form of the product needs to be considered when evaluating process choices. Like naproxen, ibuprofen is a racemic compound, this being evident from the calculated binary melting point phase diagram (Figure 3), which shows that the eutectic composition for ibuprofen is about 90% (*S*)-isomer–10% (*R*)-isomer [or 90% (*R*)-isomer–10% (*S*)-isomer].[9] The calculated diagram agrees reasonably well with results determined experimentally.[9,50]

The eutectic composition dictates that ibuprofen needs to be at least 90% pure (*S*)-isomer for crystallisation to result in optical purification. However, even

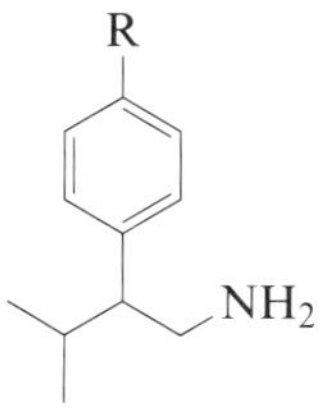

(**9**) R = Me
(**10**) R = Cl

above this composition limit, in practice, purification by crystallisation is rather inefficient. This is generally true when the eutectic is closer to pure isomer. We find that crystallisation of 96% *ee* (*S*)-(+)-ibuprofen gives a crystal crop of only about 97–98% *ee*,[9] which provides much less working margin than is available for naproxen. The result is that any asymmetric process must give nearly specification-grade (*S*)-(+)-ibuprofen, or multiple recrystallisations will be required to attain the necessary optical purity.

A method to overcome the difficult optical purification of enantiomerically enriched ibuprofen was developed at Albemarle. While ibuprofen itself exhibits an unfavourable phase diagram, that for the sodium salt of ibuprofen is much more favourable. As can be seen in Figure 4, the eutectic composition for sodium ibuprofenate is about 60% (*S*)-isomer–40% (*R*)-isomer [or 60% (*R*)-isomer–40% (*S*)-isomer], so ibuprofen produced by any method that is at least 60% optically

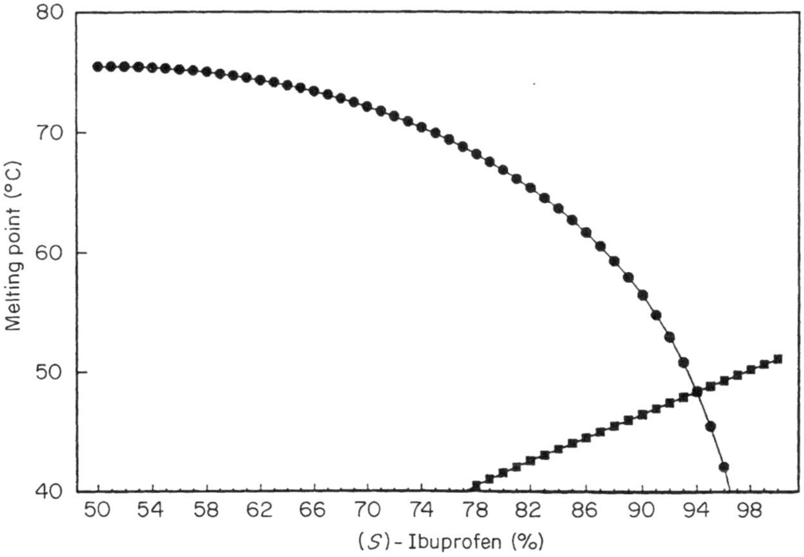

Figure 3 Calculated melting point phase diagram for ibuprofen

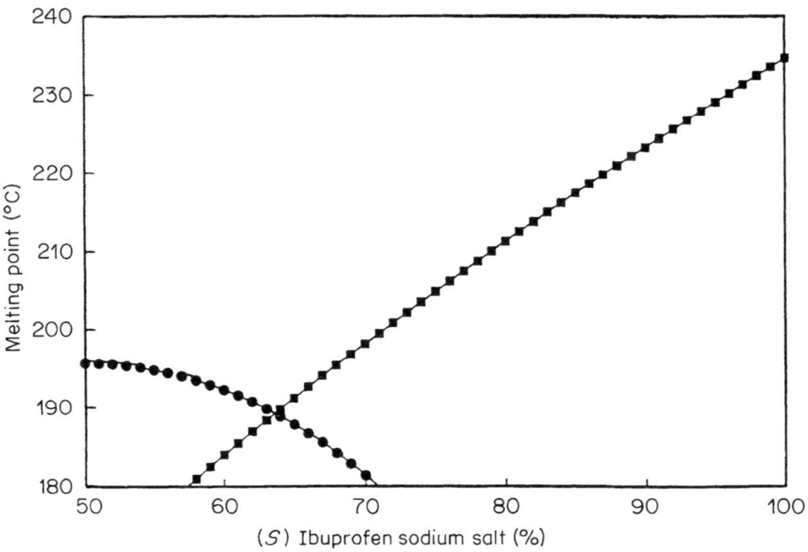

Figure 4 Calculated melting point phase diagram for sodium ibuproferate

pure can be upgraded via the sodium salt. This method works well, as shown by the one-step upgrade of 76% *ee* (*S*)-(+)-ibuprofen to 100% *ee* (*S*)-(+)-ibuprofen by treatment with sodium hydroxide in acetone.[9] The method is also of value when coupled with a diastereoisomeric resolution process, reducing the number of steps involving the resolving agent and thereby minimising its recycle and loss.

Which is the best type of process for commercial (*S*)-(+)-ibuprofen production? As it was with naproxen, the answer is greatly influenced by the plant already in place. Plants assets have been built and are currently producing racemic ibuprofen, so, with an additional capital investment, a resolution process can be added to the racemate production unit; add to this the fact that technology for ibuprofen racemisation is known,[51] the resolution route looks even better. As any unwanted isomer can be easily recycled, the yield of (*S*)-(+)-ibuprofen from racemic ibuprofen can be nearly 100%.

Another approach, taking advantage of existing racemate production capacity, might be to utilise an intermediate from the racemate process for an asymmetric synthesis. In this case, the intermediate needs to be prochiral (or easily converted into a prochiral intermediate), and any unwanted (*R*)-isomer resulting from the asymmetric step must be readily convertible back to the prochiral intermediate. This is rarely as simple in practice. Some additional questions can be raised: are the existing plants large enough to produce enough racemic isomer for sale and also for resolution, and how much of the racemic ibuprofen market will (*S*)-(+)-ibuprofen displace, if any? These questions are difficult to answer until commercial production and sale of (*S*)-(+)-ibuprofen begin. Even if it is believed that more capacity will be needed, it may be cheaper to expand a current process than to build a new plant with a more efficient process.

One consultant who has closely followed racemic switches has attempted to analyse competitive processes which he believes are being considered for (*S*)-(+)-ibuprofen production. His conclusion is that the most competitive processes today are likely to be classical resolutions carried out by racemate producers, although asymmetric synthesis and biotransformation routes are not far behind.[52]

3.4 OTHER PROFENS AND THE FUTURE

New racemic switch candidates are being aggressively sought by many companies in the pharmaceutical industry. Determination of the physiological behaviours of pure enantiomers derived from drugs currently sold as racemates is widespread.[52] The efforts of Medice and Analgesic Associates to secure and license (*S*)-(+)-ibuprofen patents have led others to pursue similar strategies for other profen drugs, including flurbiprofen and ketoprofen. Racemic flurbiprofen is sold by Upjohn as Ansaid and racemic ketoprofen by Wyeth-Ayerst as Orudis. Sepracor (Marlborough, MA, USA) has filed or received use patents on (*S*)-(+)-flurbiprofen and (*S*)-(+)-ketoprofen claiming that they have enhanced analgesic activity (including faster onset) and reduced side effects (such as gastrointestinal irritation and kidney toxicity).[53] In addition, the same drugs are claimed to inhibit bone loss due to periodontal disease.[54] Analgesic Associates has also remained active in the racemic switch area, concentrating on flurbiprofen and ketoprofen, and a series of US patents have been issued claiming advantages such as faster onset and lower dosage when the (*S*)-isomers are used.[55]

It is thus likely that the chiral profen drugs with the highest commercialisation potential, besides (*S*)-(+)-ibuprofen, are (*S*)-(+)-flurbiprofen and (*S*)-(+)-ketoprofen. Because the volumes of the racemic forms of these currently sold are relatively small, there has been correspondingly little research aimed at efficient process chemistry. Flurbiprofen and ketoprofen both contain *meta*-substituted aromatic moieties, and the former contains a fluorine atom, making their syntheses inherently more difficult and expensive compared with those of naproxen and ibuprofen.

Here, then, is the immediate future of chiral profen drugs. Methods developed for (*S*)-(+)-naproxen and/or (*S*)-(+)-ibuprofen may be readily applicable to (*S*)-(+)-flurbiprofen and/or, (*S*)-(+)-ketoprofen, or new methods may be developed. In any case (*S*)-(+)-flurbiprofen and (*S*)-(+)-ketoprofen processes based on these methods face less economic competition given the immaturity of the drugs. Of course, marketplace competition will determine whether volumes will grow, but new, efficient chemistry stands a good chance of incorporation into new plants if that growth occurs.

3.5 REFERENCES

1. For a review covering the physiology and medicinal chemistry of traditional NSAIDs, see Lombardino, J. G., (Ed.), *Nonsteroidal Antiinflammatory Drugs*, Wiley, New York, 1985.
2. Reuben, B. G., and Wittcoff, H. A., *Pharmaceutical Chemicals in Perspective*, Wiley, New York, 1989, p. 98.
3. Flower, R. J., Moncada, S., and Vane, J. R., in *The Pharmacological Basis of Therapeutics* Gilman, A. G., Goodman, L. S., Rall, T. W., and Murad, F., (Eds.) Macmillan, New York, 1985, Ch. 29.
4. Sonawane, H. R., Bellur, N. S., Ahuja, J. R. and Kulkarni, D. G., *Tetrahedron: Asymmetry*, **3**, 163 (1992).
5. Harrison, I. T., Lewis, B., Nelson, P., Books, W., Roszkowski, A., Tomalonis, A. and Fried, J. H., *J. Med. Chem.*, **13**, 203 (1970).
6. *Chemica Useful Book*, Chemica, Caterham, Surrey, 1991, naproxen profile.
7. Sheldon, R. A., *Chirotechnology*, Marcel Dekker, New York, 1993, p. 349.
8. (a) Holton, P. G., *US. Pat.*, 4246193, 1981; (b) Felder, E., San Vitale, R., Pitre, D., and Zutter, H., *US. Pat.* 4246164, 1981.
9. (a) Manimaran, T. and Stahly, G. P., *Tetrahedron: Asymmetry*, **4**, 1949 (1993); (b) Herndon, R. C., Manimaran, T., and Stahly, G. P., *US Pat.* 5248813, 1993.
10. Chan, A. S. C., *Chemtech*, **23** (3), 46 (1993).
11. Giordano, C., and Villa, M., *Synth. Commun.*, **20**, 383 (1990).
12. Ohta, T., Takaya, H., Kitamura, M., Nagai, K., and Noyori, R., *J. Org. Chem.*, **52**, 3174 (1987).
13. (a) Yoshikawa, S., Saburi, M., Ikariya, T. and Ishii, Y., (Takasago Perfumery) *US Pat.*, 4691037, (1987); (b) Takaya, H., Ohta, T., Noyori, R., Yamada, N., Takesawa, T., Sayo, N., Taketomi, T., Kumobayashi, H. and Akutagawa, S., (Takasago Perfumery) US Pat. 4739084, 1988; (c) Takaya, H., Ohta, T., Noyori, R., Sayo, N., Kumobayashi, H., and Akutagawa, S (Takasago Perfumery) US Pat. 4739085, 1988.
14. Chan, A. S. C., and Laneman, S. A (Monsanto), US Pat. 5144050, 1992.
15. Chan, A. S. C., *US Pat.* 4994607, 1991.
16. Specification is $[\alpha]_D$ = 63–68.5 °C (c = 1, $CHCl_3$), which corresponds to 95–100% *ee*: *United States Pharmacopeia*, US Pharmacopeial Convention, Rockville, MD, 1994, p. 1053.
17. (a) Wan, K. T., and Davis, M. E., *Nature* (*London*), 370, 449 (1994); (b) Wan, K. T., and Davis, M. E., *J. Catal.*, **1**, 148 (1994).
18. RajanBabu, T. V., and Casalnuovo, A. L., *J. Am. Chem. Soc.*, **114**, 6265 (1992).
19. Nugent, W. A. and McKinney, R. J., *J. Org. Chem.*, **50**, 5370 (1985).
20. Collins, A. N., Sheldrake, G. N. and Crosby, J. (Eds), *Chirality in Industry*, Wiley, Chichester, 1992, Ch. 15.
21. *Guideline for Submitting Supporting Documentation in Drug Applications for the Manufacture of Drug Substances*, Food and Drug Administration, Rockville, MD, 1987.
22. (a) Nicholson, J. S. and Adams, S. S., (Boots Pure Drug Co.) *Br. Pat.* 971700, 1964; (b) Nicholson, J. S. and Adams, S. S., (Boots Pure Drug Co.), *US Pat.* 3228831, 1966.
23. Adams, S. S., Bresloff, P., and Mason, G., *Pharm. Pharmacol.*, **28**, 256 (1976).
24. (a) Adams, S. S., Cliffe, E. E., Lessel, B. and Nicholson, and J. S., *J. Pharm. Sci.*, **56**, 1686 (1967); (b) Mills, R. F. N., Adams, S. S., Cliffe, E. E., Dickinson, W., and Nicholson, J. S., *Xenobiotica*, **3**, 589, (1973).
25. Wechter, W. J., Loughhead, D. G., Reischer, R. J., Van Giessen, G. J., and Kaiser, D. G., *Biochem. Biophys. Res. Commun.*, **61**, 833 (1974).
26. (a) Sheldon, R. A., *Chirotechnology*, Marcel Dekker; New York, 1993, p. 57; (b)

Knihinicki, R. D., Day, R. O. and Williams, K. M., *Biochem. Pharmacol.*, **42**, 1905 (1991).
27. Rieu, J.-P., Boucherle, A., Cousse, H. and Mouzin, G., *Tetrahedron,* **42**, 4095 (1986).
28. Boots Pure Drug Co., *Fr. Pat.* 1 545 270, 1968.
29. Shimizu, I., Hirano, R., Matsumura, Y., Nomura, H., and Uchida, K. (Nippon Petrochemicals) , *US Pat.*, 4 694 100, 1987.
30. (a) Shimizu, I., Matsumura, Y., Tokumoto, Y., and Uchida, K., (Nippon Petrochemicals) *US Pat.*, 5 097 061, 1992; (b) Tokumoto, Y., Shimizu, I., and Inoue, S. (Nippon Petrochemicals), *US Pat.*, 5 166 419, 1992.
31. (a) *Chem. Mkt. Rep.*, May 17 (243), 20 (1993); (b) *Chem. Mkt. Rep.*, Nov. 22 (244), 41 (1993).
32. Sunshine, A., and Laska, E. M., (Analgesic Associates) *US Pat.*, 4 851 444, 1989.
33. (Medice Chem.-Pharm.), *Eur. Pat. Appl.*, 267 321, 1986.
34. Manimaran, T., Wu, T.-C., Klobucar, W. D., Kolich, C. H., Stahly, G. P., Fronczek, F. R., and Watkins, S. E., *Organometallics*, **12**, 1467 (1993).
35. Fowler, R., Connor, H., and Bachl, R. A., *Chemtech*, 722 (1976).
36. Collman, J. P., Hegedus, L. S., Norton, J. R., and Finke, R. G., *Principles and Applications of Organotransition Metal Chemistry*, University Science Books, Mill Valley, CA, 1987, pp. 625–630.
37. Babin, J. E., and Whiteker, G. T., (Union Carbide) *PCT Int. Pat. Appl.*, WO 93/03839, 1993.
38. Takeda, M., Uchide, M. and Iwane, H., (Mitsibushi Petrochemical) *US. Pat.*, 4 329 507, 1982.
39. Alper, H. and Hamel, N., *J. Am. Chem. Soc.*, **112**, 2803 (1990).
40. (a) Larsen, R. D., Corley, E. G., Davis, P., Reider, P. J. and Grabowski, J. J., *J. Am. Chem. Soc.*, **111**, 7650 (1989); (b) Corley, E. G., Grabowski, J. J., Larsen, R. D., and Reider, P. (Merck), *US Pat.*, 4 940 813, 1990.
41. Sih, C. J., (University of Wisconsin) *Eur. Pat. Appl.*, 227 078, 1986.
42. (a) Matson, S. L. (Sepracor), *US. Pat.*, 4 800 162, 1989; (b) Wald, S. A., Matson, S. L., Zepp, C. M., and Dodds, D. R., (Sepracor) *US Pat.*, 5 057 427, 1991.
43. (a) Bertola, M. A., Marx, A. F., Koger, H. S., Quax, W. J., Van der Laken, C. J., Phillips, G. T., Robertson, B. W., and Watts, P. O., (Gist-Brocades) *US Pat.*, 4 886 750, 1989; (b) Bertola, M. A., De Smet, M. J., Marx, A. F., and Phillips, G. T., (Gist-Brocades) *US Pat.*, 5 108 917, 1992.
44. (a) Cobbs, C. S., Barton, M. J., Peng, L., Goswami, A., Malick, A. P., Hamman, J. P. and Calton, G. J., (Rhône–Poulenc) *US Pat.*, 5 108 916, 1992; (b) Goswami, A., *US Pat.*, 5 175 100, 1992.
45. *Chem. Eng. News*, Sept. 28, 67 (1992).
46. *Chem. Eng. News*, Sept. 19, 72 (1994).
47. (a) Larsen, R. D. and Reider, P. (Merck), *US Pat.*, 4 946 997, 1990; (b) Manimaran, T. and Impastato, F. J. (Ethyl Corp.), *US Pat.*, 5 015 764, 1991; (c) Choudhury, A. A., Kadkhodayan, A. and Patil, D. R. (Ethyl Corp.), *US Pat.*, 5 221 765, 1993; (d) Lin, R. W. (Ethyl Corp.), *US Pat.*, 5 278 334, 1994; (e) Trace, R. L. (Ethyl Corp.), *US Pat.*, 5 278 338, 1994.
48. Tung, H.-H., Waterson, S. and Reynolds, S. D. (Merck), *US Pat.*, 4 994 604, 1991.
49. (a) Nohira, H., *Yuki Gosei Kagaku*, **50**, 14 (1992); (b) Nohira, H. (Nagasi & Co.), *US Pat.*, 5 321 154, 1994.
50. Dwivedi, S. K., Sattari, S., Jamali, F. and Mitchell, A. G., *Int. J. Pharm.*, **87**, 95 (1992).
51. (a) Stinson, S. C., *Chem. Eng. News*, Sept. 28, 46 (1992); (b) Stinson, S. C., *Chem. Eng. News*, Sept. 19, 38 (1994).
52. DiCicco, R. L. in *Proceedings Chiral '93 USA Symposium*, Reston, VA, Spring Innovations, Stockport, 1993, p. 5.

53. (a) Barberich, T. J., and Young, J. W. (Sepracor), *PCT Int. Pat. Appl.*, WO 93 20 809, (1993); (b) Gray, N. M., Wechter, W. J., and Young, J. W (Sepracor), *US Pat.*, 5 331 000, 1994.
54. (a) Wechter, W. J., *PCT Int. Pat. Appl.*, WO 91 02 512, 1991; (b) Wechter, W. J., (Sepracor), *US. Pat.*, 5 190 981, (1993).
55. (a) Laska, E. H. (Miles and Analgesic Associates), *US Pat.*, 4 868 214, 1989; (b) Laska, E. M., Rees, J. A. and Sunshine, A. (Analgesic Associates), *US Pat.*, 4 927 854, 1990; (c) Laska, E. M., Sunshine, A. and Laska, E. (Miles and Analgesic Associates), *US. Pat.*, 4 962 124, 1990; (d) Laska, E. M. and Sunshine, A. (Analgesic Associates), *US Pat.*, 5 286 751, 1994.

4 Synthesis of Enantiomerically Pure Nucleosides

B. L. BRAY[b], M. D. GOODYEAR[a], J. J. PARTRIDGE[b] and D. J. TAPOLCZAY[a]

Glaxo Wellcome Chemical Development Division, [a]Medicines Research Centre, Stevenage, Hertfordshire, UK, and [b]Girolami Research Centre, Research Triangle Park, NC, USA

4.1 Introduction 41
4.2 GR85478 42
4.2.1 Synthesis from (−)-Carbovir 42
4.2.2 Synthesis from (+)-(1*R*, *cis*)- Cyclopent-4-ene-1,3-diol Monoacetate 47
4.3 GR95168 50
4.3.1 Resolution Approaches 54
4.3.2 Use of Palladium-catalysed Functionalisation of (+)-Hydroxyacetate **5** 60
4.4 Lamivudine 66
4.4.1 Synthesis of the Racemate 67
4.4.2 Resolution of Racemic Lamivudine 69
4.4.3 Alternative Strategies 71
4.4.4 A Novel Synthesis of Lamivudine 72
4.5 Conclusions and Summary 77
4.6 References 77

4.1 INTRODUCTION

Anti-viral research has been one of the most intensive areas of activity within both the pharmaceutical industry and the academic sector for the past 10 years, and the management of viral diseases still remains a focus for many drug companies. During this time, a wealth of chemistry has been discovered and developed; this chapter charts the development at Glaxo Wellcome of routes to the three compounds **1**–**3**, each of which has potent activity against a range of viruses.

It is the authors' aim to give the reader the background to the discovery and development of synthetic routes to these compounds, each of which contains at least two stereogenic centres. Each of these drug candidates was required *as a*

Chirality in Industry II. Edited by A. N. Collins, G. N. Sheldrake and J. Crosby

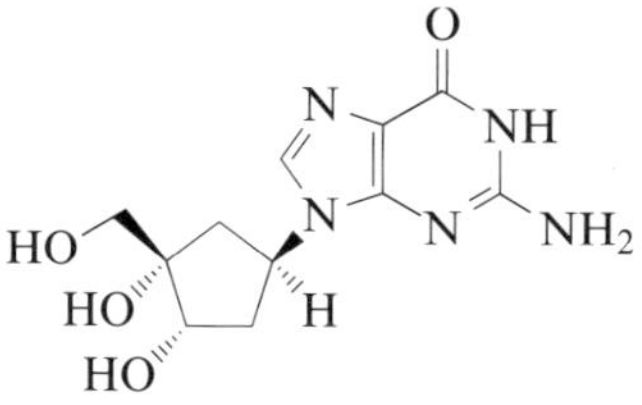

(1)

GR85478

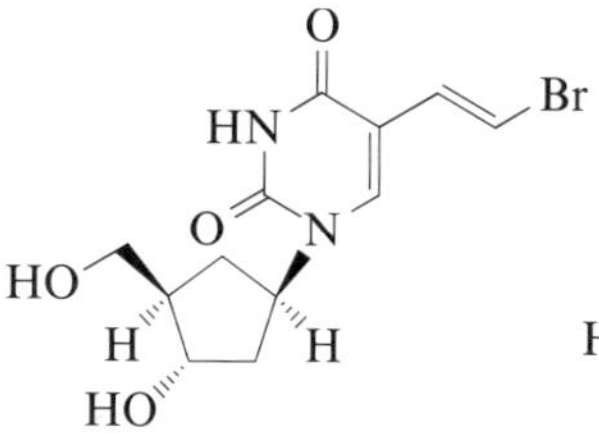

(2)

GR95168

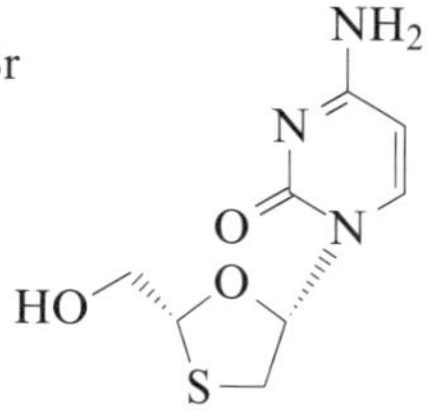

(3)

Lamivudine

single enantiomer, potentially in tonne quantities. Their syntheses are not trivial; the fact that all three are nucleoside analogues prompted some similarities in the synthetic approaches used, but each also posed its own specific challenges. These similarities and differences are explored in this chapter.

4.2 GR85478

Herpes simplex viruses exist in two major forms, known as HSV1 and HSV2; infections range from being mildly uncomfortable to extremely painful. The form of infection most people are aware of is HSV1, which manifests itself as cold sores. The compound discussed in this section [GR85478 (**1**)] showed potent activity against both forms of herpes simplex virus;[1] we describe here its synthesis by two routes, the first using a starting material from the chiral pool [the well known anti-viral agent carbovir[2] (**4**)] and the second from the readily available (+)-hydroxyacetate (**5**), obtained from an enzymatic resolution.

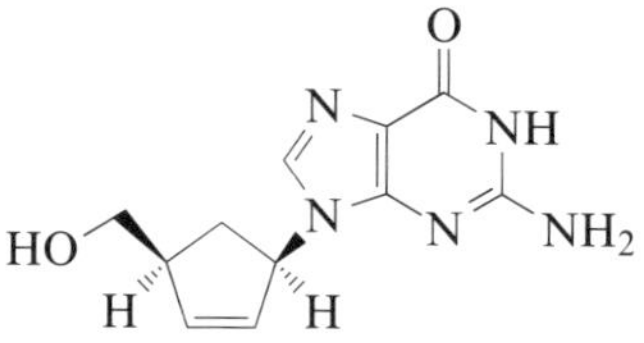

(4)

(–)-Carbovir

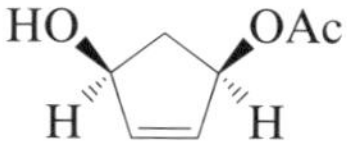

(5)

4.2.1 SYNTHESIS FROM (–)-CARBOVIR

Conversion of (−)-carbovir (**4**) to GR85478 (**1**), both of which contain a guanine fragment, requires an oxidation at C-4, migration of a double bond, reduction to the corresponding allylic alcohol and selective α-dihydroxylation (Scheme 1).[3]

Careful selection of a protecting group for the guanine moiety of (−)-carbovir (**4**) proved to be critical in this synthesis; the formamidine group seemed the best, and no problems were encountered with oxygen vs nitrogen selectivity (as is

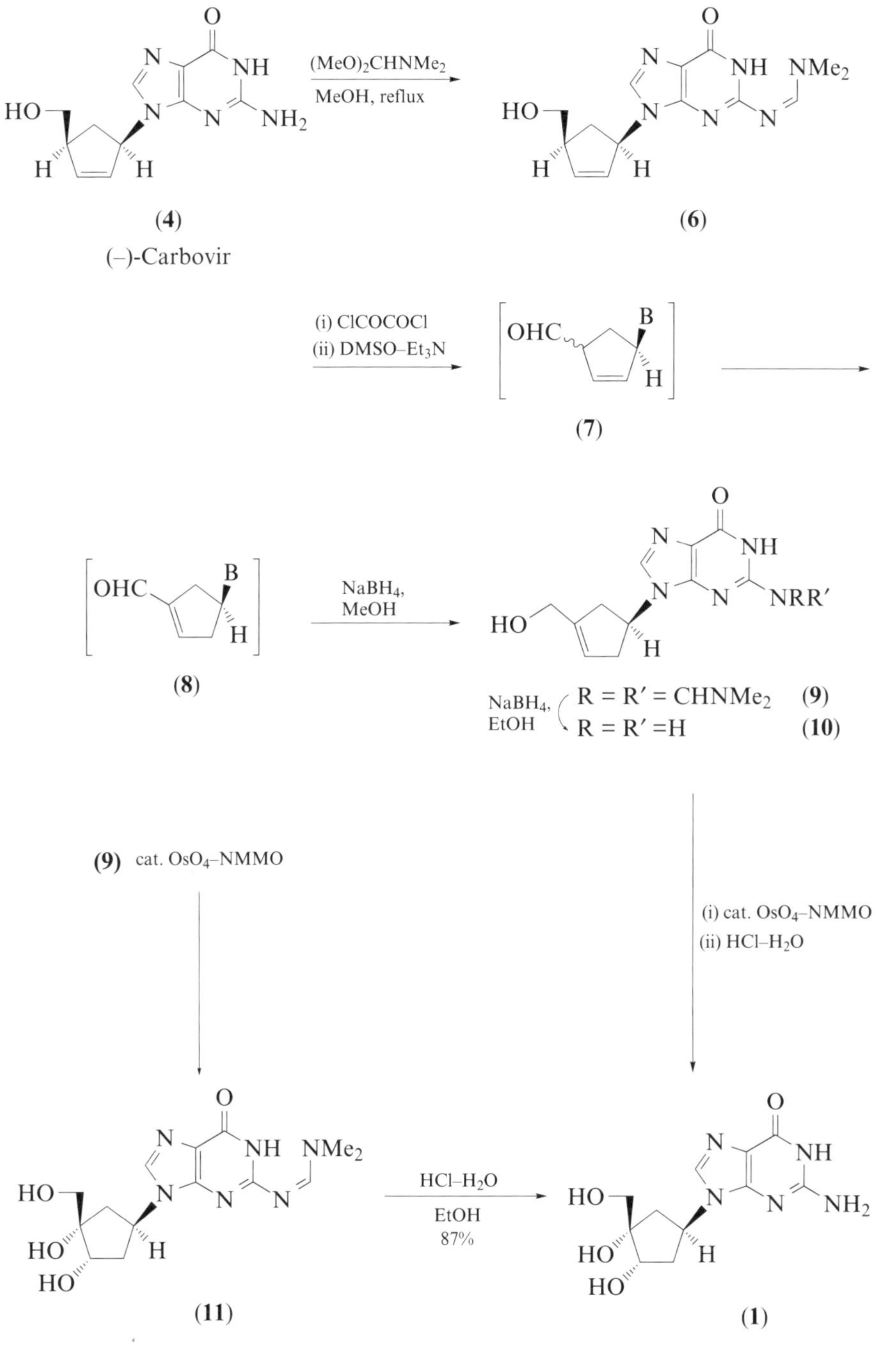
(MeO)2CHNMe2
MeOH, reflux
(4)
(–)-Carbovir
(6)
(i) ClCOCOCl
(ii) DMSO–Et3N
(7)
(8)
NaBH4,
MeOH
NaBH4,
EtOH
R = R′ = CHNMe2 (9)
R = R′ =H (10)
(9) cat. OsO4–NMMO
(i) cat. OsO4–NMMO
(ii) HCl–H2O
HCl–H2O
EtOH
87%
(11)
(1)

Scheme 1

common with protection of amide, lactam or carbamate groups). Importantly, the organic solubility of the carbocyclic nucleoside was greatly enhanced without significant mass increase using formamidine protection. Treatment of (−)-carbovir with dimethylformamide dimethylacetal in methanol at reflux gave the formamidine **6** in 99% yield. Selective oxidation of the C-4 hydroxymethylene group of **6** using standard Swern conditions[4] provided the α,β-unsaturated aldehyde **8** in 82% yield. The *cis*-aldehyde initially formed in the oxidation rapidly isomerised to a 1:1 mixture of *cis*- and *trans*-β,γ-unsaturated aldehydes **7** at −50 °C. As the solution warmed to 0 °C, the *cis–trans* mixture (**7**) was completely converted into the desired α,β-unsaturated aldehyde **8**; the oxidation and isomerisation processes were complete within 1 h. The *cis-trans* mixture of aldehydes (**7**) could be isolated at low temperature, but the mixture was found to be very air sensitive and decomposed at room temperature in a few hours. Oxidation of the C-4 hydroxymethylene group of unprotected (−)-carbovir (**4**) under the same conditions gave a low yield (<15%) of the corresponding unprotected α,β-unsaturated aldehyde. The low yield was thought to be the result of several factors, such as poor solubility, oxidation at the C-6 amino group and Schiff base formation. Reduction of aldehyde **8** with 0.25 equivalents of sodium borohydride in methanol at 0 °C gave allylic alcohol **9**; the use of a full equivalent of sodium borohydride in ethanol at 25 °C was found to reduce aldehyde **8** directly to the unprotected allylic alcohol **10**.

To complete the synthesis of **1**, a stereofacially-selective *cis*-dihydroxylation of either allylic alcohol **9** or **10** *anti* to the guanine ring was required.[5] Molecular models indicate that osmium tetraoxide-catalysed *cis*-dihydroxylation should occur selectively on the *anti* face (Figure 1); however, from literature precedent, it is clear that a combination of electronic and steric factors generally govern the selectivity of these reactions.

Surprisingly, the first reaction (Scheme 2, **12** → **13**) suggested that the osmium tetraoxide was being attracted to the more sterically congested face of the mole-

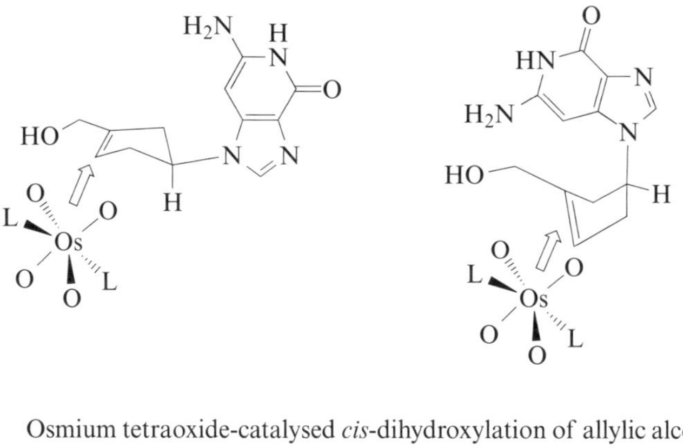

Figure 1 Osmium tetraoxide-catalysed *cis*-dihydroxylation of allylic alcohol **10**

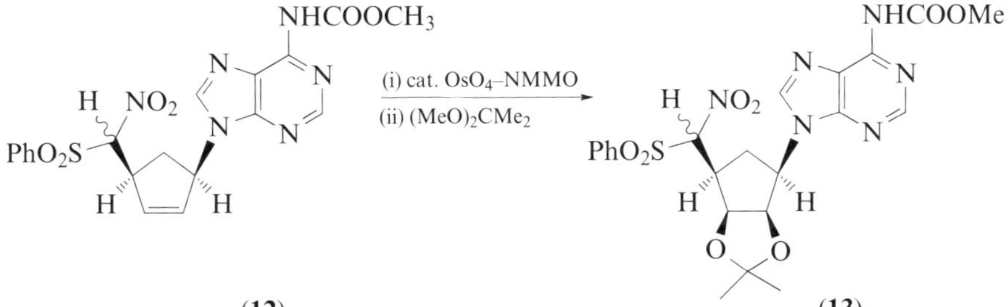

(12) (13)

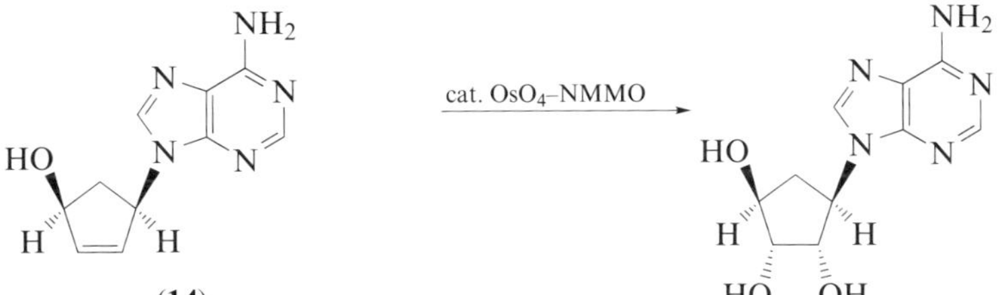

(14) (15)

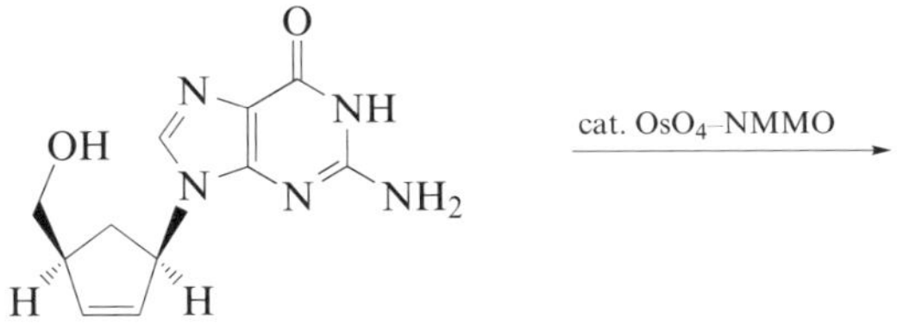

(4)

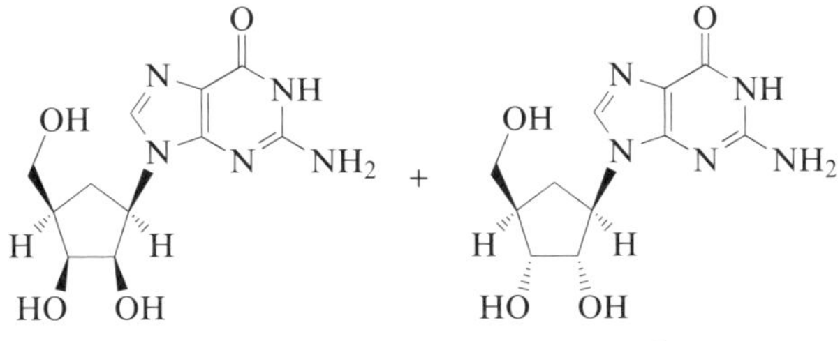

1 : 1

(16) (17)

Scheme 2 *(continued)*

Scheme 2 (*continued*)

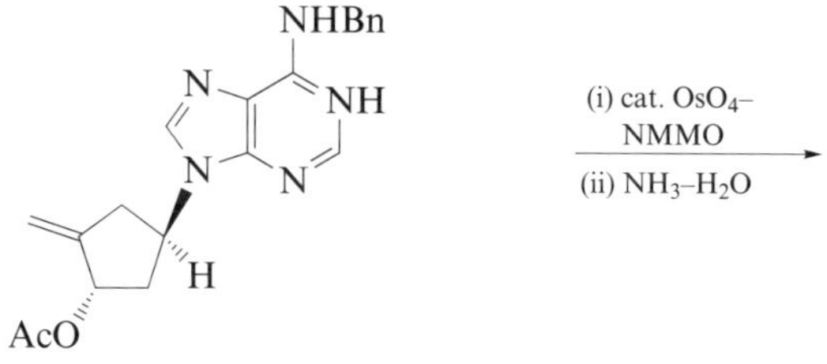

(18)

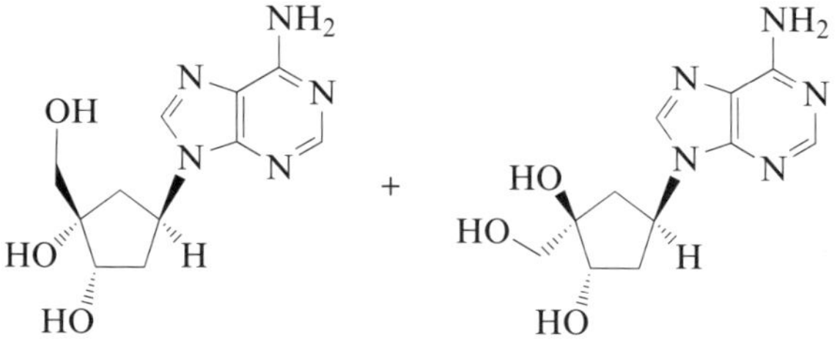

3 : 7

(19) **(20)**

Scheme 2

cule by the nitrophenyl sulphone functionality.[6] In the second reaction (**14** → **15**), selective approach of the osmium tetraoxide to the least hindered face of the molecule occurred, as would be predicted,[6] assuming no directing effect by the hydroxyl group. When (−)-carbovir (**4**) was treated with catalytic osmium tetraoxide and *N*-methylmorpholine *N*-oxide in water–acetone at 70 °C, no facial selectivity was obtained (**4** → **16** + **17**)[7]. The final example (**18** → **19** + **20**) illustrates the lack of stereofacial selectivity obtained when both faces of the cyclopentane exocyclic methylene on the cyclopentane ring are hindered.[8]

Treatment of allylic alcohol **9** with 1.5 mol% osmium tetraoxide and *N*-methylmorpholine *N*-oxide in water–acetone at 70 °C for 2 h. provided only the desired *anti*-dihydroxylation product (**11**)[3]. It is unclear why α-dihydroxylation on this allylic alcohol substrate proceeded with total *anti*-selectivity whereas, under identical conditions, the homoallylic alcohol, carbovir (**4**), displayed no selectivity at all. The formamidine protecting group on the guanine base of **9** had no effect on the rate or selectivity of the reaction; gratifyingly, the deprotected allylic alcohol **10** also afforded only the desired *anti*-dihydroxylated product (**1**), under identical conditions.

Although the catalytic *cis*-α-dihydroxylation process worked exceedingly well, removal of all traces of osmium from this water-soluble drug posed a challenge; an acceptable limit for residual osmium in the final drug **1** was not greater than

10 ppm. The guanine ring, like pyridine and other aromatic amines, binds osmium fairly strongly; in fact, osmium tetraoxide is commonly used as a biological staining tool in DNA research. Reduction of the residual osmium level to below 10 ppm required selective and efficient conditions. Pyridine was added to the reaction mixture after hydroxylation of allylic alcohol **10** was complete, then the solution was saturated with hydrogen sulphide and stirred at 70 °C for up to 70 h. Most of the osmium was precipitated as a fine, dark solid, along with sulfur, which was formed as a by-product. The solids were filtered off, the solvent volume was reduced and free base **1** was isolated by direct crystallisation. However, the solid at this stage generally still contained 50–100 ppm of osmium. To remove these final traces, the solid was dissolved in 0.5 M hydrochloric acid, warmed to 70 °C, saturated with hydrogen sulphide and stirred for 24 h. Residual osmium was precipitated and filtered off and the product (**1**) was isolated by crystallisation from the filtrate in 66% overall yield.[9] The residual osmium level after this exhaustive treatment was less than 2 ppm. Interestingly, if either the treatment with hydrochloric acid–hydrogen sulphide or with pyridine–hydrogen sulphide was performed twice, rather than once each in sequence, the residual osmium levels were not reduced below 20 ppm. Although the use of hydrogen sulphide is not desirable on a large scale, this method enabled us to prepare several kilograms of clinical-grade product **1**, essentially free from osmium contamination.

4.2.2 SYNTHESIS FROM (+)-(1*R*, *cis*)-CYCLOPENT-4-ENE-1,3-DIOL MONOACETATE

The second approach to this molecule used (+)-(1*R*, *cis*)-cyclopent-4-ene-1,3-diol monoacetate (**5**) or its enantiomer, which are both commercially available, but expensive.[10] However, either enantiomer can be obtained by an enzyme-catalysed desymmetrisation of cyclopent-4-ene-1,3-diol diacetate (**21**)[11] (Scheme 3).

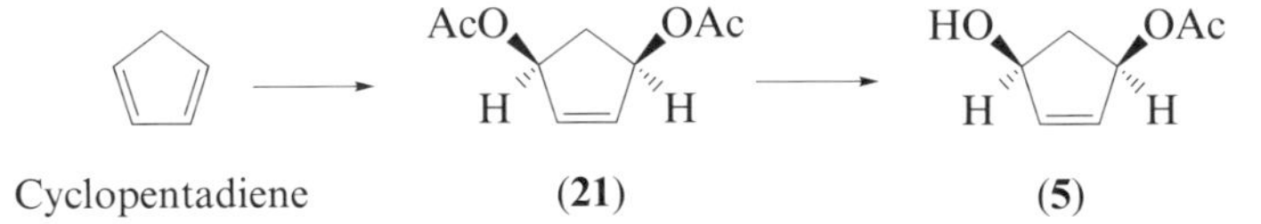

Scheme 3

In work towards an asymmetric synthesis of (−)-carbovir (**4**), we had found that the pre-formed potassium or sodium salt of 2-amino-6-chloropurine (**23**) can be efficiently alkylated at N-9 with dicarbonate **22** using palladium(0) catalysis to give the *cis*-adduct **24**.[12] Using the dihydroxylation chemistry described above, a strategy for the synthesis of the target compound **1** from (+)-hydroxyacetate **5** would simply require the addition of a one-carbon fragment to **25**, and conversion into **10**, followed by dihydroxylation giving **1** (Scheme 4).

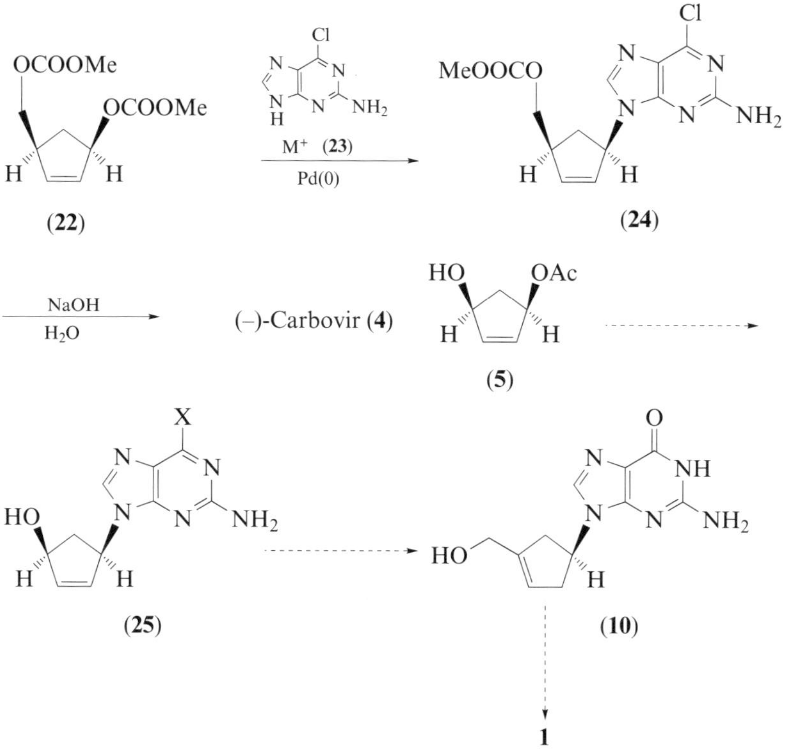

Scheme 4

Kilogram quantities of cyclopent-4-ene-1,3-diol diacetate (**21**) were prepared as described in the literature[11d,g] and desymmetrisation was accomplished using a *Pseudomonas* lipase immobilised between membranes in a Sepracor membrane bioreactor (Figure 2).[13] Using this technology, several kilograms of (+)-hydroxy-acetate **5** were prepared using less than 0.5g of purified enzyme.[14] The four-stage process, which included a final crystallisation, gave (+)-hydroxyacetate **5** in an overall yield of 30–35%, with an enantiomeric excess of >99% as determined by chiral capillary GC.

An economic plant process requires that palladium(0) is used only in catalytic quantities, and there are two means of regenerating it from the palladium–cyclopentenol π-allyl adduct **26** (Scheme 5).[15] In the presence of a soft nucleophile, palladium(0) is displaced to give the desired alkylation product **27** and palladium(0) (path A). In the presence of a strong base (or no nucleophile at all)

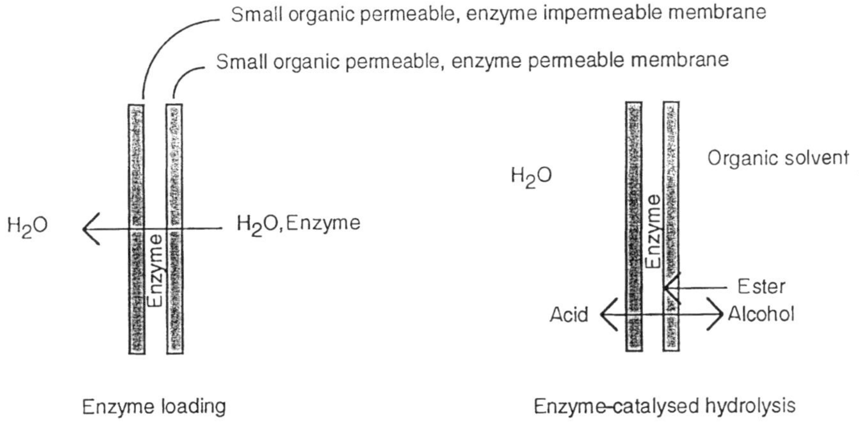

Figure 2 Membrane bioreactor

elimination is favoured to give, ultimately, the β,γ-cyclopentenone **28**, protonated base or acetic acid and palladium(0) (path B). The sodium salt of 2-amino-6-chloropurine (**23**, M = Na) is not a particularly good nucleophile and competition from path B was difficult to avoid. It was important to add (+)-hydroxyacetate **5** to a solution of pre-formed catalyst and the salt of 2-amino-6-chloropurine. On a large scale, a preformed solution of catalyst and (+)-hydroxyacetate **5** suffered significant degradation by reaction path B during the time it took to add the salt of 2-amino-6-chloropurine. The 1,8-diazabicyclo[5.4.0]undec-7-ene (DBU) salt of compound **23** gave the cleanest reaction.

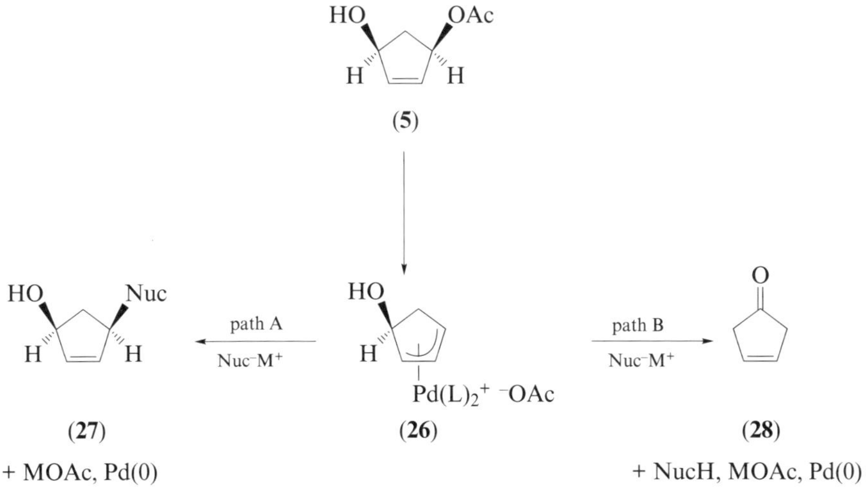

Scheme 5

The adduct of chloropurine **23** and (+)-hydroxyacetate **5** was difficult to crystallise (gels were formed), but formamidine **29**, formed by treatment of the crude product with dimethylformamide dimethylacetal, could be isolated as a crystalline solid in 45–51% yield [based on (+)-hydroxyacetate **5**] on a large scale.[16] Treatment of allylic alcohol **29** with methanesulphonic anhydride and pyridine at 0 °C afforded a reactive mesylate **30** which could not be isolated without extensive decomposition. Addition of tetraethylammonium cyanide resulted in cyanide displacement of the mesylate, giving a 1:1 mixture of *cis*- and *trans*-nitriles (**31**). Treatment of this mixture with DBU caused the β,γ-double bonds to migrate into conjugation, affording the desired α,β-unsaturated nitrile (**32**). Milder bases, such as triethylamine, did not effect this isomerisation process. The mesylate group of **30** was readily displaced with water and, as tetraalkylammonium cyanides are extremely hygroscopic salts and often contain water, the reproducibility of this cyanide displacement was poor. Several cyanide salts were evaluated, but poor solubility often resulted in slow rates of reaction. This problem was overcome by generating an anhydrous tetraalkylammonium cyanide from acetone cyanohydrin in the presence of DBU.[17] Trialkylamines, such as triethylamine, slowly liberate hydrogen cyanide from acetone cyanohydrin, but ^{1}H NMR studies showed that the reaction rate using DBU was very fast, and the equilibrium was driven by consumption of cyanide.[17] The salt of DBU and hydrogen cyanide is freely soluble in ethers, acetone and chlorinated hydrocarbons and is a powerful source of nucleophilic cyanide anion.

Treatment of the crude mesylate **30** with six equivalents of acetone cyanohydrin followed by five equivalents of DBU at 0 °C gave a 1:1 mixture of *cis*- and *trans*-nitriles (**31**); on warming to 23 °C, the double bonds spontaneously migrated into conjugation with the nitrile (**31** → **32**), even in the presence of excess acetone cyanohydrin. Treatment of the crude α,β-unsaturated nitrile **32** with 3M hydrochloric acid removed the formamidine protecting group and hydrolysed the chloropurine to a guanine ring. The resulting α,β-unsaturated nitrile (**33**) was readily isolated and purified by crystallisation. The complex series of conversions (**29** → **33**) was carried out in two stages in 60% overall yield. Conversion of the α,β-unsaturated nitrile **33** into the α,β-unsaturated *n*-propyl ester **34** was carried out in propan-1-ol in the presence of one equivalent of sulphuric acid in 65% yield after crystallisation. Reduction of this ester with DIBAL provided allylic alcohol **10** in high yield without affecting the guanine ring. The desired compound (**1**) was prepared by osmium-catalysed *cis*-dihydroxylation as previously described. This synthesis (see Scheme 6) was superior to the route from carbovir (Scheme 2), and was selected as our method of choice for large-scale manufacture.

4.3. GR95168

Varicella zoster (VZV) is the virus responsible for the diseases chickenpox and shingles. People who contracted chickenpox as children have the virus lying dor-

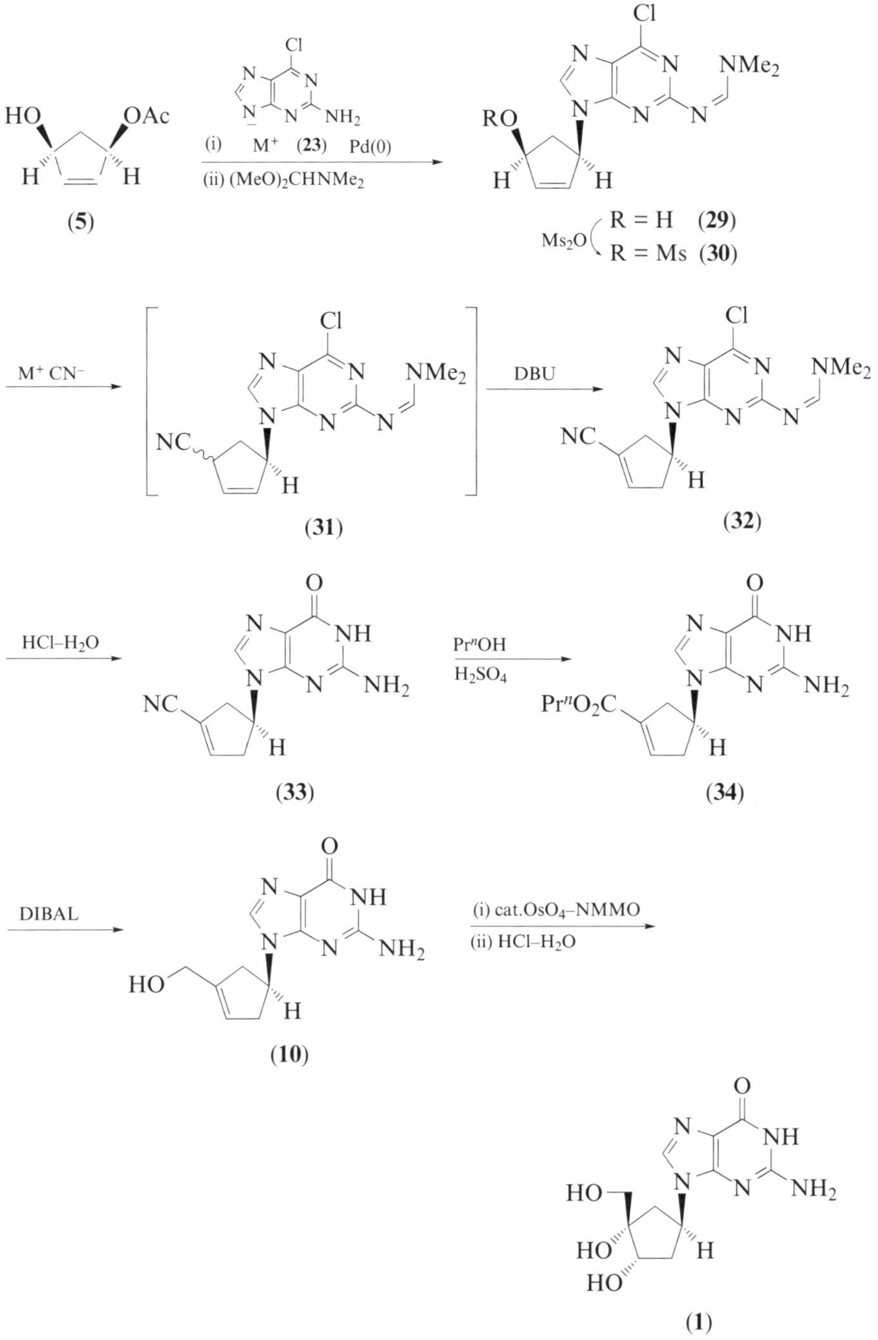
(i) M+ (23) Pd(0)
(ii) (MeO)2CHNMe2
(5)
Ms2O
R = H (29)
R = Ms (30)
M+ CN−
DBU
(31)
(32)
HCl–H2O
PrnOH
H2SO4
(33)
(34)
DIBAL
(i) cat.OsO4–NMMO
(ii) HCl–H2O
(10)
(1)

Scheme 6

mant within a large nerve bundle known as the dorsal ganglion. In adults, a mechanism which, as yet, is unknown can lead to the reactivation of this dormant virus, which manifests itself as shingles. This disease is particularly prevalent in the elderly; statistics suggest that 1 in 20 people over the age of 60 will contract it. It is an unpleasant disease that is usually accompanied by skin lesions, nerve damage and an extremely painful neuralgia that may persist long after the visible signs of the disease have disappeared. Shingles affects the quality of life of all the people who contract it; in immunocompromised or very elderly patients, the disease can be fatal. Along with other pharmaceutical companies, Glaxo Wellcome has been actively searching for a treatment for VZV, and compound **2** was shown to have outstanding activity. Two synthetic routes are discussed,[18] the first using a starting material from the chiral pool (lactam **35**) and the second from (+)-hydroxyacetate **5** using chemistry developed during work on GR85478 (described in the previous section).

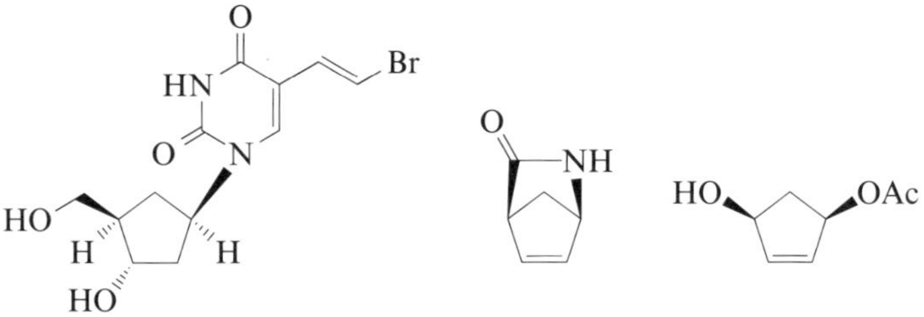

(2) GR95168 **(35)** **(5)**

The racemate of GR95168, carbocyclic bromovinyldeoxyuridine, was discovered in 1984 as a result of a collaboration between Glaxo Wellcome and Southampton University.[19] It was subsequently shown that the majority of the activity against varicella zoster virus lay in the single enantiomer **(2)** shown above.[20] This was developed as a potential therapy for shingles and, as a result, large quantities of the enantiopure drug candidate were required for safety studies. The initial synthesis used at Glaxo Wellcome is shown in Scheme 7.[20b]

There are a number of drawbacks to this route as a potential means of synthesis of large quantities of drug substance:

- Column chromatography was required to purify most of the intermediates (which were all non-crystalline).
- The use of diisopinylcamphenylborane gave separation problems during isolation of the reaction product.
- The use of vanadium in stoichiometric quantities would give significant problems in handling waste streams and ensuring that the product was free from contamination.
- The overall yield was low (owing to two low-yielding stages and the large number of steps).

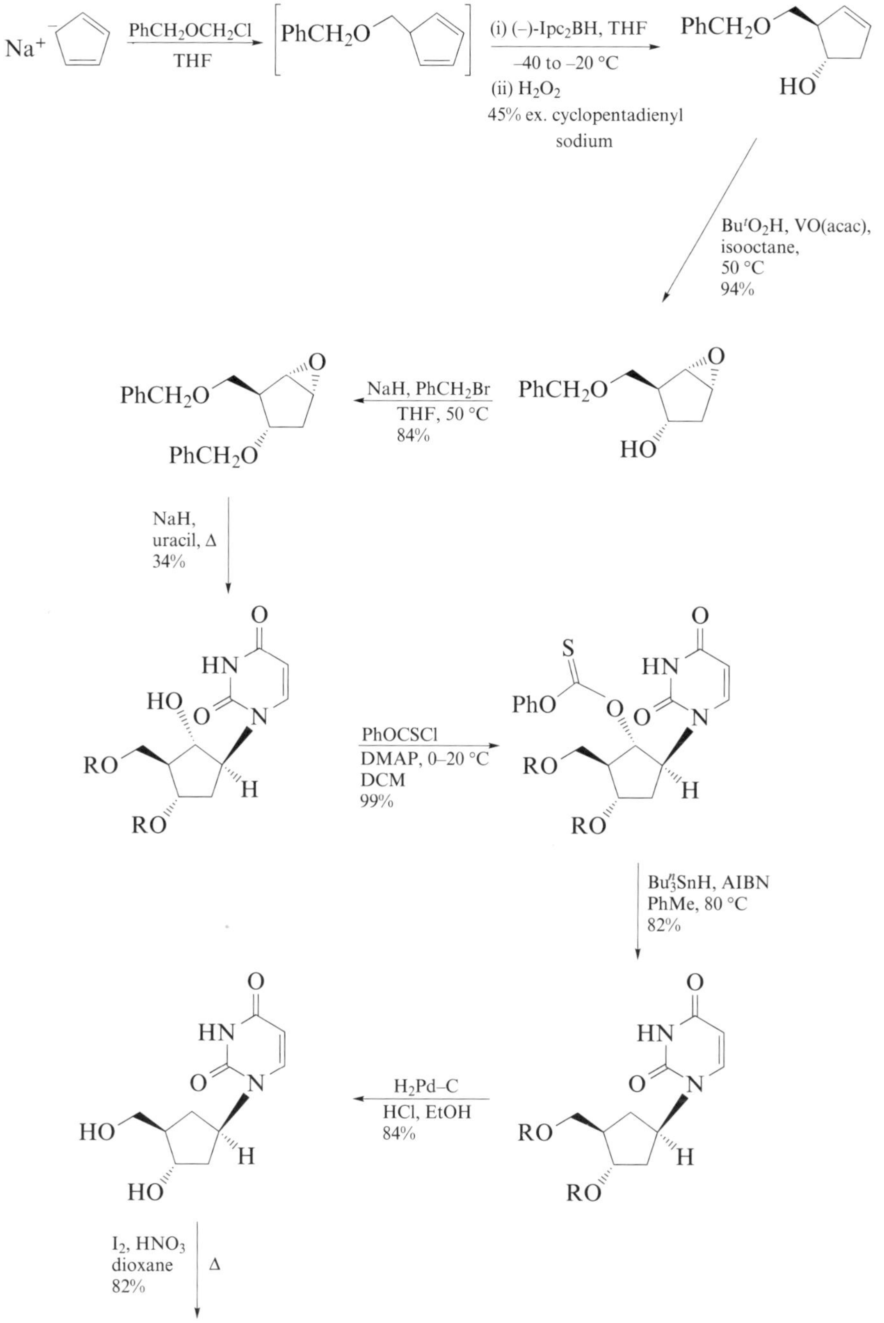

Scheme 7 *(continued)*

Scheme 7 (*continued*)

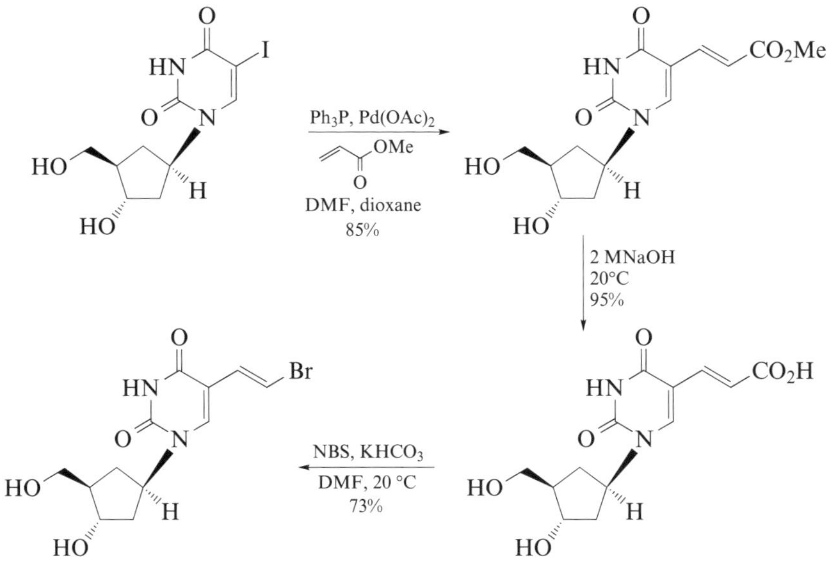

4.3.1 RESOLUTION APPROACHES

Our first approach was to use a racemic starting material (**35**),[21] containing the required relative stereochemistry, and to employ a resolution of the ring-opened derivative (**40**) to obtain the desired enantiomer. This approach is shown in Scheme 8.

The secondary hydroxyl group would then be introduced either by hydroboration or by epoxidation and regioselective ring opening. However, all attempts at the selective functionalisation of the 3-position of the cyclopentene (**38**→**37**) were unsuccessful; epoxidation of lactam **35** gave the desired α-epoxide, but all attempts at regioselective ring opening led to inseparable mixtures of isomers. Epoxidation of ring opened derivatives of **35** was less face-selective, and also not suitable. The hydroboration approach also failed, giving mixtures of 3α- and 2β-hydroboration products. Consequently, these approaches were abandoned and the alternative strategy shown in Scheme 9 was proposed.

This approach differs from that shown in Scheme 8 in the oxidation state of the substituent at C-4. A suitably protected racemic lactam is ring opened, resolved and esterified to give the unconjugated ester **43** as a single enantiomer. The double bond would then be moved into conjugation to form the α,β-unsaturated ester (**42**). This contains a trisubstituted alkene within the five-membered ring, and it is well documented that hydroboration of such systems

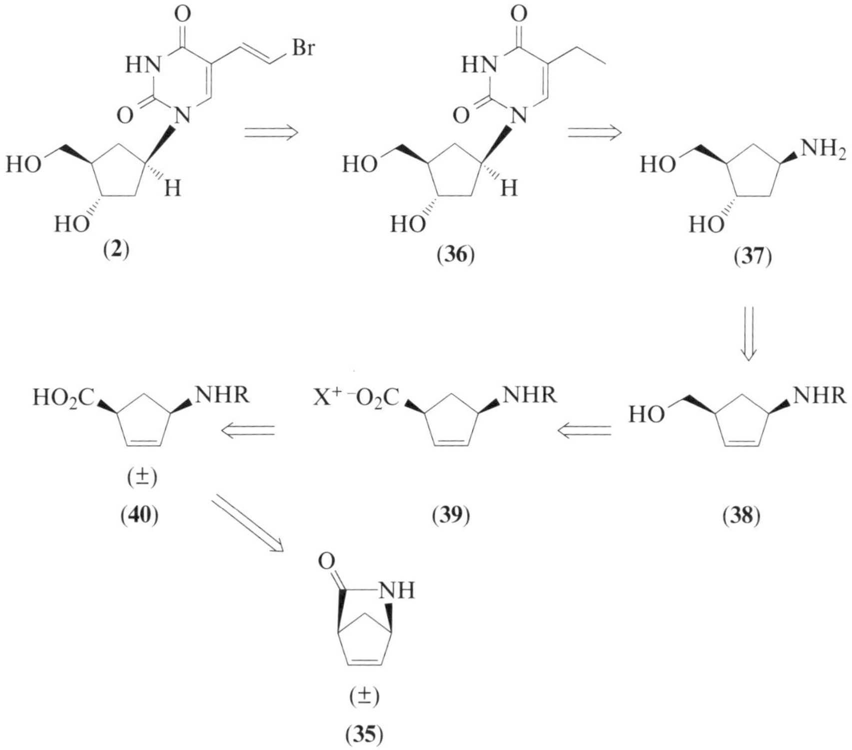

Scheme 8

can be achieved with excellent regiochemical control, although the degree of stereochemical control is less predictable.[22] By using a hindered borane, we envisaged that it might be possible to achieve some degree of selectivity via the approach of the borane from the less hindered α-face; if the amine group, suitably protected, reacted with the hydroborating agent faster than the double bond, then the steric bulk of the top face of the cyclopentane should be increased sufficiently to achieve a very high degree of face-selectivity on addition of a second equivalent of the borane. Brown *et al.*[23] have reported that tertiary amides are reduced to amines by dialkyl boranes; furthermore, protection of the lactam as an amide was likely to facilitate the ring opening prior to resolution of the free acid. Consequently, we chose to protect lactam **35** as its benzamide, as shown in Scheme 10.

Lactam **35** had previously been prepared by reaction of cyclopentadiene with *p*-toluenesulphonyl cyanide.[21b] In view of the extremely hazardous nature of the latter, and the need for a synthesis capable of operation on a large scale, we decided to switch to the safer and more readily available chlorosulphonyl isocyanate, which gave racemic lactam **35** as a crystalline solid in 35% yield. The lactam was

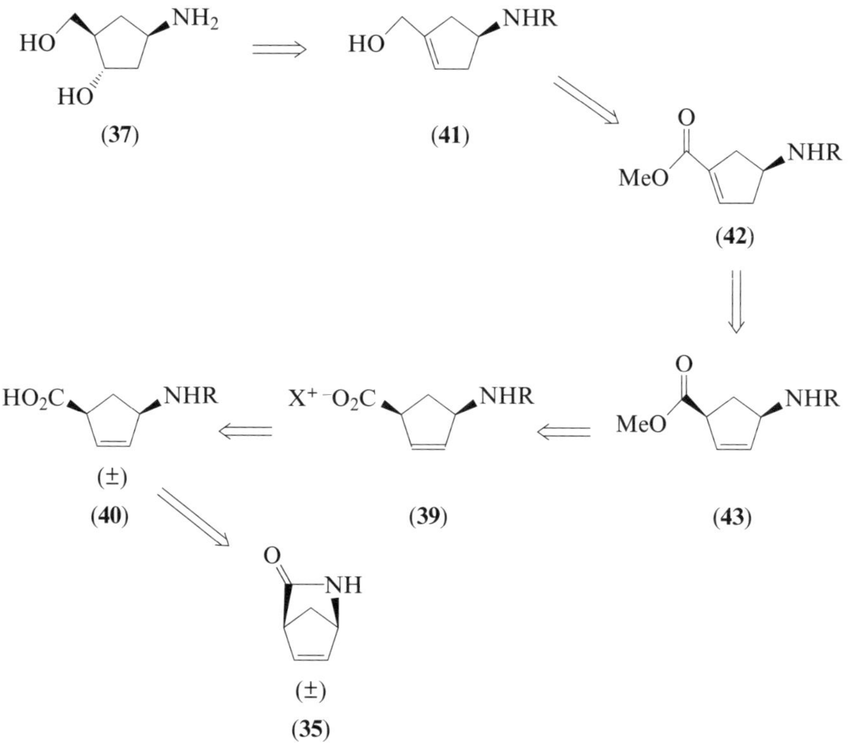

Scheme 9

protected as its benzamide (**44**) and hydrolysed with dilute sulphuric acid giving racemic acid (**40**) as a crystalline solid in 76% overall yield.

Several commercially available chiral amines were screened against the racemic acid in an attempt to obtain a diastereoisomerically pure crystalline salt; fortunately, the cheap and readily available (*R*)-α-methylbenzylamine gave the required diastereoisomer (**39**) as a crystalline salt which could be isolated directly from the reaction mixture in excellent purity (43% yield, 99% *de*, as determined by HPLC). Treatment of this salt with aqueous base followed by methanolic hydrochloric acid gave the enantiopure unconjugated benzamido ester **43** (90% yield, 99% *ee*). The double bond was moved into conjugation to give ester **42** by treatment with DBU in dichloromethane in 90% yield. The use of this solvent in production is not desirable owing to its environmental effects and possible long-term lack of availability. Ethyl acetate proved a suitable replacement.

With the conjugated ester **42** in hand it only remained to reduce the ester to the allylic alcohol and protect in order to evaluate the proposed hydroboration step. This conversion proved far more difficult than first imagined and led to the development of a novel chemoselective reduction.[24] Lithium aluminium hydride gave

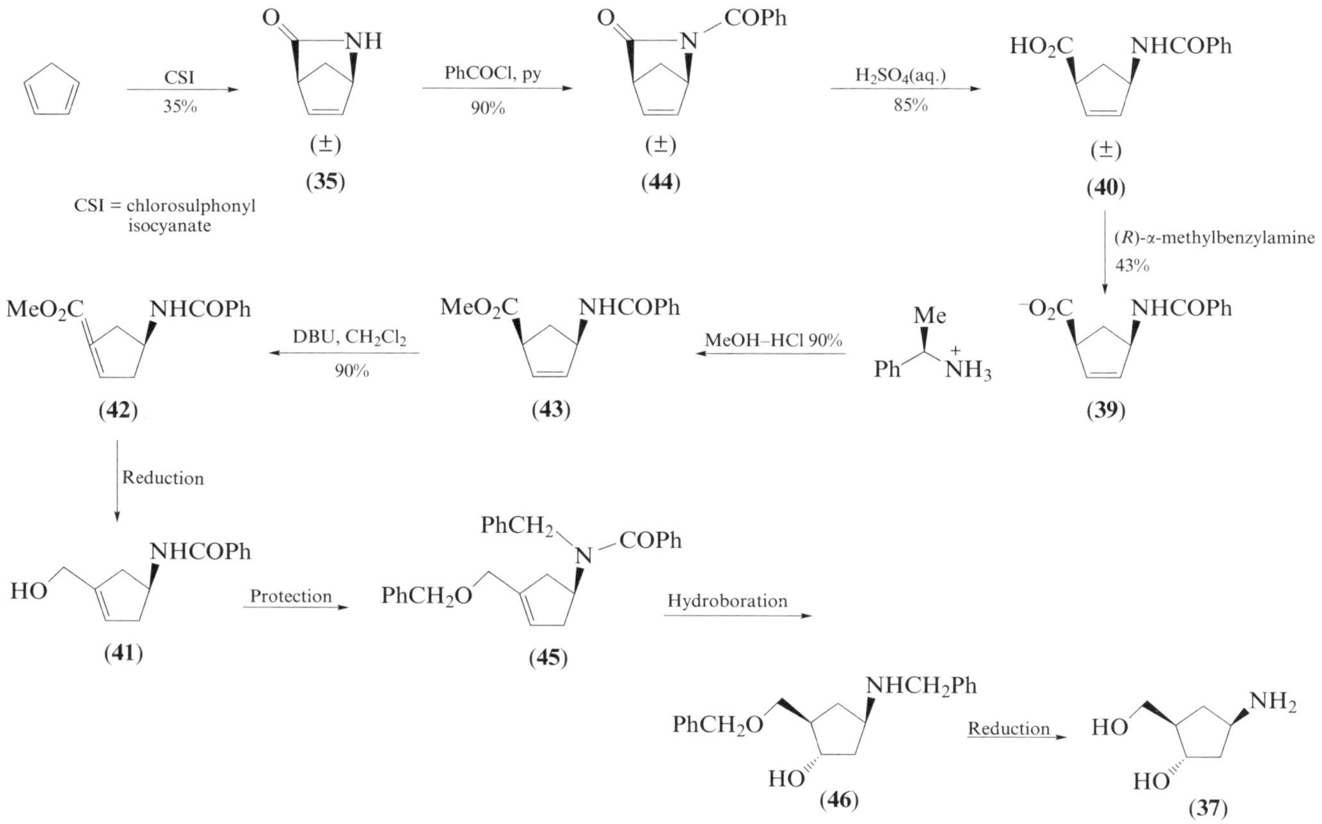
O
NH
CSI
35%
(±)
(35)
CSI = chlorosulphonyl
isocyanate
PhCOCl, py
90%
O
N
COPh
(±)
(44)
H_2SO_4(aq.)
85%
HO_2C
NHCOPh
(±)
(40)
(R)-α-methylbenzylamine
43%
^-O_2C
NHCOPh
(39)
Me
Ph
$\overset{+}{N}H_3$
MeOH–HCl 90%
MeO_2C
NHCOPh
(43)
DBU, CH_2Cl_2
90%
MeO_2C
NHCOPh
(42)
Reduction
HO
NHCOPh
(41)
Protection
$PhCH_2$
N
COPh
$PhCH_2O$
(45)
Hydroboration
$NHCH_2Ph$
$PhCH_2O$
HO
(46)
Reduction
HO
NH_2
HO
(37)

Scheme 10

a lack of selectivity between 1,4- and 1,2-reduction; the same problems occurred with modified borohydride reducing agents. However, DIBAL at −70 °C gave the required allylic alcohol **41** in reasonable yield (56%). The low yield was due to a side-reaction, in which the benzamide group was also reduced giving amine **47**.

(**47**)

Despite numerous changes in reaction conditions, the ratio of the products could not be altered in favour of the desired allylic alcohol **41**. In spite of the low yield, the product could be isolated directly from the reaction mixture, and the process was suitable for scale-up, although it required low temperatures. The main problem was the lack of chemoselectivity and resulting low yield. It was reasoned that the reduction of the amide was occurring as a result of the coordination of DIBAL to the amide as well as to the ester. Based on electronic considerations, the rate of reduction of the amide should be slower than that of the ester; it was hypothesised that selective coordination of the amide carbonyl to some other species, for example a Lewis acid, would prevent subsequent coordination of DIBAL, and thus the reduction of the amide. On adding one equivalent of aluminium chloride, ^{1}H NMR spectroscopy showed chelation of the Lewis acid mainly to the amide. On treating this pre-formed complex with DIBAL, we were pleased to find a considerable improvement in yield of **41** to 93%. Furthermore, the reaction could be carried out at 0 °C without a decrease in selectivity. We were now in a position to scale this stage up to pilot plant when required.

Protection of the alcohol and activation of the amide had been carried out using sodium hydride and benzyl bromide in dimethylformamide, the product (**45**) from this stage being isolated as a viscous oil. This procedure was not suitable for scale-up; sodium hydride in DMF is well known as a potentially hazardous mixture and, even if another solvent were used, the oil present in commercial sodium hydride dispersions can be difficult to remove and may contaminate the product. Finally, isolation of the product as a viscous oil is not practical on a large scale. Briefly, the hydroboration of **45** *in situ* was explored, but poor yields were obtained as it was difficult to estimate the correct quantity of borane required. Sodium hydride–DMF was successfully replaced by a phase-transfer system, using toluene and aqueous sodium hydroxide, from which compound **45** could be isolated as a crystalline solid in 90% yield.

We were now in a position to attempt the key stage, for which the reagent diisoamylborane was chosen, as it is known to give good regiocontrol in the hydroboration of trisubstituted alkenes and a reasonable rate of reaction.[25] The reagent is not commercially available but can be readily prepared and used *in situ*. Good control of stereochemistry requires reduction of the amide prior to hydroboration of the alkene. On adding one equivalent of the pre-formed diisoamylborane at 0 °C, we obtained a good yield of amine **48**, indicating a suitable difference in rates of reduction and hydroboration.

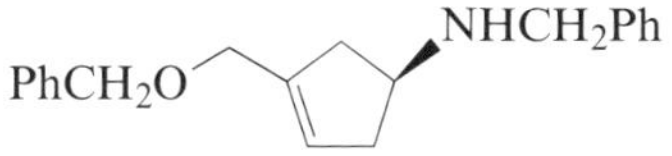

(**48**)

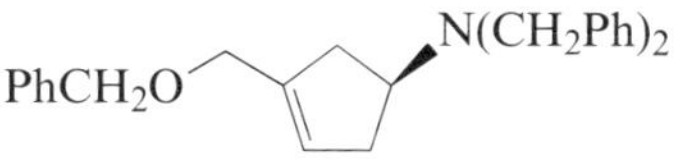

(**49**)

We repeated the reaction with two equivalents of diisoamylborane at 25 °C, and the required alcohol **46** was isolated directly from the reaction mixture as a crystalline solid in excellent yield (95%). Analysis of product remaining in the mother liquors showed the stereoselectivity of the reaction was 45:1 in favour of 3α-versus 3β-attack. The participation of the reduced benzamide group in affecting the selectivity of the reduction was subsequently demonstrated by carrying out the hydroboration of the structurally and sterically similar tribenzyl compound **49**, which gave a selectivity of only 11:1.

The benzyl groups were readily removed by hydrogenolysis, giving the key intermediate aminodiol **37**. This was successfully converted into the desired drug substance by a three-stage process in which the acryloyl isocyanate **50** was treated with the aminodiol **37** to give the cyclised unprotected uridine **51a**. This was protected as the diacetate **51b** by reaction with acetic anhydride. Conversion to **2** was achieved by using a bromination–dehydrobromination sequence with *N*-bromosuccinimide and triethylamine, followed by work-up with hydrochloric acid. After extensive development, the entire sequence (from **37** to drug substance **2**) was carried out as a 'one-pot' transformation in excellent overall yield (70%, see Scheme 11).

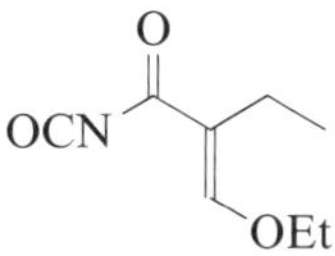

(**50**)

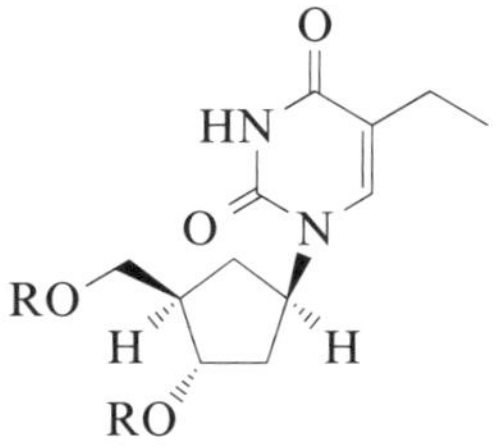

(**51a**) R = H
(**51b**) R = Ac

Subsequently, enantiopure benzamido ester **43** became commercially available from two different suppliers and, using this material, we were able to prepare several kilograms of drug substance in our pilot plant using the above route. The final reaction conditions and reagents are outlined in Scheme 11.

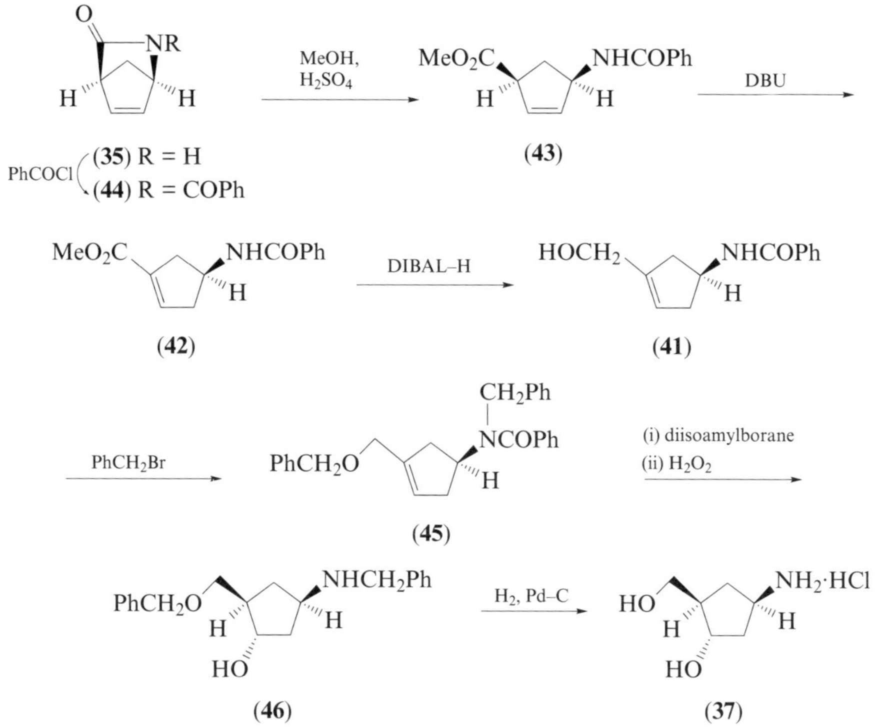

Scheme 11

4.3.2 USE OF PALLADIUM-CATALYSED FUNCTIONALISATION OF (+)-HYDROXYACETATE **5**

Following the success of a palladium-catalysed functionalisation of (+)-hydroxyacetate **5** in the synthesis of GR85478, which was discussed Section 4.2.2, we now applied that chemistry to the synthesis of GR95168. This approach has the advantage of being more convergent than the first method developed.

This strategy requires an N-3-protected 5-substituted uracil **(52)** to be coupled to **5**, followed by addition of a one-carbon fragment with requisite control of the double bond location to yield the allylic alcohol **54**. Hydroboration from the α face, as in the previously discussed route to GR95168, would give the desired *trans*-diol, which could be elaborated subsequently to give the target molecule. This approach is shown in Scheme 12.

We hoped that the palladium-mediated coupling would circumvent the problem of elimination which, in our experience, usually occurred when displacement of a leaving group (such as tosylate or mesylate) on a cyclopentane ring was attempted with a uracil derivative. The first part of the synthesis involved a palladium(0)-catalysed alkylation of a 5-substituted uracil with (+)-hydroxyacetate **5**.

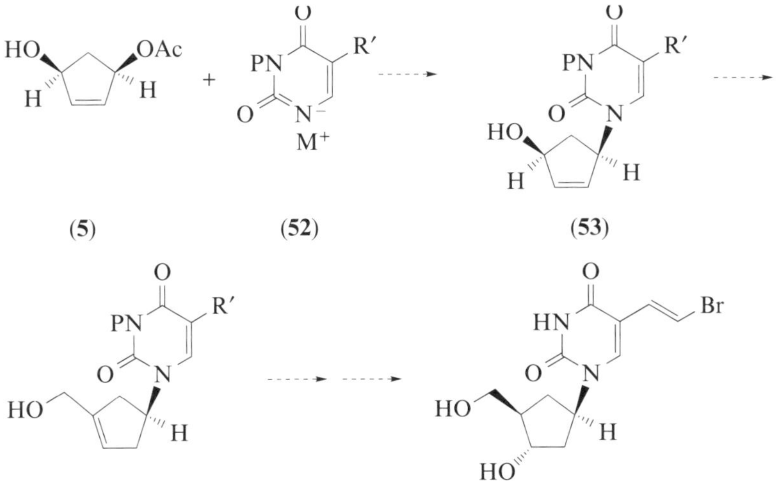

Scheme 12

A series of 5-substituted and N-3-protected, 5-substituted uracil bases was synthesised (see Scheme 13) to evaluate the efficiency of these reactions. The substituent at the 5-position was chosen such that it could be readily converted into the desired vinyl bromide moiety.

Treatment of 5-iodouracil (**55**) with two equivalents of methyl acrylate in the presence of palladium(0) provided the N-1-protected, 5-substituted uracil **56a** in nearly quantitative yield; the Michael addition to N-1 was reversible, and faster than the Heck reaction. Warming the solution of **56a** in dimethyl sulphoxide under vacuum (distilling out the methyl acrylate) removed the N-1 protecting group to give the 5-substituted uracil **57** in 90% yield. Treatment of **56a** with at least three equivalents of methyl acrylate in the presence of palladium(0) gave the *N*-,*N*-propyl ester-protected 5-substituted uracil, which, on heating in dimethyl sulphoxide under vacuum, gave **56b**; only the N-1 substituent was removed. Treatment of **56a** with benzoyl chloride afforded **58**, which also, on heating, lost the N-1-propyl ester to give the N-3-protected, 5-substituted uracil **59** in greater than 80% overall yield. N-1-Benzoyl-5-ethyluracil (**62**) was prepared by first treating 5-ethyluracil (**60**) with excess benzoyl chloride to give the N^1,N^3 dibenzoyl adduct **61**. Selective removal of the N-1 substituent with potassium carbonate in aqueous dioxane gave N-1-benzoyl-5-ethyluracil (**62**) in >70% overall yield. All of these products were successfully isolated as crystalline solids.

The reactions of these uracil derivatives are summarised in Figure 3. Disappointingly, many of the results suggested that nucleophilic substitution was not competitive with elimination (see Scheme 5, paths A and B). Dialkylation of the uracil to the bis-adduct proved difficult to control. Elimination does not appear

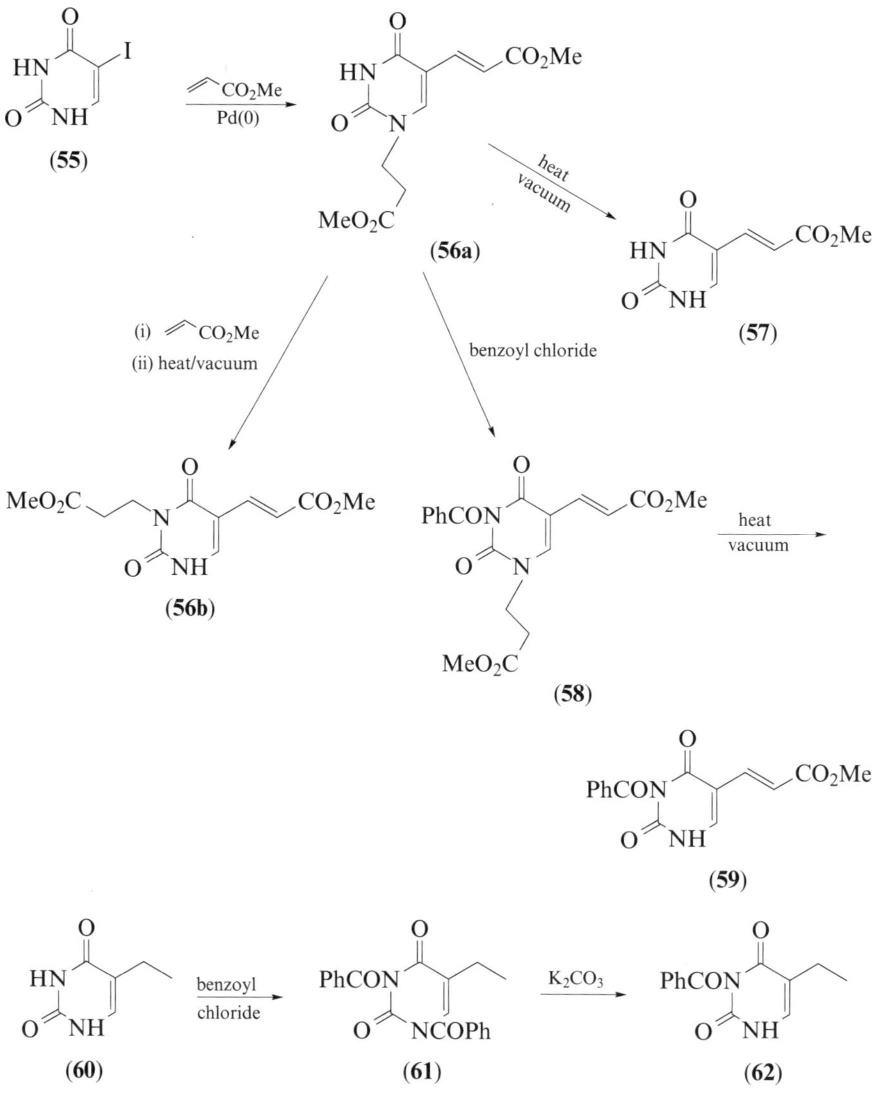

Scheme 13

to be competitive with nucleophilic substitution when the uracil base was substituted at both C-5 and N-3 with electron-withdrawing groups (entry 6); this was investigated further.

The synthesis now required a one-carbon homologation and a specific double bond migration (see Scheme 14). Nucleophilic substitution of the mesylate derivatives of allylic alcohols **63a** and **63b** with cyanide anion, as described in the

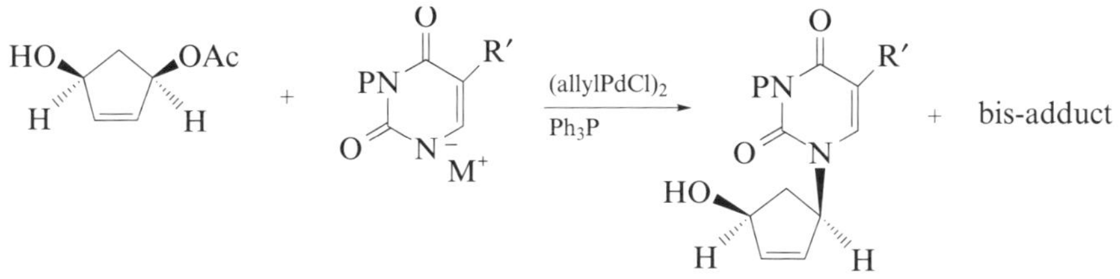

(21) **(52)** **(53)**

	P	R′	M	Catalyst (mol%)	Yield of **53**	Yield of bis-adduct
1	H	H	Na, K	5.0	20–40%	10–15%
2	H	I	Na, K	5.0	10 %	5 %
3	H	Et	Na, K	5.0	10 %	-
4	H	$CHCHCO_2Me$	Na	5.0	65%	14 %
5	$(CH_2)_2CO_2Me$	$CHCHCO_2Me$	Et_3NH	5.0	82 %	-
6	COPh	$CHCHCO_2Me$	Et_3NH	0.5	96 %	-
7	COPh	Et	DBUH	5.0	72 %	-

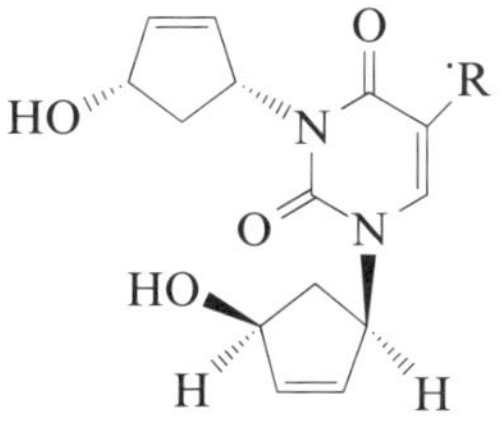

bis-adduct

Figure 3 Reactions of uracil derivatives

synthesis of GR85478 (**1**),[17] proceeded in poor yield (20–30%) and with incomplete isomerisation of the double bond; cyanide is sufficiently basic to promote extensive elimination of the sensitive allylic mesylate derivatives. Work reported by Stille and Scott[26] on the palladium(0)-catalysed carbonylation of vinyl and aryl triflates offered an attractive solution to the conversion of allylic alcohols **63a** and **63b** into α,β-unsaturated carboxylic acids **66a** and **66b**.

Oxidation of allylic alcohols **63a** and **63b** with sulphur trioxide–dimethyl sulphoxide provided the corresponding enones **64a** and **64b** in 87% yield. Selective 1,4-reduction of these enones with KS-Selectride in the presence of *N*-phenyltriflimide provided the desired $\Delta^{3,4}$-vinyl triflates **65a** and **65b** in 70–80% yield after chromatography. It was necessary to carry out the reduction in the presence of *N*-phenyltriflimide to avoid extensive elimination of the uracil base, presumably via a rapid enolate isomerisation. The vinyl triflates were converted into the α,β-unsaturated carboxylic acids **66a** and **66b** *in situ* by addition of 5 mol% of palladium(II) acetate, triphenylphosphine, triethylamine and water

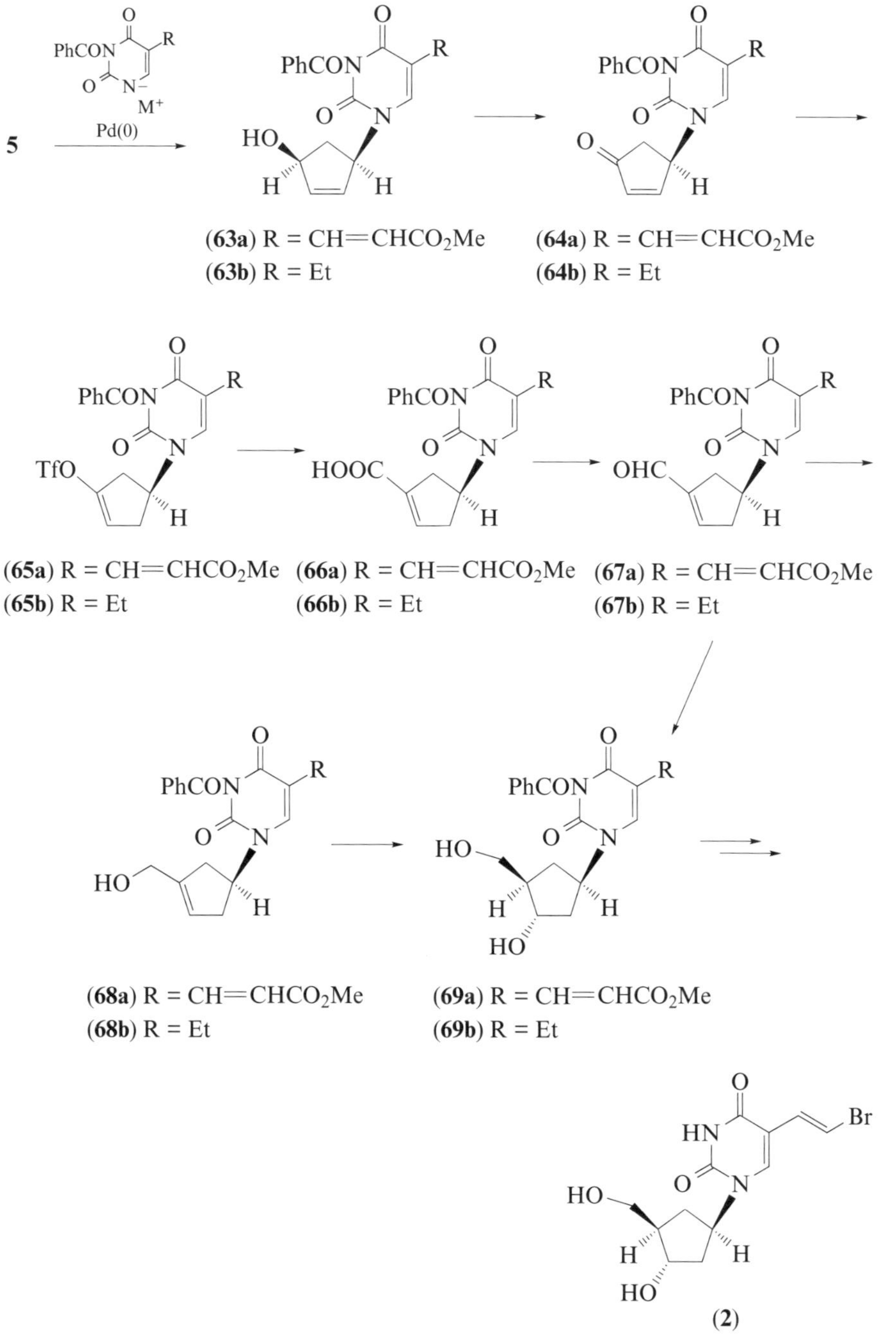
PhCON
R
M+
Pd(0)
5
HO
H
(63a) R = CH=CHCO2Me
(63b) R = Et
(64a) R = CH=CHCO2Me
(64b) R = Et
TfO
HOOC
OHC
(65a) R = CH=CHCO2Me
(65b) R = Et
(66a) R = CH=CHCO2Me
(66b) R = Et
(67a) R = CH=CHCO2Me
(67b) R = Et
(68a) R = CH=CHCO2Me
(68b) R = Et
(69a) R = CH=CHCO2Me
(69b) R = Et
HN
Br
(2)

Scheme 14

under an atmosphere of carbon monoxide. The α,β-unsaturated carboxylic acid **66a** was isolated in 65–70% yield after extractive work-up and crystallisation. The α,β-unsaturated carboxylic acid **66b**, isolated in 70% yield after chromatography, unexpectedly also contained 5–25% of the *cis*-β,γ-unsaturated isomer (see Scheme 15), although the vinyl triflate **65b** had been shown to be free of isomeric impurities. Subjecting the pure α,β-unsaturated carboxylic acid **66b** to the reaction conditions for 24 h resulted in no double bond deconjugation. It was qualitatively observed that longer carbonylation reaction times resulted in more of the *cis*-β,γ-unconjugated isomer. The problem was solved by increasing the amount of palladium(0) in the reaction mixture simply by starting with palladium(0), rather than palladium(II); this increased the rate of the reaction and reduced the amount of *cis*-β,γ-carboxylic acid to <2%.

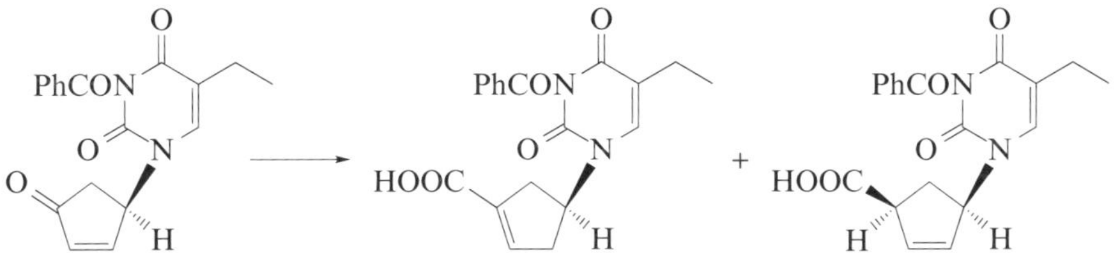

(64b) **(66b)** α,β-isomer *cis*-β,γ-isomer

Catalytic system	α,β-/β,γ-ratio
$Pd(OAc)_2$, Ph_3P, TEA, H_2O, CO	3:1
$Pd(OAc)_2$, Ph_3P, TEA, H_2O, CO, DMSO (10 vol.)	15:1
$(allylPdCl)_2$ (2 mol %), Ph_3P, TEA, H_2O, CO, DMSO (10 vol.)	30:1
$(allylPdCl)_2$ (5 mol %), Ph_3P, TEA, H_2O, CO. DMSO (10 vol.)	>50:1
$(allylPdCl)_2$(2 mol %), dppb, TEA, H_2O, CO, DMSO (10 vol.)	>50:1

Scheme 15

The choice of phosphine ligand also has an effect on the rate of the carbonylation. Substituting 1,4-bis(diphenylphosphino)butane (dppb) for triphenylphosphine reduced the reaction time by nearly 50%. Furthermore, when using this ligand, very little metallic palladium is deposited, which would cause serious problems on scale-up in the pilot plant.

Reduction of the α,β-unsaturated carboxylic acids **66a** and **66b** to either the α,β-unsaturated aldehydes **67a** and **67b** or the allylic alcohols **68a** and **68b** was envisaged as providing suitable substrates for the creation of two new chiral centres, if the previously employed hydroboration strategy was used. Reduction of the acid chloride derivative of **66b** with sodium borohydride at −78°C provided the desired allylic alcohol **68b** in 74% yield (a low temperature was needed to avoid reduction of the uracil ring). Alternatively, reduction of the acid chloride with triethylsilane with catalysis by palladium on carbon provided the α,β-unsaturated aldehyde **67b** in 45% yield after chromatography. Aldehyde **67a** was obtained by reduction of the acid chloride of **66a** with a pre-formed zinc borohydride-triethylamine adduct[27] and was isolated by crystallisation in 50–55% yield. Reduction of the acid chloride of **66a** with bis(triphenylphosphino)copper borohydride, a reagent developed to reduce acyl halides to aldehydes,[28] provided the allylic alcohol **68a** cleanly in 50–60% yield (Scheme 14).

We had now arrived at the key step in this approach, namely the chemo-, regio- and stereoselective hydroboration of allylic alcohol **68a**. We found that thexylborane added to the cyclopentene double bond *anti* to the uracil base with very high selectivity, although the hydroboration required three equivalents of thexylborane. The first equivalent complexed with the hydroxyl, the second reduced the benzamide and the third reacted as desired with the cyclopentene double bond. The resultant diol **69a** was acetylated prior to isolation owing to its instability. Hydroboration of **68b** with borane provided only a 3:1 mixture of debenzoylated *anti* and *syn* isomers in 30% yield. The dialkylboranes disiamylborane and 9-BBN reduced the benzamide, but failed to react with the cyclopentene double bond. Hydroboration of allylic alcohol **68b** with three equivalents of thexylborane provided the debenzoylated diol **69b** in 65% yield (Scheme 14).

Hydroboration of aldehydes **67a** and **67b** with three equivalents of thexylborane as described above resulted in reduction of the aldehyde, reduction of the benzamide and hydroboration of the cyclopentene double bond *anti* to the uracil base to form the desired diols **69a** and **69b** cleanly. As the methodology to convert diols **69a** and **69b** into **2** had already been established, the viability of this synthetic strategy was thus demonstrated, and it was shown that the palladium catalysis method was equally applicable to both anti-viral compounds GR85478 and GR95168.

4.4 LAMIVUDINE

The final section of this chapter deals with the synthesis of lamivudine (**3**) (also known as Epivir™ or 3TC™). This compound was first discovered by scientists at Biochem Pharma, in Montreal, Canada, as part of a programme directed towards developing therapies for AIDS[29] and is now being developed by Glaxo Wellcome as a treatment for infection by HIV and hepatitis B.

Lamivudine was initially prepared as its racemate and, in common with the previous compounds in this chapter, it was using this isomeric mixture that early

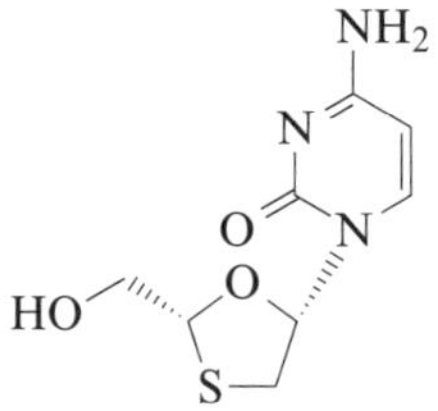

(**3**)

pharmacological and safety studies were conducted. Separation of enantiomers on a scale sufficient to determine their individual efficacies and toxicological properties is not usually a difficult problem, but the preparation of several kilograms in order to proceed with clinical studies is often a major challenge to the process research chemist. Lamivudine was no exception to this rule.

As is common with nucleosides and their analogues, it is the *cis* isomers of lamivudine which possess significant anti-viral activity. The two *cis* diastereoisomers were first separated by preparative chiral HPLC using an acetylated *β*-cyclodextrin column[30] and, although the two isomers were found to have approximately equal activities against HIV, one isomer (the laevorotatory one) was less cytotoxic than the other. When the decision to develop only the single, less toxic isomer was taken, substantial supplies of the racemate had already been prepared. From that point onward, preparation of only the laevorotatory isomer was our goal.

4.4.1 SYNTHESIS OF THE RACEMATE

The original synthesis used to prepare the racemate is shown in Scheme 16. The oxathiolane ring was constructed by reaction of hydroxyacetaldehyde [protected as its benzoate ester (**70**)] with mercaptoacetaldehyde, generated by heating its cyclic dimer, dithianediol. The resulting 5-hydroxyoxathiolane **71** was converted into acetate **72** by treatment with acetyl chloride–pyridine, then the acetate group displaced by bis-silylated *N*-acetylcytosine in the presence of titanium(IV) chloride. This coupling reaction gave many problems on scale-up; it was only marginally *cis*-selective, a *cis*:*trans* ratio of about 3:2 was obtained and a substantial amount of *N*-deacetylation occurred. The desired *cis* isomer was partially purified by column chromatography, then further purified by crystallisation. Removal of the acetyl and benzoyl groups was achieved by treatment with sodium methoxide in methanol, and the racemic product was purified by recrystallisation from water.

Although this synthesis was suitable for the preparation of small quantities, it required several modifications to enable manufacture on a large scale to be carried out safely and reliably. The protected hydroxyacetaldehyde **70** was isolated as its hydrogensulphite addition compound (a stable, crystalline solid) which needed no further purification, and distillation of the thermally unstable aldehyde was thus avoided.

The aldehyde **70** was regenerated by treatment with acid, prior to reaction with dithianediol giving hydroxyoxathiolane **71**, which, previously, had not been isolated. This compound exists in solution as a mixture of epimers, caused by rapid ring opening and closing (see Scheme 17). By carefully controlling the crystallisation conditions, the equilibrium between *cis* and *trans* isomers can be continually re-established in solution as crystallisation proceeds, enabling the product to be isolated as the pure, racemic *trans* isomer in about 70% yield.

Control over the relative stereochemistry at C-2 and C-5 of the oxathiolane ring was a major problem. Liotta's group showed[31] that if tin(IV) chloride was

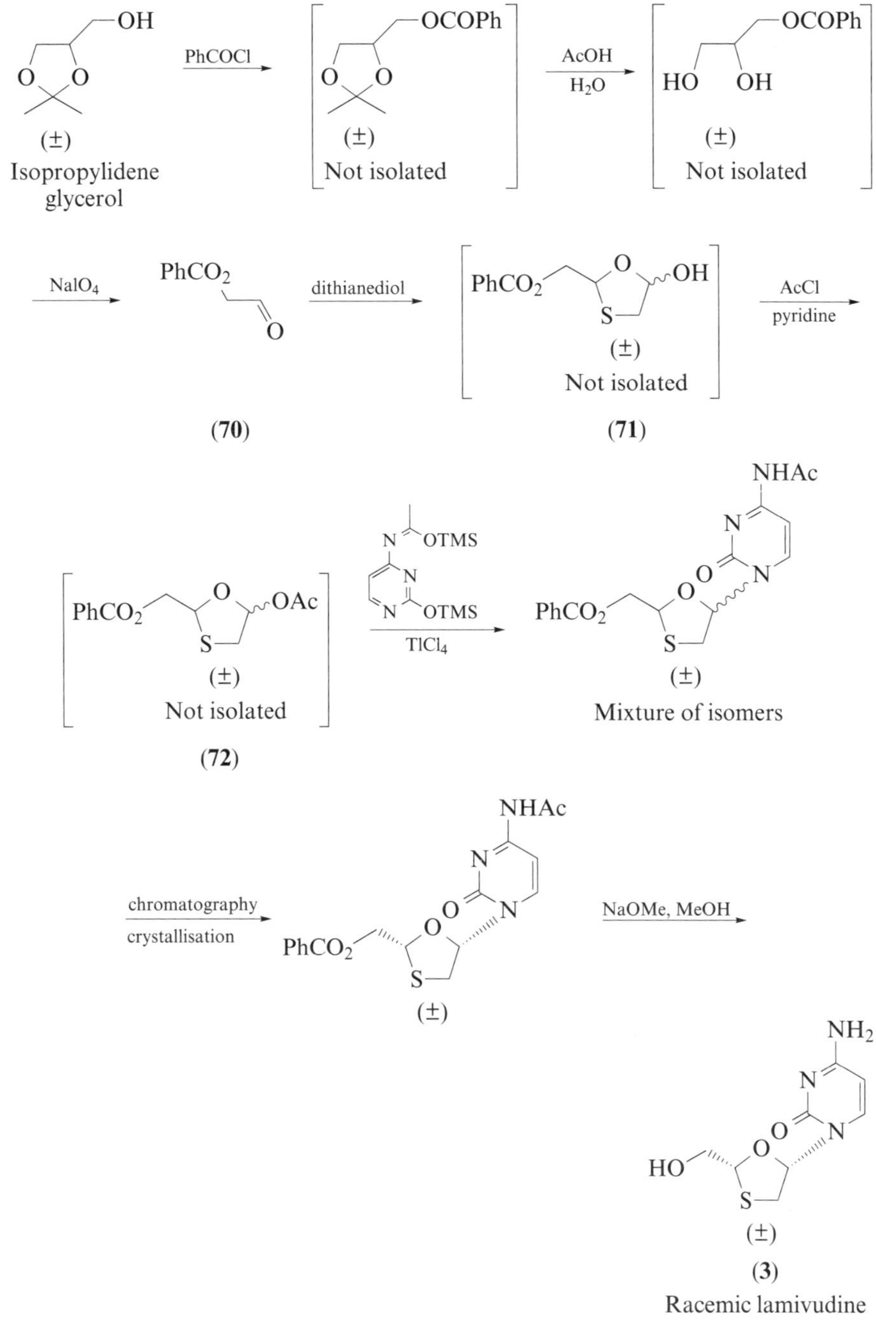
OH
PhCOCl
OCOPh
AcOH
H2O
HO
OH
(±)
Isopropylidene glycerol
Not isolated
NaIO4
PhCO2
O
dithianediol
S
AcCl
pyridine
(70)
(71)
NHAc
N
OTMS
TlCl4
OAc
OAc
Mixture of isomers
(72)
chromatography
crystallisation
NaOMe, MeOH
NH2
HO
(3)
Racemic lamivudine

Scheme 16

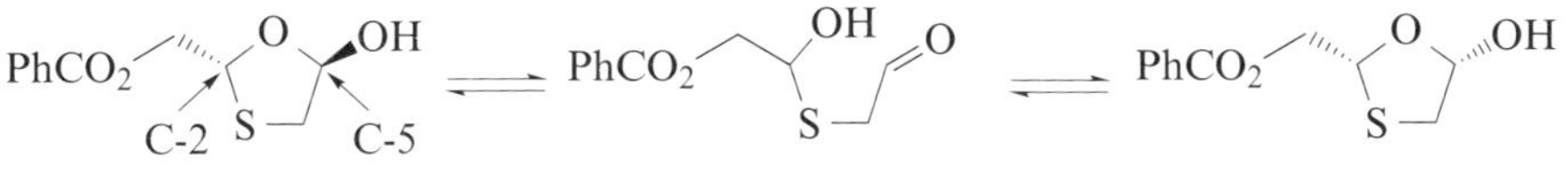

(71a) (71b)

Scheme 17

used as a Lewis acid, the *cis* selectivity of the coupling reaction was dramatically increased; no other common Lewis acid had this effect. In our hands, a selectivity of around 10:1 in favour of the *cis*-isomer could be achieved reliably. Work-up of these metal-mediated reactions is often difficult, requiring either extensive chromatography or slow filtrations to remove metal salts. We found that if the reaction mixture was acidified, we could crystallise the desired product as its hydrochloride salt merely by adding a counter solvent. The product was contaminated with only a small quantity of the unwanted *trans*-isomer and required no further purification. On scale-up, it proved necessary to recrystallise the salt to reduce the level of tin-containing impurities, but despite this the new process represented a major improvement. Finally, the benzoyl group was removed by treatment with a basic ion-exchange resin in suspension in ethanol, from which racemic lamivudine was crystallised directly in good yield and purity.

4.4.2 RESOLUTION OF RACEMIC LAMIVUDINE

As substantial supplies of the racemate were thus available, it was sensible to attempt resolution of this isomeric mixture. Formation of a diastereoisomeric salt with a chiral acid was tried with some success, and enzymic methods were also investigated. Racemic lamivudine was treated with phosphoryl chloride and trimethyl phosphate to give a mixture of the two phosphates **73a** and **73b**, which were then treated with the *Crotalus atrox* venom 5′-nucleotidase. Only the undesired, 'natural,' isomer (**73b**) was hydrolysed to the alcohol, and was easily separated from the 'unnatural' phosphate; subsequent dephosphorylation, using bacterial alkaline phosphatase, afforded enantiomerically pure lamivudine (**3**).[32] Once separated in quantity, the two isomers were compared, and the absolute configuration of the less toxic enantiomer was established by single-crystal X-ray diffraction (of the 4-bromobenzamide derivative) and found to be the '*unnatural*' (−)-isomer.[32]

Given that this 'unnatural' isomer was the required entity, a much simpler synthetic method could be developed using the enzyme cytidine deaminase.[33,34] This enzyme converts cytidine into uridine, and again only recognises the 'natural' isomer. When racemic lamivudine was treated with cytidine deaminase, the 'natural' isomer was converted selectively into the uridine analogue **74**, leaving the desired enantiomer (**3**) unchanged (see Scheme 18). This substituted uracil could easily be separated from lamivudine by selective adsorption on an ion-exchange column, exploiting the differences in acidity of the hydrogens on the heterocyclic ring. This process (which was subsequently developed into a viable plant process) was used to prepare around 3 tonnes of lamivudine, with an enantiomeric excess[35] typically greater than 99.9%.

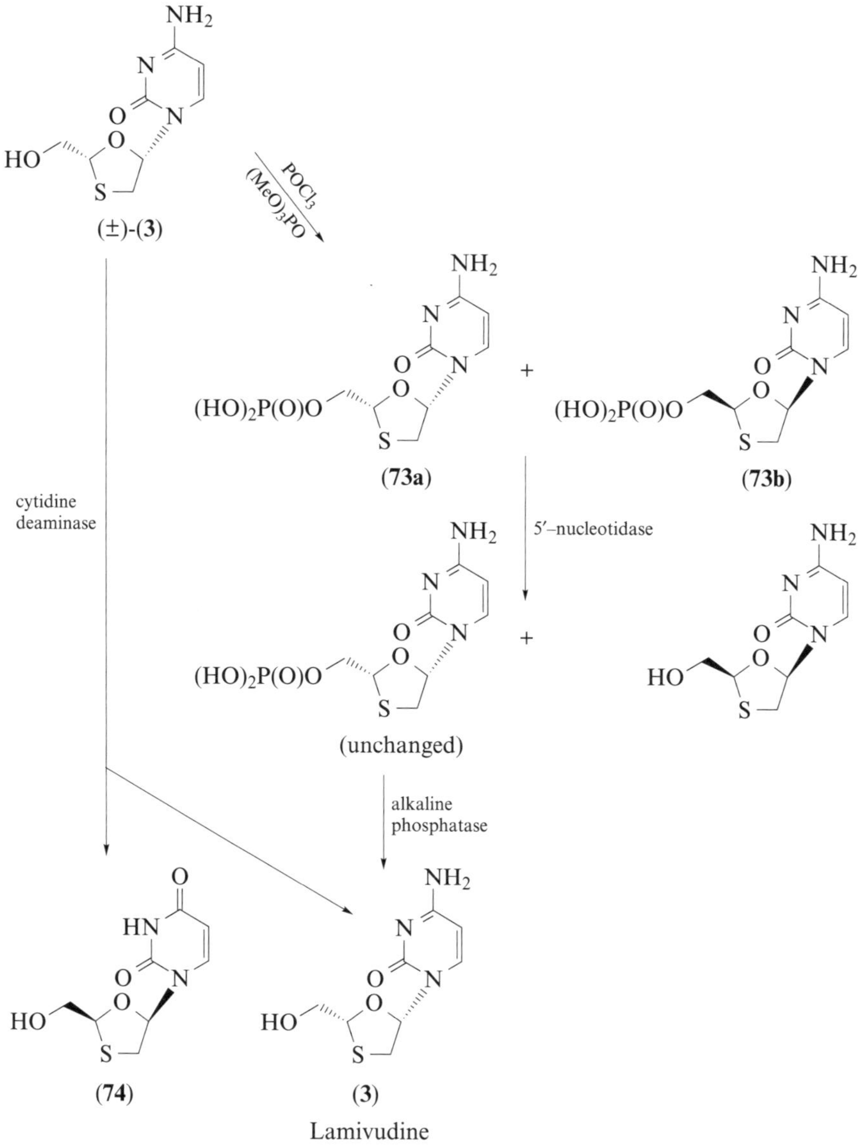

Scheme 18

4.4.3 ALTERNATIVE STRATEGIES

Although significant advances had been made, we still felt that it should be possible to find a more efficient route. Formation of an enantiopure compound by resolution as the last step of a synthesis is inefficient; at least half of all the input to each stage is lost, giving severe waste-handling problems, and the resulting process is both environmentally and economically unacceptable. We recognised the need for a new approach. An enantioselective synthesis of lamivudine had been published,[36] but this was long and unsuitable for large-scale production. Consequently, we investigated two alternative strategies:

- Synthesis of an intermediate in the current route in chiral form.
- Design of a new synthesis.

To avoid wasting large quantities of the toxic tin(IV) chloride, or the expensive starting material cytosine, the obvious target for a chiral intermediate was the stable, crystalline acetoxyoxathiolane **72**. Successful use of this intermediate requires retention of chirality under the coupling reaction conditions. The stereochemistry at C-5 of the oxathiolane ring is not important; provided that the stereogenic centre at C-2 remains intact and the mechanism of the reaction produces specifically a *cis* arrangement of the two substituents, then the desired enantiomer of lamivudine will be obtained.

Acid **76** was prepared by reaction of aldehyde **70** with mercaptolactic acid (**75**), and resolved via the brucine salt of its *S*-benzyl derivative (Scheme 19). Treatment with lead tetraacetate gave oxathiolane **72**, enantiomerically pure at C-2, although as a mixture of *cis* and *trans* isomers.[37]

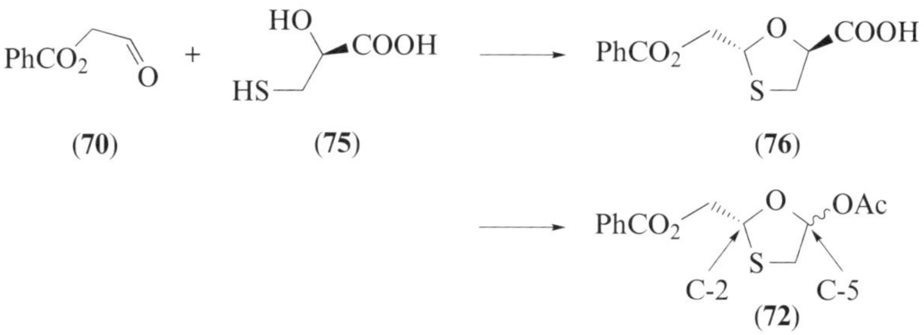

Scheme 19

Attempts were made to produce an enantiomerically pure intermediate by enzymic means with the propionate ester **77** as the preferred substrate. On treatment with *Mucor miehei* lipase only the (2*S*)-isomer (**77b**) was hydrolysed, giving a mixture from which the (2*R*)-isomer of the propionate (**77a**) could be isolated, albeit in low yield[38] (Scheme 20).

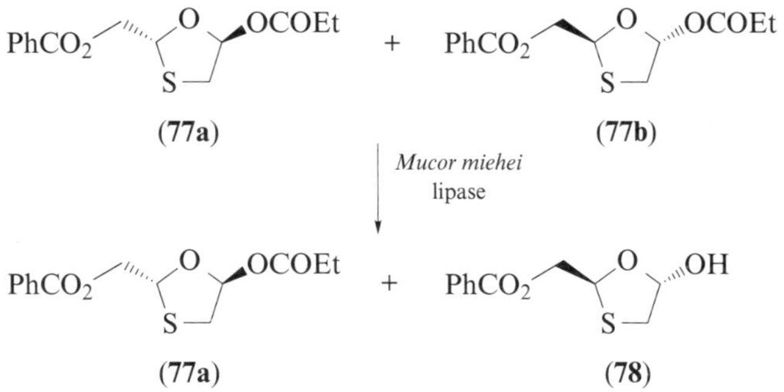

Scheme 20

Next we investigated the effect of the coupling reaction on the stereogenic centre at C-2. As the reaction tolerates a wide variety of ester and carbonate substituents at C-5, carbonate **79** was selected for investigation. The diastereoisomers were separated chromatographically, and the extra stereogenic centres present in the menthol fragment allowed easy identification of the products of the coupling reaction by either NMR or HPLC.

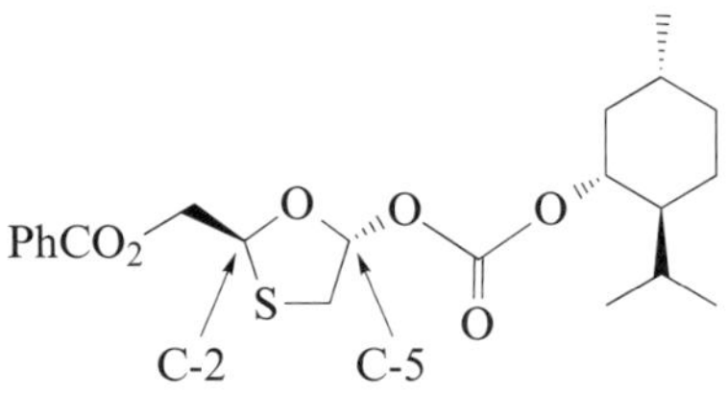

(79)

When compound **79** was treated with bis-silylated cytosine and the resulting benzoate hydrolysed, the lamivudine produced was found (by chiral HPLC) to be *racemic*; all chirality at C-2 had been lost, even though the reaction was *cis*-selective. The stereogenic centre at C-2 could be preserved if iodotrimethylsilane was used as Lewis acid, but this gave a mixture of *cis* and *trans* isomers in a ratio of approximately 1:1. Neither of these methods was attractive for potential scale-up and, with this avenue now seemingly closed, the only option left was to design a new synthesis.

4.4.4 A NOVEL SYNTHESIS OF LAMIVUDINE

The chemistry of oxathiolanes with a 2-carboxy substituent derived from glyoxylic acid was being investigated by colleagues at Biochem Pharma. In contrast to those used in the route described in Scheme 16 with a CH_2 substituent at C-2, compounds in this series reacted with silylated cytosine in the presence of iodotrimethylsilane to give the coupled products with excellent *cis*-selectivity. By making an ester with L-(−)-menthol and separating a single diastereoisomer in an analogous fashion to the menthyl carbonate described above, we showed that the coupling proceeded with *retention* of the stereochemistry at C-2.[39,40]

This chemistry therefore had the potential to form the basis of a more efficient chiral route (see Scheme 21), as the presence of an acid group at C-2 is an obvious handle by which resolution early in the synthesis could be achieved. Condensation of glyoxylic acid with dithianediol gave hydroxy acid **81**. In solution, this compound exists as a mixture of isomers, but by careful control of the crystallisation (as with the hydroxyoxathiolane **71** discussed earlier), it was isolated as the *trans* isomer, although the rate of equilibration in solution was found to be

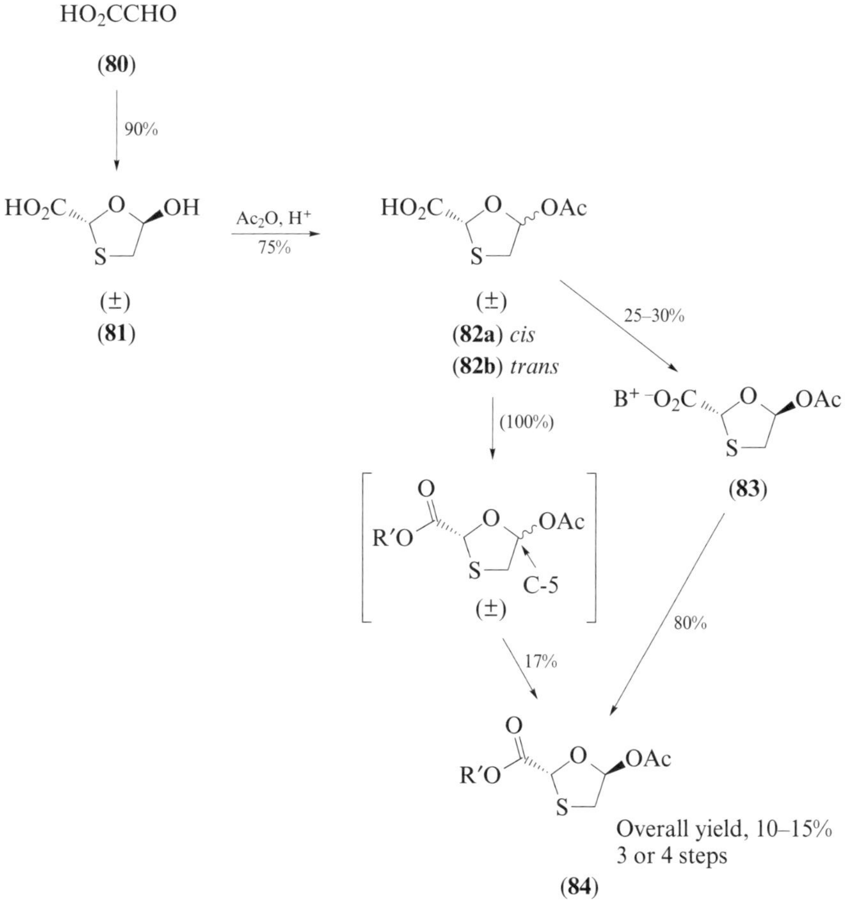

Scheme 21

solvent dependent. In DMSO the rate was slow, allowing satisfactory ^{1}H NMR spectra to be obtained. In other solvents the rate of equilibration was much faster.

For a satisfactory coupling reaction, the hydroxyl group at C-5 had to be converted into a leaving group, such as an ester or a carbonate. The simplest such group is acetoxy, and this was our first choice (although good results were achieved using other esters and carbonates). In the earlier route, it had been found that addition of acetyl chloride to a solution of hydroxyoxathiolane **71** in dichloromethane and pyridine gave a mixture of *trans* and *cis* isomers of the acetoxy derivative **72** in a ratio of about 70:30. However, if solid hydroxyoxathiolane **71** was added to acetyl chloride–pyridine, a fast acetylation took place with little epimerisation. NMR confirmed that epimerisation of **71** in chlorinated solvents was fast, but clearly not as fast as acetylation. When applied to hydroxy acid **81**, this approach was not successful, however. The acid was much less soluble in chlorinated solvents, reacted slowly and, while the product was of predominately *trans* relative stereochemistry, it was contaminated with many by-products. The major by-product was the dimer **86**, indicating that reaction of intermediate anhydride **85** was of about equal rate to reaction with acetyl chloride–pyridine (see Scheme 22).

Treatment of hydroxy acid **81** with acetic anhydride in the presence of strong acids, such as sulphuric or methanesulphonic acid, gave a clean reaction which

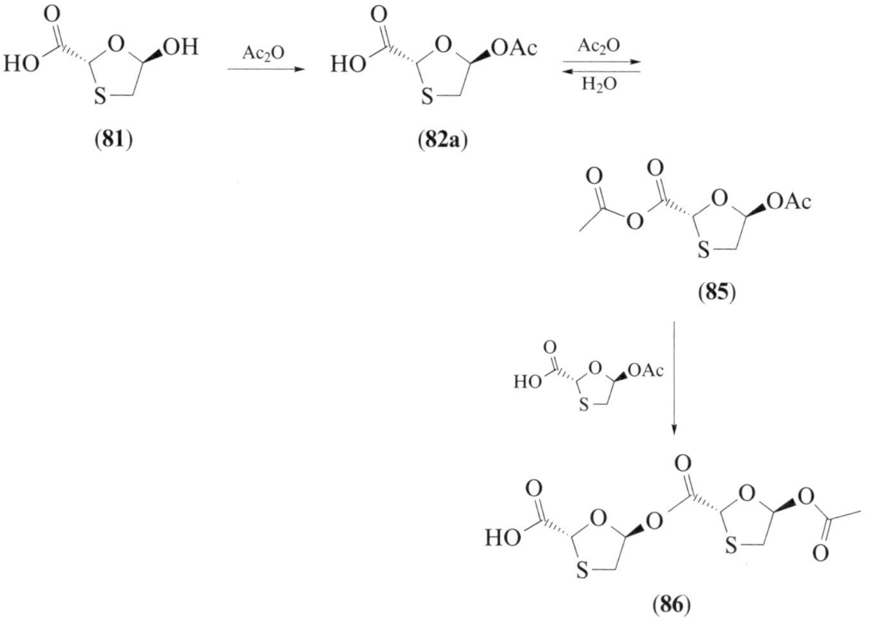

Scheme 22

was rather slow. Under these conditions the compound epimerised again to a 70:30 mixture of *trans* and *cis* isomers, before being cleanly acetylated, giving a mixture of the two acetates **82a** and **82b**. As no reliable method could be developed for the preparation of the pure *trans* isomer (nor, indeed, the *cis* isomer), we investigated methods for the resolution of the mixture which could be reliably reproduced on a large scale in a yield of about 75%.

A common way to resolve racemic acids which was applied to GR95168 (as discussed in Section 4.3) is the formation of diastereoisomeric salts with a chiral base. Several bases were screened for suitability in the resolution of acetoxy acid **82**; the best found was norephedrine. The diastereoisomerically pure salt of *trans*-acid **83** was isolated in 27% yield merely by adding a solution of norephedrine to the racemic mixture of *cis*- and *trans*-acids in isopropyl acetate. Although the yield seems low, this represents a yield of about 80% of the available diastereoisomer present in the four-component mixture.

Concurrently, we investigated the formation of diastereoisomeric esters by reaction of the crude mixture of acids with (−)-menthol. Although this was used initially as a way of both introducing crystallinity to the ester and allowing further purification of the resolved acid, we found that we could isolate a single isomer from the mixture of all four diastereoisomers by careful control of the crystallisation conditions. This represented a yield of 50% of available isomer (although only 17% of all four). Nevertheless, a simple crystallisation of the desired isomer (**84**) from a mixture of four in a ratio of 35:35:15:15 was a significant achievement, as the process could easily be run on a large scale, used no expensive resolving agents and did not require regeneration of the acid.

Following pilot-plant trials, the selective crystallisation of the desired diastereoisomer was chosen in preference to formation of the salt, and the process was developed further before being transferred to full-scale production.

Acetate **84** was treated with silyated cytosine in the presence of iodotrimethylsilane to give ester **87** (Scheme 23). Preliminary investigations showed the reaction gave a *cis*-selectivity of about 93:7, which was comparable to that in the route discussed previously using tin(IV) chloride. On isolation, most of the *trans*

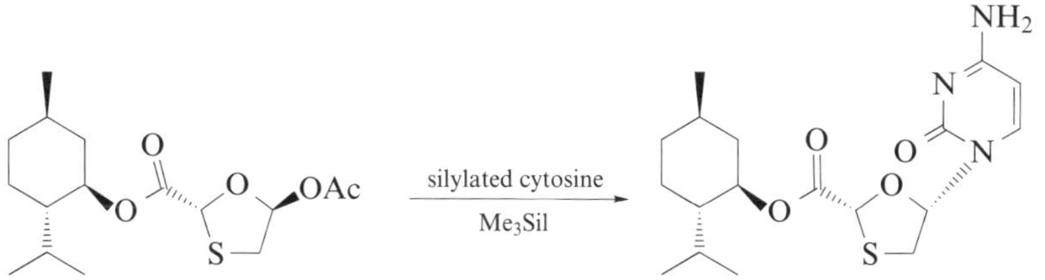

(84) **(87)**

Scheme 23

isomer could be removed, giving a product contaminated with only small amounts of unwanted isomers and no racemisation. After optimisation, the work-up was fairly simple: the pure product was isolated directly from the quenched reaction by filtration in a yield of around 75%.

The reason for the high *cis*-selectivity was not immediately apparent. Attempts to isolate intermediates from the corresponding stage of the route shown in Scheme 16 had met with failure; tin(IV) chloride caused decomposition of oxathiolane **72**, and gave a complex mixture with silylated cytosine. However, when ester **84** was treated with iodotrimethylsilane in the *absence* of silyated cytosine, a single compound was formed, which was isolated and identified as iodide **88**. NMR showed a mixture of isomers in solution, with the *trans* predominating, indicating that the reaction was merely an S_N2 displacement of iodide by silylated cytosine, and that the *cis*-selectivity resulted from the fact that the major isomer of the iodide was *trans.*

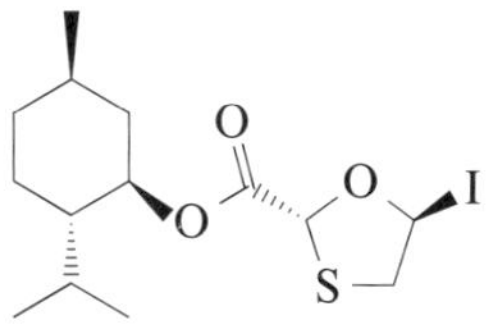

(88)

To complete the synthesis, ester **87** was reduced to give the primary alcohol, lamivudine (**3**), using any convenient reducing agent, such as lithium aluminium hydride, but isolation of the drug substance from the reaction mixture was not trivial. In the laboratory, on a small scale, column chromatography was satisfactory, but in the pilot plant a macroreticular resin was preferred.

These resins are non-functionalised and contain pores capable of accepting organic molecules. When an aqueous solution of an organic compound is passed down a column of this resin, the organic components bind to the column and the inorganics do not. On washing the column with water all inorganics are removed, leaving the desired product on the resin. If an organic solvent, such as acetone or an alcohol, is then passed down the column, the resin expands and the bound compound is released. Isolation of the product from an organic phase which is free from inorganic contamination is then possible by concentration and crystallisation. This technique was used successfully for the isolation of lamivudine, passing the solution down a column of Amberlite XAD-16 resin, washing with water, then elution with propan-2-ol. Lamivudine could be crystallised by concentration and cooling, and was obtained in good yield and high purity.

After further development and optimisation, this route was transferred to full-scale production at Glaxo-Wellcome plants, and has been used to produce several tonnes of lamivudine.

4.5 CONCLUSIONS AND SUMMARY

The synthesis on a large scale of the three anti-virals considered here posed many significant problems. The molecules all contain chiral centres which are important with respect to the activity of each drug, and levels of the undesired stereoisomers in the final drug substance had to be minimised.

In the course of developing these products, several chiral strategies have been used:

- Enzymatic resolution.
- Classical resolution with diastereoisomeric salts.
- Selective crystallisation of desired diastereoisomers.
- Asymmetric synthesis.

Practical processes have been developed which we have used to prepare several tonnes of enantiomerically pure intermediates and drugs. New routes to these molecules have also been produced, in some cases requiring the development of new methodology to solve particular problems of selectivity.

4.6 REFERENCES

1. Storer, R., Paternoster, I. L., Borthwick, A. D., and Biggadike, K (Glaxo, UK), *US Pat.* 5 155 112, 1992.
2. (a) Vince, R., and Hua, M., *J. Med. Chem.*, **33**, 17 (1990), and references cited therein; (b) Vince, R., and Hua, M., (University of Minnesota), *US Pat.*, 4 950 758, 1990, and 4 931 559, 1990.
3. Bray, B. L., Lichty, M. E., Partridge, J. J., and Turnbull, J. P. (Glaxo, UK), *US Pat.*, 5 233 041, 1993.
4. Mancuso, A. J., and Swern, D., *Synthesis*, 165 (1981).
5. Van Rheenen, V., Kelly, R. C., and Cha, D. Y., *Tetrahedron Lett.*, 1973 (1976).
6. Trost, B. M., Kuo, G. H., and Benneche, T., *J. Am. Chem. Soc.*, **110**, 621 (1988).
7. Bray, B. L., Lichty, M. E., and Partridge, J. J., unpublished results.
8. Sugiyama, H., Sera, T., Dannoue, Y., Marumoto, R., and Saito, I., *J. Am. Chem. Soc.*, **113**, 2290 (1991).
9. Bray, B. L., and Partridge, J. J., (Glaxo, UK), *US Pat.*, 5 438 132, 1995.
10. *Fluka Chemika–BioChemika Analytika 1995–96*, Fluka Chemical, Ronkonkoma, NY, 1995.
11. (a) Wang, Y., Chen, C., Girdaukas, G., and Sih, C., *J. Am. Chem. Soc.*, **106**, 3695 (1984); (b) Laumen, K., Rimerdes, E. M., and Schneider, M., *Tetrahedron Lett.*, **26**, 407 (1985); (c) Deardorff, D. R., Matthew, A. J., McKeekin, S., and Claney, C.L., *Tetrahedron Lett.*, **27**, 1255 (1986); (d) Deardorff, D. R., and Windham, C. Q., *Org. Synth.*, **67**, 114 (1988); (e) Suga, T., and Mori, K., *Synthesis*, 19 (1988); (f) Deardorff, D. R., Windham, C. Q., and Craney, C. L., *Org. Synth.*, **73**, 25 (1995); (g) Bray, B. L., Lackey, J. W., Lovelace, T. C., Mook, R. A., and Partridge, J. J., unpublished results.
12. (a) Deardorff, D. R., Linde, R. G., Martin, A. M., and Shulman, M. J., *J. Org. Chem.*, **54**, 2759 (1989); (b) Lackey, J. W., Mook, R. A., and Partridge, J. J (Glaxo, USA), *US Pat.*, 5 057 630, 1991.
13. Sepracor Inc., Marlborough, MA 01752, USA.

14. Amano International Enzyme Company, Inc., P.O. Box 1000, Troy, VA 22974, USA.
15. Trost, B. M., and Molander, G. A., *J. Am. Chem. Soc.*, **103**, 5969 (1981).
16. Partridge, J. J., and Bray, B. L., (Glaxo, USA), *US Pat.*, 5 329 008, 1994).
17. Partridge, J. J., and Bray, B. L., (Glaxo, USA), *US Pat. Appl.*, 08/271 594, 1994.
18. Coe, D. M., Myers, P. L., Parry, D. M., Roberts, S. M., and Storer, R., *J. Chem. Soc., Chem. Commun.*, 151 (1990).
19. Cookson, R. C., Dudfield, P. J., Newton, R. F., Ravenscroft, P., Scopes, D.I.C., and Cameron, J. M., *Eur. J. Med. Chem.*, **20**, 375 (1985).
20. (a) Cookson, R. C., Dudfield, P. J., and Scopes, D. I. C., *J. Chem. Soc, Perkin Trans. 1*, 399 (1986); (b) Ravenscroft, P., Newton, R. F., and Scopes, D. I. C., *Tetrahedron Lett.*, **27**, 747 (1986); (c) Ravenscroft, P., (Glaxo, UK), *US Pat.*, 4 658 044, 1987.
21. (a) Jagt, J. C., and Van Leusen, A. M., *Recl. Trav. Chim. Pays-Bas*, **92**, 1343 (1973); (b) Jagt, J. C., and Van Leusen, A. M., *J. Org. Chem.*, **39**, 564 (1974); (c) Daluge, S., and Vince, R., *J. Org. Chem.*, **43**, 2311 (1978).
22. (a) Brown, H.C., and Zweifel, G., *J. Am. Chem. Soc.*, **81**, 247 (1959); (b) Brown, H. C., and Daniels, J. J., *Org. Synth.*, 719 (1988); (c) Still, W. C., and Barrish, J. C., *J. Amr. Chem. Soc.*, **105**, 2487 (1983).
23. Brown, H. C., Bigley, D. B., Arora, S. K., and Yoon, N. M., *J. Am. Chem. Soc.*, **92**, 7161 (1970).
24. Bray, B. L., Dolan, S. C., Halter, B., Lackey, J. W., Schilling, M. B., and Tapolczay, D. J., *Tetrahedron Lett.*, **36**, 4483 (1995).
25. Brown, H.C., and Zweifel, G., *J. Am. Chem. Soc.*, **83**, 1241 (1961).
26. (a) Stille, J. K., *Angew. Chem., Int. Ed. Engl.*, **25**, 508 (1986); (b) Scott, W. J., and McMurry, J. E., *Acc. Chem. Res.*, **21**, 47 (1988).
27. Bray, B. L., unpublished results.
28. Fleet, G. W. J., Fuller, C. J., and Harding, P. J. C., *Terahedron Lett.*, 1437 (1978).
29. Soudeyns, H., Yao, X.-J., Belleau, B., Kraus, J.-L., Nguyen-Ba, N., Spira, B., and Wainberg, M. A., *Antimicrob. Agents Chemother.*, **35**, 1386 (1991).
30. Coates, J. A. V., Cammack, N., Jenkinson, H. J., Mutton, I. M., Pearson, B. A., Storer, R., Cameron, J. M., and Penn, C. R., *Antimicrob. Agents Chemother.*, **36**, 202 (1992).
31. Choi, W. B., Wilson, L. J., Yeola, S., and Liotta, D. C., *J. Am. Chem. Soc.*, **113**, 9377 (1991).
32. Storer, R., Clemens, I. R., Lamont, B., Noble, S. A., Williamson, C., and Belleau, B., *Nucleosides Nucleotides*, **12**, 225 (1993).
33. Coates, J. A. V., Mutton, I. M., Penn, C. R., Storer, R., Williamson, C. (IAF International Inc., Canada), *PCT Int. Pat. Appl.*, WO 91/17159, 1991.
34. Mahmoudian, M., Baines, B. S., Drake, C. S., Hale, R. S., Jones, P., Piercey, J. E., Montgomery, D. S., Purvis, I. J., Storer, R., Dawson, M. J., and Lawrence, G. C., *Enzyme Microb. Technol.*, **15**, 749 (1993).
35. Rogan, M. M., Drake, C., Goodall, D. M., and Altria, K. D., *Anal. Biochem.*, **208**, 343 (1993).
36. Beach, J. W., Jeong, L. S., Alves, A. J., Pohl, D., Kim, H. O., Change, C.-N., Doong, S.-L, Schinazi, R. F., Cheng, Y.-C., and Chu, C. K., *J. Org. Chem.*, **57**, 2217 (1992).
37. Humber, D. C., Jones, M. F., Payne, J. J., and Ramsay, M. V. J., *Tetrahedron Lett.*, **33**, 4625 (1992).
38. Cousins, R. P. C., Mahmoudian, M., and Youds, P. M., *Tetrahedron: Asymmetry*, **6**, 393 (1995).
39. Jin, H., Siddiqui, M. A., Evans, C. A., Tse, H. L. A., Mansour,. T. S., Goodyear, M. D., Ravenscroft, P., and Beels, C. D., *J. Org. Chem.*, **60**, 2621 (1995).
40. Mansour, T. S., Jin, H., Tse, A., and Siddiqui, A (Biochem Pharma Inc., Canada), *Eur. Pat.*, EP 0 515 157, 1992.

PHYSICAL METHODS AND CLASSICAL RESOLUTION

5 Rational Design in Resolutions

A. BRUGGINK
Chemferm, Breda, and University of Nijmegen, The Netherlands

5.1 Introduction 81
5.2 Background to a Rational Approach 81
5.3 Rational Design and Synthesis 84
5.4 Rationale of a Resolution 88
5.5 Predicting a Resolution? 92
5.6 Asymmetric Transformations, Racemisations and Recycling 94
5.7 Conclusions 96
5.8 Acknowledgements 97
5.9 References 97

5.1 INTRODUCTION

In the 100 years since the death of Louis Pasteur, the production of enantiopure compounds by 'classical' resolution has retained a prime position in the chemical industry. In terms of the *number* of products rather than tonnage, it is still the most widely used method, despite the fact that no sound theoretical basis for the method has been available; practical experience and trial and error have been the guidelines for arriving at economic resolution processes. This 'state of the art' has been well presented in two recent books.[1]

5.2 BACKGROUND TO A RATIONAL APPROACH

Around 1978, during the development of the enantiopure captopril intermediate D-(−)-*S*-acetyl-β-mercaptoisobutyric acid (**1**) (Figure 1), the need for a more rational approach to resolutions through selective crystallisations became apparent to the author and his co-workers at Océ Andeno (now DSM Andeno). Racemic acid **1**, easily obtained by Michael addition of thioacetic acid to methacrylic acid, resisted all attempts at classical resolution. Initially, chemically and optically pure **1** was not available either for reference or for investigation of the preparation of diastereoisomeric salts with resolving agents. Work with the more stable and somewhat less flexible *S*-benzoyl analogue met with more success (confirming our prejudice about the dependence of the resolution result on

Chirality in Industry II. Edited by A. N. Collins, G. N. Sheldrake and J. Crosby

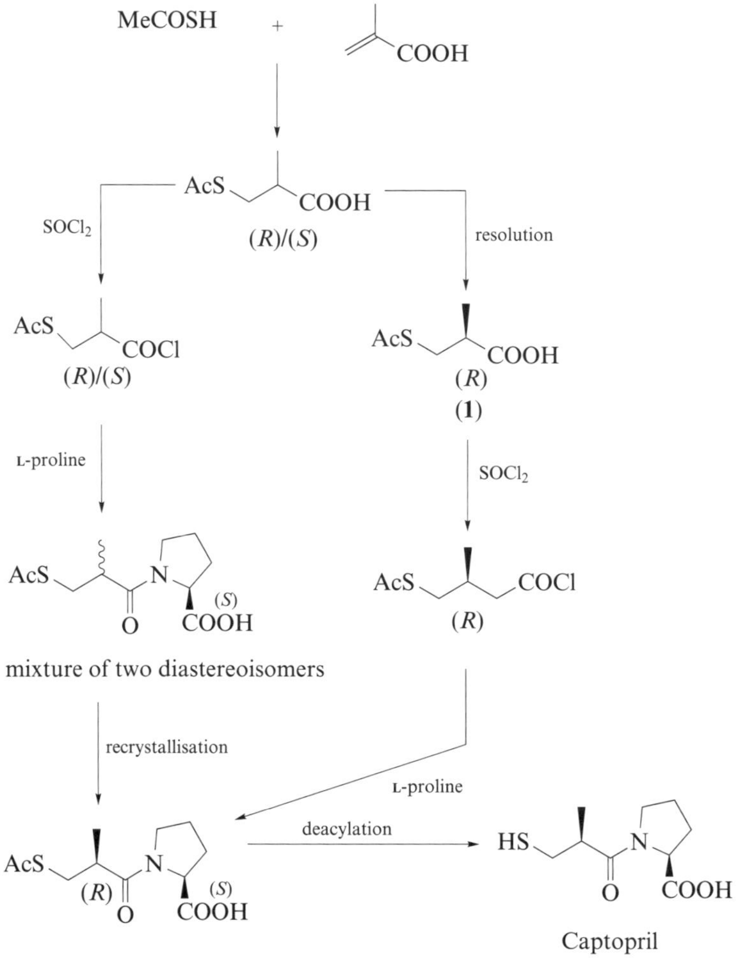

Figure 1 Captopril syntheses. Left: original Squibb process. Right: DSM Andeno process

conformational rigidity). With the traditional resolving agent cinchonidine an enantiopure product could be obtained,[2] but this did not provide a satisfactory basis for an economic industrial process. The approach suffered from several disadvantages:

- cinchonidine was expensive and difficult to recover;
- thiobenzoic acid was expensive;
- recrystallisations were required to obtain an enantiopure product;
- the unwanted enantiomer cannot be racemised for recycle in the process or recycled for other uses;

- only *S*-acetyl-derived products were registered with the FDA.

There were initial encouraging results using simple amino alcohols as resolving agents; moreover, some of these resolving agents were cheaply available on industrial scale. However, these resolutions were found to be the result of a difference in the *rate* of crystallisation of the two diastereoisomeric salts and *not based on a difference in solubility.* Experiments on a larger scale, which took longer, resulted in both diastereoisomers crystallising.

Eventually, a reliable laboratory-scale resolution was developed using D-(+)-*N*-benzyl-α-phenethylamine (not a particularly rigid molecule, given the initial preconception!) as the resolving agent.[2] At this point, only the FDA issue remained to be addressed; resolution of intermediate **1** was still required. As sufficient amounts of enantiopure D-(−)-and L-(+)-*S*-benzoyl-β-mercaptoisobutyric acid were now available, an efficient debenzoylation–reacylation procedure could, at last, be developed, giving access to the optically pure enantiomers of **1**. Systematic preparation of a range of diastereoisomeric salts and measurement of solubility differences (and other physical parameters; DSC, phase diagrams) indicated some good resolving agents for racemic **1**, leading to the present situation where:

- all captopril is produced via enantiopure **1**;
- the fact that the unwanted enantiomer L-(+)-**1** cannot readily be racemised or recycled does not preclude the economic success of the route while other potentially more economic methods for preparing **1**, e.g. asymmetric addition of thioacetic acid to methacrylic acid (requiring asymmetric protonation) are not available;
- importantly, several of the empirical rules for developing resolutions, such as the need for maximum rigidity in the starting molecules, were found not to apply.

With interest from the Dutch government and universities in the fine chemical industry, it was timely to join forces in a more fundamental research programme on resolutions and chirality. A collaboration between Andeno and the University of Groningen started in 1980 and has evolved into an almost nationwide programme between DSM Andeno, DSM Research, Chemferm and the Universities of Nijmegen and Groningen. An ambitious, long-term programme was set up aimed at:

- rational design and synthesis of new resolving agents (all the commonly used resolving agents are selected on the basis of trial and error);
- revealing, ideally common, key parameters behind successful resolutions;
- defining chirality in quantitative physical parameters and developing methods, ultimately, for predicting good resolving conditions;
- developing asymmetric transformations and more efficient methods for racemisations and recycling.

Interim results from the programme, at approximately its half way point at the time of writing, are presented in this chapter.

5.3 RATIONAL DESIGN AND SYNTHESIS

Very few approaches to the rational design of resolutions have been published. After an unsuccessful attempt by Winther in 1895[3] to classify alkaloids as resolving agents, a few publications have appeared in the last two decades. Useful compilations have been given by Wilen, Jacques and Collet,[4–6] and Fogassy and co-workers[7] have found interesting relationships between resolution results and physical parameters of the diastereoisomeric salts. Statistically relevant relationships were found between resolution results in a series of substituted phenylglycines with tartaric acid and parameters such as Taft σ-values, melting points, heats of fusion, densities and solubilities.

In addition, Fogassy and co-workers[8] have studied solvent effects. Although a lot of data were generated, basic explanations of the resolution results are lacking mainly because crystal structures of the diastereoisomeric salt pairs were not available.[8] The same applies to an extensive study by Arnett and co-workers[9] on the resolution of mandelic acid with α-phenethylamine, ephedrine and pseudoephedrine. An attempted crystal study by Gould and co-workers[10] remained incomplete because crystal structures of both partners of the diastereoisomeric pairs were not available. They did indicate, however, that small and subtle differences in interactions within the diastereoisomeric salts could have a profound effect on crystal structure and that the solvent, water, played a decisive role.

Using rationally designed molecular structures for new resolving agents to avoid preconceptions from historical studies, we embarked upon a programme of synthesis. Acidic resolving agents were chosen as the target because only a limited number, particularly of strongly acidic resolving agents, are available for industrial use (tartaric acids, mandelic acids, camphorsulphonic acids). Based on the theoretical and experimental results considered important at that time[1,11] in the design of resolving agents, cyclic phosphoric acids **(2)** were chosen. In making this selection, the following criteria (adapted from ref. 11) were considered:

(i) Strongly acidic and basic resolving agents were preferred over weakly acidic and basic reagents to increase the possibility of salt formation.
(ii) The stereocentre in the resolving agent should be close to the centre of salt formation.
(iii) Rigidity was preferred over flexibility. The incorporation of additional functionality (to increase interactions) should enhance rigidity.
(iv) Both enantiomers should be readily available.
(v) The resolving agent should be chemically and optically stable under the conditions used for the formation, separation and dissociation of diastereoisomeric salts and for recovery of the resolving agent.

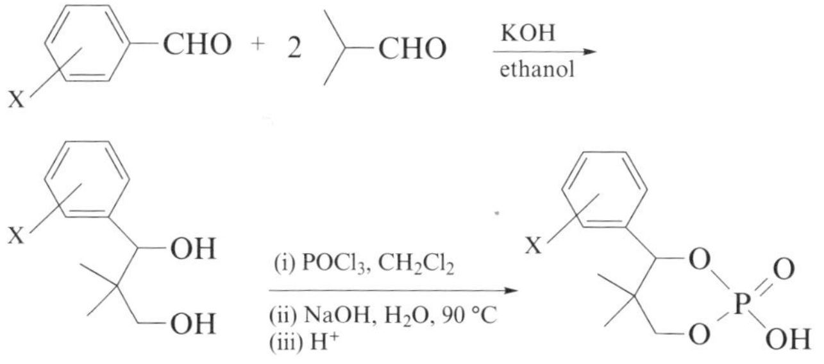

(2)

	X		X
(**a**)	H	(**i**)	4-OMe
(**b**)	2-F	(**j**)	3,4-OCH_2O
(**c**)	2-Cl	(**k**)	2-NO_2
(**d**)	2-Br	(**l**)	3-NO_2
(**e**)	2,4-Cl	(**m**)	2-thienyl
(**f**)	2,6-Cl	(**n**)	2-furyl
(**g**)	4-Cl	(**o**)	2-OEt
(**h**)	2-OMe	(**p**)	4-Me

Figure 2 Synthesis of the cyclic phosphoric acids **2**

The synthesis of compounds of the general structure **2** is straightforward and an extensive range could be prepared (see Figure 2).[11] Also, resolution was easily accomplished using D-(−)-*p*-hydroxyphenylglycine, ephedrine or 2-amino-1-phenylpropane-1,3-diol. The enantiopure phosphoric acids proved to be promising resolving agents, and several amines, alcohols and amino acids were resolved through selective diastereoisomeric salt crystallisation, as shown in Table 1. Examples of excellent resolutions were found which afforded diastereopure salts in a single crystallisation in yields of 40% or better.

Recovery of the phosphoric acids is easy owing to their very low solubility in water and common solvents, and their chemical, thermal and optical stability was excellent. Compared with camphorsulphonic acid (until then the best strong acid industrial resolving agent), a much improved strongly acidic resolving agent became available: *p*-chlorophenyl-substituted phosphoric acid. The resolution efficiency of camphorsulphonic acid for 21 bases was compared with that of 4-chlorophenylphosphoric acid, and the results (Table 2) show clearly the better resolving ability of the phosphoric acid.

Before concluding this section, a cautionary remark must be added. Using the same criteria as for the design of the phosphoric acids, the basic analogues **3**, **4** and **5** were prepared (Figure 3). Compounds **3** and **4** were readily resolved by the phosphoric acids **2**, but they did not in themselves perform as good resolving bases. Compound **5** could not even be resolved by **2**. Also, a 'designer' resolving agent for the attractive captopril precursor *β*-chloroisobutyric acid **6** did not

Table 1 Resolutions of various amines and amino acids with cyclic phosphoric acids **2**

Amine /amino acid	Phosphoric acid, **2**. X	Result (small-scale)[a]
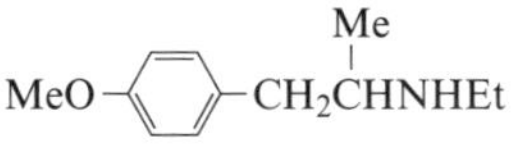	H, 4-Cl, 2-OEt, 2,4-Cl_2 2-OMe 2-Cl	– + ++
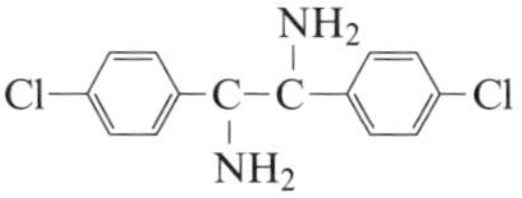	2-OMe 2-Cl H, 2,4-Cl_2	– + ++
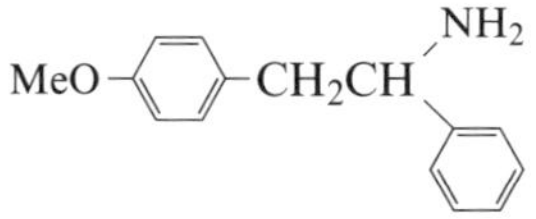	4-Cl H, 2-Cl 2-OMe, 2-OEt, 2,4-Cl_2	– + ++
	H,4-Cl, 2-OEt 2-OMe, 2-Cl, 2,4-Cl_2	– +
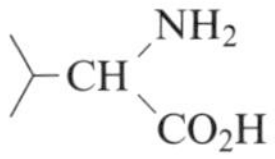	H,4-Cl, 2-OMe, 2-OEt 2-Cl	– +
Ph–CH(NH_2)CO_2H	H,4-Cl, 2-Cl, 2-OMe	–
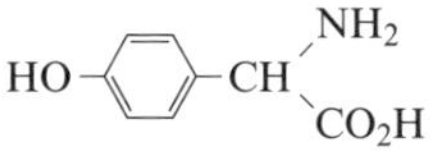	H,4-Cl 2-OMe, 2-Cl, 2,4-Cl_2	– ++
$PhCH_2CH(NH_2)CO_2H$	H,4-Cl, 2-OEt, 2,4-Cl_2 2-OMe 2-Cl	– + ++
$PhCH_2CH_2CH(NH_2)CO_2H$	H,2-OMe, 2-OEt, 2-Cl 2,4-Cl_2	– +
$MeSCH_2CH_2CH(NH_2)CO_2H$	2-OMe 2-Cl	– ++

[a] –, No or poor resolution; +, some resolution; ++, excellent resolution.

materialise from this work. Based on model studies and known preferred interactions in crystal structures, resolving agents **7** and **8** were designed and prepared. Resolutions were thwarted either at the stage of preparing enantiopure resolving agent (**8**), or in attempted resolution of acid **6** with amine **7a** or **7b**. Crystalline salts were not obtained even with enantiopure **6** and amine **7a** or **7b**.

Table 2 Resolution efficiency of 4-chlorophenylphosphoric acid (A) and camphorsulphonic acid (B)

Base	Resolution efficiency, S[a]	
	A	B
D-(−)-2-Aminobutan-1-ol	0.27	0.17
L-(+)-*threo*-2-Amino-1-phenylpropane-1,3-diol	0.35	0.33
(2*S*,3*R*)-(+)-4-Dimethylamino-1,2-diphenyl-3-methylbutan-2-ol	0.25	0.69
(−)-Ephedrine	0.50	0.24
(+)-*N*-Methylephedrine	0.31	0.66
L-(−)-Norephedrine	0.27	0.07
D-(−)-Phenylglycinol	0.34	
(+)-Pseudoephedrine		0.04
D-(+)-α-Amino-ε-caprolactam	0.18	
D-(+)-α-Dimethylamino-ε-caprolactam		0.06
(*R*)-(−)-1-Cyclohexylethylamine	0.58	0.13
L-(−)-α-Methylbenzylamine		0.05
(−)-Brucine	0.13	0.54
(+)-Dehydroabietylamine	0.58	0.20
(−)-Cinchonidine	0.79	0.18
(+)-Cinchonine		0.06
(+)-Quinidine	0.39	0.05
(−)-Quinine	0.71	0.04
(+)-Hydroquinidine	0.38	0.03
(−)-Hydroquinine		0.10
L-(−)-Strychnine	0.54	0.26
Average	0.41	0.20
Number of resolutions with efficiency >0.50	6	3

[a] Resolution efficiency *S* is defined as the difference between solubilities of the separate diastereoisomeric salts divided by the solubility of the most soluble diastereoisomer; see also ref. 14.

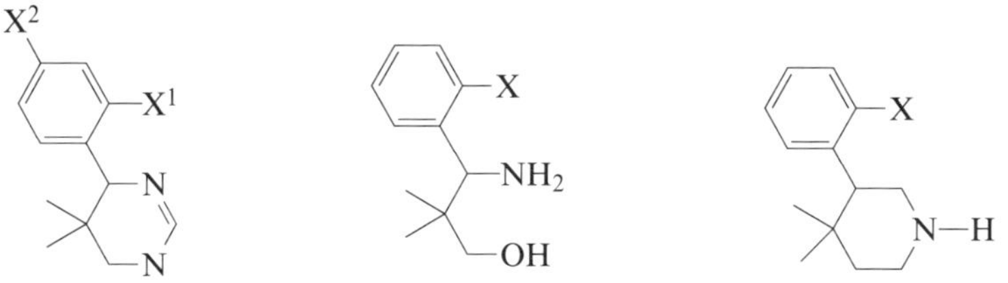

(**3**) (**4**) (**5**)

(**a**) $X^1 = X^2 = H$
(**b**) $X^1 = Cl$, $X^2 = H$
(**c**) $X^1 = X^2 = Cl$

(**a**) X = H
(**b**) X = Cl

(**a**) X = H
(**b**) X = Cl

Amidines, amino alcohols and piperidines

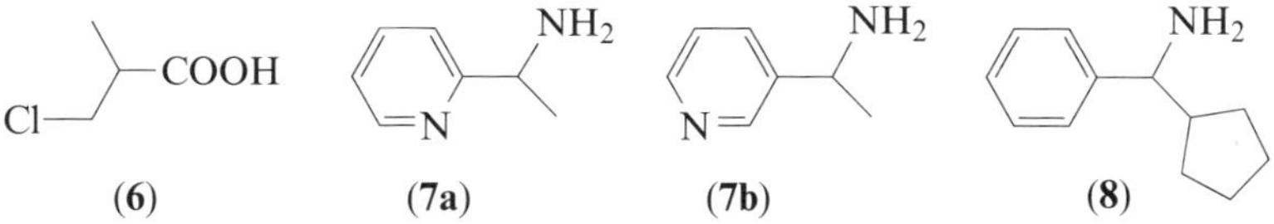

(**6**) (**7a**) (**7b**) (**8**)

Figure 3 Unsuccessful 'designer' resolving agents

5.4 RATIONALE OF A RESOLUTION

One obvious requirement for studying the basis for a resolution, namely that both salts of the diastereoisomeric pair should be available in a pure and well defined form, is not always easy to meet in practice. We found ephedrine and the cyclic phosphoric acids to be ideal pairs for salt preparations and further physical and model studies. Nineteen pairs of diastereoisomeric salts stemming from differently substituted acid–base pairs were prepared[13] and several physical parameters were determined, of which a selection are given in Table 3.

Table 3 Physical properties of cyclic phosphoric acid/ephedrine salts

Diastereoisomeric salt (X = arene subst. in acid, Y= arene subst. in ephedrine)			Solubility[a]			M.p (K)	ΔH_{fus} (kJ mol^{-1})
			EtOH	Pr^{i}OH	Pr^{n}OH		
1 X = H	Y = H	n[b]	8.8	1.37	5.33	504.1	48.5±0.4
		p	8.4	1.30	5.22	508.3	47.6+1.3
2 X = 2-F	Y = H	n	22.1	3.66		483.0	40.6±0.1
		p	46.0	8.19		474.2	37.1±0.4
3 X = 2-Cl	Y = H	n	5.7	0.86	3.34	492.7	47.7±0.8
		p	23.6	3.39	13.98	491.3	39.5±1.0
4 X = 2-Br	Y = H	n	9.3	1.48		491.6	42.6±0.6
		p	16.4	2.56		482.0	40.4±0.8
5 X = 2,6-Cl_2	Y = H	n	10.0	1.40		488.4	40.6±0.5
		p	62.6	7.56		462.9[c] 476.3[c]	31.8±4.0[d]
6 X = 2,4-Cl_2	Y = H	n	8.1	1.29	4.9	488.5	50.4±0.8
		p	49.5	7.62	30.9	457.4	31.9±0.4
7 X = 4-Cl	Y = H	n	3.32	0.45		510.7	52.0±1.1
		p	26.7	3.99		478.3	41.2±0.5
8 X = 2-Me	Y = H	n	5.75	0.95	3.57	487.4	54.3±1.0
		p	17.4	2.74	11.0	498.7	46.6+0.3
9 X = 4-Me	Y = H	n	17.5	2.82		483.0	36.4±0.9
		p	20.0	3.41		482.9	33.6+1.0
10 X = 2-OMe	Y = H	n	4.1	0.61	2.39	495.5	47.6±1.2
		p	23.1	3.04	13.9	475.4	43.9+0.4
11 X = 2-NO_2	Y = H	n	6.70	0.79	3.10	479.5	44.0±0.1
		p	4.36	0.44	2.07	486.6	48.6±0.5
12 X = H	Y =2-Cl	n		1.41		506.4	41.0±0.5
		p		1.28		499.2	47.0±0.2
13 X =2-Cl	Y = 2-Cl	n		0.21		534.0	60.2±0.7
		p		2.54		477.8	45.6+0.3
14 X = 2,6-Cl_2	Y = 2-Cl	n		0.99		507.4	45.0±0.6
		p		–[e]		–[e]	–[e]
15 X = 4-Cl	Y = 2-Cl	n		0.98		518.8	40.8±0.6
		p		–[e]		–[e]	–[e]
16 X = H	Y = 2,6-Cl_2	n		3.79		481.5	–[d]
		p		0.25		514.0	–[d]

(continued)

Table 3 *(continued)*

Diastereoisomeric salt (X = arene subst. in acid, Y= arene subst. in ephedrine)			Solubility[a] EtOH	PriOH	PrnOH	M.p (K)	ΔH_{fus} (kJ mol^{-1})
17 X = 2-Cl	Y = 2,6-Cl_2	n	–[e]			–[e]	–[e]
		p	0.13			524.4	–[d]
18 X = H	Y = 4-Cl	n	3.91			452.2	43.5±2.0
		p	1.41			492.7	45.5±1.1
19 X = 2-Cl	Y = 4-Cl	n	–[e]			–[e]	–[e]
		p	0.89			511.3	40.5±2.0

[a] Solubilities at 25 °C (g/100g).
[b] n Salt = + − or − + ; p salt = ++ or − − .
[c] Two points of fusion were observed; the first one is followed by an exothermic crystallisation.
[d] Total change in enthalpy between 460 and 480 K.
[e] Not crystalline.
[f] Reproducible values could not be obtained.

From these data the following conclusions were drawn:

- The efficiency of a resolution may be expressed as $S = (k_p - k_n)/k_p$ [where k_p and k_n are the respective solubility constants for the p and n salts of the diastereoisomeric pair (see refs 14 and 15)] and derived from the differences in solubility within the diastereoisomeric salt pair which correlates fairly well with the experimental results (yield and optical purity) of that resolution (see also Table 4 and 5).
- The resolution efficiency *S* also correlates fairly well with ΔH_f, the difference in enthalpy of fusion between a diastereoisomeric pair. Using a known thermodynamic model[14] for the crystallising diastereoisomeric pair, it is possible to arrive at a linear relationship between resolution efficiency expressed as $\ln(k_p/k_n)$ and the difference in lattice enthalpy ΔH_{solid} as shown in Figure 4.
- Useful correlations of resolution efficiency with other physical parameters could not be obtained, although low solubility and high melting point within a salt pair are generally found (see also the results of Fogassy and co-workers' work[8]).
- Effects of substitution are most profound in the phosphoric acid (see Table 4); substituents in the ephedrine moiety have a smaller influence, although they enhance the effect of the phosphoric acid substituents. In this way, very efficient resolutions for chloro-substituted ephedrines with chloro-substituted phosphoric acids were predicted and found (see Table 4).

In order to find a more definitive explanation of the successful resolutions with these phosphoric acids, the crystal structures of six diastereoisomeric salt pairs were studied,[14–16] and the results are summarised in Table 5. Analysing the crystal structures for an explanation of the efficiency of the resolution revealed no common, single cause.[17]

A subtle interplay of, mainly, hydrogen bonding, van der Waals and electrostatic forces is presumed to be responsible for the resolution results, whereby each

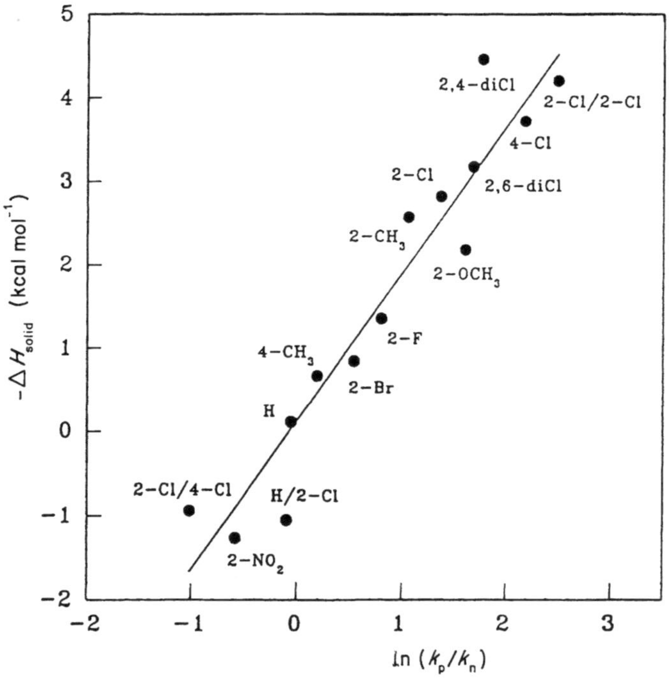

Figure 4 Plot of $-\Delta H_{solid}$ versus $\ln(k_p/k_n)$. Labels indicate substituents (with unsubstituted ephedrine) or acid–base phenyl substituents

Table 4 Calculated maximum values of *S* for the phosphoric acids/ephedrine salts, derived from the solubilities given in Figure 6

Diastereoisomeric salt		Calculated value of S^a		
Phosphoric acid	Ephedrine	EtOH(abs.)	Pr^iOH	Pr^nOH
1 X = H	Y = H	−0.05	−0.05	−0.02
2 X= 2-F	Y = H	0.52	0.55	
3 X = 2-Cl	Y = H	0.76	0.75	0.76
4 X = 2-Br	Y = H	0.43	0.42	
5 X = 2,6-Cl_2	Y = H	0.84	0.81	
6 X = 2,4-Cl_2	Y = H	0.84	0.83	0.84
7 X = 4-Cl	Y = H	0.88	0.88	
8 X = 2-Me	Y = H	0.67	0.65	0.68
9 X = 4-Me	Y = H	0.13	0.17	
10 X = 2-OMe	Y = H	0.82	0.80	0.83
11 X = 2-NO_2	Y = H	−0.35	−0.44	−0.33

(continued)

Table 4 *(continued)*

Phosphoric acid	Ephedrine			
	Y = H	Y = 2-Cl	Y = 2,6-Cl_2	Y = 4-Cl
X = H	−0.05	−0.09	–[b]	–[b]
X = 2-Cl	0.76	0.92	−0.93	−0.64
X = 2,6-Cl_2	0.81	+ +[c]	+ +[c]	+ +[c]
X = 4-Cl	0.88	+ +[c]		

[a]Positive values of *S* indicate a more soluble p salt; negative values of *S* indicate a more soulble n salt.
[b]Not determined; small-scale resolutions indicated poor resolution results.
[c]The better dissolving diastereomeric salt was obtained as an oil; small-scale resolution indicated that very efficient results are obtained.

compound, even within an isostructural series, reaches a unique optimum arrangement for crystal packing. Water or solvent molecules included in the crystals have a stabilising influence only on otherwise unstable crystal packings; inclusion of solvent does not generally lead to much more stable crystals. When inclusion solvents are not available, the non-solvated salt of the crystal pair might become the least soluble, resulting in reversal of the resolution result and, possibly, allowing selection for either enantiomer even though the resolving agent is only available as one enantiomer. Resolution of phosphoric acid **2** (X = o-Cl) with L-(+)-*threo*-2-amino-1-phenylpropane-1,3-diol in methanol vs toluene or chloroform is an example.

Table 5 Resolution results and crystal structures of six phosphoric acids with ephedrine

Phosphoric acid		Resolution efficiency, *S*		Space group	Crystal water	Mode of packing	Notes
Subst.	Salt	EtOH	Pr^iOH				
X = H	p	0.05	0.05	$P2_1$	No	I_p	Very poor resolution
	n			$P2_1$	No	I_n	
X = 2-F	p	0.52	0.55	$P2_1$	No	I_p	No difference in crystal structure; good resolution. Cause: electrostatic interactions?
	n			$P2_1$	No	I_n	
X = 2-Cl	p	0.76	0.75	$P2_1$	No	I_p	Great difference in crystal packing; very good resolution. Cause: van der Waals interactions
	n			$P2_1$	No	III	
X = 2-Br	p	0.43	0.42	$C2$	Yes	II	Mediocre resolution. No crystals of *p*-salt without water present
	n			$P2_1$	No	I_n	
X = 2,6-Cl_2	p	0.84	0.81	$P2_1$	Yes	II	Very good resolution. No crystals of *p*-salt without water present
	n			$P2_1$	No	III	
X = 2,4-Cl_2	p	0.84	0.83	$P2_12_12_1$	Yes	II	Very good resolution. No crystals of *p*-salt without water present
	n			$P2_1$	No	I_n	

5.5 PREDICTING A RESOLUTION?

In view of the results described above, this would be a brave proposition. Attempts to compute quantitatively the differences in lattice energy of the crystal pairs in Figure 5 were not successful. The results of molecular mechanics calculations supported in only a very qualitative way the correlation of resolution results and lattice energy, and also revealed that different acid–base interactions in the hydrophobic layers of the crystals influence the resolution.[14–16] Although calculating and predicting crystal packings of some rather rigid organic heterocycles are almost within reach,[18] calculation of the far more complicated diastereoisomeric salts is still very difficult (even without addressing the question of the difference in crystal packing with its diastereoisomeric counterpart or the effect of solvent inclusion).

Accepting the severe limitations of the models presented above, a different approach was taken: principal component analysis (PCA). This statistical technique[19] allows the reduction of a large number of physical parameters to a much smaller number of new parameters, the principal components (PCs), which are linear combinations of the original parameters. When a full set of physical data for a series of compounds can be reduced to two or three PCs, useful plots can be obtained of these PCs, the compounds being located on the plots according to their individual properties. Several series of amines, Lewis acids, ketones and monosaccharides have been analysed in this way, and the use of PCA for the selection of the most suitable solvent for a given synthesis has been demonstrated.[19]

A set of 34 chiral amines was characterised using 39 parameters (both measured and calculated data were used). Applying PCA to this data set showed that 90% of the original information could be described with seven PCs, and only two PCs were needed to retain 65% of the initial data. Plotting PC-1 vs PC-2 gave clear groupings of the 34 resolving agents (see ref. 17). However, attempts to correlate the locations on the plot of these resolving amines with the results of the resolving behaviour of the cyclic phosphoric acids have not been successful.

In studies which were limited to a smaller set of 23 amines, characterised by fully reliable physical parameters (26 different properties), an impressive 75% of the original information was retained using only two PCs. A plot of PC-1 vs PC-2 (see Figure 5) shows clustering in two domains: simple amines at the left-hand side and the complex alkaloids on the right. There was, however, still no correlation with the resolution results of the cyclic phosphoric acids. An obvious explanation for this lack of correlation could be the limited contribution of chirality parameters in the data set. Only optical rotations had been used up to this point, and a more structure-related description of chirality was needed. Of the various theoretical models available,[20] we chose the molecular similarity model, which determines the degree of non-superimposability of two enantiomers (i.e. very close to the original definition of chirality).[21,22]

The non-overlap can be computed in terms of molecular properties such as

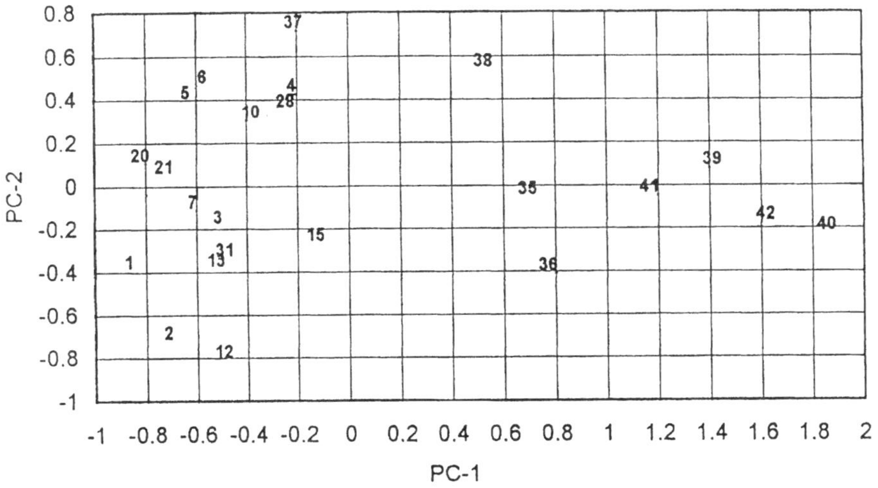

1 (*S*)-(+)-2-Aminopropanol-1(Alaninol)
2 (*R*)-(−)-2-aminobutan-1-ol
3 (1*S*,2*S*)-(+)-2-amino-1-phenylpropane-1,3-diol
4 (2*S*, 3*R*)-(+)-4-Dimethylamino-3-methyl-1,2-diphenylbutan-2-ol (Darvon alcohol)
5 (1*R*,2*S*)-(−)-2-Methylamino-1-phenyl-propan-1-ol (Ephedrine)
6 (1*S*,2*R*)-(+)-(-2-Dimethylamino-1-phenyl-propan-1-ol (*N*-Methylephedrine)
7 (1*R*, 2*S*)-(−)-2-Amino-1-phenylpropan-1-ol (Norephedrine)
8 (*S*)-(–)-2-Amino-3-phenylpropan-1-ol (Phenylalaninol)
9 (*R*)-(–)-2-Amino-2-phenylethanol (Phenylglycinol)
10 (1*S*,2*S*)-(+)-2-Methylamino-1-phenyl-propan-1-ol (Pseudoephedrine)
11 (1*R*, 2*R*-(−)-2-Amino-1-phenylpropan-1-ol (Norpseudoephedrine)
12 (*R*)-(+)-3-Aminohexahydroazepin-2-one (α-Amino-ε-caprolactam)
13 (*R*)-(+)-3-Dimethylaminohexahydroazepin-2-one (α-Dimethylamino-ε-caprolactam)
14 L-(+)-Leucinamide
15 D-(−)-Phenylglyeniamide
16 (+)-3-Aminomethylpinane
17 (*R*)-(+)-Bornylamine
18 (*R*)-(−)-Isobornylamine
19 (−)-*cis*-Myrtanylamine
20 (*R*)-(−)-1-Cyclohexylethylamine
21 (*S*)-(–)-1-Phenylethylamine
22 (*R*)-(+)-1-(1-Naphthyl)ethylamine
23 L-(+)-2,4-Diaminobutyric acid
24 L-(+)-Ornithine
25 (−)-*N*-Methyl-D-glucamine
26 D-(+)-Glucosamine
27 (*S*)-(−)-2-Hydroxymethyl-1-methylpyrrolidine (*N*-methylprolinol)
28 (*S*)-(−)-3-(1-Methyl-2-pyrrolidinyl)pyridine (Nicotine)
29 (*S*)-(+)-2-(Hydroxymethyl)pyrrolidine (prolinol)
30 (*S*)-(+)-2-(methoxymethyl)pyrrolidine (*O*-methylprolinol)
31 L-(−)-2-Pyrrolidinecarboxylic acid amide (Proliamide)
32 L-(−)-Methyl 2-pyrrolidinecarboxylate (Proline methyl ester)
33 L-(−)-α-(Methylaminomethyl)-3,4-dihydroxybenzyl alcohol (Adrenaline)
34 (*R*)-(−)-α-(aminomethyl)3,4-dihydroxybenzyl alcohol (Noradrenaline)
35 (−)-Brucine
36 (−)-Strychnine
37 (+)-Dehydroabietylamine
38 (−)-Cinchonidine
39 (+)-Cinchonine
40 (+)-Quinidine
41 (−)-Quinine
42 (+)-Hydroquinidine
43 (−)-Hydroquinine
44 (−)-Sparteine
45 (+)-Calycanthine
46 (−)-Eburnamonine
47 α-Lobeline

Figure 5 Plot of PC-1 versus PC-2 for 23 resolving bases (out of 47 studied)

electron density, electrostatic potential, electric field and van der Waals volume. An interesting application of this concept has been given by Seri-Levi and Richards,[23] who showed a correlation between the chirality indices of several chiral drugs and their eudismic ratios (i.e. a relationship between drug–receptor interaction and molecular chirality).

Whether chirality indices will turn the PCA approach into a useful tool remains to be seen, although its application in the field of drug–receptor interactions shows promise. Although drug–receptor interactions are extremely complicated, they resemble more closely the interaction of two molecules than the multi-molecule interactions of a crystallisation process. Predicting the latter is still beyond present models.

Models to describe or predict 1:1 interactions of molecules are more readily available. For example, resolution of enantiomers using chiral column chromatography can nowadays be modelled fairly well[24] using techniques which have even been used to develop new or improved homochiral stationary phases for chromatographic resolution of enantiomers. Leusen and co-workers[25] have developed a protocol to improve a chiral stationary phase in a rational way. The method starts with an existing molecular structure from which the geometry of low-energy configurations of the diastereoisomeric complexes of substrate and chiral stationary phase are computed. These calculations are repeated after small, intuitive changes towards lower energy configurations are introduced in the original molecular structure.

5.6 ASYMMETRIC TRANSFORMATIONS, RACEMISATIONS AND RECYCLING

There are two prerequisites for an industrially viable resolution: an efficient resolving agent and an efficient recycling procedure. As well as recycling the resolving agent, it is increasingly important for an economical process to recycle the unwanted enantiomer (typically obtained in 60–70% yield and in lower enantiopurity than the desired enantiomer). Other aspects of particular importance to industrial resolution processes are efficient recovery of solvents and the processing and minimisation of inorganic salts. Resolutions through selective crystallisation of diastereoisomeric salts are difficult to imagine without the production of at least one equivalent of an inorganic salt.

Considering next the reuse of the unwanted enantiomer: the principal possibilities are shown in Figure 6. Optimising the yield of the required enantiomer can be considered at three distinct levels:

- *For enantiomers:* racemisation or inversion of configuration in a given process, or use in another synthesis are possible choices.
- *For diastereoisomeric salts:* racemisation or inversion processes again offer the best prospects for industrial viability.

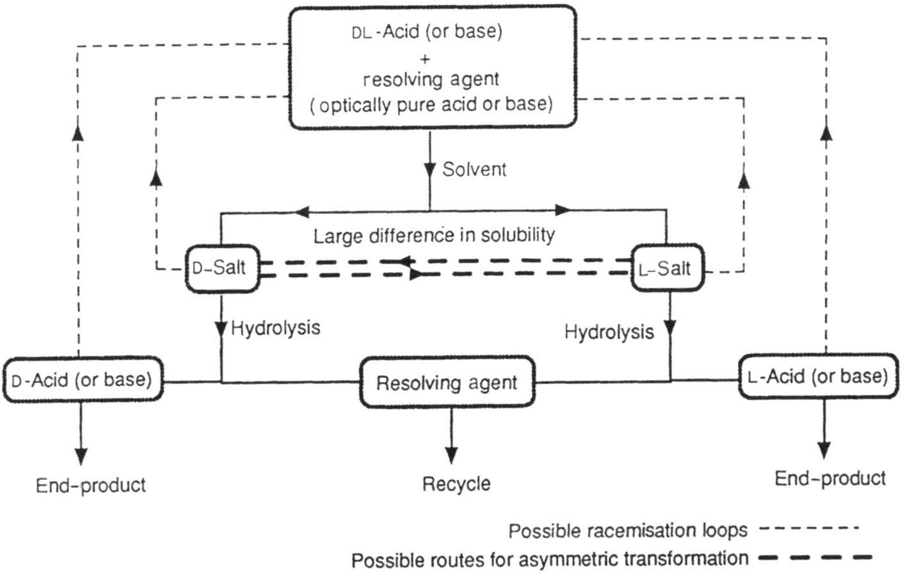

Figure 6 General scheme for resolution through diastereoisomeric salts

- *In situ racemisation*, i.e. crystallisation-induced asymmetric transformation or 'deracemisation,' is the most desirable situation, and likely to be the basis of the most economic resolution.

Unlike resolutions and resolving agents,[26] no compilation of examples of methods for racemisations and inversions is available, but a useful overview has been given by Eliel and Wilen (see ref. 6, pp. 424–440). In industrial practice, bespoke solutions are usually required.

A few guidelines can be given, of which the most general is to put the resolution step early in a multi-step synthesis. Good racemisation procedures can often be developed when the chiral centre contains a hydrogen, or is the result of a reversible synthesis. Although early resolution is essential when the co-produced enantiomer cannot be re-used, it is not important when efficient *in situ* racemisation is possible. A late resolution without racemisation can easily be 2–4 times more costly than an early resolution with recycling of the racemised co-enantiomer.[26]

In situ racemisation and recycling of the unwanted diastereoisomeric salt are preferred to racemisation of the isolated, unwanted enantiomer from an economic point of view. The great difficulty in finding a suitable resolving agent at all makes the task of finding one compatible with racemisation conditions even more difficult. Simple, stable resolving amines or acids, such as camphorsulphonic acid and the cyclic phosphoric acids, are good candidates for surviving

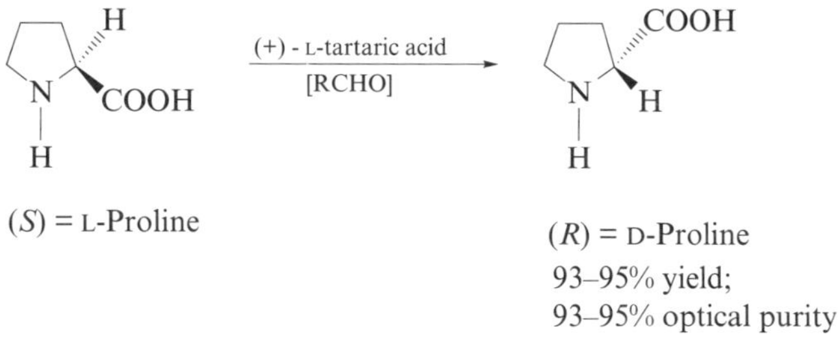

Figure 7 Transformation of L-proline into D-proline

demanding racemisation conditions. When resolving agents are stable to the racemisation conditions, the search for a process with asymmetric tranformation is more likely to succeed. This combination of selective crystallisation with racemisation in solution is receiving increasing attention and, where it can be applied, has resulted in favourable process economics. Good examples come from the work of DSM (resolution of aminoamides with mandelic acid using benzaldehyde as the racemisation catalyst[27]), Japanese work (resolution of amino acids with sulphonic acids and salicylaldehyde as racemisation promoter[28]) and the synthesis of D-proline from readily available L-proline[29] (see Figure 7). In the latter example an equilibrium is set up between D- and L-proline through racemisation catalysed by *n*-butyraldehyde, followed by precipitation of the less soluble D-proline- L-tartrate.

5.7 CONCLUSIONS

Industry today can choose from five well established methods for the production of enantiopure compounds: selective crystallisation, asymmetric synthesis, biocatalysis, fermentation and synthesis from chiral pool compounds. Although the level of development and the number of industrial applications vary greatly between methods, there is no single 'best' technology; every product has its own most economic production process governed by volumetric productivity, economy of scale, product life cycle, production location, etc. All techniques are approaching the ideal of '100% yield and 100% enantiopurity,' i.e. of furnishing, in principle, a quantitative yield of a single enantiomer or diastereoisomer.

Predictable process design is still some way off, however. Syntheses from the chiral pool and stoichiometric asymmetric synthesis are the most advanced in this respect, but methods which employ catalysis or selective crystallisations are still far from predictable. At present, chirality is still treated as a qualitative property; a description in terms of quantitative parameters related to fundamental molecular properties could be of great help in the rational design of any process for homochiral products and, in particular, for the design of classical resolutions.

5.8 ACKNOWLEDGEMENTS

Major parts of the studies presented in this chapter are the results of the research work of Dr G. J. A. Ariaans, Dr F. J. J. Leusen and Dr A. D. van der Haest. The author also acknowledges the successful design and synthesis of the cyclic phosphoric acids by Dr W. ten Hoeve at an early stage of the project and the inspiring and stimulating contributions of Professor Dr H. Wynberg, without which this project would not have been possible.

5.9 REFERENCES

1. (a) Collins, A. N., Sheldrake, G. N., and Crosby J., *Chirality in Industry*, Wiley, Chichester, 1992; see Ch. 2; (b) Sheldon, R. A., *Chirotechnology*, Marcel Dekker, New York, 1993; see Ch. 6.
2. Océ-Anderno BV, *Eur. Pat.*, EP 8.831, 8.833 and 35.811.
3. Winther, C., *Ber. Dtsch. Chem. Ges.*, **28**, 300 (1895).
4. Jacques, J., Collet, A, and Wilen, S. H., *Enantiomers, Racemates and Resolutions*, Wiley, 1981.
5. Collet, A., and Jacques, J., *Bull. Soc. Chim. Fr.*, 3857 (1972); 3330 (1973).
6. Eliel, E. L., and Wilen, S. H., *Stereochemistry of Organic Compounds*, Wiley, Chichester, 1994; see Ch. 7.
7. (a) Fogassy, E., Faigl, F, and Acs, M., *Tetrahedron*, **41**, 2837 (1985); (b) Fogassy, E., Faigl, F., and Acs, M., *Tetrahedron*, **41**, 2841 (1985); (c) Acs, M., in *Problems and Wonders of Chiral Molecules* (ed. M. Simonyi), Akadémiai Kiadó, Budapest, 1990, p.111; (d) Fogassy, E., Acs, M., Faigl, F., Simon, K., Rohonczy, J., and Ecsery, Z., *J. Chem. Soc. Perkin Trans. 2*, 1881 (1986); (e) Fogassy, E, Faigl, F., Acs, M., Simon, K., Kozsda, É., Podányi, B., Czugler, M., and Reck, G., *J. Chem. Soc. Perkin Trans. 2*, 1385 (1988), (f) Czugler, M., Csöregh, I., Kálmánm, A., Faigl, F., and Acs, M., *J. Mol. Struct.*, **196**, 157 (1989); (g) Simon, K., Kozsda, É., Böcskei, Z., Faigl, F., Fogassy, E., and Reck, G., *J. Chem. Soc., Perkin Trans. 2*, 1395 (1990).
8. (a) Kozma, D., Nyéki, A., Acs M., and Fogassy, E., *Tetrahedron: Asymmetry*, **5**, 315 (1994); (b) Kozma, D., Acs, M., and Fogassy, E., *Tetrahedron,* **50**, 6907 (1994).
9. Zingg, S. P., Arnett, E. M., McPhail, A. T., Bothner-By, A. A. and. Gilkerson, W. R., *J. Am. Chem. Soc,*. **110**, 1565 (1988).
10. (a) Gould, R. O., and Walkinshaw, M. D., *J. Am. Chem. Soc.*, **106**, 7840 (1984); (b) Gould, R. O., Gray, A. M., Taylor, P., and Walkinshaw, M. D., *J. Am. Chem. Soc.*, **107**, 5921 (1985); (c) Gould, R. O., Kelly, R., and Walkinshaw, M. D., *J. Chem. Soc., Perkin Trans. 2*, 847 (1985).
11. (a) ten Hoeve, W., and Wynberg, H., *J. Org. Chem.*, **50**, 4508 (1985); (b) Wynberg, H., and ten Hoeve, W., *Eur. Pat. Appl.*, EP 180 276; *Chem. Abstr.*, **105**, 134150u (1989); *US Pat.*, 4 814 477.
12. v. d. Haest, A. D., Wynberg, H., Leusen, F. J. J., and Bruggink, A., *Recl. Trav. Chim. Pays-Bas*, **109**, 523 (1990).
13. v. d. Haest, A. D., Wynberg, H., Leusen, F. J. J., and Bruggink, A., *Recl. Trav. Chim. Pays-Bas*, **112**, 230 (1993).
14. Leusen, F. J. J., Bruins-Slot, H. J., Noordik, J. H., v. d. Haest, A. D., Wynberg, H., and Bruggink, A., *Recl. Trav. Chim. Pays-Bas*, **110**, 13 (1991).
15. Leusen, F. J. J., Bruins-Slot, H. J., Noordik, J. H., v. d. Haest, A. D., Wynberg, H., and Bruggink, A., *Recl. Trav. Chim. Pays-Bas*, **111**, 111 (1992).

16. Leusen, F. J. J., Noordik, J. H., and Karfunkel, H. R., *Tetrahedron,* **49**, 5377 (1993).
17. Ariaans, G. J. A., Leusen, F. J. J., and Bruggink, A., in *Proceedings of the Chiral '92 Symposium*, 1992, pp. 99–106.
18. Karfunkel, H. R., Rohde, B., Leusen, F. J. J., Gdanitz, R. J., and Rihs, G., *J. Comput. Chem.*, **14**, 1125 (1993).
19. Carlson, R., *Design and Optimization in Organic Synthesis*, Elsevier, Amsterdam, 1992.
20. See, for example; Buda, A. B., a. d. Heyde, T., and Mislow, K., *Angew. Chem.*, **104**, 1012 (1992).
21. Kuz'min, V. E., Stel'makh, I. B., Bekker, M. B., and Pozigun, D. V., *J. Phys. Org. Chem.*, **5**, 295 (1992).
22. Gilat, G., *J. Phys. A*, **22**, L545 (1989); *Found. Phys. Lett.*, **3**, 189 (1990).
23. (a) Seri-Levy, A., West, S. and Richards, W. G., *J. Med. Chem.*, **37**, 1727 (1994); (b) Seri-Levy, A. and Richards, W. G., *Tetrahedron: Asymmetry.*, **4**, 1917 (1993).
24. Däppen, R., Karfunkel, H. R., and Leusen, F. J. J., *J. Chromatog.*, **469**, 101 (1989).
25. Däppen, R., Karfunkel, H. R., and Leusen, F. J. J., *J. Comp. Chem.*, **11**, 181 (1990); see also Leusen, F.J.J., PhD Thesis, University of Nijmegen, 1993, Ch. 2.
26. Bruggink, A., Hulshof, L. A., and Sheldon, R.A., *Pharm. Manuf. Int.*, 139 (1990).
27. Boesten, W. H. J., *Neth. Pat.*, 90 00386, 90 00387, 1990.
28. Yoshioka, R., Tohyama, M., Ohtsuki, O., Yamada, S., and Chibata, I., *Bull. Chem. Soc. Jpn.*, **60**, 649 (1987); *Bull. Chem. Soc. Jpn.*, **58**, 433 (1985); *Eur. Pat. Appl.*, 70114, 1983 (to Tanabe).
29. Shiraiwa, T., Shinjo, K., and Kurokawa, H., *Chem. Lett.*, 1413 (1989).

6 Resolution Versus Stereoselective Synthesis in Drug Development: Some Case Histories

B. A. ASTLEFORD and L. O. WEIGEL
Eli Lilly & Company, Indianapolis, IN, USA

6.1	Introduction	99
	6.1.1 Enantiomers and Drug Development	100
6.2	Duloxetine (LY248686)	101
	6.2.1 Enantiospecific Ketone Reduction	101
	6.2.2 Resolution–Deracemisation	103
	6.2.3 Applications to Other Systems	104
6.3	Loracarbef (LY163892)	105
	6.3.1 Synthesis from Sodium Erythorbate	107
	6.3.2 Resolution via a Phthalimido Group	110
	6.3.3 Resolution of other β-Lactam Intermediates	111
6.4	Human Steroid 5-α-Reductase Inhibitor (LY300502)	111
	6.4.1 Synthesis via a Chiral Auxiliary	111
	6.4.2 Resolution via Methanolysis of Lactams	112
6.5	Conclusion	114
6.6	Acknowledgements	115
6.7	References and Notes	115

6.1 INTRODUCTION

In the last decade, many new and exciting techniques for the production of enantiomerically pure compounds have been discovered. These methods include numerous enantiospecific reactions, asymmetric catalytic transformations, deracemisations, the use of chiral pool substrates, enzymatic methods and others.[1] Classical resolution may seem to be a 'lost art' when compared with these newer methodologies. Although apparently lacking elegance, the resolution of enantiomers by the formation of diastereoisomeric derivatives still plays a critical role in the production of non-racemic chemicals.[2] The objective of this chapter is to examine three case histories of drug development which illustrate the utility of resolutions.[3]

Chirality in Industry II. Edited by A. N. Collins, G. N. Sheldrake and J. Crosby

6.1.1 ENANTIOMERS AND DRUG DEVELOPMENT

Although numerous stereoselective and asymmetric transformations are available to chemists, relatively few are selected and developed for large scale synthesis. In reality, the manufacture of enantiomerically pure drugs is not entirely driven by overall yield efficiency, but also by the practicality of the synthesis. Issues such as speed-to-market, generally small production requirements, economics and short effective patent life are primary considerations in the route selection and the finalised manufacturing protocol. In early drug development, Chemical Process Research and Development (CPR&D) groups within the industry are invariably charged with two critical functions: rapidly providing kilograms of the target compound and early route selection for the manufacture of the drug. Enantiospecific syntheses are usually tried first because of the potential for a high yield. In many cases, this approach is more time consuming and substrates are often unsuitable. In the end, enantiomers are often obtained by classical resolution owing to time constraints, economics and reliability. However, one drawback is that resolutions are difficult without a useful 'handle' (i.e. covalent or ionic resolving agent). Even with these limitations, classical resolutions are still widely used. In 1993, a survey of the synthetic methods for the preparation of over forty enantiopure drugs in Lilly's CPR&D group showed that nearly two thirds employed ionic or covalent resolutions (Table 1).[4]

In this chapter, three enantiospecific syntheses in an early stage of development will be presented. In each case history the asymmetric strategies were

Table 1 Syntheses of Enantiopure Drugs at Lilly CPR & D in 1993[a]

'Chiral pool' or semi-synthetic	5.5 (20%)
Asymmetric synthesis	6.0 (22%)
Resolution (ionic or covalent)	15.5 (57%)

[a] Data gathered from 40 drug candiates at phase I to phase III of development

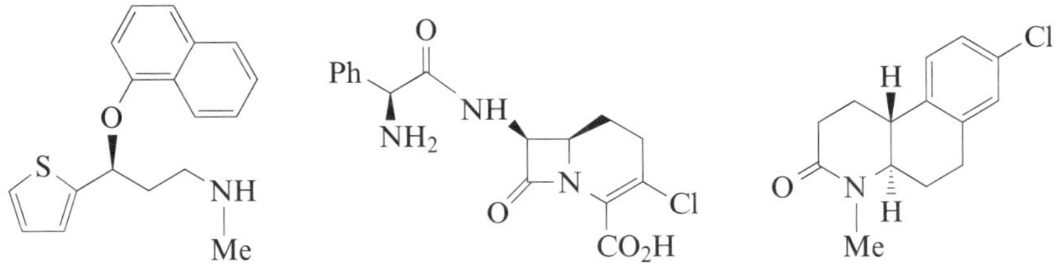

(1) (2) (3)

Figure 1 Duloxetine (**1**), Loracarbef (**2**) and LY300502 (**3**)

altered and 'specialised' resolutions were invented. The three cases are duloxetine (**1**), loracarbef (**2**) and the 5-α-reductase inhibitor LY300502 (**3**) (Figure 1).

6.2 DULOXETINE (LY248686)

Duloxetine (LY248686, **1**) is a potent inhibitor of the serotonin (5HT) and norepinephrine (NE)[5] uptake carriers. The drug is in late phase II clinical trials for depression and/or urinary incontinence, and is similar to several other drug candidates possessing the non-racemic [3-(1-aryloxy)-3-arylpropanamine] backbone. Brown, Sharpless and others[6a–c] have published enantioselective syntheses of related drugs with strategies similar to that used for duloxetine (Scheme 1).

6.2.1 ENANTIOSPECIFIC KETONE REDUCTION

Initial development efforts centred on a yield-efficient enantioselective synthesis. Duloxetine (**1**) was prepared in kilogramme quantities in about 60% overall yield from 2-acetylthiophene (**4**) *via* a four-stage sequence of Mannich reaction of 2-acetylthiophene; asymmetric reduction of ketone **5**, alkylation of alcohol **6** with 1-fluoronaphthalene and *N*-demethylation of adduct **7** to yield **1** (Scheme 1).[7]

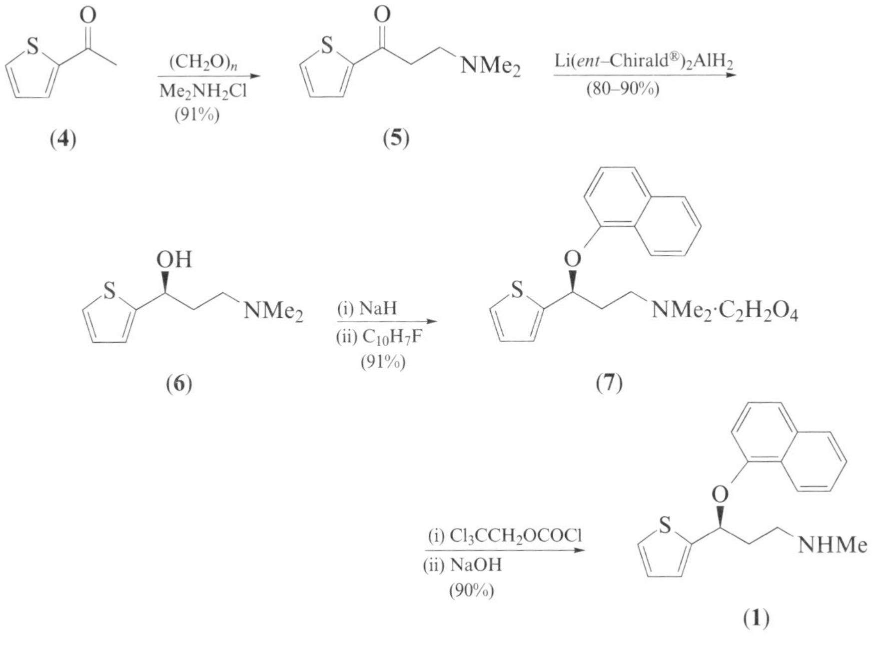

Scheme 1 Enantiospecific synthesis of duloxetine

During the course of this work, several methods for the enantiospecific reduction of ketone **5**, including reduction with yeast,[8] Corey's oxazaborolidine–borane system[9] and BINAP–LAH[10], were investigated. The reduction of the Mannich base **5**[11] with the kinetically formed Yamaguchi–Mosher–Pohland (YMP)[12] complex derived from *ent*-Chirald® (**8**) was selected for development. Thus, reaction of two equivalents of the ligand **8** with lithium aluminium hydride in toluene at low temperature in a continuous-flow reactor afforded the metastable complex **9** (Scheme 2). Immediate reduction of **5** in the flow reactor provided the amino alcohol **6** in approximately 85–90% *ee* and 90% yield.

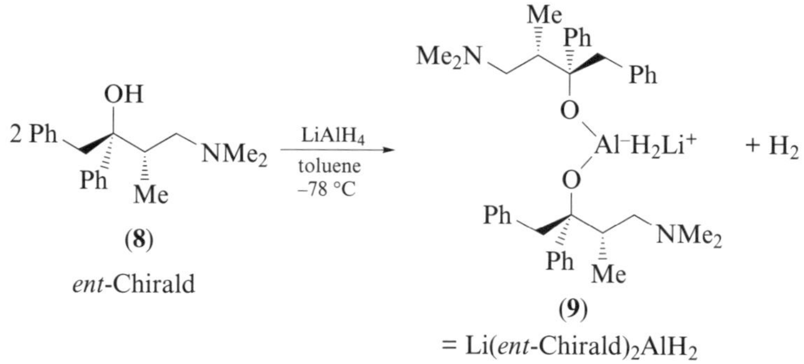

Scheme 2 Preparation of Li(*ent*-Chirald)$_2$AlH$_2$

Chirald® is a trade name for amino alcohol **12**, which is a resolved intermediate in the Lilly production of Darvon® (Scheme 3). Both Chirald® and *ent*-Chirald are available from the process. Treatment of Mannich base **10** with benzylmagnesium chloride followed by resolution of the diastereoisomeric mixture with 0.5 equivalents of camphorsulphonic acid [(−)-CSA] in ethanol yielded the resolved salt of Chirald (**11**) (Scheme 3).[13]

The CSA salt of *ent*-Chirald (**14**) was obtained from the mother liquors containing Chirald (and a small amount of the diastereoisomers **13**) by crystallisation in acetone. Neutralisation of each isolated salt and recovery of the (−)-CSA afforded Chirald (**12**) and *ent*-Chirald (**8**). Thus, only 0.5 mole equivalents of (−)-CSA was employed in the resolution of both antipodes. Since *ent*-Chirald (**8**) was a waste stream product, the net cost to the duloxetine reduction process involved only the crystallisation of *ent*-Chirald (**8**) from acetone and recovery of the (−)-CSA.[13] The free availability of the auxiliary was the primary reason for selection of this route.

Although large amounts of duloxetine (**1**) were produced, this process was *not* taken to production. A total input of nearly 5 kg of the complex **9** in 100 l of toluene was required to produce 1 kg of alcohol **6** from ketone **5**. Even with the low cost and recycle of *ent*-Chirald, the overall efficiency was, at best, comparable with that of a classical resolution.

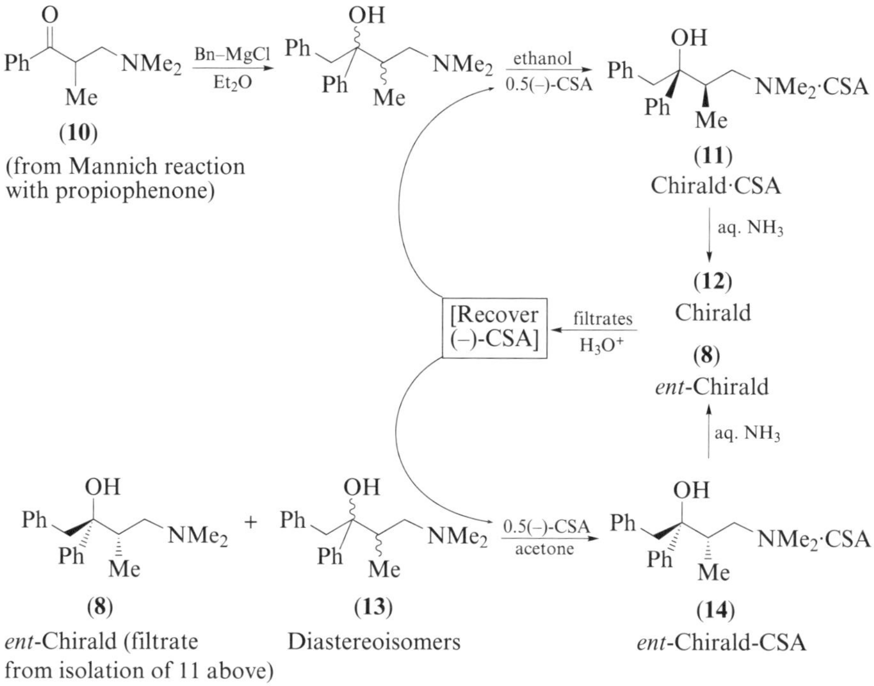

Scheme 3 Industrial source of *ent*-Chirald (**8**) and Chirald (**12**)

6.2.2 RESOLUTION–DERACEMISATION

Two critical observations were made during the development of the syntheses of duloxetine. First, classical resolution of several intermediates with (*S*)-mandelic acid (MDA) proved highly efficient. Second, racemisation of several synthetic intermediates proved extremely facile in mineral acid at 25 °C.[14] These observations were developed into a new 'net deracemisation' synthesis of duloxetine using a simple iterative sequence of resolution–racemisation–recycle (RRR) (Scheme 4).

Mannich reaction of acetylthiophene **4** (see Scheme 1) and direct sodium borohydride reduction gave the racemic amino alcohol **15.** Treatment of the latter compound with 0.45 equivalents of MDA in methyl *tert*-butyl ether (MTBE) effected immediate precipitation of salt **16** (initially 20–60% *de* at 25 °C). However, on heating this heterogeneous mixture at reflux, pure salt **16** with 95–98% *de* was isolated in 42–43% yield (95% of theory). The MTBE filtrate contained mostly the enantiomer **17** and small amounts of MDA salts. Acidification of this MTBE filtrate with hydrochloric acid effected clean racemisation of the hydrochloride salt of compound **17** in the aqueous layer. Neutralisation with sodium hydroxide allowed re-extraction of racemic amine **15** into the original

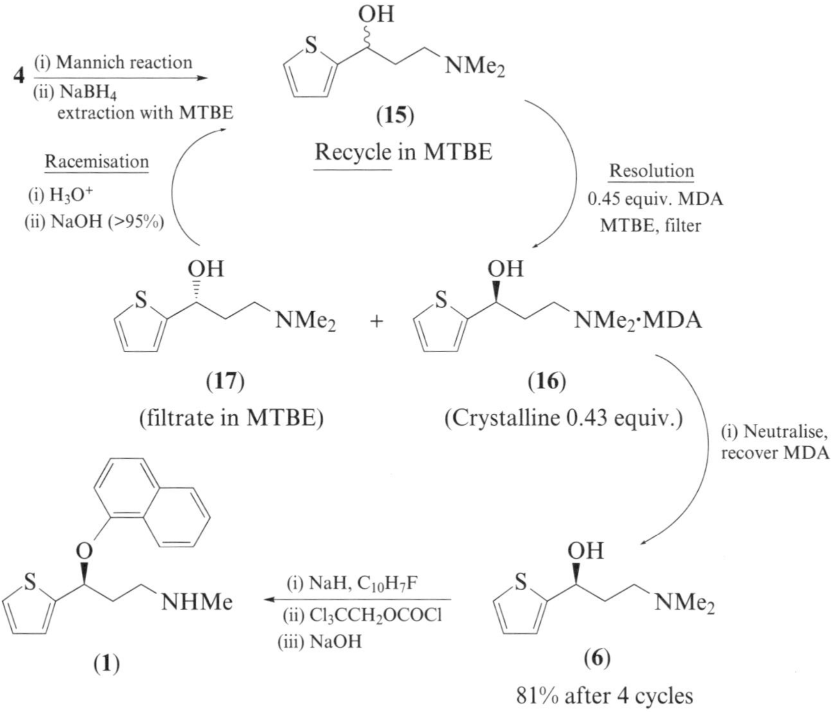

Scheme 4 Resolution–racemisation–recycle (RRR) synthesis of duloxetine (**1**)

MTBE layer. The dried solution of amine **15** was again subjected to the resolution sequence and, after four cycles, a total of 81% of alcohol **6** was obtained. Completion of the synthesis produced duloxetine (**1**) in the same overall yield as the asymmetric reduction sequence. Thus, a simple acid–base treatment allows recycle of the opposite enantiomer. The robustness and simplicity of the RRR protocol to prepare duloxetine (Scheme 4) proved to be more efficient and cost-effective than the enantioselective route (Scheme 1).

6.2.3 APPLICATIONS TO OTHER SYSTEMS

The enantiomers of 1-phenylpropane-1,3-diol (**18**) and 1-phenyl-3-chloropropan-1-ol (**19**) are starting materials in the synthesis of several drugs related to duloxetine [e.g. nisoxetine (**20**), seproxetine (**21**), tomoxetine (**22**), (*S*)-fluoxetine (**23**) and (*R*)-fluoxetine (**24**)].[6] Like the intermediate in the duloxetine synthesis, both 1-phenylpropane-1,3-diol and 1-phenyl-3-chloropropan-1-ol undergo simple acid-catalysed racemisation. When incorporated into a resolution sequence [e.g. the enzymatic resolution of 1-phenylpropane-1,3-diol (**18**)[6d]], these results provide formal RRR-based syntheses of **20**–**24** (Scheme 5).

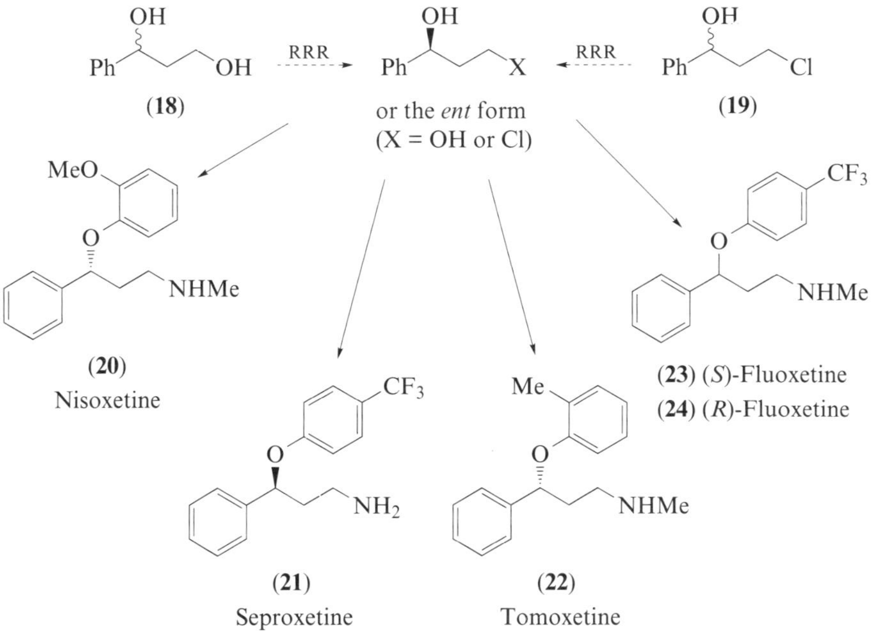

Scheme 5 Application of RRR to other drugs

6.3 LORACARBEF (LY163892)

Loracarbef (**2**) is a carbacephalosporin antibiotic with extended *in vivo* stability.[15a,b] This oral antibiotic is currently on the market and has found specialised use in the treatment of paediatric ear infections. Owing to high doses and intense competition from other β-lactam antibiotics, the cost of manufacturing was a critical issue in the development of loracarbef.

There are many elegant syntheses of loracarbef.[15c] Several critical synthetic intermediates leading to loracarbef (or closely related carbacephalosporins) are shown in Figure 2. The varying synthetic strategies from this group of intermediates are noteworthy:

(i) the racemic, bicyclic β-lactam **25**[16] is resolved via enzymatic acylation of the amino group;
(ii) the monocyclic β-lactam **26**,[17a] which is derived from a chiral auxiliary-based synthesis, is cyclised to a six-membered ring via rhodium-catalysed formation of the carbene and insertion into the N—H bond;
(iii) the amino acid **27**,[18] derived from the serine hydroxymethyl transferase (SHMT)-catalysed reaction of glycine with pentenal, is converted into a β-lactam by a Mitsunobu reaction on a β-hydroxy amide derived from the β-hydroxy acid;

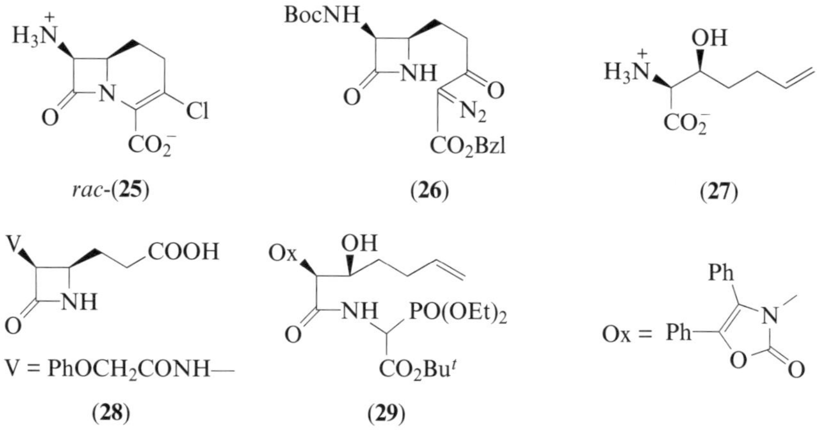

Figure 2 Selected intermediates for carbacephalosporin synthesis

(iv) the penicillin-derived acid **28**[19] is alkylated on nitrogen with a bromoglycinate, and later closed to the six-membered ring via a Dieckmann reaction;

(v) The racemic aldol-derived intermediate **29**[20] is converted into the *β*-lactam, subjected to ozonolysis [which converts the alkene into the aldehyde and cleaves the diphenyl oxazolin-2-one (Ox) protecting group] and then cyclised via an internal Wittig reaction.

Many published syntheses of carbacephalosporins utilise the classical [2 + 2] ketene–imine cyclisation (Staüdinger reaction)[15–17,21] to construct the *cis*-*β*-lactam. Lilly chemists have utilised a Dieckmann closure of Staüdinger-derived *cis*-*β*-lactams for the production process of loracarbef (**2**) (Scheme 6). Thus, the

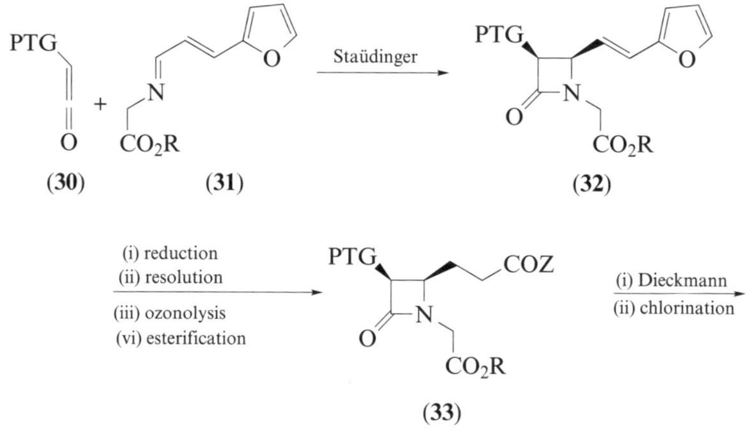

Scheme 6 Generalised Lilly syntheses of loracarbef (Lorabid) *(continued)*

Scheme 6 (*continued*)

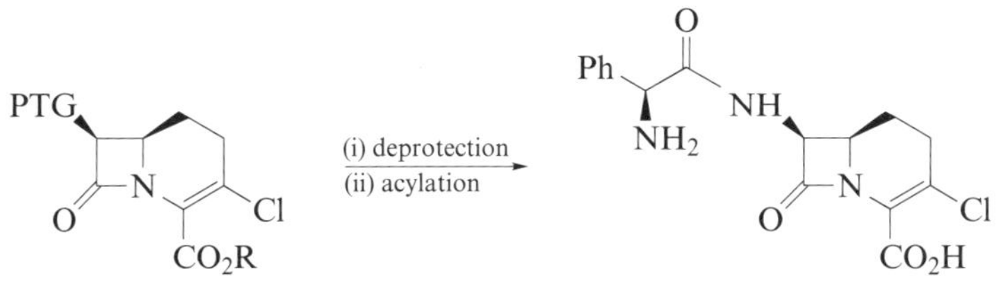

(**34**) (**2**)

PTG = PhthN— or $PhOCH_2CONH$— or EtOCOCH=C(Me)NH—;
R = Me, p-$(O_2N)PhCH_2$, Bu^t, PhO; Z = PhO, PhS

Staüdinger reaction of the protected ketene **30** (derived *in situ* from the acid chloride or anhydride and triethylamine) with the imine **31** (derived from 2-furanylacrolein and a glycinate ester) affords the corresponding β-lactam **32**. Generally, high yields are observed with appropriate protecting groups (e.g. 90% using phthalimidoacetyl chloride; PTG = PhthN, R = Me). The generalised sequence depicted in Scheme 6 affords the diester **33**, which, when subjected to a Dieckmann cyclisation and then chlorination, gives **34**. Deprotection of PTG and R followed by acylation with (*R*)-phenylglycine then affords loracarbef (**2**).

6.3.1 SYNTHESIS FROM SODIUM ERYTHORBATE

Building upon the [2 + 2] cycloaddition–Dieckmann strategy above, an alternative non-resolution-based synthesis of loracarbef (**2**) from a common food preservative has been developed. A combination of observations by Cohen *et al.*[22] on the oxidation of isoascorbic acid and the work of Hubschwerlen and Schmid[23] and Evans and Williams[24] was employed for an asymmetric ketene–imine Staüdinger reaction. The synthesis of a β-lactam with the proper framework, stereochemistry and oxidation state was rapidly realised from the very inexpensive (\$10 kg^{-1}) chiral pool intermediate, sodium erythorbate (**35**)[25] (Scheme 7).

In practice, sodium erythorbate was oxidised (hydrogen peroxide, 80%) and the intermediate lactone diol was monotosylated to afford tosylate **36** (80–88%). Treatment of the latter with sodium ethoxide afforded the crystalline, enantiomerically pure epoxide **37** (91–95%).[26] For larger scale preparations, the conversion of sodium erythorbate into the 4-hydroxyepoxycrotonate **37** was accomplished, without isolation or purification of intermediates, in approximately 75% yield. In one operation, epoxy alcohol **37** was subjected to Swern oxidation and the resulting aldehyde was condensed with *tert*-butyl glycinate. The

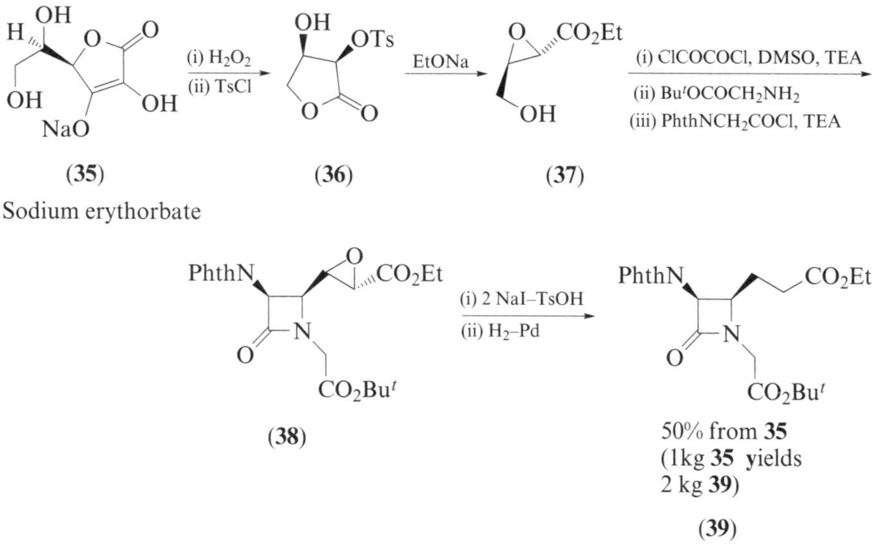

Scheme 7 Synthesis of a loracarbef precursor from sodium erythorbate

resulting unstable imine, whilst still in solution, was then treated with phthalimidoacetyl chloride in the presence of triethylamine to afford the monocyclic β-lactam **38** in 75% yield from 4-hydroxyepoxycrotonate **37** with 92–94% *de*. One recrystallisation of **38** upgraded the *de* to >99%. However, crystallisation of subsequent intermediates increased the *de* to similar levels. The epoxide **38** was easily deoxygenated by treatment with sodium iodide–tosic acid and, without purification of the intermediate alkene, was hydrogenated to afford **39** (87%).[27]

At this stage, the Dieckmann precursor **39** of loracarbef had been generated with high efficiency:

(i) four of the six carbons of sodium erythorbate (**35**) are incorporated into the nucleus of loracarbef (**2**);
(ii) two new stereogenic centres were created with high *de*;
(iii) the chiral pool material was inexpensive (< \$10 kg^{-1});
(vi) the mass conversion efficiency was very high (1 kg of sodium erythorbate gave 2 kg of **39**).

Unfortunately, the precursor **39** was not suited for a Dieckmann condensation[28] owing to unactivated esters and the reactivity of the phthalimido group. The ester and phthalimido protecting groups of the monocyclic lactam **39** were modified (without isolation) by sequential cleavage of the *tert*-butyl ester (trifluoroacetic acid, TFA), conversion of the phthalimido group to the *o*-pyrrolidinocarbonylbenzamide (OPCB)[29] group with pyrrolidine, and ethyl ester hydrolysis (sodium hydroxide, 73% from precursor **39**) to yield the diacid **40**

(Scheme 8). The diacid **40** was converted into **41** by phenol esterification (dicyclohexylcarbodiimide, phenol, 80%) and then selective transesterification (sodium *p*-nitrobenzylalkoxide, 81%). Dieckmann condensation of **41** (sodium *tert*-butoxide, 70%) and chlorination $[(PhO)_3PCl_2]$[30] with concurrent reclosure of the OPCB group to PhthN afforded **42**. Removal of the PNB group (H_2–Pd, 90%) and the phthalimido group (methylhydrazine, 90%)[18] yielded the enantiopure nucleus **43**. Acylation of **43** with the mixed carboxylic–carbonic anhydride derived from (*R*)-phenylglycine and isobutyl chloroformate (90%) provided loracarbef (**2**).

Although the cost of the starting material is low and the conversion to an advanced intermediate efficient, this sequence was never scaled up. What are the shortcomings of this synthesis? The molecular gymnastics required to convert sodium erythorbate (**35**) into intermediate **41** and the protecting group 'shuffle' resulted in a synthesis which was simply too long (20 functional group interconversions or C—C bond formations). The cumulative cost of performing multiple steps with an overall yield of about 12% quickly outweighs the cost benefit of starting with sodium erythorbate. Even if yields could be dramatically increased, the cumulative cost of processing, isolation, handling and clean-up of numerous intermediates would disfavour the development of this process for manufacture.

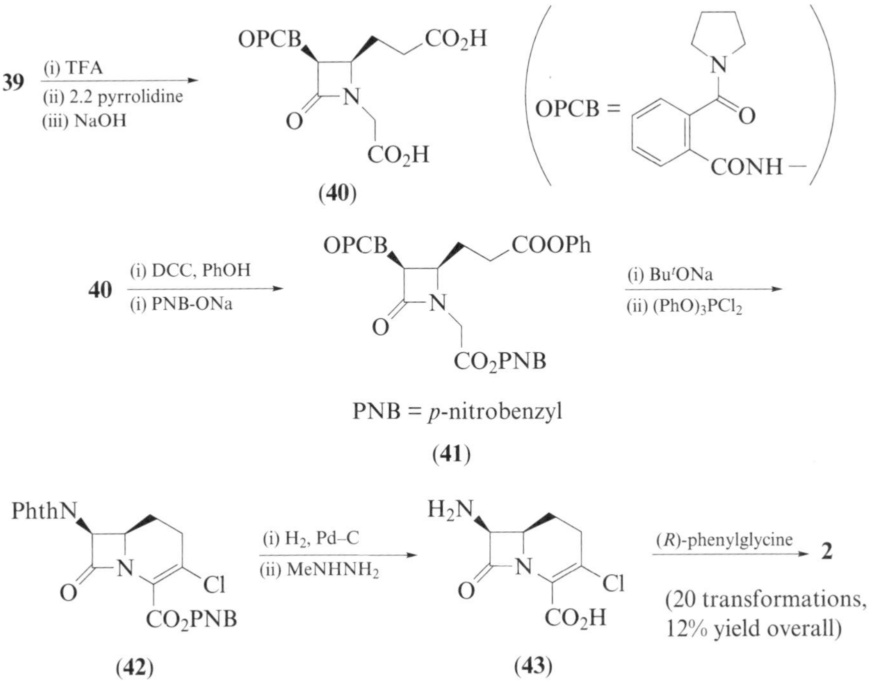

Scheme 8 Completion of sodium erythorbate route to loracarbef

6.3.2 RESOLUTION VIA A PHTHALIMIDO GROUP

Examination of protecting groups in the synthesis of loracarbef has shown that the phthalimido group could be converted into a bisamide in aprotic solvents by treatment with a primary or secondary amine. These intermediates are re-closed to the phthalimido group[29] by treatment with dilute mineral acid or with a combination of boric acid–hydrofluoric acid. Based on the classical resolution of alcohols via reaction with phthalic anhydride,[31] a similar sequence for the resolution of phthalimides with amines was devised. The sequence was demonstrated in a resolution after the first step in the loracarbef synthesis (Scheme 6).[29,31] Thus, when the racemic β-lactam **44** (obtained in 90% yield from a Staüdinger reaction) was treated with (*S*)-α-methylbenzylamine in hot THF, the pure diastereoisomer **45** crystallised directly from the reaction mixture (95% *de*). Treatment of amide **45** with mineral acid[29b] gave the resolved intermediate[29c] **46** (Scheme 9).[33] Conversion of the latter to loracarbef (**2**) was then accomplished as depicted in Scheme 6.

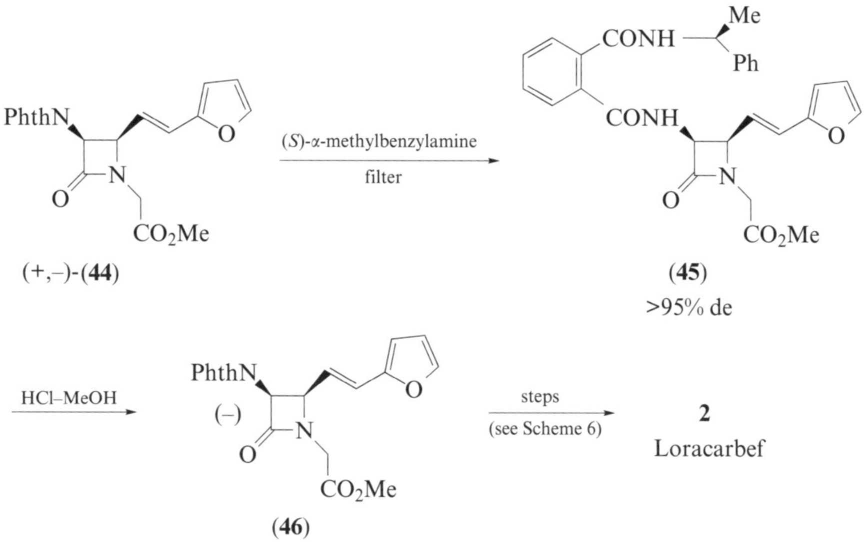

Scheme 9 Resolution of a loracarbef intermediate via a phthalimido derivative

This methodology represents a simple resolution sequence in which a racemic amine derivative (phthalimido) is resolved with a non-racemic amine. The phthalimido resolution process was replaced by a more efficient enzymatic acylation–resolution[34] sequence utilising $PhOCH_2CO_2Me$ early in the synthesis (see Scheme 6, PTG = $PhOCH_2CONH$ and R = Me).

6.3.3 RESOLUTION OF OTHER *β*-LACTAM INTERMEDIATES

This new methodology has been applied to the resolution of other *β*-lactam precursors of loracarbef. The racemic phthalimido intermediates **47**, **44** and **48** (Figure 3) have been opened with (*R*)- or (*S*)-α-methylbenzylamine or with (*S*)-2-(methoxymethyl)pyrrolidine (MMP),[29c] and the diastereoisomers separated by crystallisation or chromatography. The separation of amides derived from **47**, **44** and **48** represent an early resolution of several carbacephem[28] targets, including loracarbef.[15]

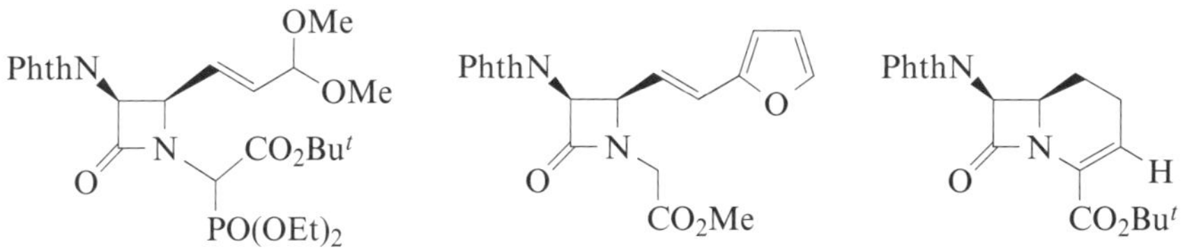

with $PhCH(Me)NH_2$ (**47**) — with MMP (**44**) — with MMP (**48**)

Figure 3 *β*-Lactams resolved by opening of phthalimido with $PhCH(Me)NH_2$ or MMP

6.4 HUMAN STEROID 5-α-REDUCTASE INHIBITOR (LY300502)

The benzoquinolinone LY300502 (**3**) has been synthesised[35] and identified as a potent, non-competitive inhibitor of 5-α-reductase in cell cultures derived from foreskin fibroblasts.[36] This activity may have implications for the treatment of acne, male pattern baldness and benign prostatic hyperplasia.

6.4.1 SYNTHESIS VIA A CHIRAL AUXILIARY

The initial synthesis of **3** was based upon an approach using a covalent chiral auxiliary (Scheme 10).[37] Friedel–Crafts reaction of ethylene and the acid chloride derived from *p*-chlorophenylacetic acid (**49**) afforded tetralone **50**. Conversion of the tetralone **50** to an enamine with (*S*)-α-methylbenzylamine and immediate treatment with acryloyl chloride afforded the aza-annulation product **51** (79% from tetralone **50**). The desired diastereoselective reduction of the endocyclic alkene in compound **51** proved difficult, since the carbon–nitrogen bond of the α-methylbenzylamine auxiliary underwent competitive reduction. Reduction of **51** with excess sodium cyanoborohydride in formic acid afforded the *trans* fused tricyclic amide **52** (admixed with the *cis* isomers) in low *de*. Attempts to

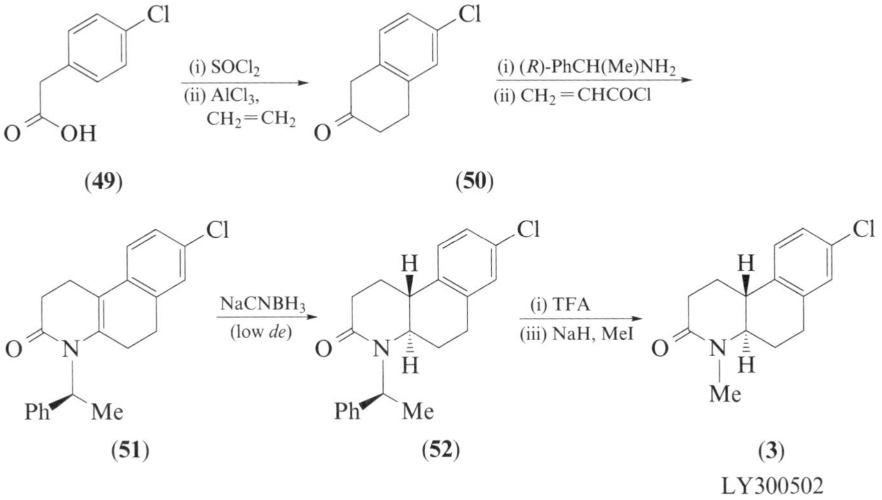

Scheme 10 Asymmetric synthesis of LY300502 via a chiral auxiliary

optimise the reduction of aza-annulation product **51**, or to obtain pure amide **52** by crystallisation, were not fruitful. Chromatographic separation of the four-component mixture afforded compound **52** in pure form. Removal of the chiral auxiliary with TFA and methylation of the intermediate amide provided LY300502 (**3**) (and its enantiomer) in about 9% overall yield from tetralone **50** (Scheme 10).

The low *de* combined with moderate yields and reliance upon chromatographic separations precluded further development of this sequence. Chiral chromatographic[38] separations of LY300502 (**3**) have been reported but production of kilogram quantities of enantiomerically pure product by this method is not yet practical.

6.4.2 RESOLUTION VIA METHANOLYSIS OF LACTAMS

Direct 'chemical' resolutions of δ-lactams similar to LY300502 are not generally viable. One tenable strategy is defined by a simplified sequence of δ-lactam opening (e.g. alcoholysis of the lactam to the ester) followed by an appropriate resolution of the amino ester (chemical, kinetic or enzymatic) and relactamisation.[39] This approach proved very successful in the development for the production of kilogram quantities of LY300502 (Scheme 11). Aza-annulation of 6-chlorotetralone (**50**) under specialised conditions using pyrrolidine enamine and acrylamide gave the ene-lactam **53** (85%). Ionic reduction of the latter with triethylsilane–trifluoroacetic acid and *N*-methylation with methyl chloride (obtained in the next step, see below) in the presence of sodium iodide–sodium

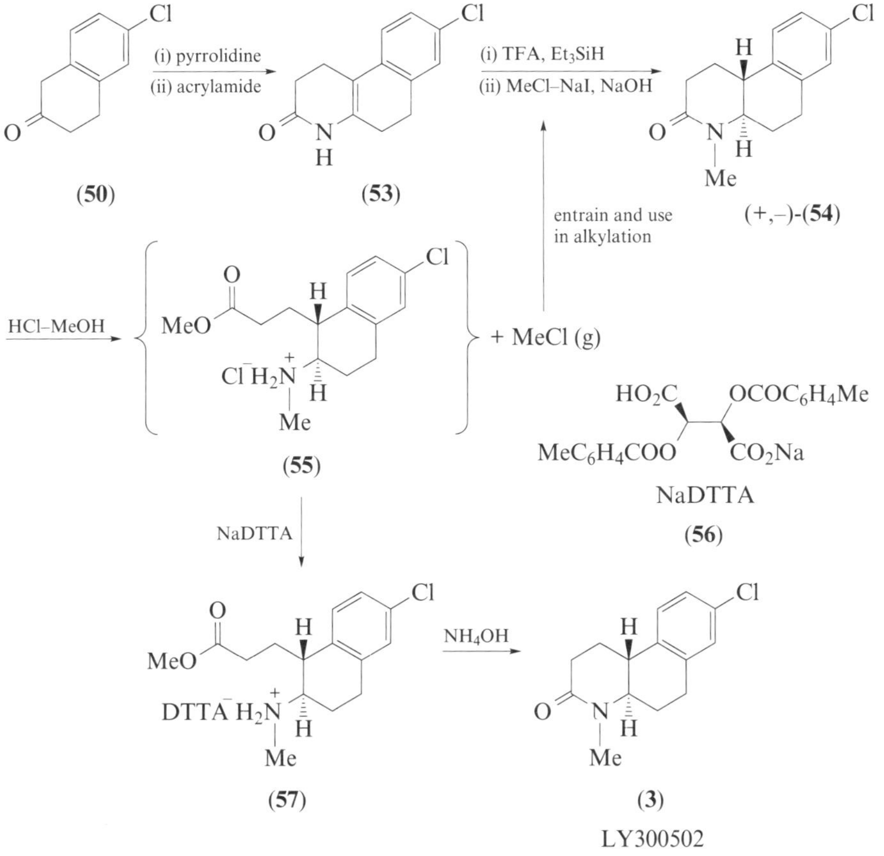

Scheme 11 Lactam resolution route to LY300502 which 'breeds' methyl chloride

hydroxide afforded the *trans*- fused bicyclic lactam (±)-**54** (90%). Methanolysis of lactam **54** with hydrochloric acid–methanol afforded hydrochloride **55** and methyl chloride. Although **55** could be isolated and characterised, this was not necessary. Treatment of the solution of **55** with sodium (*R,R*)-di-*p*-toluoyltartaric acid (NaDTTA, **56**) gave **57** in 45% yield (90% of theory, *de* > 98%). Isolation of the DTTA salt **57** also effected the removal of small amounts of the *cis* isomer of lactam **54,** as well as all other related substances. Treatment of **57** with ammonia solution and recovery of the DTTA afforded pure LY300502 (**3**) in >95% yield and 99% *ee*.

The overall process, although it contains a 'midstream' resolution, is short and efficient. Only 2 kg of 6-chlorotetralone (**50**) are required for the production of 1 kg of the enantiomerically pure product, with four pilot plant operations from **50**, in 30% overall yield. Additionally, by selection of hydrochloric acid and methanol for the alcoholysis sequence, the by-product, methyl chloride, is entrained and utilised in the alkylation sequence, obviating the need to purchase and ship methyl iodide. This new resolution sequence thus 'breeds' the *N*-methylation reagent. As with duloxetine and loracarbef, this resolution methodology is based on simple and robust acid–base chemistry. The protocol has also been utilised in the resolution of derivatives of LY300502 (**3**).[40]

6.5 CONCLUSION

These case histories illustrate resolution processes which were developed in preference to stereoselective syntheses. Some of the critical factors from these histories in route selection and development may be summarised as follows:

(i) *External ligands or auxiliaries*. The use of high molecular weight auxiliaries (ligands) is, in some cases, less effective than a simple resolution. Care should be exercised in the use of an auxiliary, and extra 'hidden' steps may also lie in recovery and re-use of the auxiliary.

(ii) *Balance of costs*. Avoid undue focus on the cost of an enantiomerically pure starting material. In the case of loracarbef, the raw material cost was quickly overtaken by cumulative processing expenses.

(iii) *Covalent auxiliaries*. The use of a covalent auxiliary generally involves introduction, transfer of chirality and removal. This approach is best when the *de* and the yields are high and the auxiliary fully complements an efficient synthesis route.

(iv) *Yields and Efficiency*. Although step yields are very important, individual yield considerations become secondary after the execution of a large number of steps.

(v) *Moderate selectivities*. In many cases, moderate *ee* (or *de*) and moderate yields cannot be tolerated (e.g. LY300502). For example, removal of an unwanted enantiomer and chemical purification from a reaction with a yield of 80% and 80% *ee* might sacrifice 25% of the desired enantiomer, netting 55% of the desired enantiomer. An efficient resolution might yield an 'uncomplicated' 45%. The 'value added' by the stereoselective transformation must be weighted against the 10% increase in yield over the resolution sequence.

(vi) '*Molecular gymnastics*'. As in the case of the synthesis of loracarbef, forcing chiral pool starting materials through circuitous routes to meet stereochemical demands of the target diminished the process efficiency.

(vii) *Speed to market* ('penny-wise, pound-foolish'). In many development circumstances, limited human resources and time-frames will not warrant a multi-year, multi-million dollar effort. This aspect of development is more pronounced as drugs evolve to be potent at submilligram doses, and production is based upon kilograms rather than tonnes.

In route selection and development, concerns for the absolute stereochemistry should not entirely outweigh the basic chemical synthesis. From the experiences above, one should consider the best, most efficient synthesis of the racemate and then superimpose the absolute stereochemical demands upon that synthetic route. In many cases, this may result in a resolution-based process. Certainly, bioorganic transformations and catalytic enantioselective reactions will become profoundly important to the single enantiomer drug industry. However, we can expect simple, direct and robust resolution processes to be of industrial importance for some time to come.

6.6 ACKNOWLEDGEMENTS

We gratefully acknowledge Professors David Evans and Barry Sharpless for the disclosure of various manuscripts in advance of publication. We also thank Professors Edward C. Taylor, Leo Paquette, William Roush, Andrew Meyers and Paul Wender for fruitful discussions throughout these programmes. We also thank our colleagues and co-authors for providing the experimental development, materials, insight and support throughout the course of this work: Marvin Hansen, Mike Martinelli, Charles Barnett, Jerry Misner, Chris Schmid, Jerry Bryant, Lowell Hatfield, Alyssa Kneisley, Gerald Thompson, Don Brannon, Jack Deeter, Gib Staten, Jeff Frazier, Mike Staszak, Jim Aikins, Jim Dunigan, John Gardner, Perry Heath, Billy Jackson, Steve Pedersen, John Rizzo, Eddie Tao, Jeff Ward, Joe Kennedy, Samantha Janisse, Jim Audia, Ken Hirsch, Dave Jones, Dave Lawhorn, Loretta McQuaid, Tom Kress, Jim Wepsiec and Bob Waggoner. We also thank Jill and Mary for their support and patience.

6.7 REFERENCES AND NOTES

1. (a) Helmchen, G., Hoffmann, R. W., Mulzer, J., and Schaumann, E. (Eds), *Stereoselective Synthesis,* Georg Thieme, Stuttgart, 1995, Vols. 1–5; (b) Atkinson, R. S., *Stereoselective Synthesis*, Wiley, Chichester, in press; (c) Norgradi, M., *Stereoselective Synthesis*, VCH, Weinheim, 1987; (d) Ojima, I. (Ed.), *Catalytic Asymmetric Synthesis*, John Wiley, New York, 1993; (e) Noyori, R., 'Enantioselective Catalysis with Metal Complexes—An Overview,' in *Modern Synthetic Methods*, Springer, Berlin, 1989; (f) Bartmann, W., and Sharpless, K. B. (Eds), *Stereochemistry of Organic and Bioorganic Transformations*, VCH, Weinheim, 1987; (g) Stille, J. R., and Barta, N. S., in *Studies in Natural Products Chemistry—Stereoselective Synthesis* (ed. Atta-ur-Rahman), Elsevier, Amsterdam, in press.
2. Collins, A. N., Sheldrake, G. N., and Crosby, J. (Eds), *Chirality in Industry*, Wiley, Chichester, 1992.
3. Weigel, L. O., Seminar presented at the 12th SCI Process Development Symposium entitled 'Racemates, Resolutions, and Results,' Cambridge, 12 December, 1994.
4. Weigel L., unpublished survey from Lilly Chemical PR&D Indianapolis, IN, 3 June, 1993.

5. (a) Robertson, D. W., Wong, D. T., and Krushinski, J. H., Jr, *Eur. Pat. Appl.*, EP 273 658, 1988; *Chem. Abstr.*, **109**, 170224n (1988); (b) Robertson, D. W., Thompson, D. C., Beedle, E. E., Reid, L. R., Bymaster, F. P., and Wong, D. T., presented at the 194th National Meeting of the American Chemical Society, New Orleans, LA, 30 Aug.–4 Sept. 1987, MEDI 49; (c) Wong, D. T., Robertson, D. W., Bymaster, F. P., Krushinski, J. H., and Reid, L. R., *Life Sci.*, **43**, 2049 (1988); (d) *Drugs Future*, **11**, 134 (1986); (e) Ankier, S. I., *Prog. Med. Chem.*, **23**, 121 (1986); (f) Robertson, D. W., Krushinski, J. H., Fuller, R. W., and Leander, J. D., *J. Med. Chem.*, **31**, 1412 (1988); (g) Fuller, R. W., *J. Clin. Psychiatry*, **47** (April Suppl.), 4 (1986).
6. (a) Srebnik, M., Ramachandran, P. V., and Brown, H. C., *J. Org. Chem.*, **53**, 2916 (1988); (b) Gao, Y., and Sharpless, K. B., *J. Org. Chem.*, **53**, 4081 (1988); (c) Mitchell, D., and Koenig, T. M., *Synth. Commun.*, **25**, 1231 (1995). (d) Boaz, N. W., *US Pat.*, 4 921 798, 1990; *Chem. Abstr.*, **116**, 105864b (1992).
7. Weigel, L. O., Deeter, J., Staten, G. S., Frazier, J., and Staszak, M., *Tetrahedron Lett.*, **31**, 7101 (1990).
8. Kumar, A., Ner, D. H., and Dike, S. Y., *Tetrahedron Lett.*, **32**, 1901 (1991).
9. Corey, E. J., and Reichard, G. A. *Tetrahedron Lett.*, **30**, 5207 (1989).
10. Noyori, R., Tomino, I., and Tanimoto, Y., *J. Am. Chem. Soc.*, **101**, 3129 (1979).
11. Weigel, L. O., Deeter, J., Staten, G. S., Frazier, J., and Staszak, M., presented at the Joint 23rd Central/24th Great Lakes Regional Meeting of the American Chemical Society, Indianapolis, IN, 29 May 1991, ORGN349.
12. (a) Yamaguchi, S., Mosher, H. S., and Pohland, A., *J. Am. Chem. Soc.*, **94**, 9254 (1972); (b) Yamaguchi, S., and Mosher, H. S., *J. Org. Chem.*, **38**, 1870 (1973); (c) Brinkmeyer, R. S., and Kapoor, V. M., *J. Am. Chem. Soc.*, **99**, 8339 (1977).
13. Carbinolamine **12** has been referred to as 'Darvon alcohol' at various points in the literature. This use of the Darvon® trademark has not been authorised by Eli Lilly and Company. For this work, non-resolved **8** was obtained from Eli Lilly Industries, Mayaguez, Puerto Rico. Resolution was effected according to Pohland, A., and Sullivan, H. R., *J. Am. Chem. Soc.*, **77**, 3400 (1955).
14. Weigel, L. O., presented at the 205th National Meeting of the American Chemical Society, Denver, CO, 31 March 1993, ORGN261.
15. (a) First public disclosure of loracarbef: Hirata, T., Matsukuma, I., Mochida, K., and Sato, K., presented at the 27th Interscience Conference on Antimicrobial Agents and Chemotherapy, New York, 4–7 October 1987; (b) Sato, K., Okachi, R., Matsukuma, I. Mochida, K., and Hirata, T., *J. Antibiot.*, **42**, 1844 (1989); (c) For a recent review of carbacephalosporins, see Zhang, T. Y., and Hatfield, L. D., in *Comprehensive Hetereocyclic Chemistry*, (ed. E. F. V. Scriven and C. W. Rees), 2nd edn, Elsevier, London, 1996, vol. 1, Chap. 20.
16. Matsukuma, I., Yoshiiye, S., Mochida, K., Hashimoto, Y., Sato, K., Okachi, R., and Hirata, T., *Chem. Pharm. Bull.*, **37**, 1239 (1989).
17. (a) Evans, D. A., and Sjogren, E. B., *Tetrahedron Lett.*, **26**, 3787 (1985); (b) for a similar approach, see Bodurow, C. C., Boyer, B. D., Brennan, J., Bunnell, C. A., Burks, J. E., Carr, M. A., Doecke, C. W., Eckrich, T. M., Fisher, J. W., Gardner, J. P., Graves, B. J., Hines, P., Hoying, R. C., Jackson, B. G., Kinnick, M. D., Kochert, C. D., Lewis, J. S., Luke, W. D., Moore, L. L., Morin, J. M. Jr., Nist, R. L., Prather, D. E., Sparks, D. L., and Vladuchick, W. C., *Tetrahedron Lett.*, **30**, 2321 (1989).
18. Jackson, B. G., Doecke, C. W., Farkas, E., Fisher, J. W., Gardner, J. P., Gazak, R. J., Kroeff, E. P., Misner, J. W., Pedersen, S. W., Staszak, M. A., and Vicenzi, J. T., presented at the 'Chiral USA 1994' Conference, Reston, VA, Spring 1994.
19. Deeter, J., Hall, D. D., Jordan, C. L., Justice, R. M., Kinnick, M. D., Morin, J. M., Jr, Paschal, J. W., and Ternansky, R. J., *Tetrahedron Lett.*, **34**, 3051 (1993).
20. Lotz, B. T., and Miller, M. J., *J. Org. Chem.*, **58**, 618 (1993).

21. (a) Cooper, R. D. G., Daugherty, B. W., and Boyd, D. B., *Pure Appl. Chem.*, **59**, 485 (1987); (b) Hegedus, L. S., Montgomery, J., Narukawa, Y., and Snustad, D. C., *J. Am. Chem. Soc.*, **113**, 5784 (1991).
22. Cohen, N., Banner, B. L., Lopresti, R. J., Wong, F., Rosenberger, M., Liu, Y. Y., Thom, E., and Liebman, A. A., *J. Am. Chem. Soc.*, **105**, 3661 (1983).
23. Hubschwerlen, C., and Schmid, G., *Helv. Chim. Acta*, **66**, 2206 (1983).
24. Evans, D. A., and Williams, J. M., *Tetrahedron Lett.*, **29**, 5065 (1988).
25. Dunigan, J., and Weigel, L. O., *J. Org. Chem.*, **56**, 6225 (1991).
26. (a) Weigel, L. O., *US Pat.*, 5 097 049, 1992; (b) Manchand, P. S., Luk, K. C., Belica, P. S., Choudhry, S. C., Wei, C. C., and Soukup, M., *J. Org. Chem.*, **53**, 5507 (1988).
27. Weigel, L. O., Frazier, J. W., and Staszak, M. A., *Tetrahedron Lett.*, **33**, 857 (1992).
28. Hatanaka, M., and Ishimaru, T., *Tetrahedron Lett.*, **24**, 4837 (1983).
29. (a) *Chem. Eng. News*, 10 September, 26 (1990); (b) Weigel, L. O., and Astleford, B., *Tetrahedron Lett.*, **32**, 3301 (1991); (c) formation of **46** was also accompanied by significant amounts of PhthNCH(Me)Ph. Use of (*S*)-2-(methoxymethyl)pyrrolidine in these resolution sequences suppresses the formation of this undesired product.
30. Hatfield, L. D., Lunn, W. H., Jackson, B. G., Peters, L. R., Blaszczak, L. C., Fisher, J. W., Gardner, J. P., Dunigan, J. M., in *Recent Advances in the Chemistry of β-Lactam Antibiotics* (ed. G. I. Gregory), Royal Society of Chemistry, London, 1980, p. 109.
31. Kenyon, J., *J. Chem. Soc.*, 2540 (1922).
32. Weigel, L.O., *US Pat.*, 5 169 945, 1992.
33. These intermediates can also be converted into the free amine by treatment with methylamine; Wolfe, S., and Hasan, S. K., *Can. J. Chem.*, **48**, 3572 (1970).
34. (a) Zmijewski, M., Briggs, B., and Thompson, A., *New J. Chem.*, **18**, 425 (1994); (b) Zmijewski, M. J., Briggs, B. S., Thompson, A. R., and Wright, I. G., *Tetrahedron Lett.*, **32**, 1621 (1991); (c) Boyer, B. D., and Eckrich, T. M., *Eur. Pat.*, EP 365 212, 1990; (d) Brennan, J., and Eckrich, T. M., *Eur. Pat.*, EP 365 213, 1990.
35. Jones, C. D., Audia, J. E., Lawhorn, D. E., Hirsch, K. S., McQuaid, L. A., Neubauer, B. L., Pike, A. J., Pennington, P. A., Stamm, N. B., and Toomey, R. E., *J. Med. Chem.*, **36**, 421 (1993).
36. Neubauer, B. L., Goode, R. L., Gray, H. M., Hanke, C. W., Hirsch, K. S., Hsiao, K. C., Jones, C. D., Lawhorn, D. E., McQuaid, L., Toomey, R. E., Valia, K., and Audia, J. E., *Drugs Future*, **20**,144 (1995).
37. Audia, J. E., Lawhorn, D. E., and Deeter, J. B., *Tetrahedron Lett.*, **34**, 7001 (1993).
38. Brandt, M., Mohammed, H., Levey, M. A. and Holt, D. A., *Bioorg. Med. Chem. Lett.*, **4**, 1365, (1994).
39. (a) Astleford, B. A., Janisse, S. K., and Weigel, L. O., presented at the 206th National Meeting of the American Chemical Society, Chicago, IL, 23 August 1993, ORGN076; (b) Astleford, B. A., Deeter, J., Janisse, S. K., and Weigel, L. O., presented at the 206th National Meeting of the American Chemical Society, Chicago, IL, 23 August 1993, ORGN075; (c) Astleford, B. A., Audia, J. E., Deeter, J., Heath, P. C., Janisse, S. K., Kress, T. J., Wepsiec, J. P., Weigel, L. O., *J. Org. Chem.*, **61**, 4450 (1996).
40. Kuo, F., and Wheeler, W. J., *J. Labelled Compd. Radiopharm.*, **34**, 915 (1994).

7 Crystal Science Techniques in the Manufacture of Chiral Compounds

W. M. L. WOOD
ZENECA Huddersfield Works, Huddersfield, UK

7.1 Introduction 119
7.2 Basic Crystal Science 120
7.2.1 Crystals, Polymorphs and Solvates 120
7.2.2 Crystal Habit 125
7.2.3 Habit Modifiers 126
7.3 Development of Crystallisation Processes from Solution 130
7.3.1 Supersaturation—the Driving Force 130
7.3.2 Choice of Solvent 131
7.3.3 Solubility Curves and Their Measurement 131
7.3.4 Nucleation, the Metastable Zone and Control of Crystal Size 133
7.3.4.1 Measurement of Metastable Zone Width 134
7.3.4.2 Control of Crystal Size and Size Distribution 137
7.3.5 Rate of Desupersaturation 140
7.3.6 Isolation of Metastable Phases 140
7.4 Application of Crystal Science to Chiral Systems 143
7.4.1 Separation of Enantiomers by Diastereoisomer Formation . 144
7.4.2 Separation of Diastereoisomers 145
7.4.3 Resolution by Direct Crystallisation of Enantiomers 148
7.4.3.1 Seeding Within the Metastable Zone 152
7.4.3.2 Nucleation Inhibition of One Enantiomer 154
7.5 Conclusion 154
7.6 References 155

7.1 INTRODUCTION

Crystal science techniques provide powerful aids in the manufacture of chiral compounds. Their applicability depends on the properties of those compounds or their intermediates, and is unpredictable unless those properties have been measured. Synthetic organic chemists often have little understanding of the factors which govern such aspects as crystal size, size distribution, the ease of separation

Chirality in Industry II. Edited by A. N. Collins, G. N. Sheldrake and J. Crosby

of the crystals from the mother liquors and the purity of the isolated crystals, all of which are important in developing a satisfactory crystallisation process.

It is intended that this chapter will help to provide that understanding, and will show how the science may be used in a practical way to achieve, and optimise, the isolation of chiral compounds by a variety of techniques, by reference to some real examples of the application of the science to the manufacture of chiral compounds. The chapter deliberately contains no mathematical equations. For those readers interested in obtaining greater detail about particular areas of the science and the underlying theory, it is suggested that they refer initially to the treatise on crystallisation by Mullin[1] and the references therein.

7.2 BASIC CRYSTAL SCIENCE

7.2.1 CRYSTALS, POLYMORPHS AND SOLVATES

A crystalline material is one in which the species present (atoms, ions, molecules) are arranged in an ordered three-dimensional structure. Under the appropriate conditions of supercooling or supersaturation, crystals are formed by nucleation followed by diffusion of the species to the surface of the nuclei and then their incorporation into the ordered structure or lattice of the crystal, causing its growth. Often there are several different possible packing arrangements of the species under different crystallisation conditions, resulting in the formation of different crystal lattices. The material is then said to be polymorphic, and the different lattices are polymorphs. A well known example is carbon, which can exist as both graphite and diamond crystalline forms. Figure 1 (from the data of Haisa and co-workers[2,3]) shows the way the molecules are arranged in the unit cell of the crystal lattices of the orthorhombic and monoclinic polymorphs of paracetamol (*p*-hydroxyacetanilide).

The lattice is not limited to only one type of molecule or set of ions. Where crystallisation is from solution, then the lattice may contain solvent molecules bound into the structure, and the resulting crystals are solvates. It is also fairly common (as with chiral racemates) for the lattice to contain more than one type of non-solvent molecule. The species co-crystallise to form a mixed crystal with a well-defined and fixed ratio of the different molecules or ions present in the lattice. Potash alum, $KAl(SO_4)_2 \cdot 12H_2O$, and sodium ammonium tartrate, $NaNH_4C_4H_4O_6 \cdot 4H_2O$, are well known inorganic and organic examples, respectively. Sometimes it is found that crystals of solute grown from solution contain variable amounts of impurities which are also present in the solution, even after washing to remove adhering solution and any material adsorbed on the crystal surface. The impurity is incorporated in the crystal lattice as a 'solid solution' and the amount present is a function of the concentration in solution.

The detailed arrangement of the species in the lattice may be determined by X-ray diffraction (XRD) studies on a single crystal. The resolution given by such

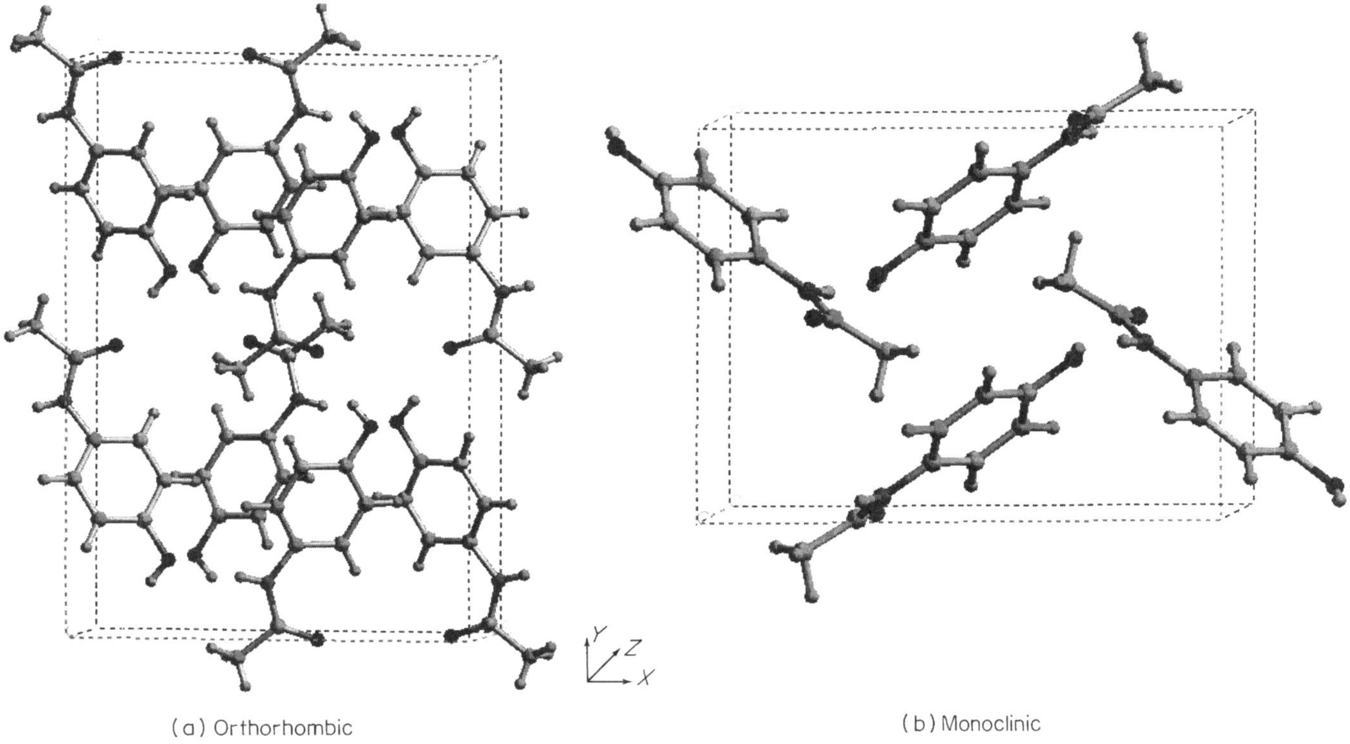

Figure 1 Arrangement of the molecules in two polymorphs of paracetamol (*p*-hydroxyacetanilide)

studies is sufficiently precise that the technique may be used to determine not only the relative position of the species in the lattice but also, where the species are molecules, the conformation of those molecules and also their molecular structure. This is particularly useful for large molecules such as proteins. Growth of suitable single crystals for these studies can be difficult; it is much simpler to produce a mass of small crystals, and the diffraction pattern produced from a sample of powder or paste is very useful for identification purposes since each polymorph or solvate gives a different pattern. Although, at present, it is not possible to obtain the detailed structural information from such samples, considerable advances are being made toward this end.[4] The use of XRD is described in detail by Glusker *et al.* [5]

The different solvates or polymorphs are known as crystal forms or phases. Generally, at a given temperature, only one phase will be stable, the others being metastable. A metastable phase can form under certain conditions, particularly in rapid crystallisations, but will tend to transform in time to a more stable phase, usually via solution. The rate-controlling processes in such transformations are discussed by Davey *et al.*[6] The stable phase at a particular temperature will always have the lowest solubility at that temperature. The exception is if a phase transition temperature exists, below which one phase is stable and above which another phase is stable. At the transition temperature, both phases have equal stability and the same solubility.

The difference in solubility of different polymorphs can alter the effectiveness of a product in end use. For example, with drugs, the lower solubility of a stable polymorph may be desirable to control the rate of release. On the other hand, the stable polymorph of a drug may have such a low solubility under physiological conditions that it is difficult to achieve a therapeutic dose. In these cases, the isolation and formulation of a metastable phase may provide the necessary improvement in solubility.

The different behaviour of polymorphs and solvates on heating provides further analytical techniques for the identification of the crystal form. On heating a solvate crystal, a temperature will be reached beyond which some or all of the solvent molecules will have sufficient energy to break free of the lattice, and the crystals desolvate. Partial desolvation can occur where some solvent molecules are held more tightly in the lattice, but these too will usually break free as the temperature is raised further. The pattern of loss in weight versus temperature is consistent for a particular solvate phase, and may be measured by thermogravimetric analysis (TGA). From a knowledge of the molecular weights of the crystal species and the solvent, and the weight loss, the ratio of solvent molecules to solute in the crystals may be obtained. Figure 2 shows the TGA trace given by gypsum ($CaSO_4{\cdot}2H_2O$). The weight loss caused by dehydration starts at about 40 °C and is complete by 280 °C. The weight loss of 20.95% compares well with the theory weight loss of 20.91%.

The loss of solvent molecules from a solvate is an endothermic process, as is the melting of crystals. Because of their different lattice energies, polymorphs

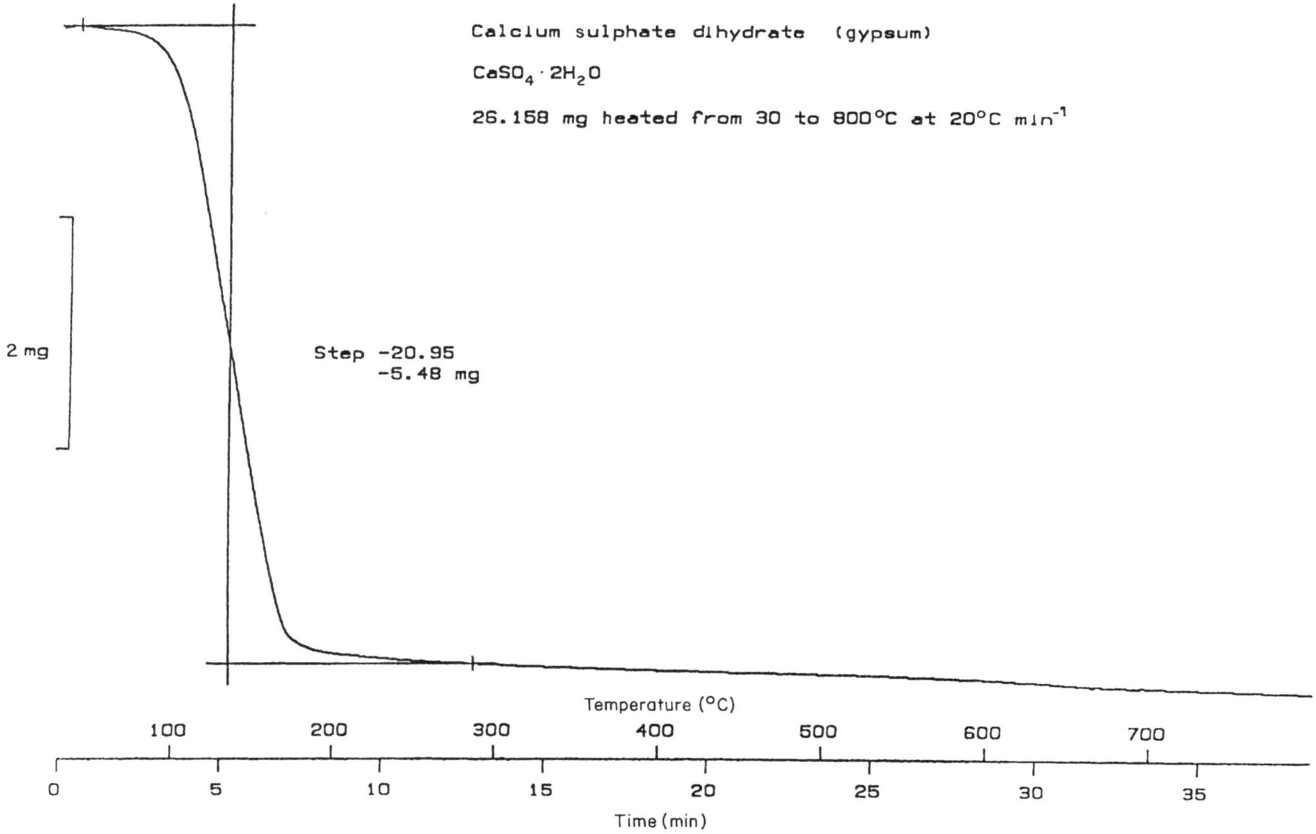

Figure 2 Thermogravimetric analysis (TGA) trace for gypsum

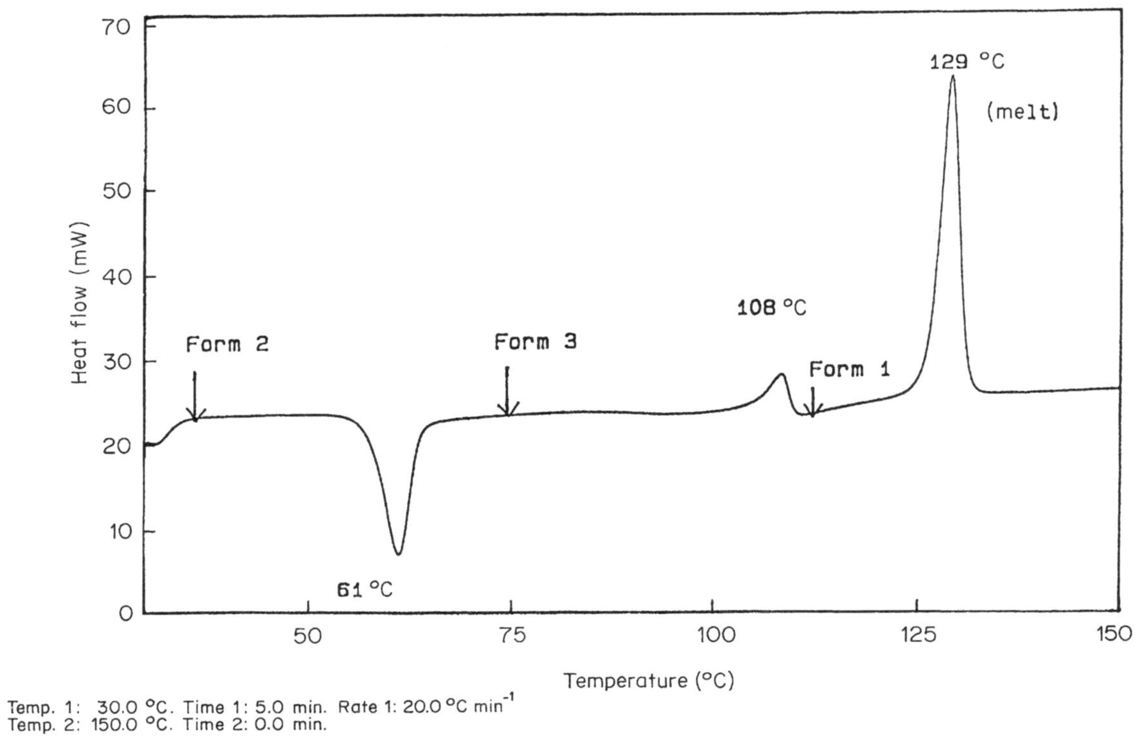

Figure 3 DSC trace for a red dyestuff (DISPERSOL Scarlet CG)

have different melting points, and therefore another technique often used to identify polymorphs and solvates is differential scanning calorimetry (DSC). A useful introduction to DSC is provided by McNaughton and Mortimer.[7]

Figure 3 shows the output for a sample of a dyestuff undergoing two polymorphic transformations to successively more stable forms before the most stable form melts. Such transformations are exothermic, producing a negative peak on the chart. Melting is endothermic, producing a positive peak. The second transformation, at 108 °C, produces a small positive peak indicating that Form 3 melts and the melt then almost immediately recrystallises as Form 1, the overall heat flux of the two processes being endothermic.

7.2.2 CRYSTAL HABIT

The shape (habit, morphology) of crystals and their size distribution are important in that they control the ease and efficiency with which the crystals are separated by filtration or centrifugation from the mother liquor (either melt or solution). Ideally, for fast and efficient separation, the crystals should have a narrow size distribution, be no smaller than about 50 μm and have a low aspect ratio, with similar length in all directions. Control of the crystal size and size distribution is discussed later in Section 7.3.4.2.

Crystal habit is important in that, while low aspect ratio crystals pack together to form a porous, easily washed filter cake, thin, plate-like crystals quickly form a thin layer with low permeability next to the filter medium, after which filtration is very slow. Their high surface area holds more mother liquor, which can be difficult to remove by washing. Needle-shaped crystals form a porous cake, but tend to block the pores of the filter medium.

The habit of an undamaged crystal is governed by a combination of its lattice symmetry and the rate at which the species is/are incorporated into the different possible sites or growth planes on the nucleus and growing crystal. For example, consider the simple case where growth rates along the *A*, *B* and *C* axes of a crystal are the same. The resulting crystal will be a cube (see Figure 4).

If the growth rates along the *A* and *B* axes are the same as before, but that along the *C* axis is only one tenth of that rate, the resulting crystal would be a thin slab. Where growth along both the *B* and *C* axes is only one tenth that along the *A* axis, the result is a flat-sided rod. The growth rate of a particular face is related to the ease with which the species become incorporated into the face, and the distance between successive layers of species perpendicular to the plane of the face.

The habit of an undamaged crystal can vary depending on the environment in which it is grown. This is because other species, such as the solvent in which the crystals form or particular impurities which may be present, can affect the growth rate of one or more faces by preferentially associating with or attaching to those faces and slowing the rate at which the solute species can be incorporated into the growth sites. If the energy with which the impurities or solvent are attached to the growing faces is relatively low, then, as the temperature of the

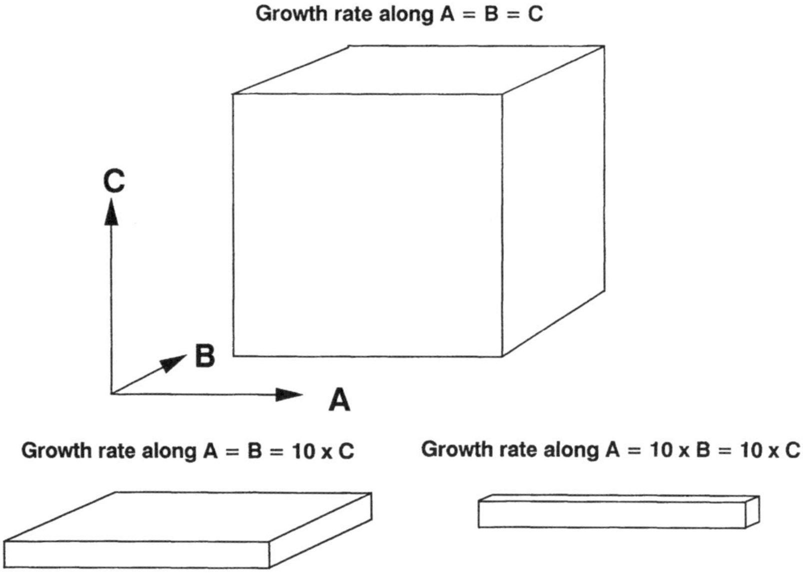

Figure 4 Effect of growth rate of different faces on crystal habit

system at which crystal growth is occurring is raised, the habit may change as growth sites become unblocked. The micrographs in Figure 5 illustrate the effect on crystal habit of crystal growth at different temperatures for the chlorophenylhydrazone $ClC_6H_4NHN{=}CHCOOH$. It is seen that a relatively small increase in reaction temperature, from ambient to 40 °C, results in a major change in the crystal habit. XRD showed that the crystals all have the same form.

The effects of different solvents on crystal habit are well known. The section on the physical constants of organic compounds in older editions of the *Handbook of Chemistry and Physics*[8] gives many examples of compounds crystallised from different solvents having different crystal habits; therefore, in the development of any crystallisation process, it should be routine to examine the crystals produced, using an optical microscope. The crystals should, wherever possible, be examined in their mother liquors, prior to any solid–liquid separation step. This will quickly give an idea of the crystal size and size distribution as well as their habit, and will also show whether they have suffered much attrition damage, for example due to agitation (such damage results in fragments which widen the distribution and can cause problems in filtration).

7.2.3 HABIT MODIFIERS

Impurities, whether present as by-products or by deliberate addition, can have a profound effect on crystal habit, even at very low concentrations. The presence of

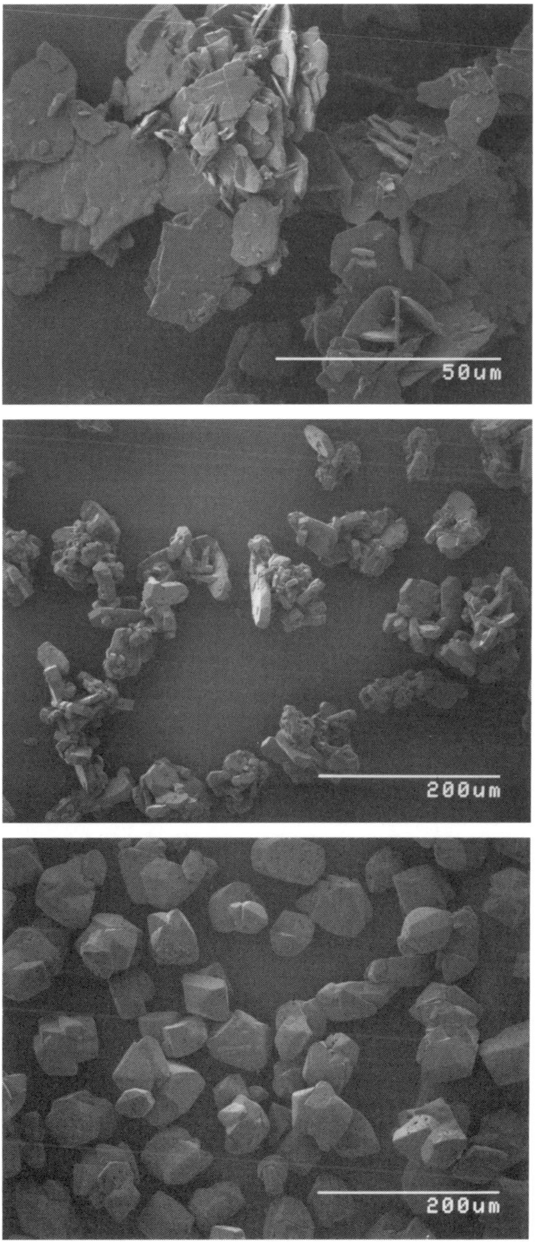

Fast addition of acid
at ambient temperature

Slow addition of acid
at 30 °C

Slow addition of acid
at 40 °C

Figure 5 Effect of reaction conditions on the habit of chlorophenylhydrazone crystals. Note the difference in magnification

Figure 6 Sodium chloride crystals of different habits

even 1 ppm of hexacyanoferrate(II) during the crystallisation of NaCl from aqueous solution is sufficient to change the habit from the normal cubes to dendrites, while higher levels inhibit nucleation. The presence of urea causes salt to grow as octagonal crystals (see Figure 6).

In the past, the experimental approach to finding an effective habit modifier to achieve a desired change in crystal habit was empirical. The effects of large numbers of materials, particularly natural products such as gums and gelatines, were examined in the hope of finding an effective modifier. With the increasing availability of single-crystal XRD and, particularly, the development of software to derive the lattice structure from the diffraction data, the precise arrangement of species in the various faces of a crystal can be determined much more easily and cheaply than before. A more structured approach to the development of effective habit modifiers to control growth at chosen faces has thus become possible. For example, consider a perpendicular slice through a particular crystal face (Figure 7). The molecules in this face are shown in diagrammatic form with an attachment end (the circle) and a receiver end (the triangle). The face grows because of the favourable energetics of fitting the attachment end of the incoming molecule

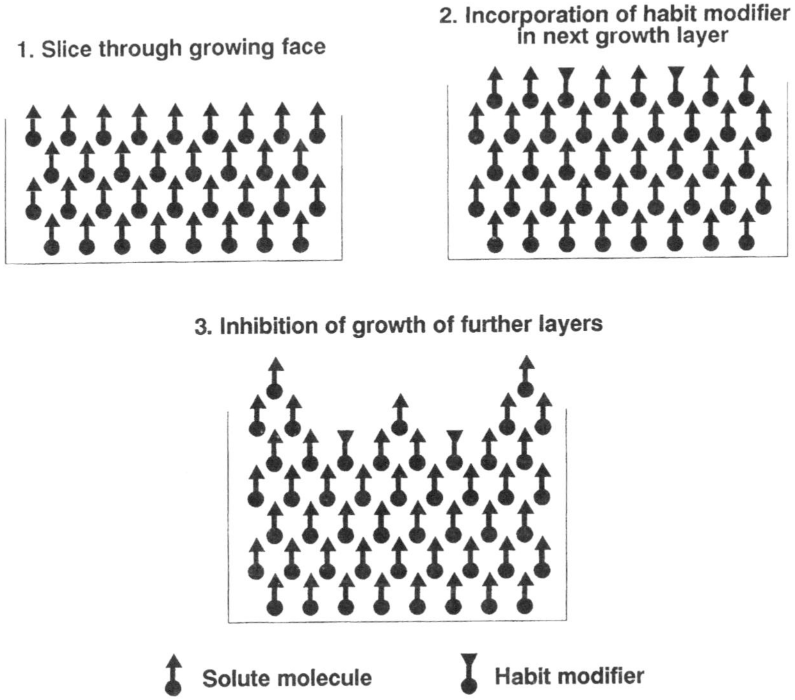

Figure 7 Effect of a habit modifier on growth rate

to the receiving ends of the molecules in the existing face, together with further interactions with its adjacent neighbours in the new face.

If impurity molecules are present of similar size and structure, with an identical attachment end, but with a different receiving end, then the likelihood of such a molecule becoming incorporated into a growth site in the crystal face may be very similar to that for the normal molecules. If it does become incorporated, then when the next layer begins to grow the energetics of the interaction between the attachment end of a normal incoming molecule and the different receiving end of the impurity will not be favourable and incorporation will not take place. Growth above and adjacent to such sites will be inhibited.

This effect is offset by the fact that the surface of a growing crystal in contact with a supersaturated solution is in a constant state of flux, with molecules continually separating from the surface and becoming incorporated. The rate of growth depends on the relative rates of separation and incorporation. Depending on the interactions with its neighbours in the new face, the impurity may, on average, stay for a longer or shorter time than a normal molecule in the growing face. If shorter, then the impurity will probably have little effect on the growth rate. To be effective, it has to stay at least as long as or preferably longer than a normal molecule, so that it is still present during the formation of the next growth layer.

7.3 DEVELOPMENT OF CRYSTALLISATION PROCESSES FROM SOLUTION

Crystallisation may occur from a melt, from solution, from the gaseous phase or from a supercritical fluid. The most common and useful method for optically active compounds is crystallisation from solution, and we shall concentrate on this.

7.3.1 SUPERSATURATION—THE DRIVING FORCE

A prime requirement for crystallisation from solution is that the concentration of the solute in the solvent exceeds its equilibrium solubility, to provide the driving force of supersaturation. Supersaturation may be created in a number of ways, for example:

—by cooling a saturated solution;
—by evaporation of solvent from a saturated solution;
—by addition of a miscible non-solvent to a saturated solution;
—for ionic salts, by use of the common ion effect;
—by salting out;
—by reaction to form the solute *in situ*.

In most crystallisations aimed at the isolation of optically active molecules, one of the first four of these methods are used, particularly the cooling of a saturated solution, and the rest of this section will concentrate on this method.

7.3.2 CHOICE OF SOLVENT

The choice of solvent for a crystallisation process from solution is governed by many considerations, depending particularly on the method of creation of supersaturation. For example, for a cooling crystallisation the solvent should provide a high concentration solution at elevated temperature and a relatively steep solubility versus temperature curve. If supersaturation is to be achieved by evaporation, the solvent should not only provide a high concentration solution but also have a relatively low boiling point and latent heat of vaporisation. If addition of a non-solvent to a solution is to be used, then the chosen solvent system should show a relatively steep relationship between solubility and solvent composition and the component solvents should be easily separated from each other after use, e.g. by distillation, to allow their recycle. On the manufacturing scale, the choice of solvent is also often limited on the grounds of the toxicity of any solvent residues which may be retained in the isolated product, and/or fire and explosion hazards.

7.3.3 SOLUBILITY CURVES AND THEIR MEASUREMENT

The yield of crystals given by a single crystallisation step not involving a reaction depends on the solubility of the solute at the initial and final conditions of temperature and/or solvent composition in the crystallisation, and the relative volumes of solvent at the start and end of the crystallisation. For example, in a simple cooling crystallisation a high yield depends on a steep relationship between solubility and temperature. In order to be able to design a crystallisation process with high yield, it is important to know how the solubility varies with the changing conditions during crystallisation, and so the solubility curve must be determined.

A further important factor which must be considered is the solubility of impurities or by-products present in the system. Since a prime objective of a crystallisation process is to isolate a pure product from an impure starting material, the choice of final conditions in a crystallisation which would lead to impurities also crystallising should be avoided. Therefore, the solubility of the impurities should also be known. In principle, a solubility curve is measured by holding stirred dispersions of the solute crystals at a number of different conditions (fixed temperatures, fixed solvent compositions) for sufficient time to reach equilibrium, and the concentration of solute in solution is then measured. The equilibrium may be approached either from a higher temperature, by cooling a suspension and crystallising, or by heating a suspension and dissolving. There are a number of practical factors to consider. It is always best to use the representative, impure system if possible, rather than pure solute and solvent, since impurities can sometimes affect the solubility of the solute. If the impurities form solid solutions in the product crystal lattice, then the solubility of the product can be increased significantly above that of pure product; a doubling or more of solubility is not uncommon in such cases. In addition, the presence of impurities may inhibit crystal growth, as discussed in Section 7.2.2, and this effect can be so strong that, if equilibrium is approached by cooling/crystallisation, growth can cease when the solution is still significantly supersaturated. If such behaviour is suspected, it is therefore good practice to measure initially the equilibrium solubility by cooling/crystallisation from solution, and then use the crystals produced to measure their solubility by dissolution. If the result by dissolution is significantly smaller than that by crystallisation, then crystal growth is ceasing when the solution is still supersaturated. Such growth inhibition can have a major effect on the yield from a crystallisation, and further studies may be justified to investigate which species may be the inhibitors. It may be possible to avoid their presence by altering the upstream chemistry or by removing them by a pre-treatment.

There are potential problems which can complicate the solubility measurements, the most common being the crystallisation of the solute as different polymorphs or solvates. If a phase transition temperature (PTT) occurs within the solubility curve temperature range, then the equilibrium solubility curve should display a discontinuity, as shown in Figure 8.

However, there is no guarantee that, as the PTT is crossed, the phase transition actually takes place. What was the stable phase becomes metastable beyond the PTT, but if the new stable phase does not nucleate, the crystals present will

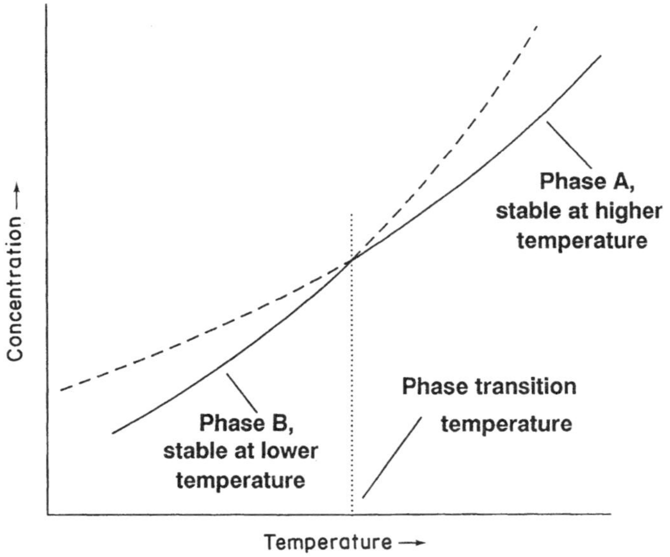

Figure 8 Solubility curve showing a phase transition

remain as the metastable phase, and the measured solubility will be of that phase. Therefore, instead of showing the discontinuity, the measured solubility curve will be smooth (dotted line in Figure 8). For this reason, it is important that the nature of the crystalline phase present at equilibrium at the extremes of high and low temperature likely to be used in practice are established by powder XRD. If the XRD patterns show different phases to be present at the different temperatures, then it is good practice to conduct two series of solubility curve measurements, one with rising temperature and the other with cooling. If, as sometimes happens, the phase transition does not occur, the PTT will be where the two solubility curves intersect. There may be more than one PTT within the temperature range.

The solubility curve defines the equilibrium concentration of solute at a given temperature and solvent composition. However, during a crystallisation process the system is not at equilibrium; supersaturation is being created, and crystal nucleation and growth occur towards equilibrium.

7.3.4 NUCLEATION, THE METASTABLE ZONE AND CONTROL OF CRYSTAL SIZE

A simple solubility versus temperature curve is shown in Figure 9. Consider a solution at the concentration and temperature represented by point A. This point lies beneath the solubility curve and so represents an undersaturated solution. If this solution is now cooled, the system will travel along a horizontal line on the

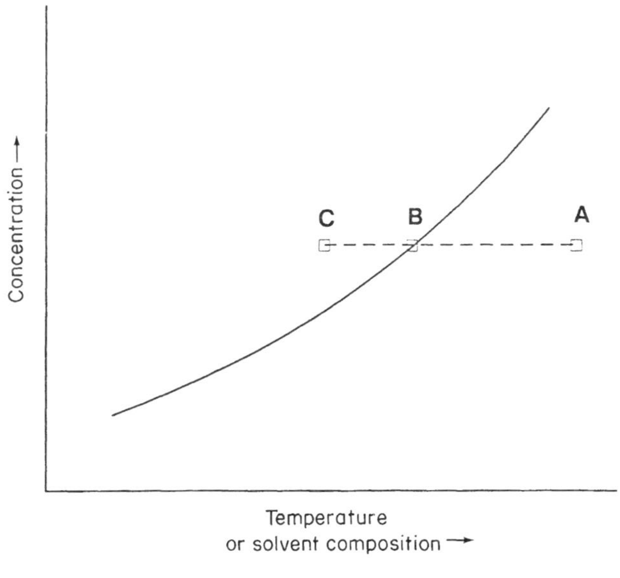

Figure 9 A simple solubility curve

diagram until, at point B, the system becomes just saturated. Further cooling causes the system to become supersaturated, but nucleation to form crystals does not usually occur immediately the system becomes supersaturated.

Only when cooling is continued to some point C does sufficient supersaturation exist for nucleation to occur. By analogy with reaction kinetics, in which there is an activation energy, there is an 'activation supersaturation' which has to be exceeded before nucleation occurs. The gap between points B and C (the 'metastable zone') may be small, perhaps less than 1 °C in a cooling crystallisation, but may be large, particularly for large molecules. Cases where the gap is many tens of degrees are not uncommon. It may take some hours (the induction time) for nucleation to take place at point C, but as the supersaturation is increased still further, so the induction time normally reduces rapidly. Assuming the solubility curve has been determined, the position of point C relative to B in a cooling crystallisation may be established.

7.3.4.1 Measurement of Metastable Zone Width

In the first method, a slightly undersaturated solution is prepared, for example by filtering the equilibrium solution from a dissolution solubility curve experiment. The solution is then re-heated to several degrees above its original equilibrium temperature, and maintained at this temperature for a period of several hours to ensure that any nuclei which may have passed through the filter have redissolved, then cooled to within a few degrees of the saturation temperature. At elevated temperatures or with volatile solvents, great care must be taken to minimise any loss of solvent, for example by using pressure rather than vacuum filtration.

The undersaturated solution is then slowly cooled in a stirred vessel and the temperature at which the first sign of nucleation is observed is noted. This is tedious to do visually, and a preferred method is to use a fibre-optic colorimeter (Figure 10), which detects the onset of nucleation by attenuation of the light beam. If the nucleation temperature is reached quickly, then the experiment should be repeated using a slower cooling rate. The same solution can be re-used, provided that the temperature is raised again and held at several degrees above the saturation temperature to ensure dissolution of all the nuclei. It is generally found that the points at which nucleation occurs lie on a curve which is roughly parallel to the solubility curve (Figure 11), the region between the curves being the metastable zone. Within this zone any crystals present, for example by the addition of seed crystals, will grow but the solution will not self-nucleate.

The second method for measuring the metastable zone width is more time consuming but gives additional information. The same equipment is used, but the temperature of the solution is quickly reduced to a chosen value and then maintained. Again, the time taken for the first nuclei to appear is noted. If no nucleation occurs within several hours, the solution is reheated to above its equilibrium solubility temperature, then re-cooled to a slightly lower temperature. A graph is then plotted of the induction time at each temperature before nucleation

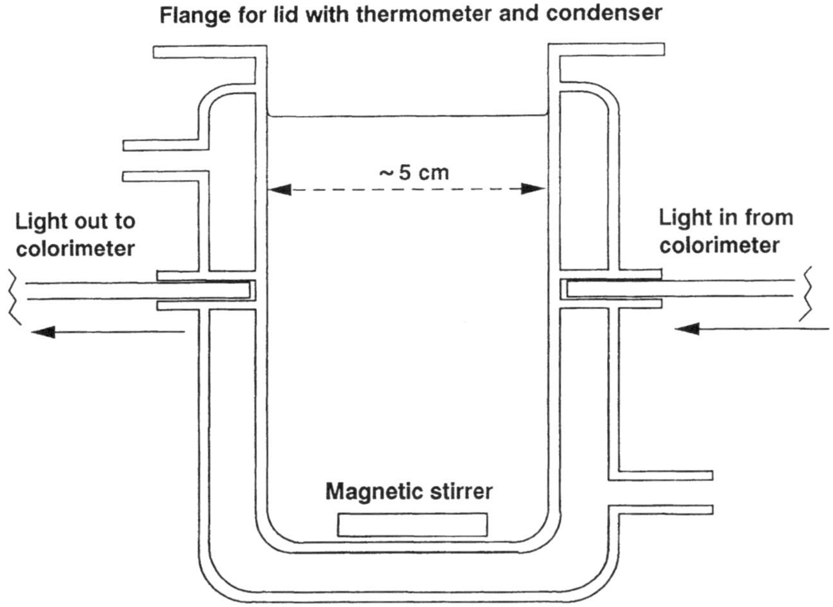

Figure 10 Jacketed vessel for metastable zone measurement. The side tubes are used for fitting fibre-optic cables from a Brinkmann fibre-optic colorimeter

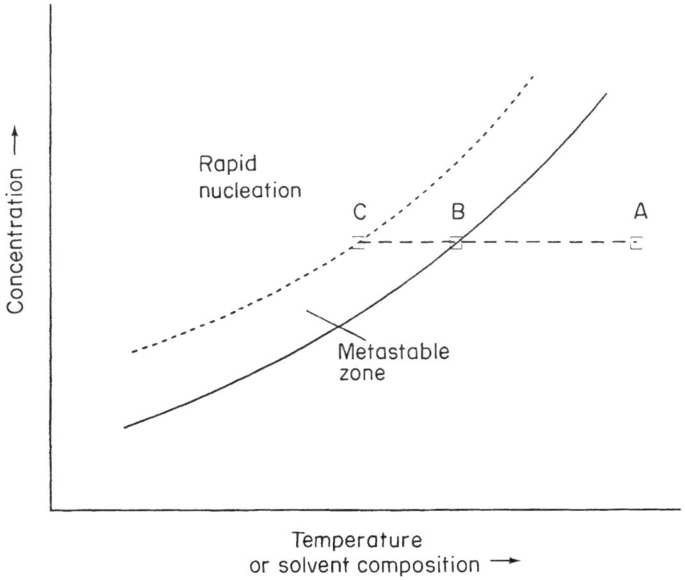

Figure 11 Solubility curve and metastable zone

occurs versus the degree of supersaturation created at that temperature, which is obtained from the solubility curve. A typical graph is shown in Figure 12.

This graph is particularly useful in that it indicates how long a solution can withstand a given level of supersaturation without nucleating, which is important in designing the combination of the area of heat transfer surface and the temperature difference (ΔT) across that surface to be used in a cooling crystallisation. The solution close to the heat transfer surface will cool quickly to the surface temperature, and then agitation will cause re-heating by mixing with the bulk. A knowledge of the induction time versus supersaturation curve allows the calculation of the minimum surface temperature allowable if crash nucleation is not to occur at, or close to, the heat transfer surface.

In Figure 11, the nucleation point curve is deliberately shown as a dotted line. This curve is in practice rather variable for a particular solvent–solute system, its position depending particularly on the scale of the experiment. Nucleation is a statistical process: the larger the volume of a supersaturated solution, the more likely it is that nucleation will occur and the narrower is the metastable zone.

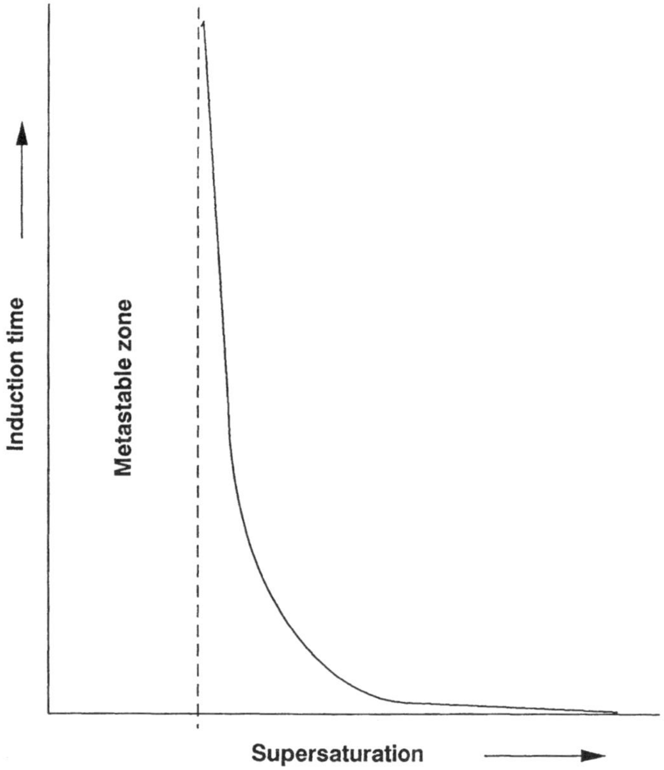

Figure 12 Graph of induction time versus supersaturation

In the case of crystallisation by addition of a miscible non-solvent, the metastable zone can be measured in similar fashion to that for cooling crystallisation but, in this case, the temperature is held constant and the solvent composition is changed by addition of the non-solvent. It is important that the non-solvent is dispersed as quickly as possible to avoid localised high supersaturations and consequent crash nucleation.

7.3.4.2 Control of Crystal Size and Size Distribution

The number of nuclei formed per unit volume of solution during a nucleation event is controlled primarily by the rate at which supersaturation is created and the nucleation point curve is crossed. Crossing it very rapidly, e.g. by very rapid cooling, can increase the number of nuclei formed by up to three orders of magnitude compared with slow cooling. If each nucleus grows to a crystal then, since a fixed amount of the solute will crystallise from the solution, a larger number of nuclei must result in each crystal being smaller.

Figures 13–15 show how the concentration of solute in solution can vary during a cooling crystallisation. Figure 13 shows the case where the cooling rate is

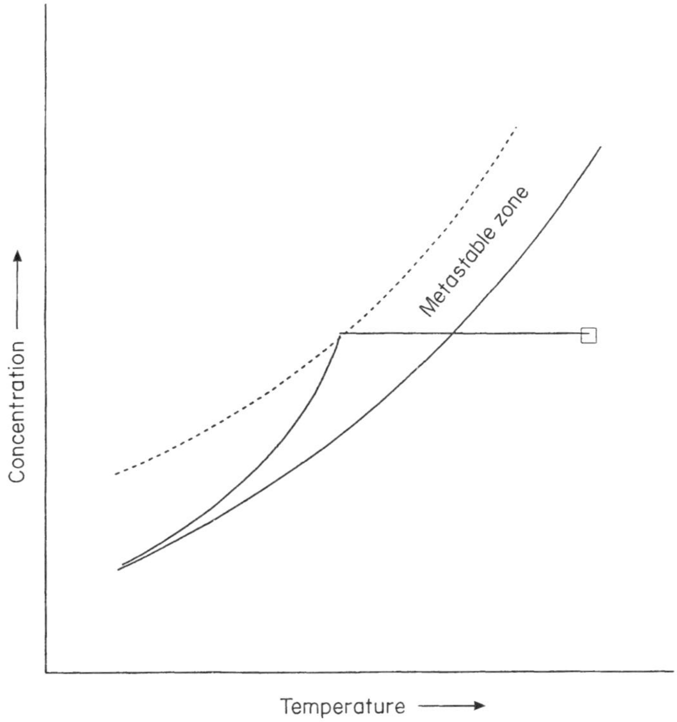

Figure 13 Supersaturation path for unseeded system with slow cooling. One nucleation event gives small crystals of uniform size

relatively slow. After the nucleation event, the concentration of solute falls as growth occurs on the nuclei. As cooling continues, the concentration remains within the metastable zone, although its position relative to the solubility curve will depend on the relative rates of cooling, and hence creation of supersaturation, and of crystal growth, which discharges that supersaturation. The resulting crystals will be of relatively uniform size which, as described above, will be governed by the number of nuclei initially formed.

If cooling is very rapid, as in Figure 14, the rate at which supersaturation is created may exceed the rate at which it is discharged via crystal growth on the nuclei created at the first nucleation event, particularly if habit modifiers are also present which reduce the crystal growth rate. In this case, further nucleation events occur as shown, the second and subsequent crops of nuclei giving rise to successively smaller final crystals. When crystallisation is complete, the crystal size distribution will be wider than if there had been only one nucleation event, which can result in problems with the subsequent isolation of the crystals from the mother liquor. The wider size distribution gives rise to a less permeable cake

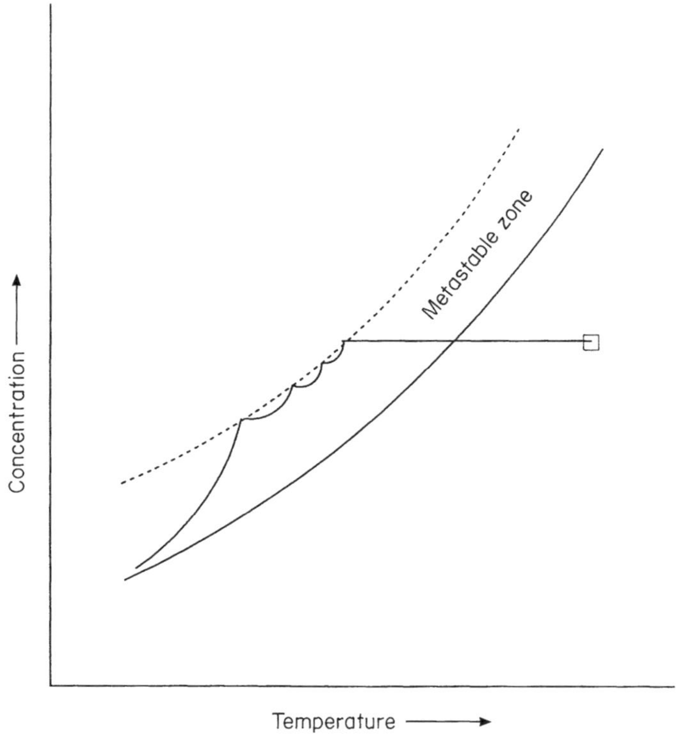

Figure 14 Supersaturation path for unseeded system with fast cooling. Several nucleation events give a wide distribution of very small crystals

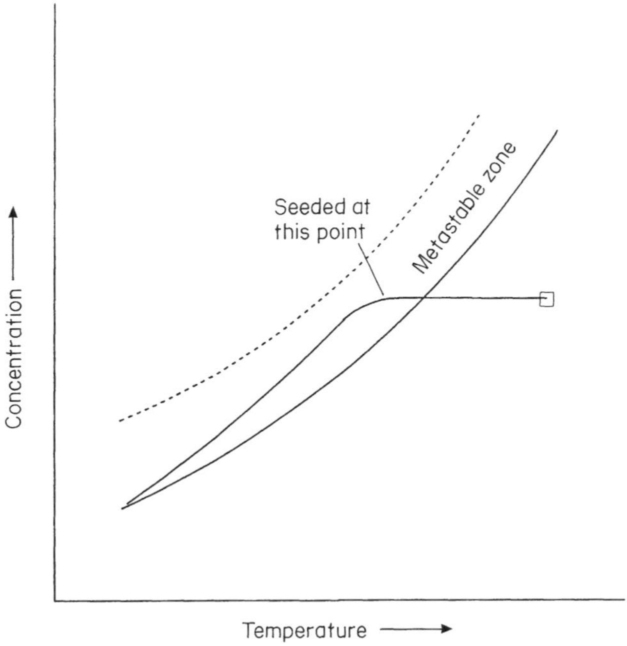

Figure 15 Supersaturation path for seeded system with slow cooling. No nucleation event; the crystal size depends on the number and size of seeds added

or blocks the filter medium, and such multiple nucleation should generally be avoided.

Figure 15 shows the case where the solution, cooled until it is supersaturated but remaining within the metastable zone, is then seeded with solute crystals and cooled slowly so that the system never self-nucleates. The resulting crystals will be of a uniform size which is dependent on the number of seed crystals added. In this way, much larger crystals may be formed than achievable by self-nucleation. Seeding may also be used to induce crystallisation of a particular metastable polymorph which would not normally nucleate under the crystallisation conditions. For this to be feasible, the metastable zones for both polymorphs must overlap and seeding must be carried out within this overlap region. This is discussed further in Section 7.3.6.

While addition of seeds on the laboratory scale is usually straightforward, on the plant scale this can be difficult. Also, the window of metastable zone where seed must be added is often rather narrow. Where control of crystal size rather than polymorph is desired, an alternative to the addition of seed crystals is the use of ultrasound. This has been shown to induce nucleation at lower supersaturations than necessary for natural nucleation, i.e. it narrows the metastable zone.[9] An example of the practical use of ultrasound in a chiral system is given in ref. 10.

The curves in Figures 13–15 are for a cooling crystallisation, but similar arguments can be made for systems where the supersaturation is created by addition of a non-solvent or by a pH change, for example the crystallisation of a phenol by addition of acid to a solution of the phenolate salt. In such systems it is very important to control the rate of addition of the non-solvent or the acid/alkali used to adjust the pH, and to ensure good agitation to obtain rapid mixing.

Where habit modifiers are being used, their effect on crystal size is not easily predictable. As they act by slowing the growth rate of different faces, after a given growth time the crystals will tend to be smaller than in the absence of modifiers. However, the modifiers may be so efficient at blocking the growth sites that they may inhibit the initial nucleation of the crystals. The resulting increase in supersaturation may then result in crash nucleation, producing very many more nuclei than from a normal nucleation event and smaller final crystals. Alternatively, the modifier may moderate the number of nuclei formed at nucleation which are capable of growth, leading to fewer, larger crystals.

Thus, the nucleation conditions can be used to control the crystal size and size distribution. Such control depends not only on measuring the solubility curve and the width of the metastable zone as described above, but also the rate of desupersaturation via crystal growth.

7.3.5 RATE OF DESUPERSATURATION

Desupersaturation of a supersaturated solution by crystal growth takes time, and this has to be taken into account when controlling a crystallisation process. In a given system, the rate of desupersaturation of the solution in a supersaturated dispersion of crystals depends both on the temperature and on the surface area of the crystals present on which growth is occurring. For a given degree of supersaturation, (the percentage by which the concentration exceeds the equilibrium solubility at a given temperature), as the temperature is raised the rate of crystal growth generally increases because the rates both of diffusion of solute species to the crystal surface and their re-orientation to fit into the growth sites increase. In addition, as the surface area of crystals present increases so the mass of solute that can crystallise in unit time also increases. Both of these effects increase the rate of desupersaturation.

Where the crystals have low solubility and the solute is created by reaction, as in the precipitation of a simple ionic salt, the rate of desupersaturation can be very fast. However, with molecular crystals, particularly where the molecules are large, half-lives for desupersaturation of several hours are known, even where there is a large area available for growth.

7.3.6 ISOLATION OF METASTABLE PHASES

If, in a polymorphic system, it is desired to isolate a phase which is metastable at ambient temperature, then there are several possible ways of achieving this depending on the system and, in particular, the solubility curves of the phases and their metastable zone widths. Where nucleation occurs via the creation of high supersaturations, it is often found that the predominant nucleating phase is a metastable phase. This empirical observation was first noted by Ostwald and was the basis of his 'Rule of Stages'.[11] In practice, the rule does not always hold,[12] and where it does it is because the kinetics of nucleation and crystal growth of the metastable phase are faster than those of the stable phase.

Even where it does hold, some nuclei of the stable phase are likely to be created. If nuclei of the stable crystal phase exist in a dispersion of a metastable phase, then in time the system will change so that only crystals of the stable phase remain. Therefore, while crystallisation via the creation of very high supersaturation may be worth trying, it does not guarantee lasting success. In addition, the crystals formed are likely to be very small. There is, however, an approach which is likely to be more lastingly successful and give larger crystals.

Consider first a system with a phase transition temperature (PTT) within normal operating temperatures, up to the boiling point of the solvent. Let us refer to the stable phase at lower temperatures as S and the metastable phase as M. The solubility curves and metastable zone widths versus temperature are shown qualitatively in Figure 16.

If controlled nucleation is made to occur above the PTT at temperature T_1, then the crystals which form will be the stable phase at that temperature, i.e. M. Phase S is very unlikely to nucleate because the solute concentration stays within

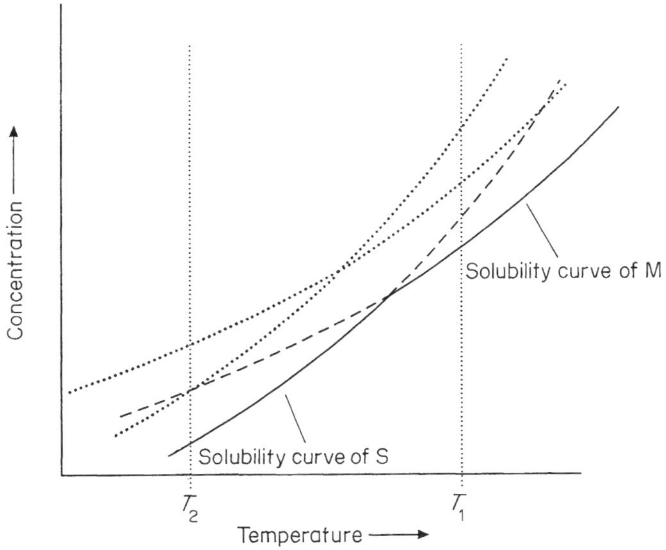

Figure 16 Solubility curves and metastable zones of a polymorphic system with a phase transition temperature

its metastable zone. If the system is then cooled slowly, at such a rate that the solute concentration remains only slightly above the solubility curve of M, then the temperature could be reduced to T_2 before the solution becomes sufficiently supersaturated with respect to S that S nucleates. Therefore, if the crystals are isolated above T_2, they should consist of pure M.

Consider now a system which does not have a PTT at a useful temperature. In order to isolate M from such a system, its solubility curve must lie within the metastable zone of S, as shown in Figure 17. So long as this is so, then it should be possible to seed the system with M seeds within the metastable zone of S and within the narrow band of conditions between the solubility curve of M and the metastable zone limit of S, and then apply controlled cooling to achieve growth of M while avoiding nucleation of S. This of course assumes a source of M seeds is available.

If the metastable zone of the stable phase S in a particular solvent does not overlap the solubility curve of M, as in Figure 18, then any saturated solution of M will lie outside the metastable zone of S, and so S will nucleate. Therefore, it is impossible to isolate only phase M from such a system. The only options are to investigate different solvent systems, to try to find one in which the solubility curve of M does lie within the metastable zone of S, or possibly search for, or design, a nucleation inhibitor for phase S, and hence widen its metastable zone.

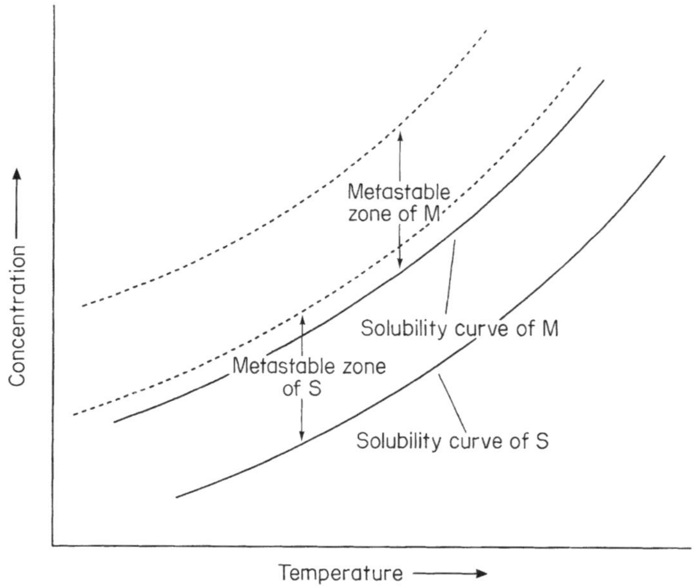

Figure 17 Case where solubility curve of M lies within the metastable zone of S. M may be isolated by seeding within the narrow region between the solubility curve of M and the metastable zone limit of S

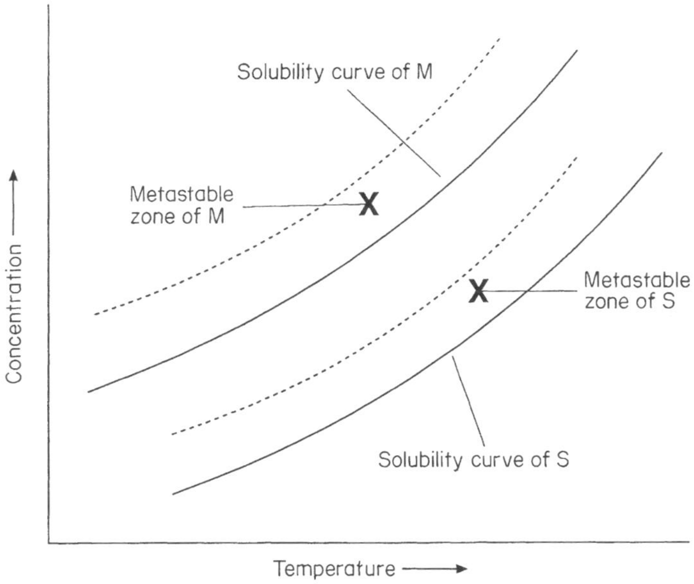

Figure 18 Case where solubility curve of M lies outside the metastable zone of S. M cannot be isolated by seeding

Occasionally, what was thought to be the stable phase at a given temperature turns out to be metastable with respect to a new phase which spontaneously nucleates for the first time, and thereafter isolation of the original metastable phase may prove difficult if not impossible. This can be a very serious problem, particularly in the pharmaceuticals area, where physical form may dramatically affect the bioavailability of the drug. The phenomenon of 'disappearing polymorphs' is discussed in ref. 13.

7.4 APPLICATION OF CRYSTAL SCIENCE TO CHIRAL SYSTEMS

As in the manufacture of non-chiral molecules, crystallisation is used routinely to separate chiral molecules from impurities formed during their synthesis. In addition, crystal science may be applied in the isolation of individual enantiomers from a racemic starting material, and this subject is covered in depth in the excellent book of Jacques *et al.*[14] This chapter is intended as an introduction, illustrating the various ways in which the science discussed above may be applied to chiral systems, with some specific examples.

A successful crystallisation process to separate a mixture of solutes depends on utilising or controlling one or more of the following properties of those solutes:

—different solubility profiles versus temperature or solvent composition;
—different metastable zone widths;
—their different rates of crystal growth.

For molecules with a single chiral centre, the opportunities to achieve a good separation of the enantiomers from a racemate by application of crystal science appear at first sight to be limited, their individual physical properties (melting points, latent heats of fusion and solubility in non-chiral solvents) being the same, as are their metastable zone widths and rates of crystal growth. If, however, a second chiral centre is introduced to form a diastereoisomer, then the opportunities increase.

7.4.1 SEPARATION OF ENANTIOMERS BY DIASTEREOISOMER FORMATION

The basis for the majority of racemate resolutions is fractional crystallisation of diastereoisomers, and the subject is reviewed in depth by Jacques *et al.*,[14] Bayley and Vaidya[15] and Bruggink in Chapter 5 of this volume.

In favourable cases, where the metastable zone width for the more soluble diastereoisomer is wide, the yield of the less soluble diastereoisomer need not be

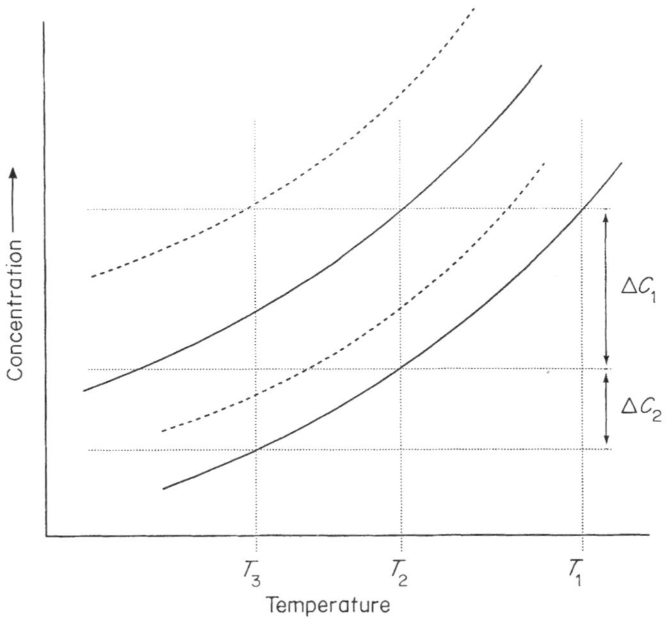

Figure 19 Solubility curves and metastable zones of two diastereoisomers. On crystallising a 50:50 mixture of diastereoisomers, the yield of the less soluble diastereoisomer can be increased by making use of the wide metastable zone of the more soluble diastereoisomer

limited solely by the difference in the solubilities of the two diastereoisomers. It should be possible to drive the crystallisation further, creating conditions where the more soluble diastereoisomer is supersaturated, but still within the metastable zone, so that it does not nucleate. Figure 19 shows the solubility curves and metastable zones of two diastereoisomers. For a 50:50 mixture of the diastereoisomers, a solution can be made which is just saturated in the less soluble diastereoisomer at temperature T_1. Cooling to T_2 will cause nucleation and growth of that diastereoisomer and give a crystallised yield equivalent to the concentration difference ΔC_1. It will also make the solution just saturated in the more soluble diastereoisomer. However, the solution can be cooled further to T_3, still within the metastable zone of the more soluble diastereoisomer, further increasing the yield of crystals of the other diastereoisomer by an amount equivalent to ΔC_2 and still avoiding nucleation of the more soluble diastereoisomer.

7.4.2 SEPARATION OF DIASTEREOISOMERS

An interesting example of this approach from the author's experience concerned the separation of the diastereoisomers of a molecule with two chiral centres. The structure of the molecule cannot be disclosed, for commercial reasons, but nevertheless it provides a good illustration of the application of crystal science in several ways.

The relative solubilities of two diastereoisomers [(*RR*′) and (*RS*′)] versus temperature in the chosen solvent are shown in Figure 20. From this solvent the diastereoisomers crystallise as solvates. Given the difference in solubility, by

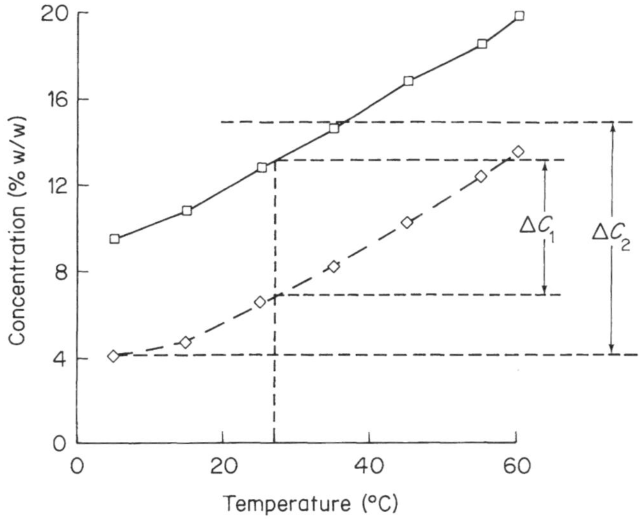

Figure 20 Solubility curves for (□)(*RR*′) and (◇)(*RS*′) diastereoisomers

Table 1 Yield and purity of (RR′) and (RS′) diastereoisomers from first crystallisation sequence

Form	Yield (%)	Purity (%)
(*RR′*)	77	98
(*RS′*)	49	98

making a solution at 60 °C containing 26.5% of the 50:50 mixture of the diastereoisomers, that solution will be just saturated in the (*RS′*)-isomer but well undersaturated in the (*RR′*)-isomer. If the solution is then cooled to 27 °C the solution at equilibrium will have become just saturated in (*RR′*), and (*RS′*) will have crystallised to the equivalent of ΔC_1. At temperatures lower than 30 °C, the (*RR′*) form will also be supersaturated.

However, the solution was prepared from non-solvated crystals of the mixed diastereoisomers and, since these are metastable with respect to the solvated crystals, they have a higher solubility. Therefore they continue to dissolve when added to a solution saturated with respect to the solvated (*RS′*) phase, and so a significantly supersaturated solution in (*RS′*) can be prepared containing 30% of mixed diastereoisomers. Also, it was found that the width of the metastable zone for (*RR′*) was so wide that the resulting solution could safely be cooled to 5 °C and held without (*RR′*) nucleating. The yield of (*RS′*) crystals could therefore be increased to the equivalent of ΔC_2. Desupersaturation of (*RS′*) from the solution proceeded with a half-life of 100 min, and so, once nucleated, the system had to be held at 5 °C for about 7–8 h before the concentration of (*RS′*) in solution reached equilibrium solubility.

The crystals of (*RS′*) were then filtered off and the liquors at 5 °C seeded with (*RR′*) seeds. The half-life for desupersaturation of (*RR′*) was found to be 10 h, so the system required holding for about 50 h to reach equilibrium before filtration! The isolated yields and purities of the products from this first crystallisation sequence are shown in Table 1.

If the final filtrates were then re-heated to 60 °C and further mixed diastereoisomers and solvent added to raise the concentration of (*RR′*) to its original value, the whole process could be repeated. Since the concentrations of (*RR′*) and (*RS′*) in the final filtrates will be the same as at the end of the first crystallisation sequence, it follows that, in theory at least, the isolated yields of both isomers should be 100%, based on the weights of isomers added at the start of the second crystallisation sequence. In practice, the yields were found to be (*RR′*) 98% and (*RS′*) 95%, the main causes of yield loss being handling and washing of the filtered crystals.

There is a limit to the number of times that the filtrates may be recycled in such a process before undesirable side-effects are observed. This is because impurities build up in the filtrates, which may begin to interfere with the habit or size of the crystals or their growth rates. Even where these effects are absent, the concentration of the impurities will eventually rise to a point where they themselves crystallise, reducing the purity of the isolated products.

A more sophisticated example of the isolation of a desirable diastereoisomeric pair with biological activity from a pair with no activity is provided by the process for the synthetic pyrethroid cyhalothrin (Figure 21). The synthetic route[16] produces four of the eight possible isomers of this molecule (those with a *cis* configuration at the cyclopropane ring), which comprise two diastereoisomeric pairs known as *cis* A and *cis* B. The insecticidal activity resides only in the *cis* B pair.

During the development of this product, it was found that on cooling a solution of all four isomers, crystals were obtained each of which contained equal amounts of the *cis* B pair of enantiomers (an example of a racemic compound crystal; see below). The crystal growth rate was very slow, and the process was complicated by the fact that the non-crystallising pair of isomers acted as a solvent for the crystallising pair, leading to the observation that the crystallisation rate fell on increasing the initial concentration.

However, the hydrogen atom on the chiral carbon bearing the cyano group is fairly acidic and proton exchange occurs readily in the presence of base. As reprotonation can occur from either side of the planar carbanion, the presence of

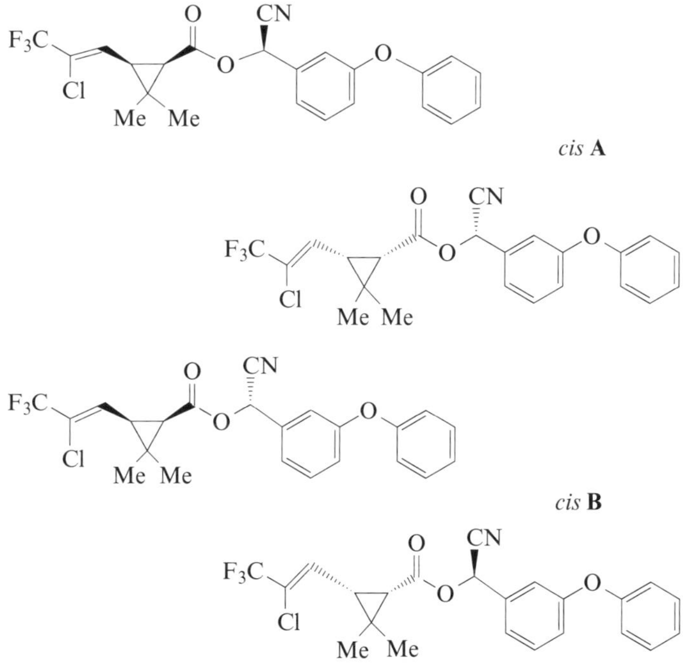

Figure 21 The *cis* A and *cis* B diastereoisomeric pairs of isomers of cyhalothrin

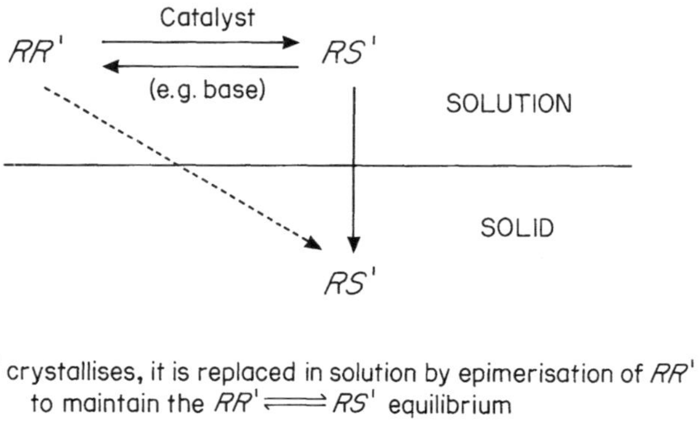

Figure 22 Second-order asymmetric transformation

base causes epimerisation of the diastereoisomers at this chiral centre. If crystallisation is carried out in the presence of a base, then, as the active diastereoisomer pair *cis* B crystallises, so the remaining diastereoisomers in solution epimerise to maintain the equilibrium and help to maintain the concentration of *cis* B in solution. The process is known as a second-order asymmetric transformation (see Volume I, p. 26) and is shown diagramatically in Figure 22.

This approach may also be used in resolutions for diastereoisomeric salts, where the chemistry of the target chiral molecule allows *in situ* racemisation during crystallisation. An example is described by Yoshioka *et al.*,[17] involving the resolution of DL-*p*-hydroxyphenylglycine with (+)-1-phenylethanesulphonic acid. Epimerisation was achieved by holding the solution/dispersion of the diastereoisomers at 100 °C in acetic acid containing some salicylaldehyde. After 5 h the slurry was cooled and filtered, giving a yield of 85% of the D-(+)-diastereoisomer, based on the initial weight of the DL-(+)-form.

7.4.3 RESOLUTION BY DIRECT CRYSTALLISATION OF ENANTIOMERS

Let us now consider the use of crystal science to obtain single enantiomers from racemates directly. As mentioned above, enantiomers have identical solubilities and metastable zone widths in non-chiral solvents. Therefore, if one is to crystallise one enantiomer but not the other, one has either to work within the limits of the metastable zone, seeding with crystals of an individual enantiomer, or else use chiral nucleation inhibitors to widen the metastable zone of one of the enantiomers.

An additional complication is that when a racemate crystallises, the crystals formed can be of two types (Figure 23). The more common is where the enantiomers co-crystallise to form a mixed crystal containing both enantiomers in equal proportions, known as a racemic compound.

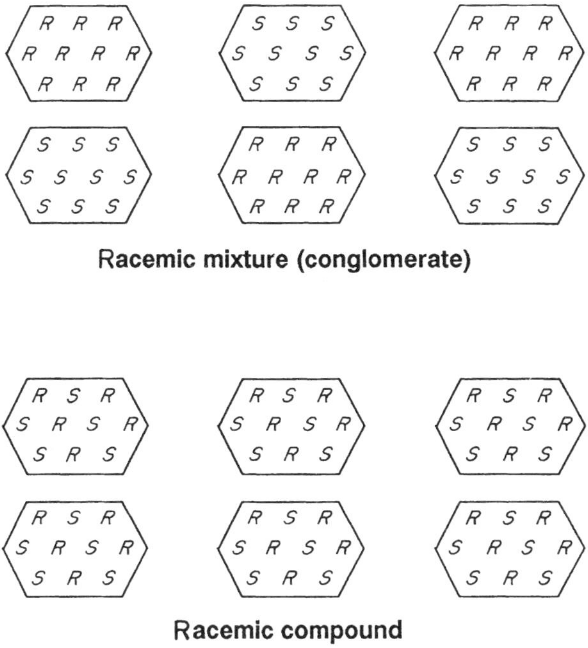

Figure 23 Crystal forms of a racemate

If a racemate produces only this type of crystal, crystallisation may not be used to separate the enantiomers except by formation of a diastereoisomer. Seeds of pure enantiomer added to a supersaturated solution of such a system within the metastable zone will simply dissolve.

Crystal science may be applied, however, in a minority of systems (perhaps 10%) in which the racemate crystallises to give crystals in which the lattice contains either one or the other enantiomer. These are referred to as racemic conglomerate (or sometimes racemic mixture) crystals. The crystals may have an asymmetric habit (although not always[18]), and where this is the case the (*R*) and (*S*) crystals will be mirror images of each other. Louis Pasteur performed the first resolution of a racemate in 1848[19] by manually sorting out the (*R*) and (*S*) crystals of sodium ammonium tartrate, although this approach is hardly an industrial procedure!

It is important, therefore, to determine whether a racemate crystallises as racemic compound or conglomerate crystals. Several methods may be used, but all rely on having a sample of one of the enantiomers in a relatively pure crystalline form, ideally recrystallised from the same solvent system and under similar conditions as the racemate. Probably the most widely used method is to compare the infrared spectra of the crystals of the racemate and of a pure enantiomer in the solid state. If the spectra are identical, then the racemate is a con-

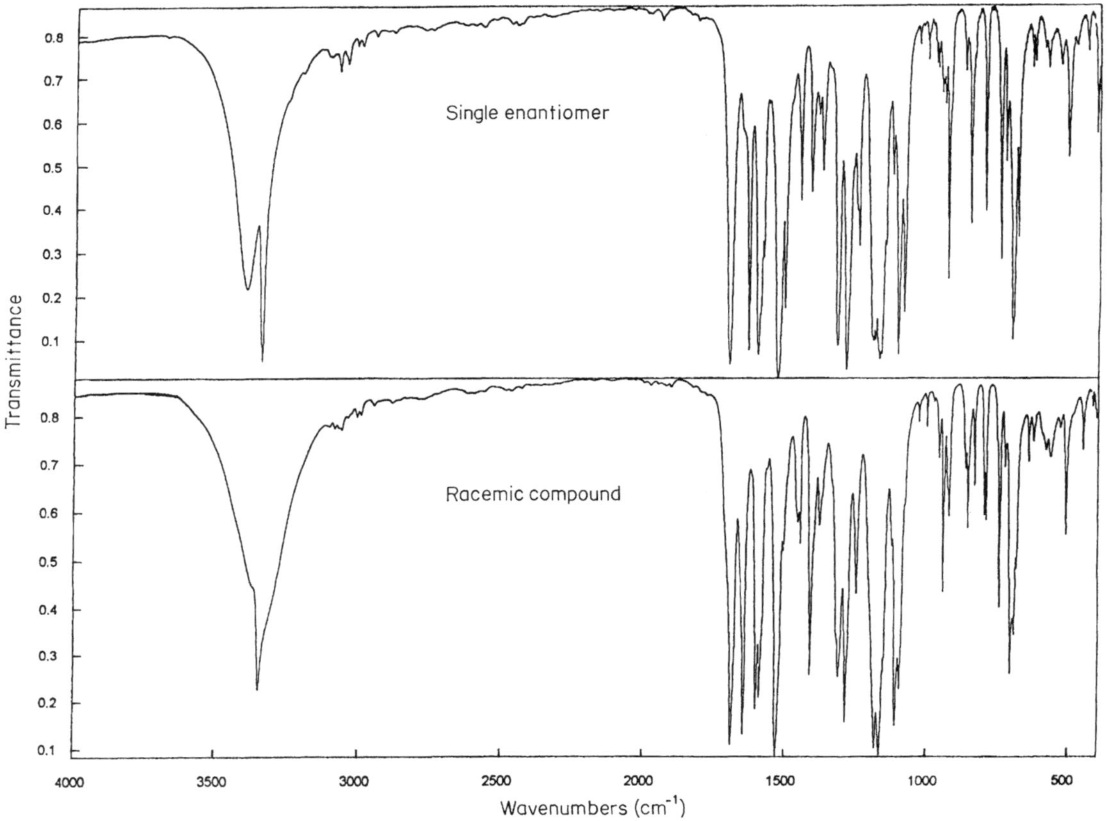

Figure 24 IR spectra of single enantiomer and racemic compound crystals of a development drug

glomerate. If the spectra are different, then it is most likely that the racemate forms compound crystals (although it is also a possibility that the racemate is a conglomerate, but in a different polymorph or solvate phase from the pure enantiomer). To minimise this possibility it is important that the crystal samples should be isolated under similar conditions. Figure 24 shows the IR spectra of crystals of a single enantiomer and of a racemic compound of a development pharmaceutical. Although the spectra are very similar, there are several significant differences.

Another method is to compare the powder XRD patterns of the racemate and pure enantiomer crystals. If they are identical, then the racemate is a conglomerate. If they are different, then again it is most likely, although not certain, that the racemate gives compound crystals.

If the crystals melt without decomposing, another method which is particularly useful for compounds that are liquid at room temperature, and which has not been reported before now, is to measure the heat of fusion of the racemate and single enantiomer crystals using DSC. For example, the latent heat of fusion of the pure L-enantiomer of 2-chloropropionic acid, MeCHClCOOH, was measured in a sub-ambient DSC and found to be about 84 $J\,g^{-1}$. The value for the racemate was about 149 $J\,g^{-1}$. Now, if the racemate was a racemic conglomerate, its latent heat of fusion in $J\,g^{-1}$ would be expected to be equal to that of the single enantiomer, less some heat of mixing of the enantiomers in the liquid phase. That its heat of fusion is almost double that of the single enantiomer indicates that the racemate cannot be a conglomerate, but rather a racemic compound.

It should be noted that there may be a phase transition temperature on one side of which the racemate crystallises as a conglomerate and on the other as compound crystals. Sodium ammonium tartrate has such a transition temperature at 28 °C. Below this temperature, it forms conglomerate tetrahydrate crystals. Between 28 °C and 35 °C it forms compound monohydrate crystals, while above 35 °C it forms separate anhydrous compound crystals of sodium tartrate and ammonium tartrate. It was fortuitous that Pasteur's final crystallisation temperature was below 28 °C! There are also cases[14] where conglomerate crystals are the stable phase above a transition temperature, specifically, where the conglomerate crystals have a lower degree of solvation than the compound crystals.

As described earlier, the existence of such a phase transition temperature may be detected from discontinuities in the solubility versus temperature curve, and confirmed by XRD, DSC and TGA on the crystals present at equilibrium above and below the transition temperature.

It is good practice to resolve the racemate at as early a stage as possible in the synthesis, assuming the subsequent chemistry is stereospecific. The intermediates formed in the steps after the chiral centre has been introduced into the molecule, as well as the final product, can all be checked to see whether they form racemic conglomerate crystals. Assuming a racemate is found which gives conglomerate crystals, crystallisation techniques may be applied to resolve it.

7.4.3.1 Seeding Within the Metastable Zone

The basis of one approach, which has been used commercially by Merck to resolve an α-methyldopa intermediate, was first described by Jungfleisch[20] in 1882. He showed that on careful seeding of a supersaturated solution of sodium ammonium tartrate (at a temperature below 2 °C) within the metastable zone (which is wide for this system) with one seed crystal each of the pure enantiomers, these grew without further nucleation to give two large crystals which were enantiomerically pure. Modern practice is to create a supersaturated solution within the metastable zone and then seed with the pure enantiomers in separate crystallisers using the equipment shown diagramatically in Figure 25.

In each crystalliser the supersaturation is controlled so that it remains within the metastable zone and nucleation of the other enantiomer does not occur. Care must be taken that seeds of the unwanted enantiomer are not inadvertantly introduced into the crystallisers via the feed solution. This approach works particularly well for the α-methyldopa intermediate because the enantiomers have a very wide metastable zone. The temperature difference between the dissolver and the crystallisers is quoted as 60 °C.[21] Where the metastable zone is narrow, the control of supersaturation becomes very critical.

While it is not feasible on an industrial scale to sort mirror image racemic conglomerate crystals of the same size, they might be sorted by sieving if the crystals differ significantly in size. This is the basis of a process first described by Dowling,[22] where the crystals of different sizes are obtained by seeding a supersaturated solution simultaneously with large crystals of one enantiomer and very small crystals of the other. Dowling applied the method to glutamic acid salts, while Watanabe and Noyori[23] applied it to acetylglutamic acid.

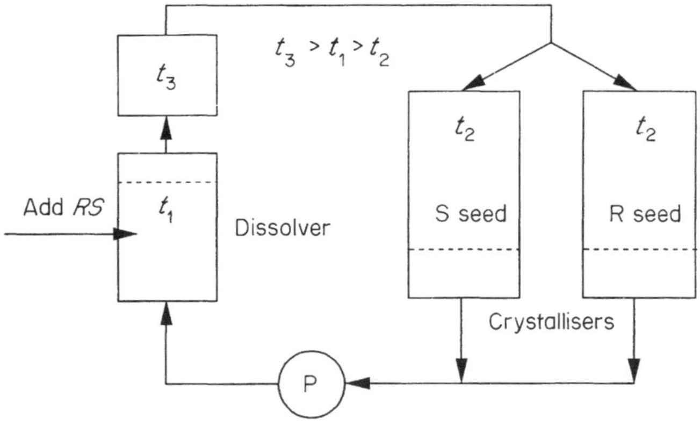

Figure 25 Process for resolution by seeding with individual enantiomers within metastable zone (after Merck process for an α-methyldopa intermediate)

If the chemistry of the enantiomers permits an easy *in situ* racemisation under crystallisation conditions, then the opportunity exists to use a second-order asymmetric transformation process as described above for diastereoisomers. For example, in the synthesis of the agrochemical paclobutrazol (Figure 26), the penultimate stage involves the formation of a racemic chiral ketone. Since the reduction of this ketone to the final product using borohydride is completely stereoselective, then if the ketone is resolved to obtain the (*S*) enantiomer, the final product will be the desired (2*S*,3*S*)-diastereoisomer.

It was found that crystals grown from aqueous methanol were conglomerate crystals[24] and that the metastable zone width was about 5 °C. In addition, in the presence of base, the ketone is in equilibrium with the enolate form and racemisation is very rapid. Therefore, the system is a good candidate for seeding to produce either desired enantiomer.

Although the ability to form racemic conglomerate crystals is a necessary requirement for successful resolution of a racemate by direct crystallisation, it

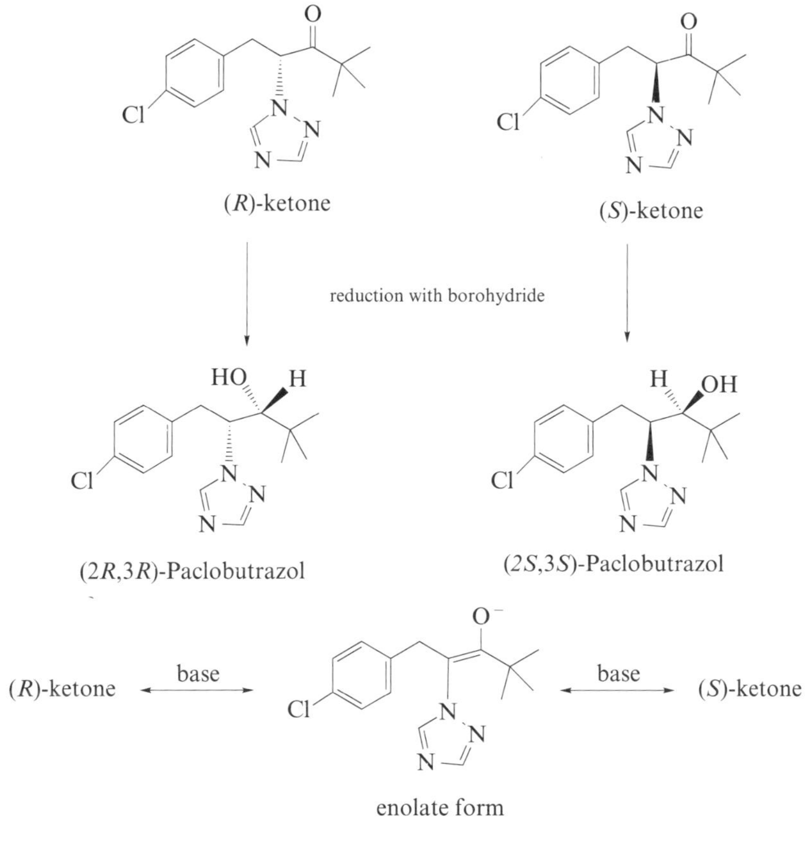

Figure 26 Paclobutrazol chemistry

does not guarantee it. Some conglomerates, while giving crystal lattices containing only one enantiomer, form crystals which contain domains of both lattices in a single crystal. This was found to be the case with the above ketone, if high supersaturations within the metastable zone were used. Apparently, enantiomerically pure seed was nucleating the other enantiomer at its surface. Only when a low supersaturation, given by 1 °C of undercooling, was used did this effect disappear to give the product with a high enantiomeric purity. This behaviour at higher supersaturation is consistent with observations on β-phenylglyceric acid.[25]

When this aspect became understood, the process was changed from a cooling crystallisation to an evaporative crystallisation, allowing much closer control of the supersaturation during the process.

7.4.3.2 Nucleation Inhibition of One Enantiomer

The last crystallisation technique for resolving a racemate that will be discussed involves the use of additives to inhibit selectively the nucleation and growth of one of the enantiomers. As discussed in Section 7.2.3, such inhibitors or habit modifiers should be similar in structure to the target crystallising molecule. Given that it should affect one enantiomer but not the other, it follows that the agent must itself be chiral.

This approach has been investigated in detail[26] for certain amino acids. Thus, if a racemic solution of DL-asparagine is crystallised in the presence of D-glutamic acid, the first crystals to appear are L-asparagine. The metastable zone for the D-asparagine has effectively been widened due to the nucleation-inhibiting effect of the D-glutamic acid. The design of more potent polymeric chiral inhibitors has been described.[27]

It has been mentioned several times that the solubility of enantiomers is the same in non-chiral solvents. In a single enantiomer chiral solvent, however, there is the possibility for the solubilities to be slightly different. More importantly, chiral solvents can themselves sometimes inhibit nucleation and/or growth of one enantiomer relative to the other, and the successful resolution of enantiomers by direct crystallisation has been reported using isopropyl D-tartrate[28] and (–)-α-pinene[29] as the chiral solvents.

7.5 CONCLUSION

The successful application of crystal science in the isolation of chiral compounds depends very much on the whims of Nature in determining, for example, whether diastereoisomeric salts can be prepared with suitably large differences in solubility and/or metastable zone width, or whether a racemate crystallises as racemic compound or conglomerate crystal forms from a chosen solvent system. The likelihood of success may be improved by judicious choice of the stage in the route at which resolution is attempted, bearing in mind particularly the advantages offered by second-order asymmetric transformations, where they are feasible.

Development of a successful process depends on finding out just how generous Nature has been, by measuring solubility curves in a range of solvents, the metastable zone widths, and the crystal phase(s) present at various temperatures and by checking the crystal habits. Crystal science can then be applied to optimise the process to ensure the isolation of the desired crystal phase as uniform, easily isolated crystals. Where Nature is kind, or accepts a helping hand, crystal science can provide the basis for very effective processes in the manufacture of chiral compounds.

7.6 REFERENCES

1. Mullin, J. W., Crystallisation, 3rd edn, Butterworth-Heinemann, Oxford, 1993.
2. Haisa, M., Kashino, S., and Maeda, H., *Acta Crystallogr., Sect. B*, **30**, 2510 (1974).
3. Haisa, M., Kashino, S., Kawai, R., and Maeda, H., *Acta Crystallogr., Sect. B*, **32**, 1283 (1976)
4. Fagan, P. F., Roberts, K. J., Docherty, R., Charlton, A. P., Jones, W., and Potts, G. D., *J. Chem. Soc., Chem. Commun.*, in press.
5. Glusker, J. P., Lewis, M., and Rossi, M., *Crystal Structure Analysis for Chemists and Biologists*, VCH, New York, 1994.
6. Davey, R. J., Cardew, P. T., McEwan, D., and Sadler, D. E., *J. Cryst. Growth*, **79**, 648 (1986).
7. McNaughton, J. L., and Mortimer, C. T., in *IRS*; *Physical Chemistry Series 2*, Butterworths, London, 1975, Vol. 10; subsequently reprinted by Perkin-Elmer Corpo., Norwalk, CT, USA, from whom it may be ordered, code PE 0993–8669.
8. See, e.g. Weast, R. C. (Ed.), *Handbook of Chemistry and Physics*, 58th edn, CRC Press, Cleveland, OH, 1977.
9. Martin, P. D., Phillips, E. J., and Price, C. J., in *1993 I. Chem. E. Research Event*, University of Birmingham, Birmingham, 1993, pp. 516–518, ISBN 0 852 95308 9).
10. Noguchi Research Foundation, *Fr. Pat.*, 1 389 840, 1965. *US Pat.*, 3 365 492, 1968.
11. Ostwald, W., *Z. Phys. Chem.*, **22**, 289 (1897).
12. Cardew, P. T., and Davey, R. J., in *Proceedings of Symposium on Tailoring of Crystal Growth*, (ed. by J. Garside), I. Chem. E. NW Branch, Manchester, 1982, pp. 1.1–1.7.
13. Dunitz, J. D., and Bernstein, J., *Acc. Chem. Res.*, **28**, 193 (1995).
14. Jacques, J., Collet, A., and Wilen, S. H., *Enantiomers, Racemates and Resolutions*, Wiley–Interscience, New York, 1981.
15. Bayley, C. R., and Vaidya, N. A., in *Chirality in Industry* (ed. A. N. Collins, G. N. Sheldrake and J. Crosby), Wiley, Chichester, 1992, pp. 69–77.
16. Robson, M. J., Cheetham, R., Fettes, D.J., and Crosby, J., *1984 British Crop Protection Conference Proceedings*, BCPC Publications, Croydon, 1984, Vol. 3, pp 853–857.
17. Yoshioka, R., Tohyama, M., Ohtsuki, S., Yamada, S. and Chibata, I., *Bull. Chem. Soc. Jpn.*, **60**, 649 (1987).
18. Addadi, L., Berkovitch-Yellin, Z., Weissbuch, I., Lahav, M., and Leiserowitz, L., *Top. Stereochem.*, **16**, 1 (1986).
19. Pasteur, L., *C. R. Acad. Sci.*, **26**, 535 (1848).
20. Jungfleisch, M. E., *J. Pharm. Chim., 5ème Ser.*, **5**, 346 (1882).
21. *Chem. Eng.*, 247 (1965).
22. Dowling, B. B., *US Pat.* 2 898 558, 1959.

23. Watanabe, T., and Noyori, G., *Kogyo Kagaku Zasshi*, **72**, 1083 (1969).
24. Black, S. N., Williams, L. J., Davey, R. J., Moffatt, F., Jones, R. V. H., McEwan, D. M., and Sadler, D. E., *Tetrahedron*, **45**, 2677 (1989).
25. Furberg, S., and Hassel, O., *Acta Chem. Scand.*, **4**, 1020 (1950).
26. Addadi, L., Berkovitch-Yellin, Z., Domb, N., Garti, E., Lahav, M., and Leiserowitz, L., *Nature* (*London*), **296**, 21 (1982).
27. Zbaida, D., Weissbuch, I., Shavit-Gati, E., Addadi, L., Leiserowitz, L., and Lahav, M., *React. Polym.*, **6**, 241 (1987).
28. Luttringhaus, A., and Berrer, D., *Tetrahedron Lett.*, 10 (1959).
29. Groen, M. B., Schadenberg, H., and Wynberg, H., *J. Org. Chem.*, **36**, 2797 (1971).

8 Membrane Separations in the Production of Optically Pure Compounds

J. T. F. KEURENTJES and F. J. M. VOERMANS
Akzo Nobel Central Research, Arnhem, The Netherlands

8.1 Introduction . . . 157
8.2 Enantioselective Membranes . . . 158
  8.2.1 Selective Polymers . . . 158
  8.2.2 Selective Liquid Membranes . . . 162
8.3 Membrane-assisted Separations . . . 168
  8.3.1 Liquid–Liquid Extractions . . . 168
  8.3.2 Liquid-membrane Fractionation . . . 171
  8.3.3 Micellar-enhanced Ultrafiltration (MEUF) . . . 174
  8.3.4 Membrane-assisted Kinetic Resolutions . . . 175
8.4 Conclusion . . . 178
8.5 References . . . 179

8.1 INTRODUCTION

In the past two decades, the range of conventional separation techniques, such as distillation, crystallisation and extraction, has been extended by a wide range of membrane separation processes. Traditionally, membranes are found in biomedical applications (e.g. blood dialysis and plasmaphoresis[1]) and reverse osmosis (e.g. desalination[2]). However, material developments over the past 15 years have opened the way to many industrial applications. Membrane separations often provide opportunities as cost-efficient alternatives to separations which are troublesome or impossible using classical methods and, since most membrane processes are performed at ambient temperature, they can offer clear advantages compared with conventional separation processes.[3,4]

There are two general types of membrane processes for enantiomer separations: either direct separation using an enantioselective membrane, or separation in which a non-selective membrane assists an enantioselective process. Obviously, the most direct method is to apply enantioselective membranes which provide a selective barrier, allowing preferential transport of one of the enantiomers of a

Chirality in Industry II. Edited by A. N. Collins, G. N. Sheldrake and J. Crosby

racemic mixture. These membranes can be of either the dense polymer or liquid type. In the latter case a selective carrier is usually added to the liquid membrane. Non-selective membranes can also provide essential non-chiral separation characteristics, thus making a chiral separation based on enantioselectivity outside the membrane technically and economically feasible. Several configurations can be envisaged for this purpose:

(1) hollow-fibre membrane fractionation (liquid–liquid extraction);
(2) liquid-membrane fractionation;
(3) micellar-enhanced ultrafiltration (MEUF);
(4) membrane-assisted kinetic resolution.

In this chapter, an overview is given of the (potential) application of membrane separation processes in the separation of racemic mixtures; although not exhaustive, it should give an impression of the developments in this field. Application of the different process options will be discussed with special regard to scale operations. Many of the systems described in this chapter still are in an early stage of development whereas others are currently being scaled up to pilot plant or are already applied at full scale.

8.2 ENANTIOSELECTIVE MEMBRANES

8.2.1 SELECTIVE POLYMERS

Enantioselective polymer membranes usually consist of a non-selective porous support coated with a thin layer of an enantioselective polymer. The selective layer is made from a chiral polymer or a polymer containing chiral side groups, such composite materials having the mechanical strength to withstand a pressure gradient over the membrane. The selectivity of such a membrane is determined by enantiospecific interactions between the isomers to be separated and the top-layer polymer matrix, resulting in one enantiomer being selectively dissolved in the membrane. This separation mechanism is termed 'solution/diffusion,' which is similar to the mechanism found in membrane processes like reverse osmosis and pervaporation.[5] In principle, these membranes can be used in a similar process configuration to reverse osmosis (RO) or ultrafiltration (UF) membranes, and a pressure difference across the membrane is the driving force for transport. Scale-up of the system to the desired productivity is relatively easy, as RO and UF are well established technologies with well known guidelines for operation on a large scale. When a membrane does not yield the required optical purity in a single step, a cascade of membrane units can result in a feasible process.[6]

The performance of an enantioselective membrane is dependent on two factors: the permeability P (flux) and enantioselectivity (α), the latter being defined as the ratio of the permeabilities of the L- and D-enantiomers:

$$\alpha = \frac{P_{\mathrm{L}}}{P_{\mathrm{D}}} \tag{1}$$

The search for enantioselective polymers has been mainly in the context of the separation of racemic amino acids. The polymers commonly used as stationary phases in chromatography can be used as starting points, e.g. polysaccharides (especially cellulose derivatives), acrylic polymers, poly(α-amino acids) and poly-acetylene-derived polymers.

Poly-L-glutamate with amphiphilic side-chains (**1**) forms an α-helix, which is selectively permeable for water-soluble optical isomers, such as α-amino acids.[7–10] Dialysis experiments (with no trans-membrane pressure) showed that the resolution of LD-tryptophan can be almost complete (Figure 1), but that the flux

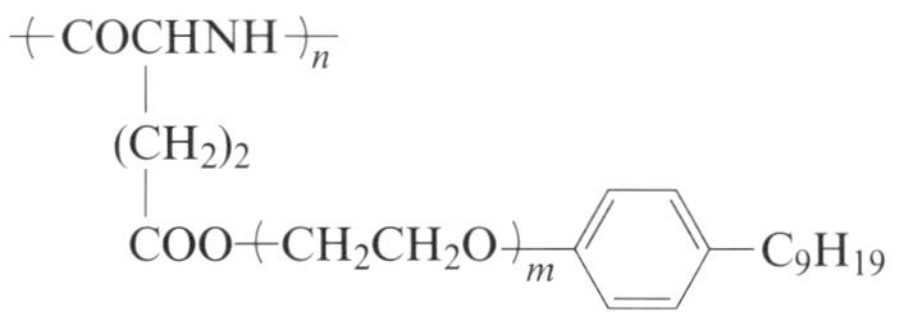

(**1**)

n-Nonylphenoxyhexa(oxyethylene)-1-ol poly(γ-methyl-L-glutamate) (NON6-PLG)

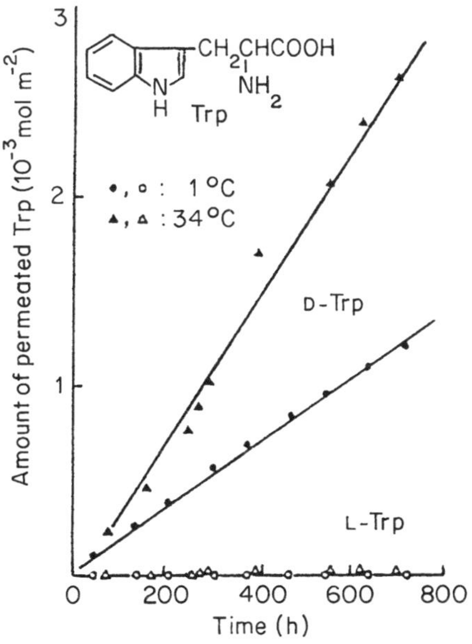

Figure 1 Permeation behaviour of D- and L-tryptophan from the racemic mixture through poly-L-glutamate-derived membranes (NON-6-PLG) in a dialysis configuration. Reproduced with permission from ref. 9

through these membranes is extremely low (of the order of 3×10^{-6} mol m^{-2}h^{-1} at 2 mmol l^{-1} in the feed). Similar fluxes, but lower selectivities, were reported for the separation of DL-tyrosine ($\alpha = 6.4$[7]) and DL-serine ($\alpha = 3$[9]). Polypropene membranes with (−)-β-pinene side groups [PDPSP (**2**)][11] exhibit a higher selectivity ($\alpha = 13.4$) for the resolution of DL-tryptophan than the poly-L-glutamate-derived membranes when a dialysis configuration is used. An increase in the trans-membrane pressure to 8 bar results in a significant increase in the permeation rate (*ca.* 7×10^{-5} mol m^{-2} h^{-1}, assuming the membrane is 10 μm thick), but the selectivity values approach unity. When a membrane is made from a blend of PDPSP and PTMSP {poly[1-(trimethylsilyl)prop-1-yne]} a pressure-resistant membrane results, yielding $\alpha = 2.3$ at 8 bar (flux *ca.* 5.4×10^{-4} mol m^{-2}h^{-1}, assuming a 10 μm thick membrane).

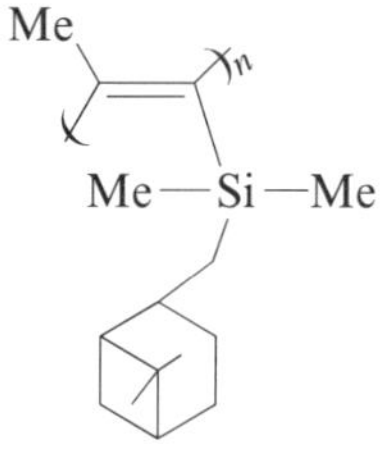

(**2**)

Poly{1-[dimethyl(10-pinanyl)silyl]prop-1-yne} (PDPSP)

To improve the permeability of the derivatised poly-L-glutamate membranes[7] and the PDPSP membranes,[11] alternative membranes consisting of a chiral backbone [poly(methyl-L-glutamate)] and flexible short siloxane side-chains have been synthesised by Aoki *et al.*[12] Permeation rates increase significantly to 5×10^{-4}–6×10^{-3} mol m^{-2} h^{-1} at 10 bar trans-membrane pressure (assuming a 10 μm thick membrane), although the selectivity values drop to 1.2–1.35.

A different approach is to immobilise the chiral component in a porous membrane. For example, an amino acid can be immobilised in the pores of a polysulphone UF membrane (Figure 2a).[13] Depending on the trans-membrane pressure, selectivities for the separation of DL-phenylalanine were between 1.25 and 4.1, at permeabilities between 10^{-6} and 10^{-7} m s^{-1}, respectively (Figure 2b). A rather dilute feed stream is used (1–4 mol m^{-3}), and this results in low solute fluxes. Although with these selectivity values many cycles are required to obtain high enantiomeric ratios, fluxes are at a level commonly found for reverse osmosis.[5] In another case, bovine serum albumin (BSA) has been immobilised on a polysulphone UF membrane which is selective for amino acids, and selectivities of 1.2 have been obtained with a corresponding flux of 0.016 mol m^{-2}h^{-1}.[14]

Only a few non-amino acids have been resolved using enantioselective polymer membranes. Examples are the β-blocker oxprenolol (**3**),[15] stilbene oxide (**4**)[16] and butan-2-ol,[11] using cellulose tris(3,5-dimethylphenylcarbamate) (**5**), tri-(*p*-chlorophenyl)carbamate (**6**) and PDPSP (**2**) membranes, respectively.

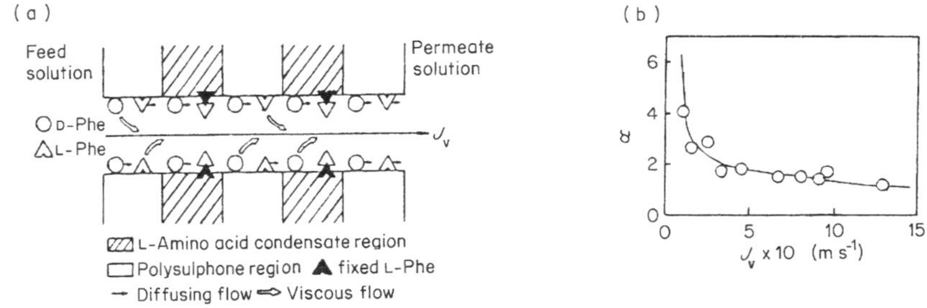

Figure 2 (a) Transport mechanism for the separation of DL-phenylalanine by an enantioselective UF membrane. (b) Effect of volume flux on the separation factor when DL-phenylalanine is separated by an enantioselective UF membrane (T = 37 °C). J_v = volume flux.

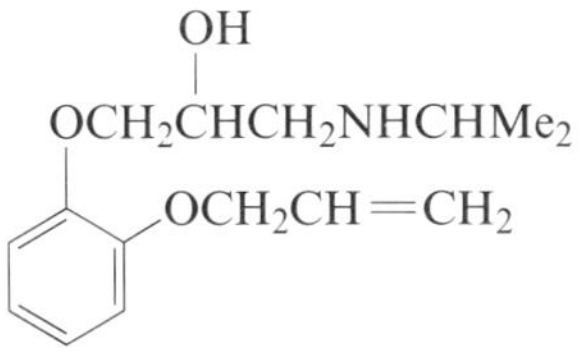

(3)
Oxprenolol

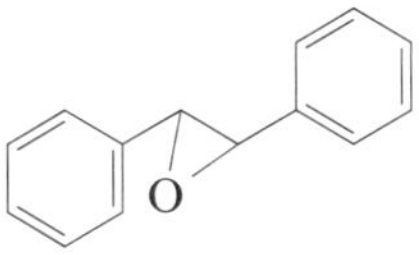

(4)
Stilbene oxide

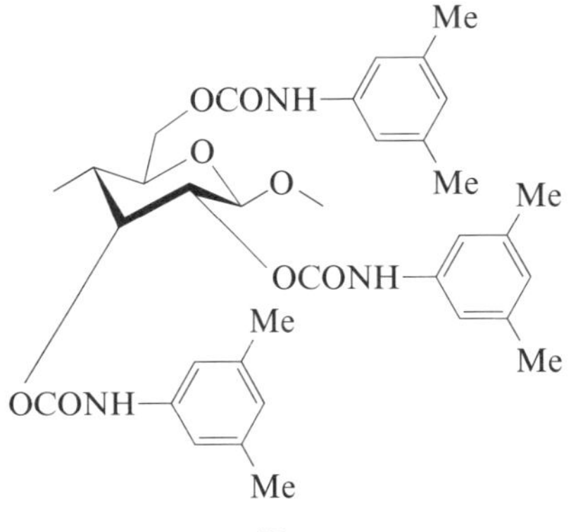

(5)
Cellulose tris(3,5-dimethylphenylcarbamate)

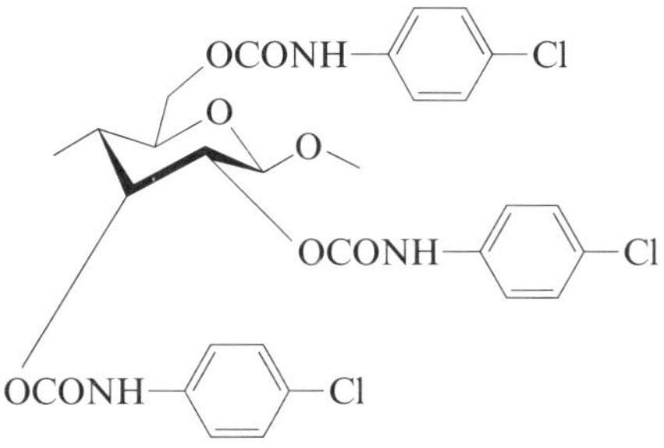

(6)
Cellulose tris(*p*-chlorophenylcarbamate)

On examining the data for enantioselective polymer membranes, it is clear that most experiments have been performed on the laboratory scale, using membrane surface areas typically in the range 1–100 cm^2. Most experiments have been performed using a dialysis type of experiment, and the selectivity thus obtained represents an 'intrinsic' property of the membrane. When the flux is increased by increasing the trans-membrane pressure, this usually results in a sharp decrease in selectivity. Although the selectivities reported so far are encouraging, large-scale application of these membranes does not seem very likely in the short term.

8.2.2 SELECTIVE LIQUID MEMBRANES

A liquid film can be immobilised in the pores of a membrane by capillary forces. Two miscible liquids that do not wet the porous membrane can then be kept apart by the immobilised film. This liquid membrane may contain an enantioselective carrier which selectively forms a complex with one of the enantiomers of a racemic mixture at the feed side, and transports it across the membrane, where it is released into the receptor phase (Figure 3). To avoid leakage of the carrier from the liquid membrane, it must not dissolve in the feed liquid or the receptor phase, and to achieve sufficient selectivity it is necessary to minimise non-selective transport through the bulk of the membrane liquid. Transport of the enantiomers to be separated can be driven by a gradient in concentration or pH.

The simplest way to apply enantioselective liquid membranes is to use a chiral liquid, e.g. chiral alcohols [nopol (**7**) and (*S*)-(−)-2-methylbutan-1-ol] immobilised in the pores of a polyethylene film for the separation of amino acid hydrochlorides.[17] Nevertheless, most methods for enantioselective transport across a

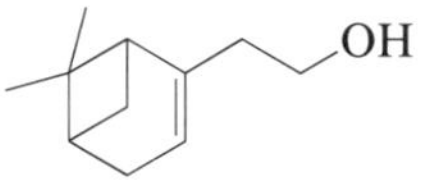

(**7**)

Nopol

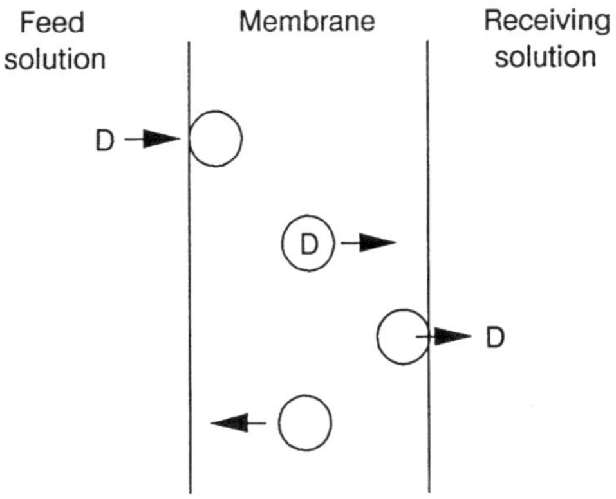

Figure 3 Schematic representation of a liquid membrane. The carrier selectively forms a complex with one enantiomer at the feed side and releases it in the receiving solution

liquid membrane make use of chiral selector molecules added to an achiral liquid membrane, and an obvious advantage of these systems is the possibility of tailoring the selector molecules to the application.

Despite the fact that cyclodextrins (**8**) are used extensively in gas chromatography,[18] liquid chromatography[19] and capillary electrophoresis,[20] they have been little used as selective carriers in liquid membranes. Cyclodextrins are expected to be selective in transporting cyclic and heterocyclic compounds,[19] e.g. racemic 1-ferrocenylethylphenylsulphide (**9**), which has been resolved using β-cyclodextrins at a permeation rate of 10^{-3} mol $m^{-2}h^{-1}$.[21]

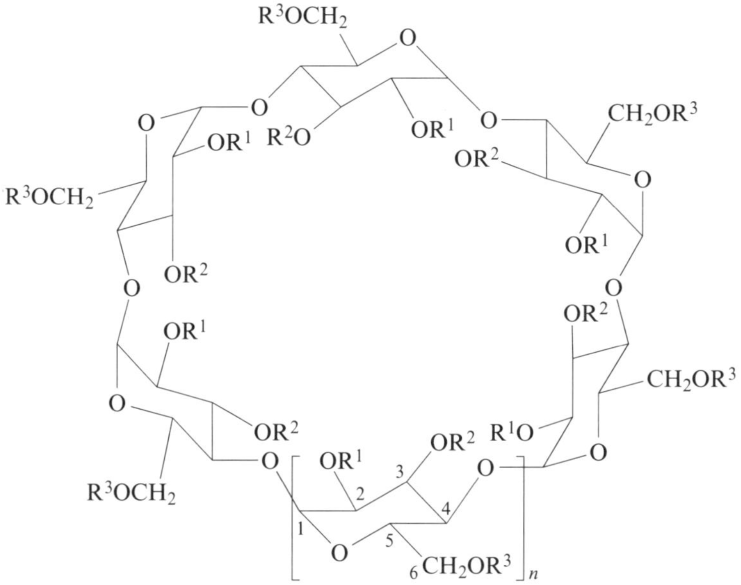

(**8**)

Cyclodextrin

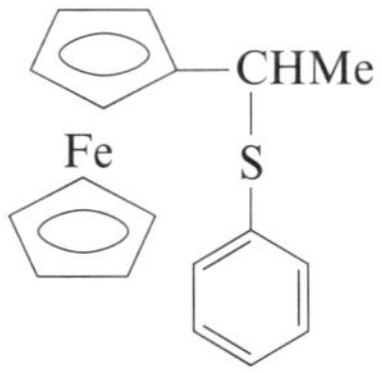

(**9**)

1-Ferrocenylethylphenylsulphide

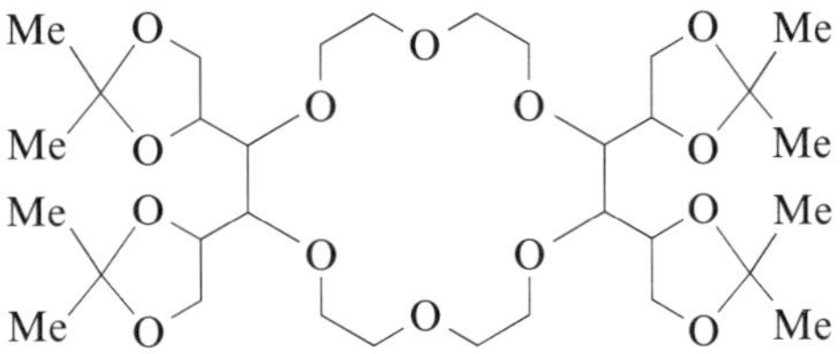

(**10**)

1,2:1′,2′:5,6:5′,6′-Tetra-*O*-isopropylidene-3,3′:4,4′-bis-*O*-oxydiethylenedi-D-mannitol

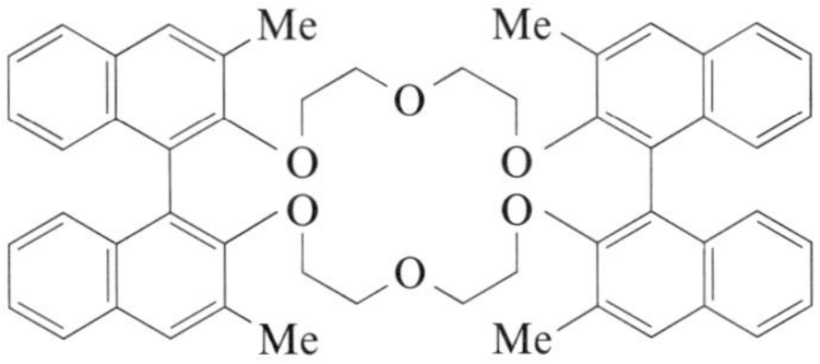

(**11**)

3,3′-Dimethylbis(α,α′-dinaphthyl)-22-crown-6

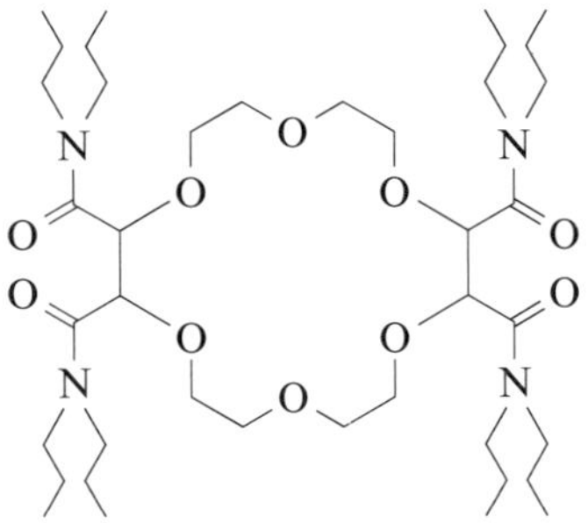

(**12**)

N,N,N′,N′,N″,N″,N‴,N‴-Octapropyl-1,4,7,10,13,16-hexaoxacyclooctadecane-2,3,11,12-tetracarboxamide

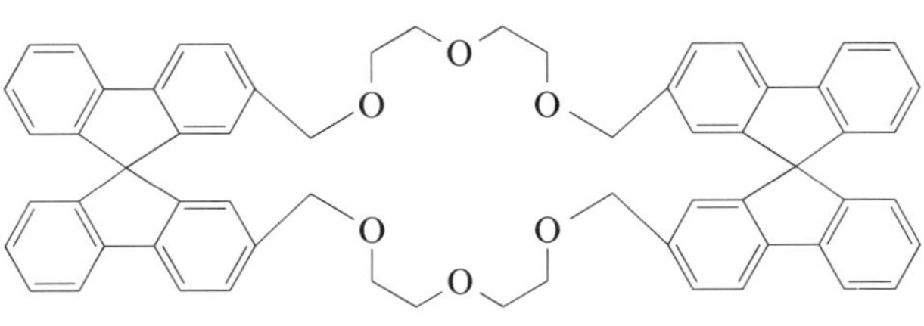

(**13**)

2,2:2′,2′-Bis(2,5,8-trioxanonamethylene)di(9,9′-spirobifluorene)

Chiral crown ethers are lipophilic compounds, showing selectivity towards chiral amine salts. Chirality is introduced by chiral side groups, such as tartaric acid derivatives (e.g. **10**),[22,23] dinaphthyl compounds (**11**),[24,25] and carboxamides (**12**).[26,27] 9,9′-Spirobifluorene crown ethers (**13**) sometimes exhibit high enantioselectivities for α-amino alcohols;[28] enantioselectivies of up to 9.5 have been reported. Binding of the OH group of the α-amino alcohol is probably the key factor in chiral recognition. Selectivities up to 19.5 have been achieved with binaphthyl crown ethers for the resolution of phenylglycine,[25] in which binding of the aromatic moiety is essential for obtaining selectivity.

Surfactant solutions have also been used for chiral separations in liquid membrane systems. Bile salt micelles, immobilised in a cellulose support, have been used for the enantioselective transport of 1,1′-binaphthol ($\alpha = 1.9$).[29] In a similar experiment, dansylphenylalanine has been separated using ionene (OIM-6) micelles, resulting in enantioselectivities of 2.1, whereas without the micelles no transport occurs.

Amino acid derivatives have been used as carriers in liquid membrane transport[30,31] and in chiral liquid–liquid extractions[32,33] for the separation of amino acids. However, other compounds can also be transported selectively. Examples include the separation of propranolol (**14**) and bupranolol (**15**),[31] using *N*-*n*-alkyl-hydroxyprolines (**16**) as the selective carriers, resulting in enantioselectivities of 1.24 and 1.16, respectively. The transport mechanism is ion pairing (Figure 4). *N*-(3,5-Dinitrobenzoyl)leucine can be transported selectively through a liquid membrane containing fatty esters or amines of (*S*)-*N*-(1-naphthyl)-leucine in hexane.[34] Here, a bulk liquid membrane system was used (Figure 5), in

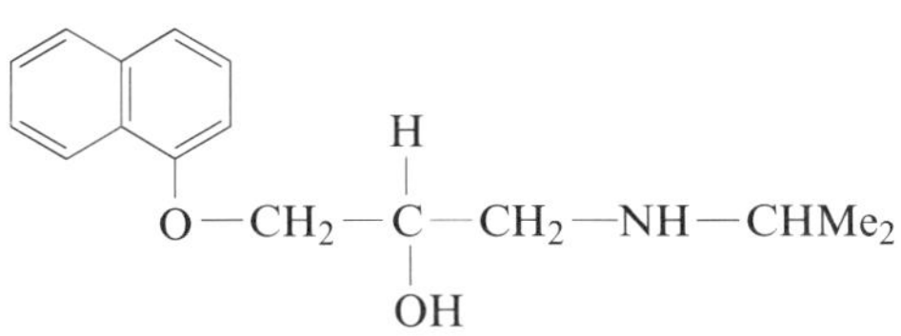

(**14**)

Propranolol

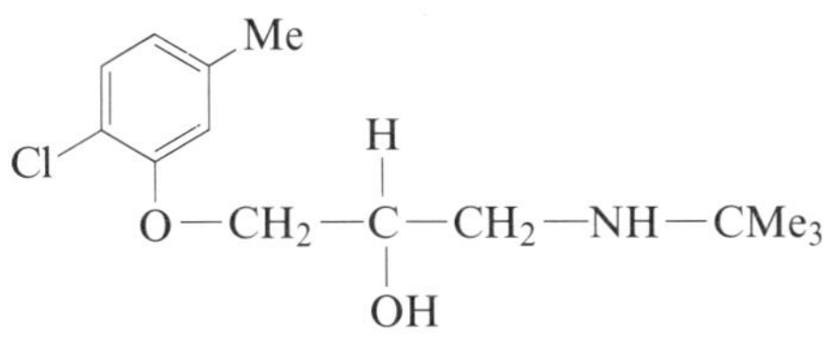

(**15**)

Bupranolol

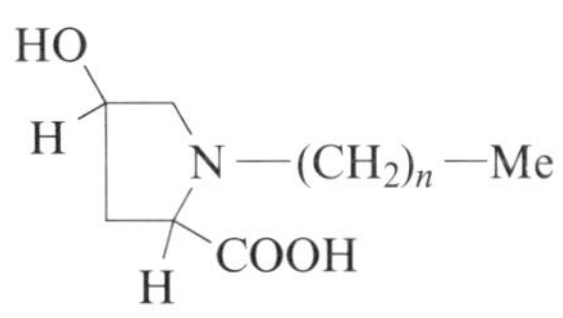

(**16**)

N-*n*-Alkylhydroxyproline

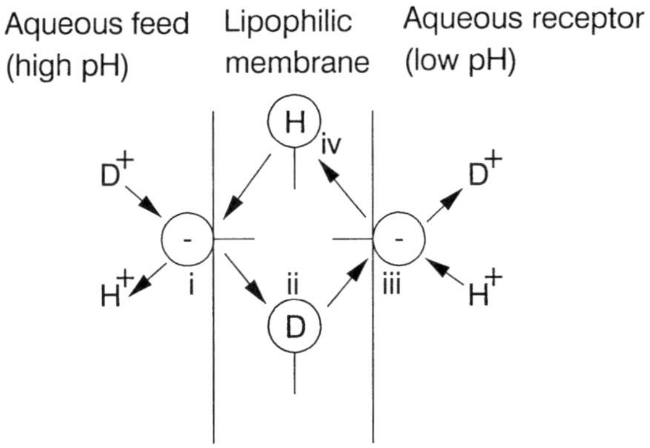

Figure 4 Principle of ion pairing for selective transport across a liquid membrane. (i) Carrier head group deprotonates at donor/membrane interface and undergoes ion-pair interaction with positively charged drug; (ii) neutral ion pair diffuses through the lipophilic membrane; (iii) at the membrane–receptor interface the carrier re-protonates owing to the lower pH, thereby releasing drug into receptor phase; (iv) carrier diffuses back

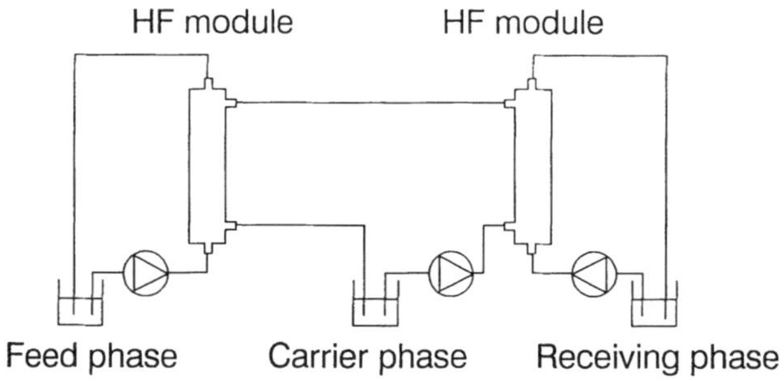

Figure 5 Principle of a bulk liquid membrane. In the first hollow-fibre membrane module enantioselective extraction from the feed phase occurs to the carrier phase. The carrier phase is passed to the second hollow-fibre unit, where the enantiomer is released into the receiving phase

which the membrane liquid is circulated between an absorption and a desorption hollow-fibre unit. Formation and dissociation of the chiral complex take places at different locations, so that appropriate process conditions for absorption and desorption (such as temperature) can be readily achieved. Using this configuration, *ee* values in excess of 95% were obtained in a single step.

Enantioselective supramolecular transport of complexes of lasolocid A (**17**) with metal ammines and amines resulted in enantioselectivities of 1.22 for 1,2-diaminoethane, 1.74 for diethylenetriamine and 1.38 for the complex of

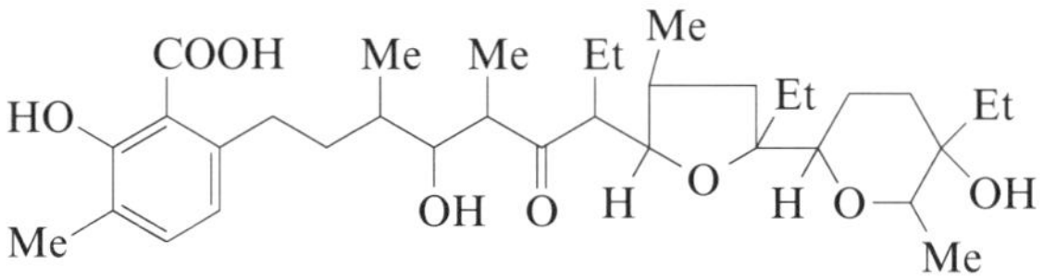

(**17**)

Lasolocid A

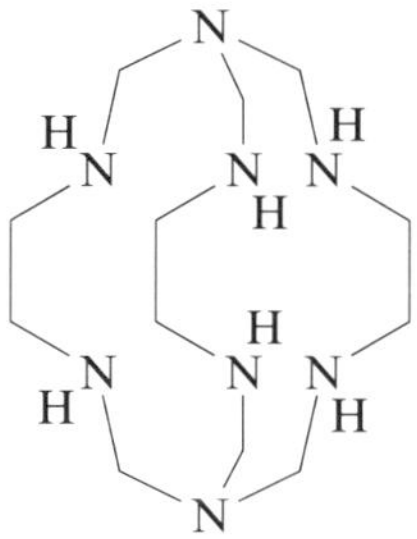

(**18**)

1,3,6,8,10,13,16,19-Octaazabicyclo[6.6.6]eicosane

1,3,6,8,10,13,16,19-octaazabicyclo[6.6.6]eicosane (**18**), using a counter gradient of ammonium ions as the driving force. Permeation rates through this bulk liquid membrane containing 1 mM of **17** in chloroform are of the order of 5×10^{-3} mol $m^{-2}h^{-1}$.[35]

Although high selectivities can be achieved in liquid membrane separations, there are potential drawbacks. When the feed stream is depleted in one enantiomer, the enantiomer ratio in the permeate decreases rapidly (Figure 6). This is similar to a kinetic resolution,[36] where the initial *ee* of the desired product can be high, but decreases as the feed mixture is depleted in the more rapidly converted component. To avoid this either requires racemisation of the feed or, alternatively, a system containing the two opposite selectors can be used,[25] so that the feed stream remains virtually racemic. This was illustrated by using binaphthyl crown ethers as carriers in a W-tube configuration for the separation of phenylglycine.

Although selectivities reported for enantioselective transport through liquid membranes can be high, several passes are required to obtain purities in excess of 99%. The problems encountered in this type of multi-stage process are comparable with those in crystallisation processes, which, owing to the many recycles required, lead to a relatively complicated flow scheme and expensive process equipment. In the authors' opinion, the use of supported enantioselective liquid membranes is relatively complicated for large-scale separations unless the product has the required purity after a single step.

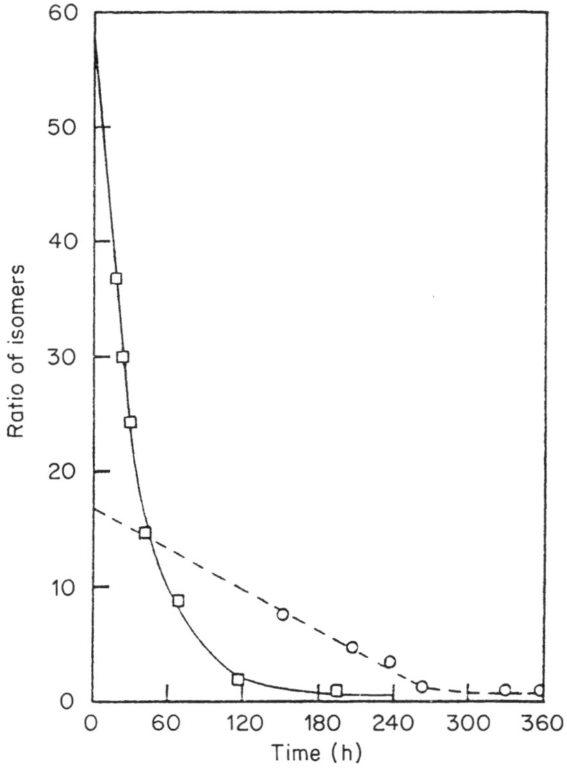

Figure 6 Enantiomer ratio of isomers penetrating a 0.7 M β-CD liquid membrane *vs* time. The solid line represents the ratio of *p*- to *o*-nitroaniline. The dashed line represents the ratio of (+)-(*R*)- to (−)-(*S*)-1-ferrocenylethylphenylsulphide. Reproduced with permission from ref. 21

8.3 MEMBRANE-ASSISTED SEPARATIONS

8.3.1 LIQUID–LIQUID EXTRACTIONS

Enantiomers can be separated using conventional liquid–liquid extraction, provided that enantioselectivity is incorporated into one or both of the liquids. Usually, enantioselectivity is achieved by adding a chiral additive (the chiral selector) to just one of the two phases. In principle, the selectors applied in liquid–liquid extraction can be the same as the selectors applied in supported liquid membranes. Examples are the resolutions of DL-valine and DL-isoleucine in an extraction system consisting of *n*-butanol and water containing *N*-*n*-alkyl-L-hydroxyproline and copper(II) ions.[32,33] Chiral ammonium salts, based for example on norephedrine (**19**), can be separated by partitioning between an aqueous

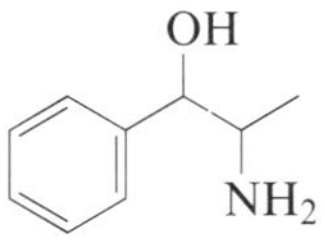

(19)

Norephedrine

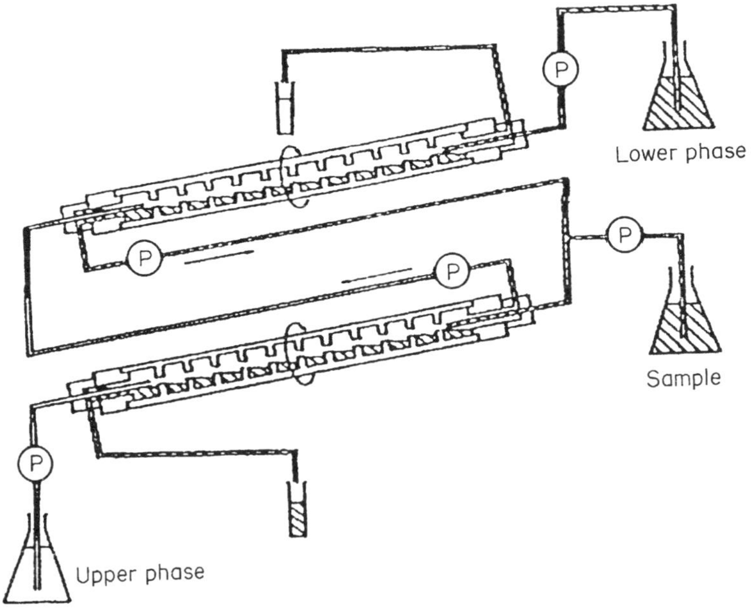

Figure 7 Schematic representation of the countercurrent extraction apparatus used by Takeuchi *et al.*[33] for the resolution of amino acids.

and a lipophilic phase containing esters of tartaric acid.[37–39] Nevertheless, despite the relatively high separation factors obtained in these systems (1.25–3.54), a significant number of stages are required to obtain high optical purities, and to achieve this number of stages a good countercurrent flow is essential. This can be achieved relatively easily in an arrangement described by Takeuchi *et al.*[33] (Figure 7); however, scale-up of this type of equipment is expected to be difficult.

Equipment for liquid–liquid extractions is often described in terms of a required number of transfer units (NTU) for a given degree of separation:[40]

$$\mathrm{NTU} = \frac{\Lambda}{\Lambda - 1}\ln\left[\left(1 - \frac{1}{\Lambda}\right)\cdot\frac{x_\mathrm{i}}{x_\mathrm{o}} + \frac{1}{\Lambda}\right] \tag{2}$$

where x_i/x_o is the ratio between the inflow and outflow concentrations in the feed stream and Λ is the extraction factor, defined as

$$\Lambda = \frac{mF_e}{F_f} \tag{3}$$

F_e and F_f are the flow rates of the two liquids, respectively, and m is the distribution ratio over the two phases. The apparatus characteristics are combined in the height of a transfer unit (HTU):

$$\mathrm{HTU} = \frac{v}{K_o A} \tag{4}$$

where v is the superficial flow velocity, K_o is the overall mass-transfer coefficient and A is the specific surface area of the equipment. The height of the separation apparatus (H) now equals

$$H = \mathrm{NTU} \times \mathrm{HTU} \tag{5}$$

Back-mixing and flooding often limit the performance of conventional extraction equipment, and this can be avoided when the liquid–liquid interface is immobilised in the pores of a membrane (Figure 8). This type of process is called 'hollow-fibre membrane extraction' (and is also referred to as 'pertraction' or 'dispersion-free extraction'). The advantage of hollow-fibre membrane extraction does not result from large mass-transfer coefficients (which are similar to those obtained with conventional equipment), but from the extremely high surface area per unit volume, typically 30 times larger for fibres than for packed towers, and 100 times larger than for conventional extractors,[41] thus reducing HTU accordingly. Extensive literature is available on the (potential) applications of hollow-fibre extractions.[42–45]

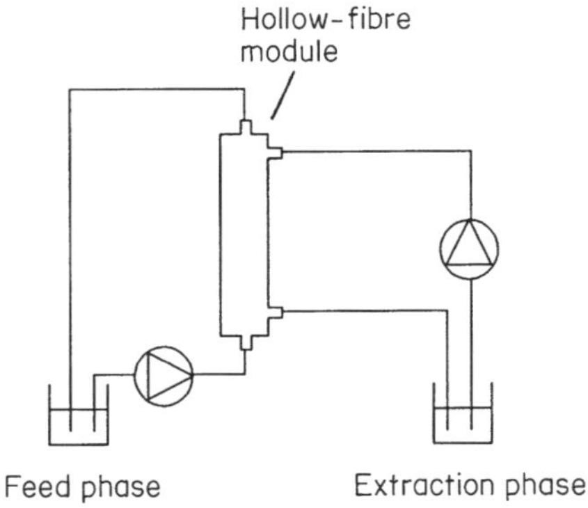

Figure 8 Schematic representation of hollow-fibre membrane extraction

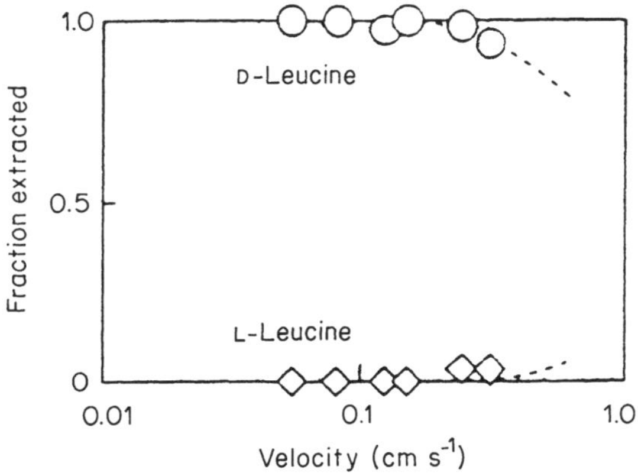

Figure 9 Separation of DL-leucine in hollow-fibre membrane extraction using a *N-n*-dodecyl-L-hydroxyproline solution in octanol as the enantioselective extraction liquid. The modules used were 32 cm long and contained 96 Celgard X-20 fibres. Reproduced with permission from ref. 46

As an example, DL-leucine has been separated using *N-n*-dodecyl-L-hydroxyproline in octanol as the extraction phase flowing countercurrently with an aqueous phase[46]. Polypropylene hollow fibres (Celgard X-20) were used, in which the pores were filled with a cross-linked poly(vinyl alcohol) gel. In experiments with an organic to aqueous flow ratio of 4 (so that Λ_D and Λ_L were 1.32 and 0.64, respectively), DL-leucine was separated almost completely (Figure 9).

When complete resolution is required, it follows from equation 2 that Λ_D and Λ_L have to be on the opposite sides of unity. Because the gross distribution over the two phases usually deviates substantially from unity, the flow ratios have to be adjusted accordingly, which may result in excessive phase volumes. In addition to extreme dilution of the separated enantiomers, this may also lead to significant losses of valuable material.

8.3.2 LIQUID-MEMBRANE FRACTIONATION

To avoid the extreme flow ratios often encountered in liquid–liquid extractions, a completely symmetrical system has been developed at Akzo Nobel in which two miscible chiral liquids with opposite chiral selectors are used, separated by a non-miscible liquid membrane (Figure 10[47]). The selector has to be chosen to be incapable of passing the liquid membrane, i.e. highly lipophilic in the case of an aqueous liquid membrane and charged or highly polar in the case of a non-polar liquid membrane. Using an aqueous liquid membrane, enantioselectivities have been determined for several racemic drugs using a range of chiral selectors (Table 1).

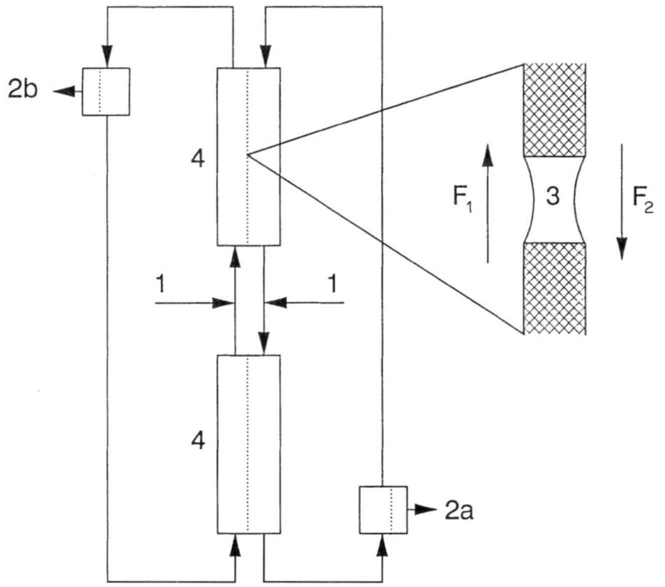

Figure 10 Schematic representation of the Akzo Nobel enantiomer separation system. Two liquids containing opposing enantiomers of the selector (F_1 and F_2) flow counter-currently through the column (4). The racemic mixture is added in the middle of the column (1) and the enantiomers are recovered from the outflows (2a and 2b)

Table 1 Enantioselectivities[47] determined for several drugs in the liquid-membrane fractionation system[a]

Compound	Selector	Solvent	Anion	α
Norephedrine (**19**)	0.25 M DHT	Heptane	–	1.19
	0.25 M DHT	Heptane	–	1.50*
	10 wt% PLA	Chlorofom	–	1.07
	10 wt% PLA	Chloroform	–	1.10*
Ephedrine (**20**)	0.10 M DHT	Heptane	0.06 M PF_6^-	1.30
Mirtazapine (**21**)	0.10 M DHT	Heptane	–	1.05
	0.25 M DHT	Heptane	–	1.06
	0.50 M DHT	Heptane	–	1.08
	0.25 M DBT	Decanol	–	1.06
	10 wt% PLA	Chloroform	–	1.04
	0.10 M DHT	Heptane	0.01 M PF_6^-	1.07
	0.50 M DHT	Heptane	0.03 M PF_6^-	1.16
Phenylglycine	0.25 M DHT	Heptane	–	1.06
Salbutamol (**22**)	0.25 M DHT	Heptane	0.018 M BPH_4^-	1.06
	0.25 M DHT	Cyclohexane	0.0035 M BPh_4^-	1.04
Terbutaline (**23**)	0.43 M DHT	Dichloroethane	0.5 M PF_6^-	1.14
	2.15 M DHT	Heptane	0.0045 M BPh_4^-	1.05
Propranolol (**14**)	0.25 M DHT	Heptane	–	1.03
Ibuprofen (**24**)	0.10 M DHT	Heptane	–	1.10

[a] All experiments were performed at room temperature, except those marked with asterisks, which were performed at 4°C. In some cases a lipophilic anion was used to increase the solubility of the drug in the organic phases. PF_6^- = hexafluorophosphate; BPh_4^- = tetraphenylborate; DHT = dihexyl tartrate; DBT = dibenzoyl tartrate; PLA = poly (lactic acid)

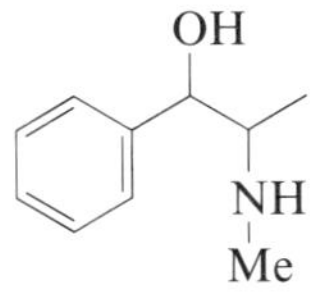

(20)
Ephedrine

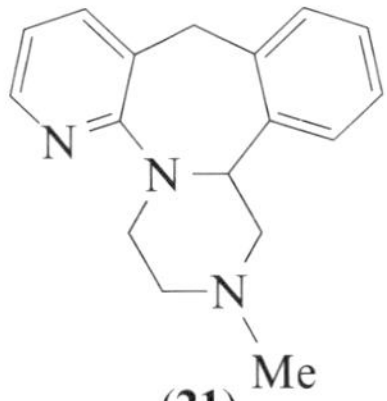

(21)
Mirtazapine

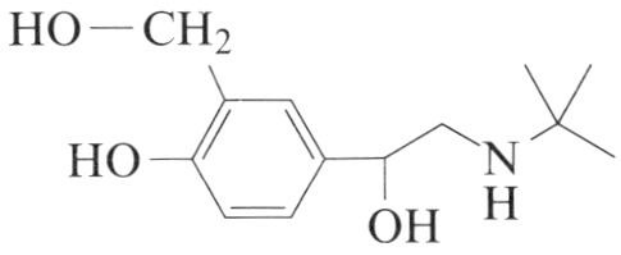

(22)
Salbutamol

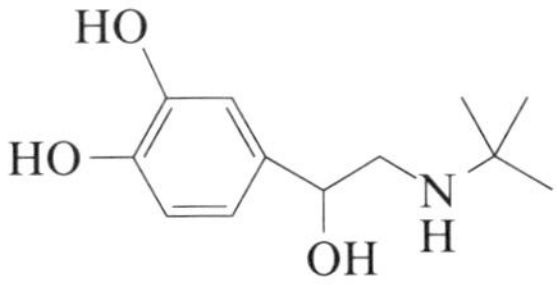

(23)
Terbutaline

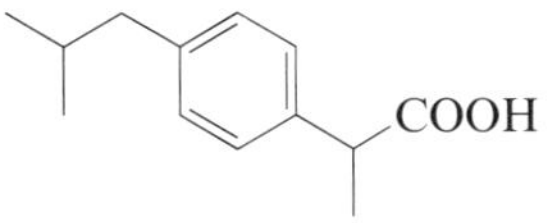

(24)
Ibuprofen

Using this configuration, both enantiomers can be recovered completely with an enantiomer ratio, L/D, in the outflow given by

$$\frac{L}{D}=\frac{\dfrac{\Lambda_D}{\Lambda_D-1}\left[exp\left(\dfrac{\Lambda_D-1}{2\Lambda_D}NTU\right)-\dfrac{1}{\Lambda_D}\right]}{\dfrac{\Lambda_L}{\Lambda_L-1}\left[exp\left(\dfrac{\Lambda_L-1}{2\Lambda_L}NTU\right)-\dfrac{1}{\Lambda_L}\right]} \tag{6}$$

Owing to the extremely high surface area per unit volume in the hollow-fibre membrane modules, relatively low selectivities are sufficient to separate a racemic mixture into its component enantiomers. Typical values for HTU have been determined experimentally,[48] and range from 2 to 6 cm at flow velocities between 0.01 and 2 mm s^{-1}. From equation 6 the required NTU can be calculated for 99% pure products: these values are 190, 100 and 50 for enantioselectivities of 1.05, 1.1 and 1.2, respectively. From equation 5, it then follows that the total length of the apparatus typically is of the order of 2–5 m.

To demonstrate the utility of this method for obtaining both enantiomers with high purity, experiments were performed using racemic norephedrine (**19**). In Figure 11 the enantiomer ratio of norephedrine in the two outflows is given during start-up, and it can be concluded that the system reaches equilibrium within 24 h, both enantiomers being recovered with >99% purity.

The successful application of the separation system described above depends strongly on the ability to find an efficient chiral selector molecule, and molecular mechanics methods have proved a versatile tool for this purpose.[49,50] Currently, this system is being scaled up, and application on a large scale should be possible in the near future.

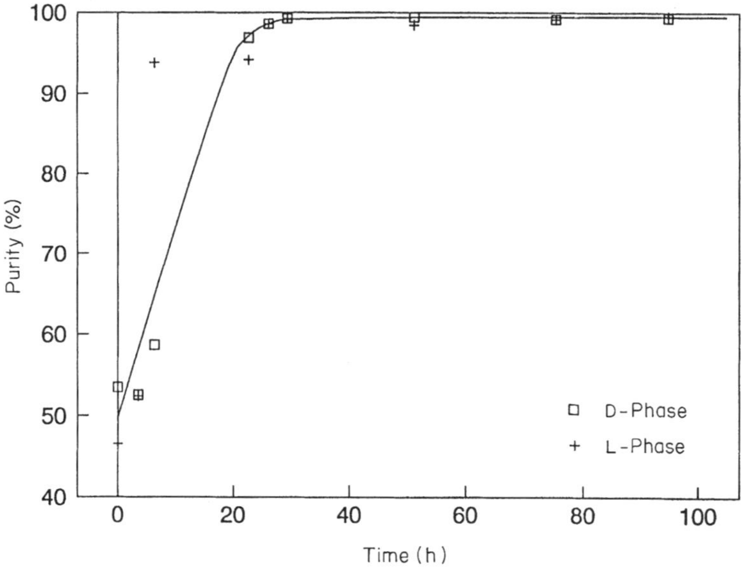

Figure 11 Enantiomer ratio in the outflows versus time for the separation of racemic norephedrine using the system in Figure 10

8.3.3 MICELLAR-ENHANCED ULTRAFILTRATION

Micellar-enhanced ultrafiltration (MEUF), a micelle-based membrane separation system, has been used for the separation of amino acids.[51] An amphiphilic amino acid derivative, L-5-cholesteryl glutamate (**25**), is used as a chiral co-surfactant in micelles made with the non-ionic surfactant Serdox NNP™ 10. Cu(II) ions are added for the formation of ternary complexes between phenylalanine and the amino acid co-surfactant. The basis for the separation is the difference in stability between ternary copper complexes formed with D- or L-phenylalanine

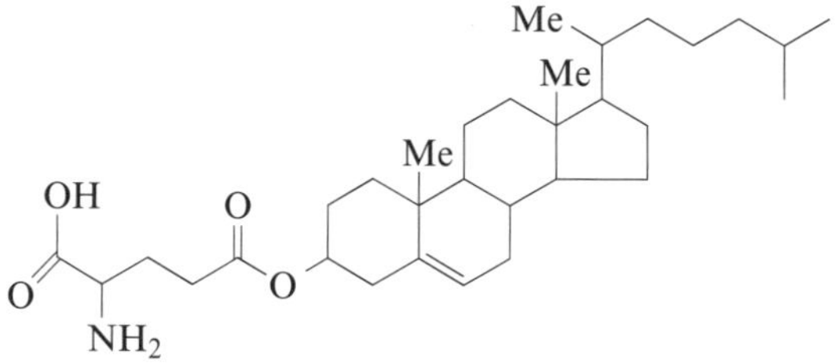

(**25**)

L-5-Cholesteryl glutamate

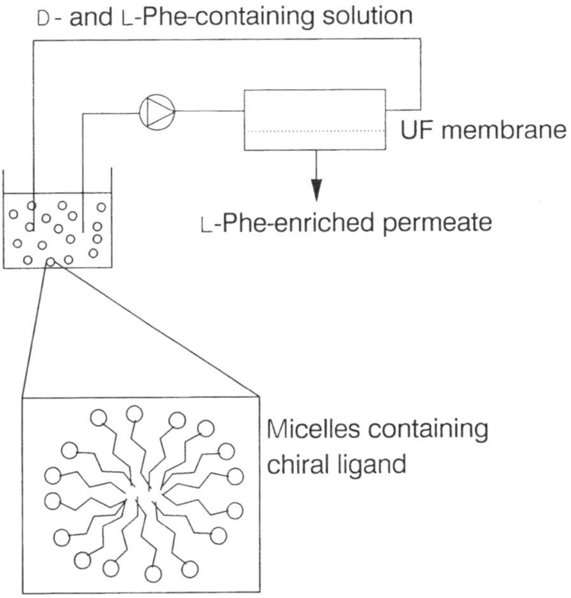

Figure 12 Principle of micellar-enhanced ultrafiltration (MEUF). The D-enantiomer of a racemic mixture is preferentially bound to the micelles, which are retained by the membrane. The bulk liquid containing the L-enantiomer is separated through the membrane [51]

and L-5-cholesteryl glutamate. The micelles (and the enantiomer preferentially bound to them) are retained by an ultrafiltration membrane which is permeable to the uncomplexed phenylalanine enantiomer (Figure 12).

The 'operational' selectivity of this system reaches values of around 4 when a 2:1 ratio between the chiral co-surfactant and phenylalanine is used. A single-stage MEUF system can, therefore, serve as an effective low-grade separation method, although complete resolution requires a multi-stage MEUF apparatus. The industrial potential of MEUF obviously lies in the possibility of processing large volumes of dilute racemic mixtures inexpensively.

8.3.4 MEMBRANE-ASSISTED KINETIC RESOLUTIONS

Until now, the use of membranes to facilitate kinetic resolutions has been limited to enzymatic resolutions, which can be classified into two main types. In one system the enzymes are immobilised inside or on the membrane, and in the other the enzymes are kept in solution and are retained by the membrane. According to Matson and Quinn,[52] the potential benefits of these systems compared with conventional process technology (such as stirred tank reactors) are as follows:

— Membrane bioreactors can improve the efficiency of enzyme-catalysed bioconversions by integrating the conversion and product purification steps and by eliminating certain pieces of process equipment and attendant energy requirements.
— Membrane bioreactors increase yields in bioconversions that are subject to by-product inhibition (for reasons of unfavourable thermodynamics or enzyme inhibition).
— Membrane bioreactors will prove useful in extending the utility of enzyme and whole-cell catalysis to organic chemical syntheses that are not currently amenable to biocatalysis because of various process engineering limitations.

This combination of an enzymatic conversion and product separation using ultrafiltration is being applied to the production of optically pure α-amino acids,[53] and has been scaled up to a level of 100 tonnes annually. For this purpose, a hollow-fibre polysulphone or regenerated cellulose membrane with a molecular weight cut-off of 10 000 is used to retain the enzyme (acylase I). The substrate (*N*-acetyl-DL-amino acids) and product (L-amino acid) are capable of passing through the membrane for downstream processing. A similar process has been described for the production of menthyl-(−)-laurate from racemic menthol in a non-aqueous solvent using ceramic membranes to retain lipase from *Candida cylindracea*.[54] Other examples of kinetic resolutions using membrane-retained enzymes are the production of *N*-acetyl-L-phenylalanine[55] and *N*-acetyl-L-methionine.[56]

It has already been shown for liquid–liquid extractions that hollow-fibre membranes can be used efficiently to create intensive contact between two immiscible phases. Using a membrane in which the enzyme is immobilised, a lipophilic product can effectively be separated from an aqueous-soluble reactant, and vice versa. Examples are the enzymatic resolution of glycidyl butyrate[57] and the enantiospecific hydrolysis of *trans*-3-(4-methoxyphenyl)glycidic acid methyl ester [MPGM (**26**)], a key intermediate in the synthesis of diltiazem[58]. For the hydrolysis of MPGM, lipase from *Serratia marcescens* was immobilised in the pores of an asymmetric polyacrylonitrile UF membrane [59]. (+)-MGPM dissolved in toluene is selectively hydrolysed to (2*S*,3*R*)-(+)-3-(4-methoxyphenyl)glycidic acid and methanol. Enzyme inhibition caused by *p*-methoxyphenylacetaldehyde derived from the glycidic acid could effectively be suppressed by the addition of sodium hydrogensulphite to the aqueous phase (Figure 13).

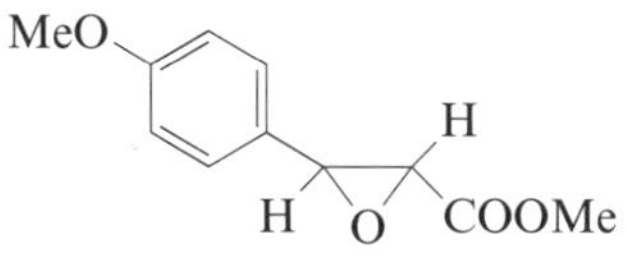

(**26**)

3-(4-Methoxyphenyl)glycidic acid methyl ester (MPGM)

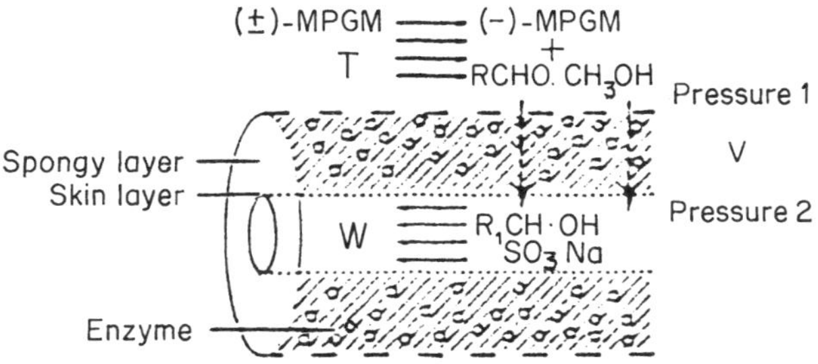

Figure 13 Behaviour of the products in the membrane bioreactor used for the production of (−)-MPGM. Reproduced with permission from ref. 59

Subsequently, (−)-MPGM is recovered from the toluene solution by crystallisation. The productivity of the membrane reactor is *ca* 40 kg of (−)-MPGM per m^2 annually. This system has been developed by Sepracor Inc., and has been in operation as part of the process for the production of diltiazem hydrochloride since 1993.

A similar system has been developed by Matsumae *et al.* for the separation of (*R,S*)-ibuprofen **(24)** using proteases which enantioselectively hydrolyse the (*R*)-ibuprofen esters[60] (Figure 14). A 100% yield of (*S*)-ibuprofen is achieved by racemisation of the (*R*)-ibuprofen and, in principle, this arrangement can be used for the separation of other acids, such as naproxen and 2-chloropropionic acid (see also Chapters 3 and 10). Another process, comprising a membrane extraction and a permselective membrane, has been used by van Eikeren *et al.*[61]

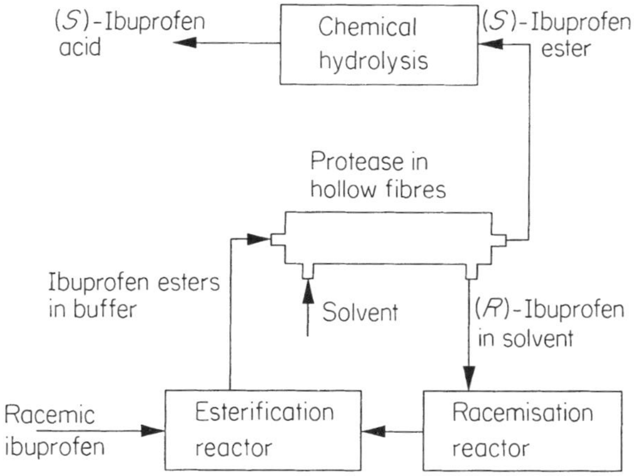

Figure 14 Membrane bioreactor for the production of (*S*)-ibuprofen[60]

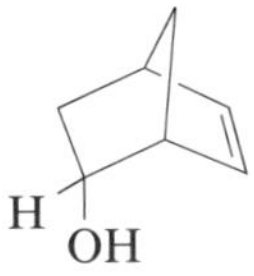

(27)
(*R*)-*endo*-Norbornen-2-ol

to produce (*R*)-*endo*-norbornen-2-ol (**27**) from the racemate at a scale of tens of kilograms.

The membrane-assisted enzymatic resolutions are particularly useful technologies which are amenable to large-scale use; the membrane separation required to enhance the enzymatic process is often established technology, which makes the process relatively easy to operate on a multi-kilogram scale. A potential drawback, often encountered when enzymes with a wide range of substrate specificity (such as lipases and esterases) are used, is the fact that the conversion rate can decrease the less the substrate resembles the natural substrate. Moreover, at the moment the use of isolated enantiospecific enzymes on a large scale is operationally easier for those enzymes not requiring a cofactor such as NADH or $FADH_2$. Once simpler cofactor-regeneration technologies have been developed, the large-scale use of redox enzymes will be extended substantially. Although it is sometimes difficult to obtain products in the 99% enantiomeric excess range using conventional process equipment, the combination of the enzymatic conversion with a membrane separation often results in very high optical purities.

8.4 CONCLUSION

One of the key advantages of membrane separation processes for the production of optically pure products is their ready applicability to large-scale use. This is particularly important for hybrid processes when the effectiveness of a new enantioselective technique can be enhanced with membrane technology in which scale-up procedures are well established, and can be implemented rapidly and efficiently. For extraction processes, only a moderate enantioselectivity is required to achieve high optical purities, and process configurations are similar to conventional unit operations such as extraction and distillation.

Enantioselective membranes are still at an early stage of development and large-scale implementation does not look likely in the short term. In the authors' opinion, considerable improvements will have to be made in the separation factors before this technology will be viable on a manufacturing scale. Nevertheless, enantioselective polymer membranes have considerable potential for large-scale operation because of the precedence of techniques such as reverse osmosis and ultrafiltration.

8.5 REFERENCES

1. Dorson, W. J. and Pierson, J. S., *J. Membr. Sci.*, **44**, 35 (1989).
2. Sourirajan, S. and Matsuura, T., *Reverse Osmosis/Ultrafiltration, Process Principles*, National Research Council Canada, Ottawa, 1985.
3. Strathmann, H., in *Synthetic Membranes: Science, Engineering and Applications* (ed. P. M. Bungay, H. K. Lonsdale and M. N. de Pinho), NATO ASI Series C: Mathematical and Physical Sciences, Vol. 181, Reidel Dordrecht, 1986.
4. Hwang, S. T., and Kammermeyer, K., *Membranes in Separations*, Wiley–Interscience, New York, 1975.
5. Rautenbach, R., and Albrecht, R., *Membrane Processes*, Wiley Chichester, 1989.
6. Keurentjes, J. T. F., Linders, L. J. M., Beverloo, W. A. and van 't Riet, K., *Chem. Eng. Sci.*, **47**, 1561 (1992).
7. Maruyama, A., Adachi, N., Takatsuki, T., Torii, M., Sanui, K., and Ogata, N., *Macromolecules*, **23**, 2748 (1990).
8. Ogata, N., in Proceeding of the 1st International Conference on Frontiers in Polymer Research (ed. P. N. Prasad and J. K. Nigam) Plenum Press, New York, 1991, p. 103.
9. Ogata, N., *Macromol. Symp.*, **77**, 167 (1994).
10. Ogata, N., *Polym. Prepr. Am. Chem. Soc. Div. Polym. Chem.*, **34**, 96 (1993).
11. Aoki, T., Shinohara, K., and Oikawa, E., *Makromol. Chem. Rapid. Commun.*, **13**, 565 (1992).
12. Aoki, T., Tomizawa, S., and Oikawa, E., *J. Membr. Sci.*, **99**, 117 (1995).
13. Masawaki, T., Sasai, M., and Tone, S., *J. Chem. Eng. Jpn.*, **25**, 33 (1992).
14. Higuchi, A., Hara, M., Horiuchi, T., and Nakagawa, T., *J. Membr. Sci.*, **93**, 157 (1994).
15. Yashima, E., Noguchi J., and Okamoto, Y., *J. Appl. Polym. Sci.*, **54**, 1087 (1994).
16. Linder, C., Nemas, M., Perry, M., and Ketraro, R., *Br. Pat. Appl.*, 2 233 248, 1991.
17. Bryjak, M., Kozlowski, J., Wieczorek, P., and Kafarski, P., *J. Membr. Sci.*, **85**, 221 (1993).
18. Venema, A., Henderiks, H. H., and van Geest, R., *J. High Resolut. Chromatogr.*, **14**, 676 (1991).
19. Armstrong, D. W., Ward, T. J., Armstrong, R. D., and Beesley, T. E., *Science*, **232**, 1132 (1986).
20. Nielen, M. W. F., *J. Chromatogr.*, **637**, 81 (1993).
21. Armstrong, D. W., and Jin, H. L., *Anal. Chem.*, **59**, 2237 (1987).
22. Curtis, W. D., Laidler, D. A., Stoddart, J. F., and Jones, G. H., *J. Chem. Soc., Chem. Commun.*, 835 (1975).
23. Behr, J. P., Girodeau, J. M., Hayward. R. C., Lehn J. M., and Sauvage, J. P., *Helv. Chim. Acta*, **63**, 2096 (1980).
24. Yamaguchi, T., Nishimura, K., Shinbo, T., and Sugiura, M., *Chem. Lett.*, 1549 (1985).
25. Newcomb, M., Toner, J. L., Helgeson R. C., and Cram, D. J., *J. Am. Chem. Soc.*, **101**, 4941 (1979).
26. Bussmann, W., Lehn, J.-M., Oesch, U., Plumeré, P., and Simon, W., *Helv. Chim. Acta*, **64**, 657 (1981).
27. Holy, P., Morf, W. E., Seiler, K., Simon, W., and Vigneron, J. P., *Helv. Chim. Acta*, **73**, 1171 (1990).
28. Prelog, V., and Mutak, S., *Helv. Chim. Acta*, **66**, 2274 (1983).
29. Hinze, W. L., Williams, R. W., Fu, Z. S., Suzuki, Y., and Quina, F. N., *Colloids Surf.*, **48**, 79 (1990).
30. Pirkle, W. H., and Bowen, W. E., *Tetrahedron: Asymmetry*, **5**, 773 (1994).
31. Heard, C. M., Hadgraft, J., and Brain, K. R., *Bioseparation*, **4**, 111 (1994).

32. Takeuchi, T., Horikawa, R., and Tanimura, T., *Anal. Chem.*, **56**, 1152 (1984).
33. Takeuchi, T., Horikawa, R., and Tanimura, T., *Sep. Sci. Technol.*, **25**, 941 (1990).
34. Pirkle, W. H., and Doherty, E. M., *J. Am. Chem. Soc.*, **111**, 4113 (1989).
35. Chia, P. S. K., Lindoy, L. F., Walker, G. W., and Everett, G. W., *J. Am. Chem. Soc.*, **113**, 2533 (1991).
36. Sheldon, R. A., *Chirotechnology, Industrial Synthesis of Optically Active Compounds*, Marcel Dekker, New York, 1993.
37. Prelog, V., and Dumic, M., *Helv. Chim. Acta*, **69**, 5 (1986).
38. Prelog, V., Mutak, S., and Kovacevic, K., *Helv. Chim. Acta*, **66**, 2279 (1983).
39. Prelog, V., Stojanac, Z., and Kovacevic, K., *Helv. Chim. Acta*, **65**, 377 (1982)
40. Beek, W. J. and Muttzall, K. M. K., *Transport Phenomena*, Wiley, London, 1975.
41. Dahuron, L., and Cussler, E. L., *AIChE J.*, **34**, 130 (1988).
42. D'Elia, N. A., Dahuron, L., and Cussler, E. L., *J. Membr. Sci.*, **29**, 309 (1986).
43. Yang, M. C., and Cussler, E. L., *AIChE J.*, **32**, 1910 (1986).
44. Prasad, R., and Sirkar, K. K., *AIChE J.*, **34**, 177 (1988)
45. Keurentjes, J. T. F., Sluijs, J. T. M., Franssen, R. J. H., and van 't Riet, K., *Ind. Eng. Chem. Res.*, **31**, 581 (1992).
46. Ding, H. B., Carr P. W., and Cussler, E. L., *AIChE J.*, **38**, 1493 (1992).
47. Keurentjes, J. T. F., *PCT Int. Pat. Appl.*, WO 94/07814, 1994.
48. Keurentjes, J. T. F., Nabuurs, L. J. W. M., and Vegter, E. A., in *Proceedings of Chiral USA '94 Symposium*, Spring Innovations, Stockport, 1994, p. 39.
49. Lipkowitz, K. B., and Baker, B., *Anal. Chem.*, **62**, 770 (1990).
50. Aerts, J., *J. Comput. Chem.*, in press.
51. Creagh, A. L., Hasenack, B. B. E., van der Padt, A., Sudholter, E. J. R., and van 't Riet, K., *Biotechnol. Bioeng.*, **44**, 690 (1994).
52. Matson S. L., and Quinn, J. A., *Ann. N.Y. Acad. Sci.*, **469**, 152 (1986).
53. Bommarius, A. S., Drauz, K., Groeger, U., and C. Wandrey, in *Chirality in Industry* (ed. Collins, A. N., Sheldrake, G. N., and Crosby, J)., Wiley, Chichester, 1992, p. 371.
54. Rios, G. M., Lambert, F., and Jallageas, J. C., *Entropie*, **27**, 31 (1991).
55. Chen, S., *Eur. Pat. Appl.*, EP 273 679, 1988.
56. Schmidt-Kastner, G., and Goelker, C., *GBF Monogr. Ser. 9 (Techn. Membr. Biotechnol.)*, 201 (1986).
57. Wu, D. R., Cramer, S. M., and Belfort, G., *Biotechnol. Bioeng.*, **41**, 979 (1993).
58. Matsumae, H., Furui, M., and Shibatani, T., *J. Ferment. Bioeng.*, **75**, 93 (1993).
59. Matsumae, H., Furui, M., Shibatani, T., and Tosa, T., *J. Ferment. Bioeng.*, **78**, 59 (1994).
60. Matson, S. L., Wald, S. A., Zepp, C. M, and Dodds, D., *PCT Int. Pat. Appl.*, WO 89 09765, 1991.
61. Van Eikeren, P., Muchmore, D. C., Brose, D. J., and Colton, R. H., *Ann. N.Y. Acad. Sci.*, **672**, 539 (1992).

BIOLOGICAL METHODS AND CHIRAL POOL SYNTHESES

9 Four Case Studies in the Development of Biotransformation-based Processes

R. McCAGUE and S. J. C. TAYLOR
Chiroscience Ltd, Cambridge, U.K.

9.1 Introduction 184
9.2 Resolution of the Carbocyclic Nucleoside Synthon 2-Azabicyclo[2.2.1]hept-5-en-3-one with Lactamases 184
9.2.1 Routes to Carbocyclic Nucleosides 184
9.2.2 Identification of Microbial Strains for the Resolution of the γ-Lactam 2-Azabicyclo[2.2.1]hept-5-en-3-one 186
9.2.3 Development of a Whole-Cell Bioresolution Process for the (–)-γ-Lactam 186
9.2.4 Development of an Immobilised Lactamase Resolution Process for the (+)-γ-Lactam and the Ring-opened (–)-Amino Acid 187
9.2.5 Downstream Chemistry from the Resolved γ-Lactam or Amino Acid 188
9.3 Resolution of a Versatile Hydroxylactone Synthon 4-*endo*-Hydroxy-2-oxabicyclo[3.3.0]oct-7-en-3-one by Lipase De-esterification 190
9.3.1 Substrate Synthesis 190
9.3.2 Development of a Biotransformation for Resolution of the Hydroxylactone 191
9.3.3 Esterification versus Hydrolysis 192
9.3.4 Downstream Chemistry from the Resolved Hydroxylactone 192
9.4 Integration of an Acylase Biotransformation with Process Chemistry: a One-pot Synthesis of *N*-BOC-L-3-(4-thiazolyl)alanine and Related Amino Acids 194
9.4.1 Uses of Unnatural Amino Acids 194
9.4.2 Overall Synthetic Route to *N*-BOC-3-(4-thiazolyl)alanine . . 195
9.4.3 An Ester-based Route 195
9.4.4 Development of a One-pot Process Based on Aminoacylase Resolution 197
9.4.5 Recycle of the D-Enantiomer 199

Chirality in Industry II. Edited by A. N. Collins, G. N. Sheldrake and J. Crosby

9.4.6 Comparison of Different Biotransformation Processes . . . 200
9.4.7 Application of the Aminoacylase Process to Other Unnatural Amino Acids 200
9.5 Dynamic Resolution of an Oxazolinone by Lipase Biocatalysis: Synthesis of (*S*)-*tert*-Leucine 201
9.5.1 Use of (*S*)-*tert*-Leucine 201
9.5.2 Azlactones for the Dynamic Resolution of Amino Acids . . 201
9.5.3 Development of a Lipase-catalysed Dynamic Resolution . . 202
9.6 Conclusions 203
9.7 Acknowledgements 205
9.8 References 205

9.1 INTRODUCTION

Biotransformations are playing an increasing role in the manufacture of single-enantiomer intermediates, especially for pharmaceutical agents. Advantages of biotransformations are (i) they are catalytic and take place under mild, generally near ambient, conditions, (ii) when the biocatalyst is derived from a microbial strain the scaleability is unlimited, as the biocatalyst can be provided by growing the microbial strain to whatever biomass is necessary, and (iii) there is much scope to develop a transformation that is both highly enantioselective and highly economic.

In this chapter, a selection of biotransformation-based processes are described that afford intermediates for pharmaceutical agents. In each case the optimisation of the biotransformation with respect to enantiospecificity, conversion rate and enzyme utilisation is described. Generally, a biotransformation cannot stand on its own; it needs to be integrated within the total process. Frequently, the biotransformation chosen for development will be governed by the chemical route, so consideration is given here to the integration of the biotransformation within the overall process chemistry, including the development of the product isolation and discussion of means of utilising the unwanted enantiomer. Finally, we draw some broad principles from the specific cases studied that should be of assistance to process scientists in the future development of improved biotransformation-based processes.

9.2 RESOLUTION OF THE CARBOCYCLIC NUCLEOSIDE SYNTHON 2-AZABICYCLO[2.2.1]HEPT-5-EN-3-ONE WITH LACTAMASES

9.2.1 ROUTES TO CARBOCYCLIC NUCLEOSIDES

Carbocyclic nucleosides are analogues of the natural nucleosides in which the ribosyl oxygen has been replaced by a methylene linkage. They have attracted much interest as therapeutic agents as the absence of the labile ribosidic linkage results in compounds with greatly improved metabolic stability to nucleases and correspondingly greater bioavailability. Two therapeutic applications of carbocyclic nucleosides are (i) as antiviral agents (e.g. carbovir for the treatment of HIV[1] and c-AFG and c-BVDU for Herpes simplex[2,3]) and (ii) as adenosine agonist cardiac vasodilators (e.g. c-NECA[4]) (Scheme 1). For use as therapeutic agents, the carbocyclic nucleosides need to be available as single enantiomers, which, for the above mentioned examples, is the configuration corresponding to that of the natural nucleosides. However, while analogues of natural nucleosides are generally prepared by synthetic manipulation of natural nucleosides or built up from ribose, which is available in single-enantiomer form, there is no such ready access to carbocyclic nucleosides. There is a carbocyclic nucleoside natural product, aristeromycin (Scheme 1), which may be obtained from fermentation of strains of *Streptomyces*;[5] however, the synthetic elaboration into the target compounds might be tortuous. Nevertheless, a clever procedure has been devised by Glaxo to convert the adenine base of aristeromycin into the guanine base required for the antiviral agents.[5] Apart from this, carbocyclic nucleosides are usually constructed by total synthesis (a review of the various approaches has appeared[6]). Of the various options, a particularly versatile approach starts with the bicyclic γ-lactam synthon 2-azabicyclo[2.2.1]hept-5-en-3-one (**1**), first described by Jagt and van Leusen,[7] and prepared by cycloaddition of cyclopentadiene with chlorosulphonyl isocyanate, *p*-toluenesulphonyl cyanide or methanesulphonyl cyanide.[8] The application of the resulting racemic lactam **1** to the

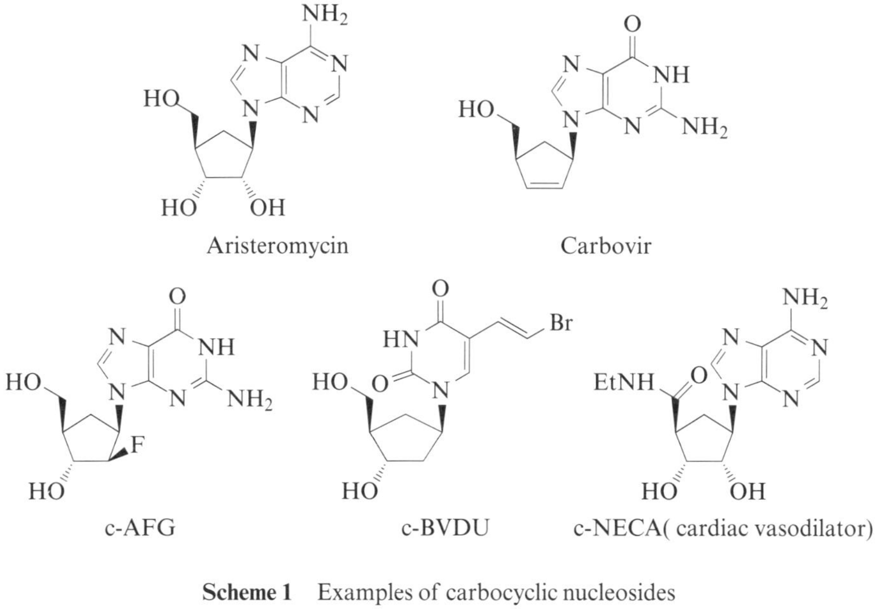

Scheme 1 Examples of carbocyclic nucleosides

synthesis of carbocyclic nucleosides has been demonstrated by the group of Vince *et al.*; generally, the lactam ring is opened and the heterocyclic base is constructed from the resulting amine.[9] However this route leads to a racemic carbocyclic nucleoside, and, while methods to resolve such products exist, it would be more convenient to have access to the single enantiomer of the lactam. Thus, a synthesis can be envisaged (Scheme 2) whereby lactam **1** is resolved and the appropriate enantiomer is carried forward into the remainder of the synthesis.

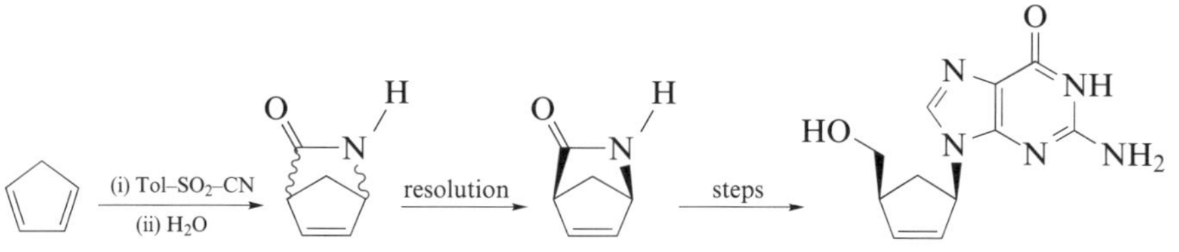

(1)

Scheme 2 Overall route to a carbocyclic nucleoside

9.2.2 IDENTIFICATION OF MICROBIAL STRAINS FOR THE RESOLUTION OF THE γ-LACTAM 2-AZABICYCLO[2.2.1]HEPT-5-EN-3-ONE

With the aim of obtaining a single enantiomer of the lactam, efforts were made to locate a biocatalyst capable of cleaving the amide linkage of the lactam enantiospecifically. While available isolated enzymes (such as proteases) failed to act on the lactam, enzymes were discovered in microbial strains that were screened which did possess such activity. Screening turned out to be relatively straightforward owing to the very high optical rotation of the enantiopure lactam $\{[\alpha]_D = 557$ $(c = 1, CH_2Cl_2)\}$, and a number of potentially useful strains were identified quickly. Moreover, strains were found with enantiocomplementary lactamases that cleave either the (+)- (**1b**) or the (–)-enantiomer (**1a**) of the lactam[10] (Scheme 3). In the former case, the (–)-lactam **1a**, which has the required configuration for the synthesis of carbocyclic nucleosides of the natural configuration, is not consumed. This was verified by conversion to (–)-carbovir,[11] the absolute configuration of which had already been established by stereochemical correlation to aristeromycin.[5]

9.2.3 DEVELOPMENT OF A WHOLE-CELL BIORESOLUTION PROCESS FOR THE (–)-γ-LACTAM

Following an initial laboratory resolution of the lactam **1**, work was undertaken to establish a viable manufacturing process. Cells of suitable strains, for instance *Pseudomonas* spp., were grown in an appropriate medium comprising glucose as carbon source, yeast extract and mineral salts. Cells grown in this way and resuspended in a neutral buffer gave a rapid transformation of racemic lactam at up to

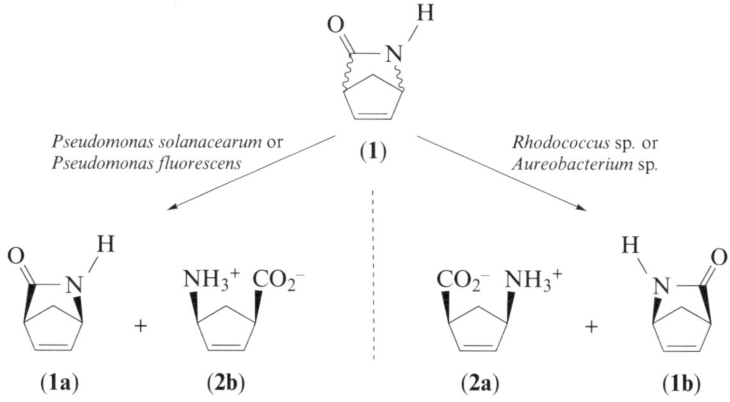

Scheme 3 Enantiocomplementary lactamase resolutions

100 g l^{-1} in aqueous solution such that bioconversion was essentially complete in about 3h at ambient temperature. Whilst the isolated microbial enzyme was found to be of limited stability, it was relatively stable within the cells. Furthermore a convenient procedure has been found in which the cells are grown by fermentation and the resulting cell mass is frozen and stored. The biotransformation can then be conducted at a later time, and addition of the frozen cell mass to an aqueous lactam solution is all that is needed to effect the biotransformation. This procedure has the advantage that the biotransformation can be performed at a different site from the fermentation, thereby improving manufacturing flexibility. Also, the freeze–thaw process causes some degree of cell lysis, which liberates the enzyme and increases the rate of the transformation. A further feature of this biotransformation is that control of the pH is not required, since the amino acid formed acts as its own buffer, maintaining the mixture at around neutral pH. After the biotransformation, the resulting aqueous solution contains the required (–)-lactam (**1a**), the amino acid (**2b**) of opposite configuration and cell debris. The lactam can be selectively extracted into an organic solvent, such as dichloromethane, leaving the amino acid behind. Such a procedure lends itself to scale-up because of its simplicity and the fact that the microbial strain can be grown to the necessary biomass to provide sufficient catalyst. At the time of writing, the process has been used to produce enantiopure (–)-**1a** on a 1 tonne scale.

9.2.4 DEVELOPMENT OF AN IMMOBILISED LACTAMASE RESOLUTION PROCESS FOR THE (+)-γ-LACTAM AND THE RING-OPENED (–)-AMINO ACID

As indicated above, the enzymes that provide (–)-lactam **1a** are of limited stability outside of cells, so the transformation is more practicable with whole cells. It was discovered, however, that a lactamase present in an *Aureobacterium* species, which has the opposite stereoselectivity, is remarkably robust.[12] It is stable when isolated and, moreover, it maintains its activity up to 70 °C. These characteristics are ideal for establishing an immobilised enzyme system (see Volume I, Chapter 19). The lactamase was purified by a procedure which included ammonium sulphate fractionation and anion-exchange chromatography, and was immobilised directly on a glutaraldehyde-activated solid support. The biotransformation was then operated simply by cycling an aqueous solution of racemic lactam (up to 200 $g\,l^{-1}$) through the immobilised enzyme reactor until all the (–)-enantiomer (**1a**) was hydrolysed. At this stage there was no evidence of enantiomer contamination in either the remaining substrate or the product; an enantiospecificity value, *E*, of >7000 was calculated, which is at the limits of sensitivity of the assay methods used. The immobilised lactamase proved to be extremely stable; a continuous transformation using a small reactor revealed little loss of activity during 8 months, after which time the conversion was discontinued owing to mechanical failure of the support.

Clearly, with such a high level of catalyst re-use, the contribution of the cost of the catalyst to the cost of the single enantiomer product becomes minimal. Indeed, the volume productivity (amount of substrate converted by a given amount of biomass in a given time) has been increased some 20 000-fold by the development. A feature of biotransformation processes is that with development, substantial economic improvements can often be realised. Another feature of an immobilised enzyme biotransformation process is that the downstream work-up is greatly simplified; there is no interference from proteinaceous matter during the isolation and large volumes of solvent are not needed. After the transformation the mixture is simply slurried in acetone, which leaves the (+)-lactam (**1b**) in solution, while the (–)-amino acid (**2a**) crystallises, and can be collected by filtration. Noteworthy, also, is that this biotransformation provides the amino acid directly in its neutral (zwitterionic) form, which cannot be achieved by chemical cleavage of the lactam (in which acid hydrolysis would produce the amine salt of the acid).

9.2.5 DOWNSTREAM CHEMISTRY FROM THE RESOLVED γ-LACTAM OR AMINO ACID

Given the availability of the single enantiomer of the lactam 2-azabicyclo[2.2.1]hept-5-en-3-one (**1a**), established routes based on the racemate may be used to provide enantiopure carbocyclic nucleosides such as (–)-carbovir. Thus, acid hydrolysis of the lactam (**1a**), esterification, *N*-acetylation and calcium borohydride reduction give the amino alcohol (**3**), which is then converted into the carbocyclic nucleoside carbovir[11,12] (Scheme 4).

Carbocyclic nucleosides in which the ring has all the functionality of the ribose, required for the cardiac vasodilator adenosine agonists (e.g. c-NECA), are accessed through osmium tetraoxide-catalysed dihydroxylation. Here, the rigid bicyclic framework ensures that the dihydroxylation is almost completely

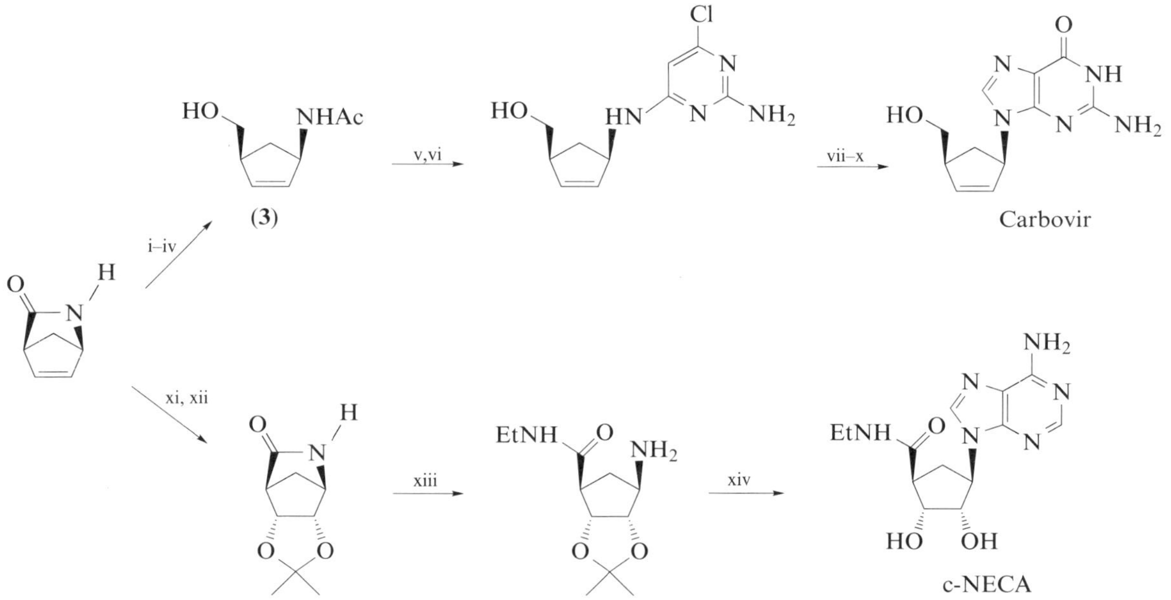

Reagents: i, aq. HCl; ii, $(MeO)_2CMe_2$; iii, Ac_2O; iv, $Ca(BH_4)_2$; v, HCl, aq. EtOH; vi, 2-amino-4,6-dichloropyrimidine; vii, $4\text{-}ClC_6H_4N_2^+Cl^-$; viii, Zn, HOAc; ix, $(EtO)_3CH$; x, NaOH; xi, *N*-methylmorpholine *N*-oxide, OsO_4 (cat.), aq. acetone; xii, $(MeO)_2CMe_2$; xiii, $EtNH_2$; xiv, standard steps.

Scheme 4 Conversion of the single-enantiomer bicyclic lactam into carbocyclic nucleosides stereoselective for the *exo*-face whereas, when carried out on ring-opened lactam substrate, such alkene functionalisation tends to give isomer mixtures.[13] In the case of the immobilised enzyme biotransformation described in the previous section [which gives the (–)-amino acid **2a** and the (+)-lactam **1b**], it is necessary to use the amino acid product (**2a**) to obtain carbocyclic nucleosides of the natural configuration. The amino acid **2a** is converted easily into its benzoylamide methyl ester derivative which is an established precursor of certain carbocyclic nucleosides. One drawback of any such biocatalytic kinetic resolution is that it leaves an unwanted isomer. The inversion of the (+)-lactam into the (–)-lactam requires a multi-step sequence in which the key step is a skeletal rearrangement;[14] this does not seem an economic option at present, however.

9.3 RESOLUTION OF A VERSATILE HYDROXYLACTONE SYNTHON 4-*ENDO*-HYDROXY-2-OXABICYCLO[3.3.0]OCT-7-EN-3-ONE BY LIPASE DE-ESTERIFICATION

9.3.1 SUBSTRATE SYNTHESIS

As discussed in the previous section, conversion of the bicyclic lactam **1a** into a carbocyclic nucleoside usually requires the heterocyclic base to be built on to the amine nitrogen (although displacement of the nitrogen, via the *N,N*-ditosylate, has recently been reported[15]). In seeking a route whereby the base could be introduced intact as a nucleophile, we discovered that the bicylic hydroxylactone **4** was suitable for this purpose.

The racemic hydroxylactones **4** and **5** are synthesised simply by the cycloaddition of glyoxylic acid (as a 50% aqueous solution) and cyclopentadiene, which is effected by mixing the neat components at ambient temperature (Scheme 5). The

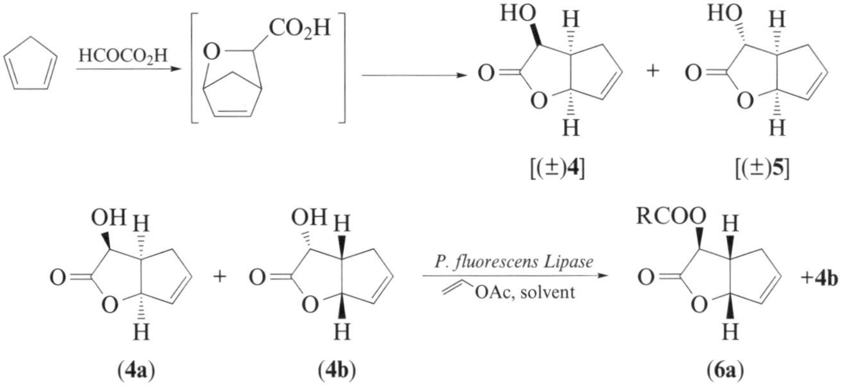

Scheme 5 Products from the cycloaddition of cyclopentadiene with glyoxylic acid

reaction was first described by Lubineau *et al.*[16] as an example of a reaction accelerated by water. The reaction is presumed to proceed initially by a [4 + 2]-cycloaddition of glyoxylic acid and cyclopentadiene in a hetero-Diels–Alder reaction to form a bicyclo[2.2.1] system. This adduct then undergoes rapid electrocyclic rearrangement to the more stable [3.3.0] system, resulting in a mixture of the *endo* (**4**) and *exo* (**5**) isomers in a ratio of about 4:1. On a small scale these diastereoisomers can be separated by column chromatography or by fractional crystallisation; however, the former cannot be scaled up easily and the latter gives poor yields. Hence the challenge was to develop a resolution with selectivity for the *endo*-isomer in addition to enantiocontrol.

9.3.2 DEVELOPMENT OF A BIOTRANSFORMATION FOR RESOLUTION OF THE HYDROXYLACTONE

The first resolution[17] identified in the laboratory employed *Pseudomonas fluorescens* lipase to catalyse the acetylation of the hydroxylactone using vinyl acetate in an organic solvent. Here, the lactone isomer (**4a**) with the correct absolute configuration for synthesis of the natural nucleoside was converted into the corresponding acetate product (**6a**; R = Me). The acetate and remaining alcohol enantiomer could be separated by chromatography, but this procedure was not convenient on a large scale. For this reason, we examined the corresponding hydrolytic biotransformation.

Screening for hydrolysis of the acetate, propionate and butyrate esters of a mixture of the diastereoisomers **4** and **5** by commercially available lipases soon revealed hydrolysis of the butyrate ester by *Pseudomonas fluorescens* lipase as a method of choice (Scheme 6).[18] Thus, a crude mixture of the *endo*- and *exo*-butyrate esters (**6** and **7**; R = Pr^n) was resolved efficiently in aqueous buffer, with continuous addition of NaOH to maintain neutral pH. The hydrolysis typically gave *endo* product of >92% *ee*, with minimal hydrolysis of the *exo*-isomer. The enantiomeric ratio (*E* value) for the butyrate ester hydrolysis was calculated as >200, and this process was subsequently scaled for use with multi-kilogram quantities of substrate, with an isolation procedure which avoids chromatography. Initially, solvent extraction studies (heptane–water) showed that if a suffi-

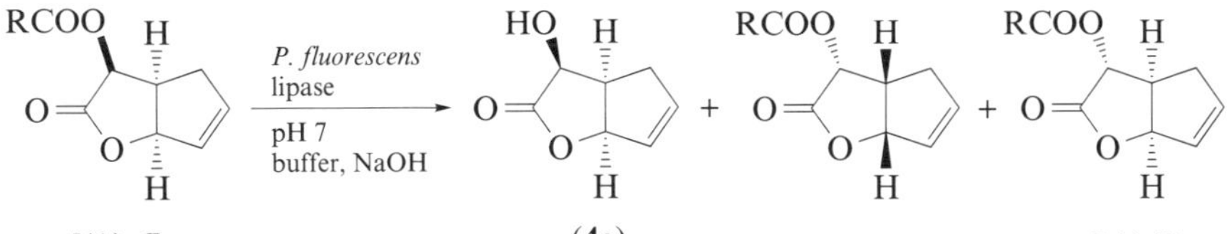

[(±)-**6**]

(**4a**)

[(±)-**7**]

(R = Me, Et, Pr^n)

Scheme 6 Resolution of 4-*endo*-hydroxy-2-oxabicyclo[3.3.0]oct-7-en-3-one [formulae follow the recommendation of Maehr, H., *J. Chem. Educ.*, **62**, 114 (1985)]

ciently hydrophobic ester was used (e.g. the butyrate), the ester could be selectively partitioned into the organic phase while the alcohol formed in the biotransformation was retained in the aqueous phase. Furthermore, the neat butyrate ester was immiscible with the aqueous phase, so centrifugation of the biotransformation mixture separated most of the unwanted ester impurities as an oily phase. The alcohol (**4a**) could then be recovered by simple solvent extraction, followed by a recrystallisation which enhanced the *ee* to 99%.

There was one complication: both acetate and butyrate esters were found to undergo spontaneous pH-dependent ring opening of the lactone to give a carboxylate salt, with faster decomposition at higher pH. This resulted in some loss of product for the butyrate ester (10%), but a much higher loss for the more soluble acetate ester, providing further justification for the choice of butyrate ester as substrate.

9.3.3 ESTERIFICATION VERSUS HYDROLYSIS

In view of the loss of some of the product to ring hydrolysis, the enzymatic esterification approach in which, hopefully, such ring opening should not occur, was re-investigated. Furthermore, conducting biotransformations in non-aqueous media frequently offers benefits such as increased enzyme stability or easier work-up. A biocatalytic, irreversible transesterification seemed, at first, to be the preferred route for resolution of the lactone **4**. However, trials showed that acylation of the hydroxylactone mixture by *Pseudomonas fluorescens* lipase, using vinyl acetate or vinyl butyrate in toluene, was slow and not as selective, with E values reduced to 30–40; (it is fairly common to find such differences in enantioselectivity between biocatalytic acylations and aqueous hydrolyses). Consequently, the acylation reaction required considerably more enzyme than the hydrolysis, and gave a less pure product after isolation.

9.3.4 DOWNSTREAM CHEMISTRY FROM THE RESOLVED HYDROXYLACTONE

The optically pure (–)-hydroxylactone **4a** proved to be a versatile building block for key chiral compounds, as, for example, in an alternative synthesis of the anti-HIV agent carbovir[18] (Scheme 7). This was achieved by reduction of the lactone to a triol, followed by oxidative cleavage of the vicinal diol and another reduction to a diol. Selective protection of the diol, then nucleophilic displacement via a π-allyl palladium complex intermediate using the intact heterocyclic base 2-amino-6-chloropurine and, finally, a deprotection yielded (–)-carbovir.

The (+)-hydroxylactone **4b**, on the other hand, can be used to provide intermediates for hypocholesteremic reagents,[19] where a key step is a Baeyer–Villiger oxidation of a cyclopentanone to a δ-lactone (Scheme 8). (+)-Hydroxylactone **4b** has also been used as a precursor to the antifungal agent brefeldin A, in what is probably the shortest synthesis of this compound to date (Scheme 9).[20]

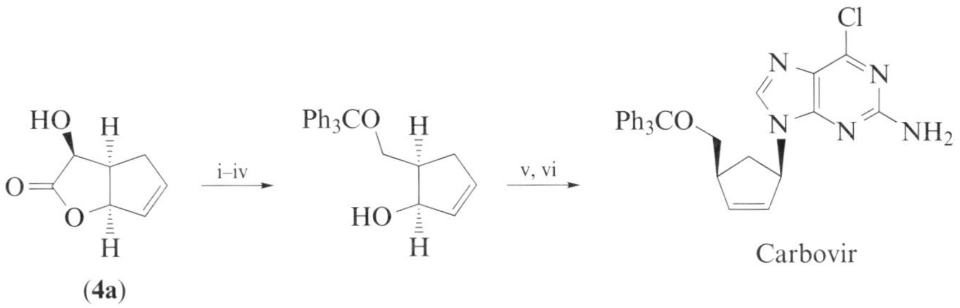

Reagents: i, $LiAlH_4$; ii, $NaIO_4$; iii, $NaBH_4$; iv, Ph_3CCl; v, Ac_2O; vi, 2-amino-6-chloropurine, $Pd(PPh_3)_4$ (cat.)

Scheme 7 Conversion of the resolved hydroxylactone into a carbocyclic nucleoside

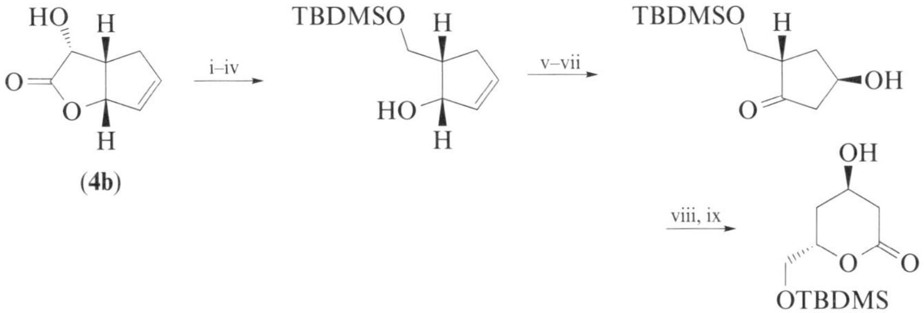

Reagents: i, $LiAlH_4$; ii, $NaIO_4$; iii, $NaBH_4$; iv, TBDMSCl; v, PCC; vi, H_2O_2, NaOH; vii, Al–Hg; viii, MCPBA; ix, AcOH

Scheme 8 Conversion of the resolved hydroxylactone in a precursor of hypocholesterolaemic agents

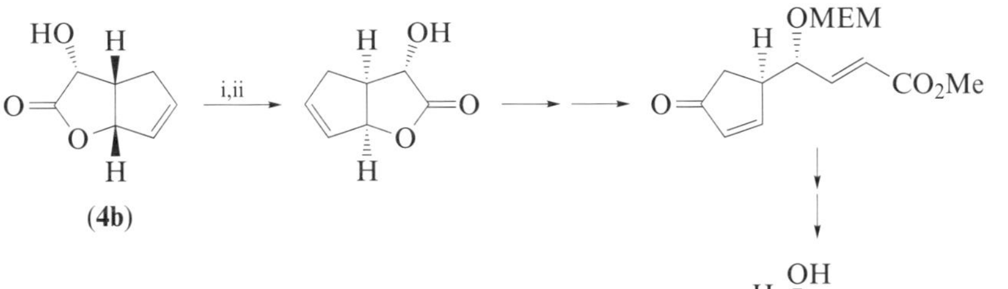

Reagents: i, Ph_3P, $ClCH_2CO_2H$, DEAD; ii, thiourea, EtOH

Scheme 9 Conversion of resolved hydroxylactone **4b** into brefeldin A

9.4 INTEGRATION OF AN ACYLASE BIOTRANSFORMATION WITH PROCESS CHEMISTRY: A ONE-POT SYNTHESIS OF *N*-BOC-L-3-(4-THIAZOLYL)ALANINE AND RELATED AMINO ACIDS

9.4.1 USES OF UNNATURAL AMINO ACIDS

Unnatural amino acids are of growing importance as components of therapeutic oligopeptides, where replacement of a natural amino acid in an enzyme or receptor binding agent may provide a required inhibitory or antagonist activity or, alternatively, may improve properties such as oral bioavailability or resistance to metabolic degradation. In addition, unnatural amino acids may themselves be versatile synthetic intermediates or have potential for use as chiral auxiliaries (see also Volume I, Chapters 8 and 20). One such unnatural amino acid is *N*-BOC-3-(4-thiazoyl)alanine (BOC-TAZ) (**9**). This has been used as a component of antihypertensive inhibitors of the enzyme renin, where it acts as a mimic of histidine.[21,22] The complete structures of two such renin inhibitors are shown in Scheme 10. The following sections describe the synthesis of the single enan-

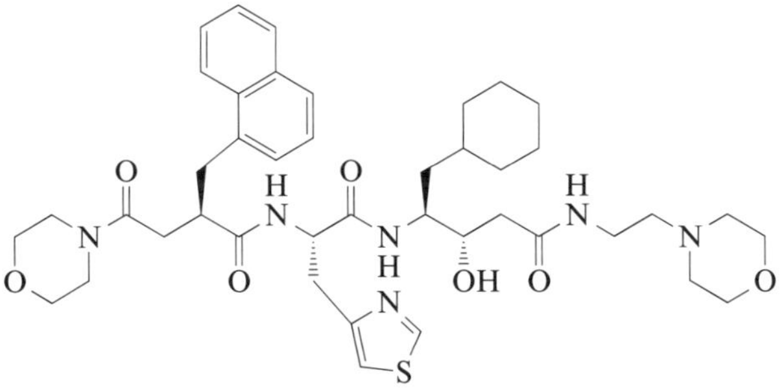

Sankyo

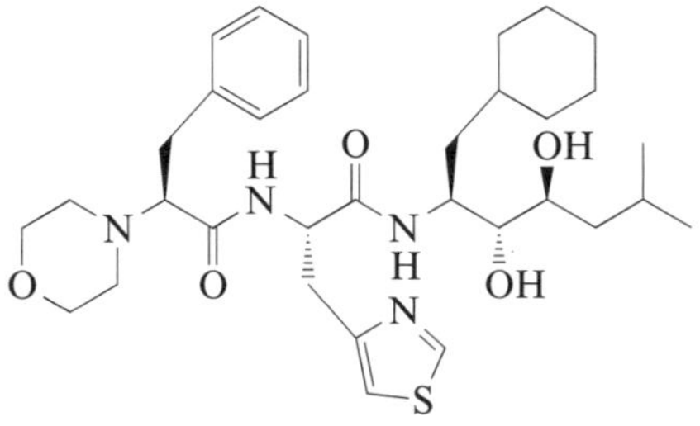

Abbott

Scheme 10 Examples of renin inhibitors containing L-thiazolylalanine

tiomer L-BOC-TAZ, although it should be noted that the method is applicable to a range of different amino acids.

9.4.2 OVERALL SYNTHETIC ROUTE TO *N*-BOC-3-(4-THIAZOYL)ALANINE

Today there are a great many synthetic approaches to amino acids.[22] One very effective synthesis of phenylalanine derivatives starts with a condensation reaction of a glycine anion equivalent with the appropriate benzaldehyde derivative; however, this requires the corresponding aldehyde to be available. In the case of L-BOC-TAZ (**9**) the appropriate aldehyde is not readily available, necessitating synthesis from 4-chloromethylthiazole hydrochloride (**8**), which, in turn, is prepared by the condensation of thioformamide with 1,3-dichloroacetone.[24] This approach, which utilises deprotonated diethyl acetamidomalonate as a glycine anion equivalent, is shown in Scheme 11. The process development steps used to obtain the single enantiomer of the protected amino acid from the acetamidomalonate condensation product are described below. The initial target was to produce 20 kg of single enantiomer product, and this was accomplished within a three month timescale.

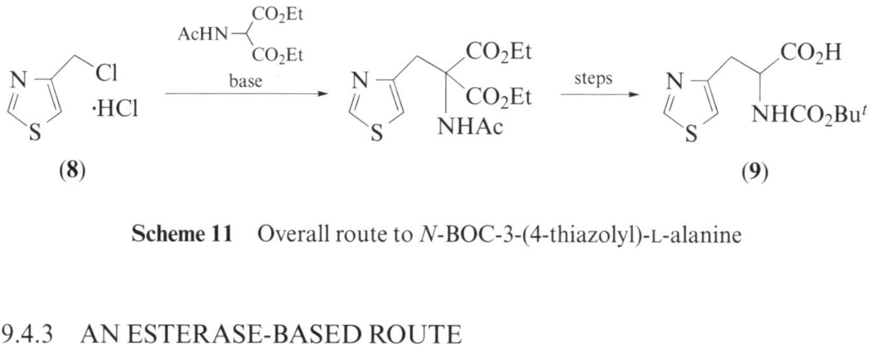

(**8**) (**9**)

Scheme 11 Overall route to *N*-BOC-3-(4-thiazolyl)-L-alanine

9.4.3 AN ESTERASE-BASED ROUTE

Our first experimental approach was based on the use of an esterase[25,26] to effect the key resolution, as outlined in Scheme 12. This initially appears to be an attractive route, since the unwanted D-enantiomer (**10b**), which remains as the ester after the transformation, is readily extracted from the mixture and racemised with ethoxide to recycle to the process.

However, from a process point of view the approach was very cumbersome and inefficient. The condensation step was carried out in ethanol with ethoxide as base, which was problematic because of the need to use ethanol free of other alcohols in order to avoid mixed ester products. In a typical procedure, the anion of the acetamidomalonate was generated by treatment with one equivalent of ethoxide in ethanol at about 60 °C. To this solution was added the chloromethylthiazole, generated as a solution of free base by the prior addition

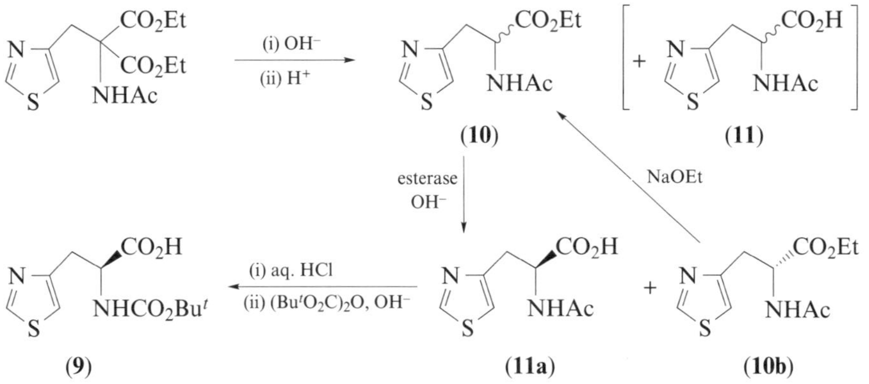

Scheme 12 Esterase-based route to *N*-BOC-3-(4-thiazoyl)-L-alanine

of 1 equivalent of ethoxide to a suspension of the hydrochloride in ethanol. The chloromethylthiazole, which is a potent vesicant, had to be used quickly since it polymerises readily. After the condensation reaction, which proceeds smoothly at 60–80 °C, the diester condensation product may be isolated by drown-out into cold water, in which it is only slightly soluble. The next step is selective hydrolysis and decarboxylation to give the monoester. On the laboratory scale, this can be effected fairly easily by adding sufficient aqueous lithium hydroxide (ca two equivalents) to a solution of the diester and warming until hydrolysis is complete, then acidifying and heating to promote decarboxylation. On the plant, it would not be safe to add all the lithium hydroxide at once, and it was added portionwise at 55 °C. HPLC analysis mid-way through the addition showed that a substantial amount of monoester **10** had already formed, indicating that decarboxylation was occurring directly under these conditions. Moreover, when a sample of the mixture was treated with more hydroxide, it was evident that the monoester hydrolysed faster than the diester. Whatever the method used to carry out the diester hydrolysis/decarboxylation, at least 10% of the racemic *N*-acetylamino acid **11** was formed.

To avoid the unwanted D-enantiomer being carried through the biotransformation step, it was necessary to remove the racemic acid **11** from the ester **10**. This was done by extraction of the ester **10** into an organic solvent at neutral pH, then solvent removal prior to dissolution in water for the biotransformation. It quickly became apparent that such a process was highly inefficient. The ester **10** was produced in a predominantly aqueous solution (from the hydrolysis), and extracted into organic solvent only to be redissolved in water. Also, a significant loss of yield resulted from chemical 'over-hydrolysis.' Biotransformation of the ester **10** took place readily with proteases such as α-chymotrypsin or subtilisin, to give the L-*N*-acetylamino acid **11a** and residual D-ester **10b**, which could be extracted from the resulting mixture and recycled. Concentrated acid had to be

Table 1 Steps in the ester resolution process to L-BOC-TAZ

Step		Solvent	pH
1	Condensation	Ethanol	Basic
2	Diester crystallisation	None	
3	Ester hydrolysis	Aq. ethanol	Basic
4	Decarboxylation	Water	Acidic
5	Isolation of racemic ester	Dichloromethane	Neutral
6	Biotransformation	Water	Neutral
7	Removal of D-ester		
8	Amide hydrolysis		Acid
9	Butoxycarbonylation	Aq. methanol	Basic
10	Product extraction	Dichloromethane	Acid

used for the hydrolysis of the L-acetylamino acid **11a**, and, hence, a lot of base was required for neutralisation prior to the butoxycarbonylation. However, a fundamental problem with the esterase bioresolution was the lack of enantiospecificity for the thiazoylalanine substrate (**10**); the product was only around 92% *ee*, which was not sufficient for its purpose. The *ee* of the BOC-protected thiazoylalanine **9** could be raised by formation of the salt with (*S*)-α-methylbenzylamine, allowing completion of the synthesis (note that this procedure only enhances the *ee* of pre-enriched material, and cannot be employed for *de novo* classical resolution of the racemate).

A complete synthesis of the *N*-BOC-L-thiazoylalanine **9** had thus been achieved, but the process was far from satisfactory. The process steps are summarised in Table 1, and the problems are (i) too many extraction steps, (ii) too many pH changes, (iii) yield loss from the chemical formation of racemic amide acid, (iv) use of pure ethanol as solvent (which requires an excise license) and (v) lack of enantiopurity in the product from the biotransformation, which entails an extra step to improve the *ee*.

9.4.4 DEVELOPMENT OF A ONE-POT PROCESS BASED ON AMINOACYLASE RESOLUTION

The problems described above prompted the development of an alternative process. Effective resolution of *N*-acetylamino acids had previously been shown for a wide range of phenylalanine analogues by Whitesides *et al.*[27] and it was discovered that racemic amide acid **11** (a by-product from the esterase-based process) readily underwent stereoselective amide hydrolysis with aminoacylase from pig kidney or *Aspergillus niger*. Moreover, the L-amino acid **12** formed proved to be of high enantiomeric excess (> 98% ee). This bioresolution formed the basis of the one-pot process to L-BOC-TAZ (**9**) which is outlined in Scheme 13.

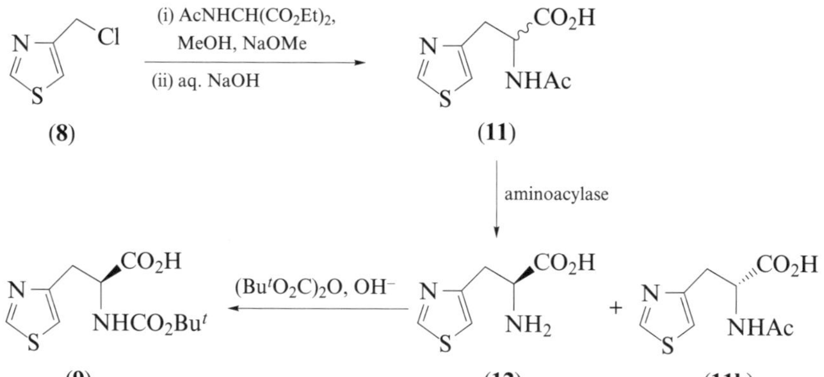

Scheme 13 Aminoacyclase-based one-pot process to *N*-BOC-3-(4-thiazoyl)-L-alanine

The first advantage in carrying out the resolution at the amide stage is that the initial acetamidomalonate condensation could use methanol instead of ethanol. Thus, while the condensation in methanol at 60 °C resulted in a product mixture of ethyl and methyl esters, these were hydrolysed in the next step. The hydrolysis was performed by slow addition of hydroxide solution, maintaining the mixture at pH 9–10. At this pH, not only does hydrolysis occur at a reasonable rate, but also the initial product undergoes decarboxylation. The decarboxylation causes a reduction in pH and, by careful control of the base addition, the resulting mixture can be kept approximately neutral; in this way a specific acidification step for the decarboxylation is not necessary.

A further feature of this hydrolysis method is that it was carried out with addition of the aqueous base, with concurrent removal of the methanol by distillation. The remaining aqueous solution could be used directly for the biotransformation, for which an acylase had been identified that was sufficiently robust to perform the hydrolysis in what was, essentially, a crude reaction mixture. After the biotransformation the mixture of L-amino acid (**12**) and D-acetylamino acid **11b** was in aqueous solution, thus avoiding the acidic hydrolysis step used in the esterase process. The butoxycarbonylation of the amino acid could be carried out directly with di-*tert*-butyl dicarbonate upon basification and addition of methanol as co-solvent. The BOC-amino acid **9** could then be extracted directly from the aqueous reaction mixture with *tert*-butyl methyl ether (TBME) and was of high enantiomeric excess. Thus, the sequence from chloromethylthiazole (**8**) to L-BOC-TAZ had been accomplished without removal of the mixture from the reaction vessel. This combines optimum vessel utilisation and minimum solvent waste and provides a demonstration of the effective integration of the biotransformation into the chemical process. The complete process, summarised in Table 2, is much simplified compared with the previous esterase-based process.

Table 2 Steps in the one-pot aminoacylase-based process

Step		Solvent	pH
1	Condensation	Methanol	Basic
2	Hydrolysis/decarboxylation		Basic
3	Biotransformation	Water	Neutral
4	Butoxycarbonylation		Basic
5	Product extraction	TBME	Acidic

9.4.5 RECYCLE OF THE D-ENANTIOMER

In the esterase-based process, the unwanted enantiomer is readily extracted as unreacted ester and is racemised with base. It is less straightforward with the shorter, acylase-based route, but the D-acetylamino acid **(11b)** can be recovered after concentration of the aqueous solution and then racemised by heating with acetic anhydride. The mechanism of this racemisation involves activation of the acid function, which causes cyclisation to an oxazolinone (azlactone) (**13**), which tautomerises to the enol **14** (Scheme 14).

Hydrolysis of the azlactone under alkaline conditions then provides the racemised acetylamino acid, which can be recycled to the biotransformation stage. In principle, the azlactone (**13**) could be a substrate for biotransformation; this approach has been used for a synthesis of L-*tert*-leucine, as described in Section 9.5.

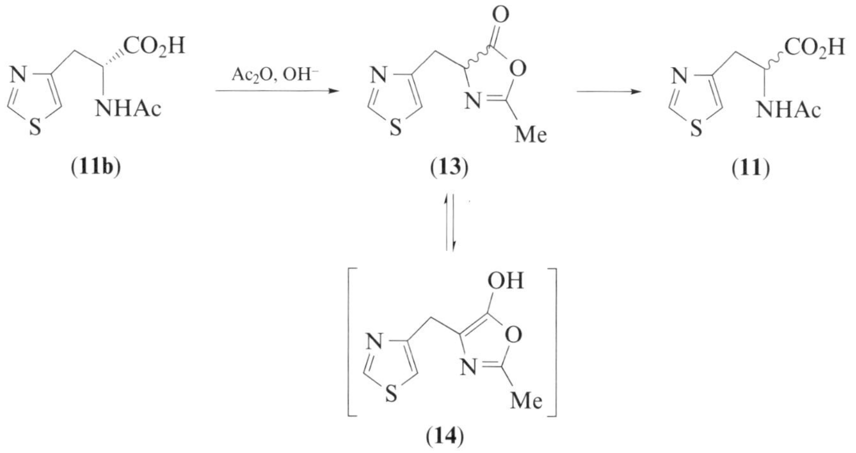

Scheme 14 Racemisation of the D-isomer via an oxazolinone (azlactone)

9.4.6 COMPARISON OF DIFFERENT BIOTRANSFORMATION PROCESSES

Another route worthy of consideration combines the esterase and aminoacylase reactions. Here, the racemic ester **10** is prepared, as described in Section 9.4.3, but is not separated from the racemic amide acid by-product (**11**). The esterase biotransformation reaction is then carried out on this mixture, and the product is subjected to the aminoacylase reaction which provides the L-amino acid **12**. The D-ester (**10b**) left behind from the tandem biotransformations can be selectively extracted and racemised with base for recycle. In terms of the number of solvent changes, a comparison of the three possible routes shows six changes for the esterase-based process, two changes for the aminoacylase-based process and four changes for the combined enzyme process. Owing to its greater simplicity and the higher biocatalyst costs in the combined method, the aminoacylase route is preferred, even though recycling of the unwanted enantiomer is more difficult.

9.4.7 APPLICATION OF THE AMINOACYLASE PROCESS TO OTHER UNNATURAL AMINO ACIDS

Initially, work described in Section 9.4.4 used an aminoacylase powder in solution that could not be recovered after use. More recently, an acylase has been identified and immobilised that performs the transformation when the acid amide solution is simply passed through a column of the supported enzyme. This has the double benefit of giving a protein-free solution for the downstream steps and allowing enzyme re-use, which markedly improves the process economics. Additionally, the process has been found to be applicable to other unnatural alanine derivatives such as L-2-naphthylalanine (**15**), L-4-cyanophenylalanine (**16**), L-2-pyridylalanine (**17**) and L-allylglycine (**18**) (Scheme 15), with appropriate process adjustments to account for the different solubilities of these compounds.

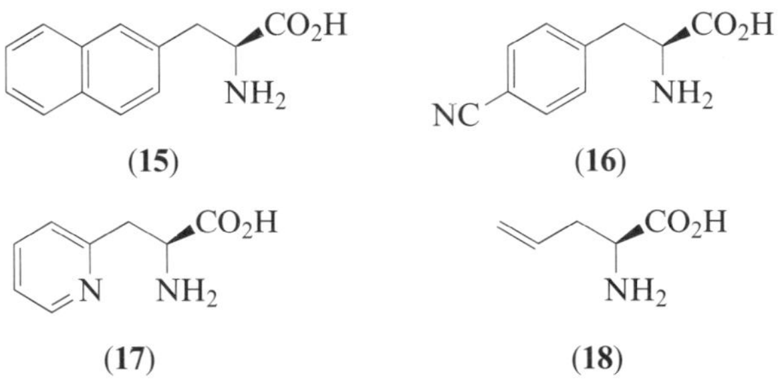

Scheme 15 Other amino acids prepared by the acylase-based resolution process

9.5 DYNAMIC RESOLUTION OF AN OXAZOLINONE BY LIPASE BIOCATALYSIS: SYNTHESIS OF *(S)-tert*-LEUCINE

9.5.1 USES OF *(S)-tert*-LEUCINE

The unnatural amino acid (*S*)-*tert*-leucine (*tert*-butylglycine) is useful as a lipophilic, hindered component of peptides[28] where, because of its bulky nature, cleavage by peptidases is disfavoured, giving peptides of improved metabolic stability. *tert*-Leucine is also useful as the building block for a number of chiral auxiliaries and ligands, such as those shown in Scheme 16, where the presence of the bulky *tert*-butyl group makes these compounds particularly effective for asymmetric synthesis. However, the *tert*-butyl group also makes the synthesis of the optically active amino acid more challenging.

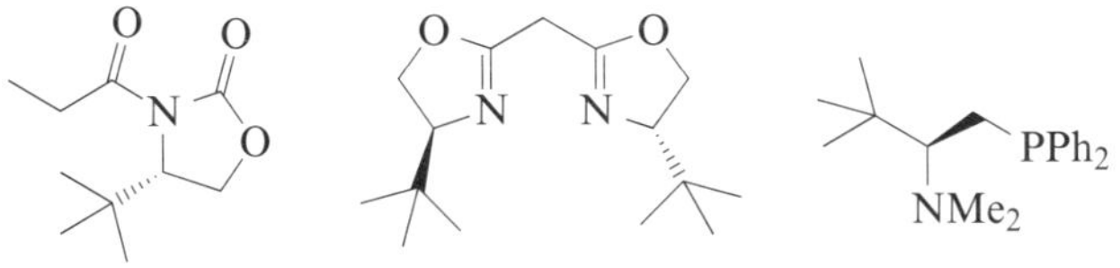

Scheme 16 Products derived from L-*tert*-leucine

9.5.2 AZLACTONES FOR THE DYNAMIC RESOLUTION OF AMINO ACIDS

One previously reported solution to the synthesis of (*S*)-*tert*-leucine involved overall asymmetric reductive amination of the corresponding prochiral keto acid (3,3-dimethyl-2-oxobutanoic acid) with a coupled enzyme system developed by Degussa comprising a dehydrogenase and enzymic cofactor recycling.[29] Chiroscience have developed an alternative approach based on a hydrolase resolution of a 5-(4*H*)oxazolinone (azlactone) intermediate in which a cofactor is not required. Appropriate azlactones are prepared by cyclodehydration of an *N*-acylated amino acid with acetic anhydride (cf. Section 9.4.5). Generally, azlactones are readily ring opened by a variety of nucleophiles to amino acid derivatives, and, because of the relatively acidic C-4 hydrogen (p*K*a 8.8–9), are easily racemised. These compounds have received a lot of attention because of their transient formation during peptide synthesis, leading to partial racemisation. Lipases and proteases have been used to effect asymmetric ring opening of azlactones using either water or alcohols as nucleophiles;[30] facile racemisation of the less reactive enantiomer permits a dynamic resolution and the complete conversion of a racemate to an optically pure amino acid derivative.

9.5.3 DEVELOPMENT OF A LIPASE-CATALYSED DYNAMIC RESOLUTION.

An azlactone (**19**) of racemic *tert*-leucine is readily prepared by acetic anhydride treatment of the *N*-benzoyl derivative (Scheme 17). Whilst water can be used as the nucleophile for enzyme-catalysed ring opening, unselective, non-enzymatic hydrolysis of the azlactone reduces the *ee* of the resulting acid. This problem can be overcome by using an alcohol as the nucleophile in a water-free system. Lipozyme (an immobilised *Mucor miehei* enzyme) was found to be the enzyme of choice and, using *n*-butanol as the nucleophile, a highly selective reaction was possible, giving the ester **20** in 97% *ee* and 90% yield (Scheme 17); lower alcohols give lower rates in the resolution reaction and lower enantiomeric excesses.

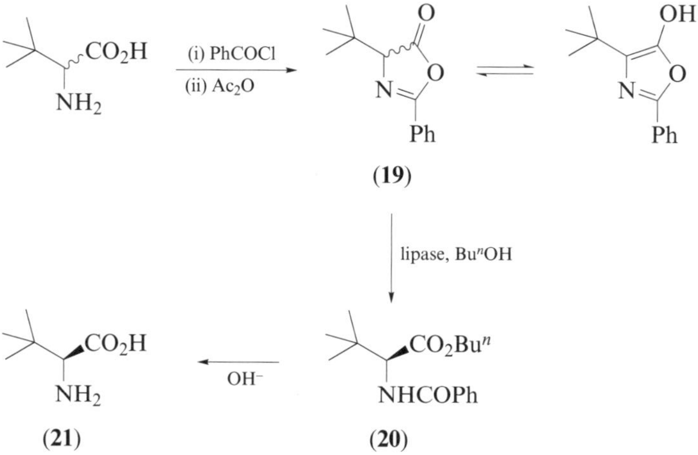

Scheme 17 Dynamic biocatalytic resolution process for L-*tert*-leucine

The initial laboratory procedure for the resolution[31] was not considered suitable for manufacture since it was performed at a high dilution, with a high loading of enzyme (substrate-to-enzyme ratio 10:1) and still required 5 days for complete conversion. Also, in the early experiments it was necessary to include triethylamine, otherwise the yield and product *ee* were reduced; presumably the triethylamine effected the racemisation of the (*R*)-azlactone. However, during development it was found that the concentration of the substrate could be increased to 20%, and conditions were established whereby the reaction was complete within 24 h; it was also found that the triethylamine was no longer necessary to obtain a high yield and enantiomeric excess. It seems that the higher concentration of the substrate is sufficient to promote racemisation of the (*R*)-azlactone.

Importantly, the enzyme could be reused, although this resulted in decreased yields and *ee*. It is possible that the azlactone somehow modifies the lipase to a form which has reduced enantiospecificity, but which is strongly inhibited by triethylamine. Deprotection of the ester amide **20** to the amino acid **21** is not as straightforward as might be expected. Normally, an amino acid protected in this way would be cleaved by acid hydrolysis but, in this case, such conditions led to partial racemisation caused by transient recyclisation to the stereochemically labile azlactone **19**. This recyclisation probably reduces adverse steric interactions caused by the bulky *tert*-butyl group. A solution to this problem was to perform a sequential deprotection, hydrolysing the ester first using enzymatic (Alcalase) or chemical (NaOH/water/methanol) methods. Debenzoylation, with minimal racemisation, was effected using caesium hydroxide, and it was shown later that the complete hydrolysis could be carried out with potassium hydroxide without racemisation. Note that the *tert*-butyl group probably reduces the risk of base-induced racemisation by hindering abstraction of the α-hydrogen.

Overall, while the process may formally be described as a resolution, by virtue of the *in situ* recycling it is a deracemisation, and is as effective as an asymmetric synthesis from a prochiral starting material. The viability of the process is dependent on the economics of supply of the racemate. However, it is possible to make the azlactone directly from the inexpensive *N*-benzoylglycine (hippuric acid) (**22**) via the glycine azlactone (**23**) as shown in Scheme 18,[32] thus providing a particularly attractive overall route.

Clearly, while the above case history has focussed on L-*tert*-leucine, the methodology could be applied equally to other *tert*-alkyl glycine derivatives. For example, using the route shown in Scheme 18, condensation of the glycine azlactone **23** with different ketones would be followed by introduction of the final alkyl group using an appropriate copper-coupled Grignard reagent.

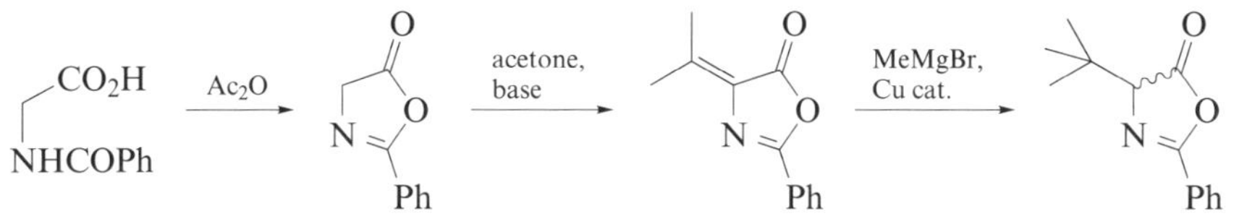

(**22**) (**23**) (**19**)

Scheme 18 Alternative route to the azlactone bioresolution substrate for L-*tert*-leucine

9.6 CONCLUSIONS

Four examples have been described of development work carried out on biocatalytic resolution procedures to provide scaleable processes. Combining the learning from these case studies, it is possible to highlight aspects that are of general importance when establishing bioresolutions.

First, it is important to consider the most appropriate substrate, not only for the biotransformation itself (e.g. in respect of enantiospecificity, conversion rate, etc.) but also to facilitate product isolation. Thus, in the case of the bicyclic hydroxylactone (Section 9.3), it was found that the butyrate ester rather than the acetate allowed solvent partitioning as a scaleable means of separation of the biotransformation product. It is better still if the substrate can be chosen such that the product crystallises directly during the work-up. Also, isolation of the product can be facilitated when the biotransformation has been developed, as in the case of the immobilised lactamase (Section 9.2), to operate at a high concentration (200 g l^{-1}).

It is important when developing a biotransformation-based step to consider its fit within the overall synthetic pathway, as exemplified by the process development for the *N*-BOC-TAZ (Section 9.4). Here, the preferred biotransformation led to simplification of the overall process, taking into consideration issues such as the number of pH changes and solvent exchange steps. It is also of note here that a bioresolution involving amide hydrolysis has a significant advantage over an esterase resolution in having less, or no, requirement for pH control during the transformation, which facilitates operation in non-specialised equipment. This is illustrated well by the case of the lactamase process (Section 9.2), in which the transformation required only an aqueous solution of the substrate and the biocatalyst. To integrate a biotransformation with the total process, it can be valuable to combine, in the earlier stages of a project, a bioresolution with a method for enhancing the enantiomeric purity of the product, typically a crystallisation technique. This gives an interim process with the capability of providing enantiomerically pure product during the development and optimisation of the biocatalytic resolution (see Section 9.4).

A choice which frequently has to be made in a biocatalytic resolution is whether to work in the hydrolytic direction, in aqueous solution, or in the esterification (or transesterification) mode in a low-water system. The hydrolytic approach frequently has the advantages of higher biotransformation rates and, often, greater enantioselectivity, but can suffer from competing chemical hydrolysis; these were considerations for both the bicyclic hydroxylactone (Section 9.3) and the *tert*-leucine (Section 9.5), which are, preferably, carried out in aqueous and organic solvent systems, respectively.

Lastly, a problem with all resolutions is the utilisation of the unwanted enantiomer—a most important consideration in providing the best economics. In the case of the lactamase work (Section 9.2), and the bicyclic hydroxylactone (Section 9.3), utilisation of the unwanted enantiomer is not practical, but the resolution takes place early in the total synthesis, so lessening the economic impact. In the case of the *N*-BOC-thiazoylalanine acylase resolution approach (Section 9.4), a racemisation procedure for the unwanted D-enantiomer was established via an oxazolinone (azlactone) intermediate. This concept was extended in a biotransformation of the azlactone itself in a process for (*S*)-*tert*-leucine where, owing to *in situ* racemisation of the unwanted enantiomer, the process is as efficient as an asymmetric synthesis.

9.7 ACKNOWLEDGEMENTS

We are grateful to colleagues and collaborators at Chiroscience and the University of Exeter whose efforts have provided the results discussed in this chapter.

9.8 REFERENCES

1. Coates, J. A. V., Ingall, H. J., Pearson, B. A., Penn, C. R., Storer, R., Williamson C., and Cameron, J. M., *Antiviral Res.*, **15**, 161 (1991).
2. Borthwick, A. D., Butt, S., Biggadike, K., Exall, A. M., Roberts, S. M., Youds, P. M., Kirk, B. E., Booth, B. R., Cameron, J. M., Cox, S. W., Marr C.L.P., and Shill, M., *J., Chem. Soc., Chem. Commun.*, 656 (1988).
3. Scopes, D. I., Newton, R. F., Ravenscroft, P., and Cookson, R.C., *Eur. Pat.*, 0 104 066, 1984.
4. Chen, J., Grim, M., Rock, C., and Chan, K., *Tetrahedron Lett.*, **30**, 5543 (1989).
5. Exall, A. M., Jones, M. F., Mo, C.-L., Myers, P. L., Paternoster, I. L., Singh, H., Storer, R., Weingarten, G. G., Williamson, C., Brodie, A. C., Cook, J., Lake, D. E., Meerholtz, C. A., Turnbull, P .J. and Highcock, R. M., *J. Chem. Soc., Perkin Trans. 1*, 2467 (1991).
6. Borthwick, A.D., and Biggadike, K., *Tetrahedron*, **48**, 571 (1992).
7. Jagt, J. C., and van Leusen, A.M., *J. Org. Chem.*, **39**, 564 (1974).
8. Griffiths, G.J., and Previdoli, F.E., *J. Org. Chem.*, **58**, 6129 (1993).
9. Daluge, S., and Vince, R., *J. Org. Chem.*, **43**, 2311 (1978); Vince, R. and Hua, M., *J. Med. Chem.*, **33**, 17 (1990).
10. Taylor, S. J. C., Sutherland, A. G., Lee, C., Wisdom, R., Thomas, S., Roberts, S.M., and Evans, C., *J. Chem. Soc,. Chem. Commun.*, 1120 (1990).
11. Evans, C. T., Roberts, S.M., Shoberu K.A., and Sutherland, A.G., *J. Chem. Soc., Perkin Trans. 1*, 589 (1992).
12. Taylor, S. J. C., McCague, R., Wisdom, R., Lee, C., Dickson, K., Ruecroft, G., O'Brien, F., Littlechild, J., Bevan, J., Roberts, S. M., and Evans, C.T., *Tetrahedron: Asymmetry*, **4**, 1117 (1993).
13. Vince, R., and Daluge, S., *J. Org. Chem.*, **45**, 531 (1980).
14. Palmer, C., and McCague, R., *J. Chem. Soc., Perkin Trans. 1*, in press.
15. Jung, M. E., and Rhee, H., *J. Org. Chem.*, **59**, 4917 (1994).
16. Lubinau, A., Augé, J,. and Lubin, N., *Tetrahedron Lett.*, **32**, 7529 (1991).
17. MacKeith, R. A., McCague, R., Olivo, H. F., Palmer, C. F., and Roberts, S.M., *J. Chem. Soc., Perkin Trans. 1*, 313 (1993); *ICT Int. Pat. Appl.*, WO 93/00826, 1993.
18. MacKeith, R. A., McCague, R., Olivo, H. F., Roberts, S. M., Taylor, S. J. C., and Xiong, H., *Bioorg. Med. Chem.*, **2**, 387 (1994).
19. McCague, R., Olivo, H. F., and Roberts, S. M., *Tetrahedron Lett.*, **34**, 3785 (1993).
20. Casy, G., Gorins, G., McCague, R., Olivo, H. F., and Roberts, S.M., *J. Chem. Soc., Chem. Commun.*, 1085 (1994); Carnell, A.J., Casy, G., Gorins, G.A., Kompany-Saeid, A., McCague, R., Olivo, H.F., Roberts, S.M., and Willetts, A.J., *J. Chem. Soc., Perkin Trans. 1*, 3431 (1994).
21. Rosenberg, S. H., Spina, K. P., Woods, K.W., Polakowski, J., Martin, D. L., Yao, Z., Stein, H. H., Cohen, J., Barlow, J. L., Egan, D. A., Tricarico, K. A., Baker, W. R., and Kleinert, H. D., *J. Med. Chem.*, **36**, 449 (1993).
22. Nishi, T., Saito, F., Nagahori, H., Kataoka, M., Morisawa, Y., Yabe, Y., Sakurai, M., Higashita, S., Shogi, M., Matsushita, Y., Ijima, Y., Ohizumi, K., and Koike, H., *Chem. Pharm. Bull.*, **38**, 103 (1990).
23. Duthaler, R.O., *Tetrahedron*, **50**, 1539 (1994).

24. Caldwell, W. T., and Fox, S. M., *J. Am. Chem. Soc.*, **73**, 2935 (1951).
25. Miyazawa, T., Iwanaga, H., Ueji, S., Yamada, T., and Kuwata, S., *Chem. Lett.*, 2219 (1989).
26. Hsiao, C.-N., Leanna, M. R., Bhagavatula, L., and De Lara, E., *Synth. Commun.*, **20**, 3507 (1990).
27. Chenault, H. K., Dahmer, J., and Whitesides, G. M., *J. Am. Chem. Soc.*, **111**, 6354 (1989).
28. Fauchere, J.-L., and Petermann, C., *Helv. Chim. Acta,* **63**, 824 (1980).
29. Drauz, K., Bommarius, A, Knaup, G., Kottehahn, M. and Schwarm, M., presented at 'Chiral Europe '94' Symposium 19–20 Sept. 1994, Nice, (Spring Innovations).
30. Crich, J. Z., Brieva, R., Marquart, P., Gu, R.-L., Fleming, S., and Sih., C. J., *J. Org. Chem.*, **58**, 3252 (1993).
31. Turner, N. J., Winterman, J. R., McCague, R., Parratt, J. S. and Taylor, S. J. C., *Tetrahedron Lett.*, **36**, 1113 (1995).
32. Miyazawa, T., Nagai, T., Yamada, T., Kuwata, S. and Watanabe, H., *Mem. Konan Univ., Sci. Ser.*, **23**, 51 (1989).

10 (*S*)-2-Chloropropanoic Acid: Developments in Its Industrial Manufacture

S. C. TAYLOR
ZENECA LifeScience Molecules, Billingham, UK

10.1 Introduction 207
10.2 Market Requirements for (*S*)-2-Chloropropanoic Acid 208
10.3 (*S*)-2-Chloropropanoic Acid Synthesis Options 210
10.4 The ZENECA Approach to (*S*)-2-Chloropropanoic Acid 211
10.4.1 Biological Dehalogenation 211
10.4.2 Identification of an (*R*)-2-Chloropropanoic Acid Dehalogenase 212
10.4.3 Properties of (*R*)-2-Chloropropanoic Acid Dehalogenase and *Pseudomonas putida* NCIMB 12018 213
10.4.4 Downstream Processing Requirements 213
10.4.5 Process Development Overview 214
10.4.6 Developments Through Molecular Biology 219
10.4.7 Dehalogenation as a General Route to Chiral Molecules . 220
10.5 Conclusions 221
10.6 Acknowledgements 222
10.7 References 222

10.1 INTRODUCTION

(*S*)-2-Chloropropanoic acid [(*S*)-CPA] was one of the first commercially valuable single-enantiomer materials to be identified and has, as a consequence, received considerable attention from industrial and academic groups exploring synthesis options. Although structurally a relatively simple molecule, its importance as a large-volume homechiral building block and the cost sensitivity associated with its major agrochemical end markets have posed a difficult challenge to businesses and groups seeking to exploit the market opportunity. Numerous options for its production have been proposed and investigated at laboratory and pilot scale. During the 1980s, ZENECA (formerly ICI) invented and developed a novel bio-technological process which, since 1989, has been operated in a full-scale commercial plant.

Chirality in Industry II. Edited by A. N. Collins, G. N. Sheldrake and J. Crosby

It is the purpose of this chapter to explore the background interest in (*S*)-CPA and its end markets, to review the chemical and biological production options, to present a historical picture of how the current ZENECA bio-process has evolved and, finally, to draw some conclusions as to the application of biological stereoselective dehalogenation to other chiral molecules.

10.2 MARKET REQUIREMENTS FOR (*S*)-2-CHLOROPROPANOIC ACID

The opportunity for (*S*)-CPA falls into two distinct commercial areas: an undeveloped market as a pharmaceutical intermediate and a growing agrochemical market where it is a key building block for a major group of herbicides.

The importance of chirality to the pharmaceutical industry is now well recognised, reflecting the common situation where the desired biological activity resides only in one of a pair of enantiomers or, as is often the case, an undesirable activity is associated with the other enantiomer. The principal driving forces for making pharmaceutical compounds as individual enantiomers rather than racemates have been product efficacy and regulatory. Cost reduction, through the raw materials savings that can come from moving to a single enantiomer, has generally been of lesser importance. (*S*)-CPA has not been widely used as a chiral building block in the pharmaceutical sector, partly reflecting its lack of availability in high enantiomeric purity at the time when processes for the production of many of today's active molecules were being established. The non-steroidal anti-inflammatory group of therapeutics can be synthesised from 2-halopropanoates, e.g. 2-bromopropanoic acid or 2-chloropropanoic acid (CPA) and their esters. Products such as naproxen, ibuprofen and flurbiprofen have historically been made as racemates, but it is now clear that single-enantiomer forms will be marketed, creating an opportunity for (*S*)-CPA and its analogues from which they can be produced.[1]

Chirality is also important in the agrochemical industry, with pesticidal activity often localised in single enantiomers. Here, regulatory issues have been, and remain, less profound than in pharmaceuticals, and the principal reason for making products as single active enantiomers rather than racemates has been to achieve greater cost effectiveness in use. This has not been the sole reason, however, and some companies have recognised the opportunity to gain product differentiation and possibly lead regulatory change in favour of the more environmentally sound single-enantiomer forms. This is well illustrated by the group of herbicides based on the phenoxypropanoic acid structure.

There are many examples of the phenoxypropanoic acid herbicides such as mecoprop and dichlorprop[2] which have been produced and sold since the 1950s. All products are based on the same core structure (Figure 1) but with differing substitution on the aromatic ring. This can range from simple chloro or methyl substitution as in dichlorprop and mecoprop through to complex heteroaromatic systems as in fluazifop.

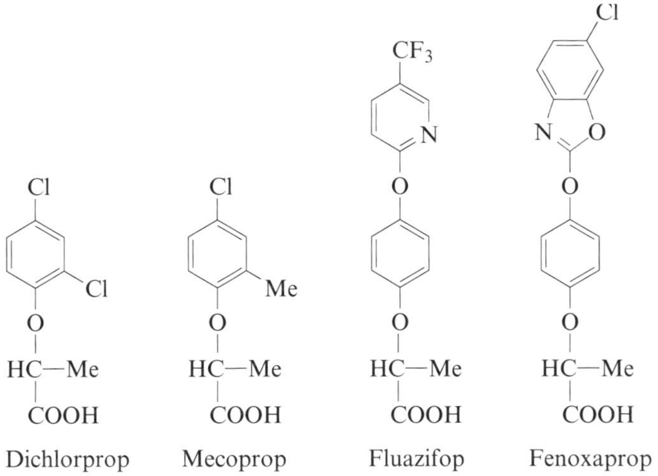

Figure 1 Examples of phenoxypropanoic herbicides

Products such as fluazifop are particularly valuable herbicides giving excellent post-emergent control of annual and perennial grass weeds in broadleaved crops at low dose rates of <0.5 kg a.i.ha^{-1} even as the racemic form. The older products such as mecoprop are used to control broadleaved weeds in cereals, but at much higher application rates of 1.5–3 kg a.i. ha^{-1} in the racemic form. Thus, very substantial volumes of these herbicides are applied worldwide and, although they are readily biodegradable, the presence of 50% inactive isomer is becoming an environmental issue. All of these products show herbicidal activity in the (*R*)-enantiomer, with the (*S*)-enantiomer having much reduced or no activity. Until recently, all major products were made as racemates with, in most cases, the chiral centre being derived from low-cost, racemic CPA, a salt or an ester derivative. In order to produce the single active (*R*)-enantiomers, the existing chemistry from CPA to final herbicide may be used, as it does not cause significant loss of enantiomer purity through racemisation during processing when used with (*S*)-CPA with minor proprietary process modifications. (*S*)-CPA, or in some instances the ester, can thus be substituted as the starting material, allowing the use of existing manufacturing plant and hence minimising new capital requirements.

One of the earliest phenoxypropanoate herbicides to be converted from racemic to single-enantiomer form was the ZENECA product fluazifop. Similar strategies were adopted by A. H. Marks in the UK and BASF in Germany with regard to mecoprop and other older, commodity herbicides, where the lower specific activity of the racemic products has meant a significant environmental load. Conversion to a single-isomer product allows for an effective doubling in manufacturing capacity for the final product and offers cost saving through reduced raw material use. Coupled with the environmental and potential regulatory benefits, this has given a strong case for moving such products to single-enantiomer form. In recent years, most of the major agrochemical companies have either initiated the switch of racemic phenoxypropanoate herbicides to single-enantiomer products or introduced completely novel single-enantiomer compounds, both approaches creating a growing market for (*S*)-CPA.

Synthetic approaches to phenoxypropanoic acid herbicides not based on the use of (*S*)-CPA have also been proposed. These include direct microbiological resolution of the final racemic herbicide, which was demonstrated by ZENECA with fluazifop,[3] and also, more recently, by esterase hydrolysis.[4] However, these do not appear to compete economically with the (*S*)-CPA option, indicating the effectiveness and high productivity of the chemistry from (*S*)-CPA and the cost implication of yield loss when working with the final active molecule. This substantiates the general rule that resolutions should take place as early as possible to minimise the total quantity of material produced.

The herbicide market needs described above require (*S*)-CPA to be available at a low price, but with good chemical and enantiomer purity (>92% *ee*) and on a scale of several thousand tonnes per annum.

10.3 (*S*)-2-CHLOROPROPANOIC ACID SYNTHESIS OPTIONS

The principal route options to (*S*)-CPA have all included a biological stage. The oldest established route is based upon the fermentative production of (*R*)-lactic acid from glucose using selected strains of *Lactobacillus*. Although the (*R*)-isomer is not considered to be the natural lactic isomer, bacteria which are able to produce this isomer almost exclusively are now well characterised.[5,6] However, they do require a much higher degree of fermentation control than is common with *Lactobacillus* species, in order to prevent contamination with (*S*)-isomer producing species. Conversion of (*R*)-lactic acid to (*S*)-CPA can be done by various chemical methods, e.g. esterification followed by chlorination with thionyl chloride.

The second main synthesis option utilises the now well known ability of esterase and lipase enzymes to catalyse the selective hydrolysis or transesterification of esters of acids such as CPA. The early work on CPA resolution with esterases was done by Klibanov and co-workers in the USA [7] simultaneously with studies by the Stauffer Chemical Company.[8] However, extended reaction times, a carefully controlled ratio of organic and aqueous phases and a degree of conversion in excess of 60% were necessary to obtain good enantiomer purity. This could be a result of the reaction taking place at a site distant from the chiral centre on a small molecule. More recently this reaction has been further developed by BASF (using the isobutyl ester of CPA and an immobilised *Pseudomonas* esterase[9]), Chemie Linz[10] and others[11] (Figure 2).

Other routes have been investigated in the laboratory, including stereospecific enzymic hydrolysis of racemic 2-chloropropionamide,[12] hydroformylation with

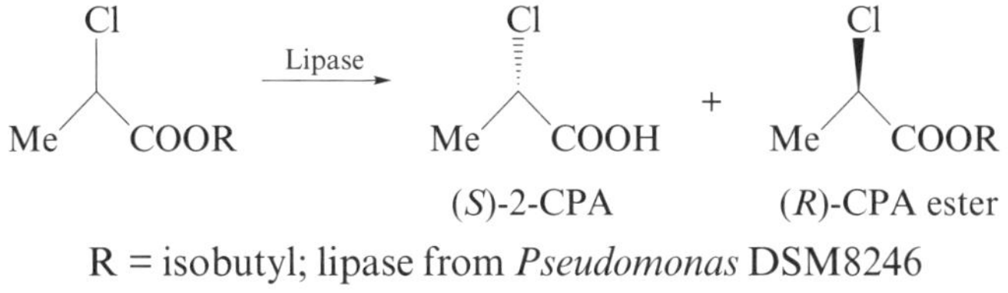

Figure 2. Proposed use of ester hydrolysis as a route to (*S*)-CPA

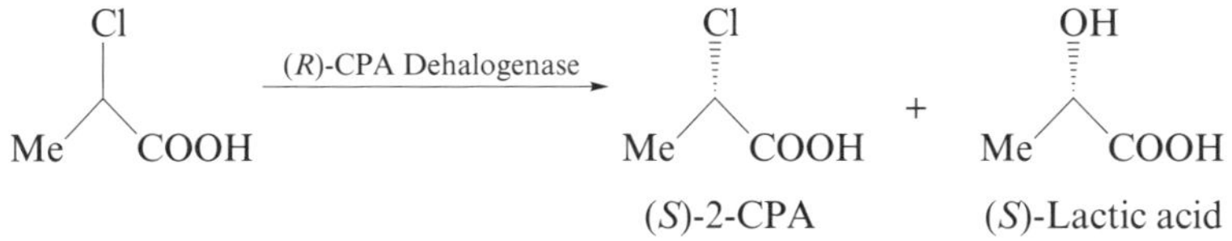

Figure 3. Proposed use of stereoselective enzyme-catalysed dehalogenation as a route to (*S*)-CPA

metal–ligand complexes [13] and diastereoselective halogenation of silyl ketene acetals.[14]

The development of a ZENECA process was aimed at overcoming the performance shortcomings indicated above with the lactic acid and esterase routes, and achieving a cost base in advance of other more speculative alternatives. The low cost of racemic CPA made resolution approaches potentially economic, and novel enzyme-catalysed dehalogenation, where the reaction would occur directly at the chiral centre and thus, potentially, offer high enantiomer specificity, was recognised by ZENECA as an innovative approach. The principle was to use a hydrolytic, non-cofactor-requiring enzyme that catalysed the removal of chlorine from (*R*)-CPA but not from (*S*)-CPA. Thus, from racemic CPA, reaction products would include unreacted (*S*)-CPA, which should be readily separated from the expected hydroxy acid (lactic acid) co-product of resolution (Figure 3).[15]

10.4 THE ZENECA APPROACH TO (*S*)-2-CHLOROPROPANOIC ACID

10.4.1 BIOLOGICAL DEHALOGENATION

Enzymes that catalyse the dehalogenation of molecules are widespread in nature and occur in both prokaryotic and eukaryotic organisms. Microbes that live in environments contaminated with halogenated compounds, particularly *Pseudomonas* bacteria, are good sources of enzymes that catalyse the elimination of halogen from organic molecules and many organisms are able to use halogenated aliphatic and aromatic compounds as sole sources of carbon and energy for growth. Many biochemical studies on CPA, 2,2-dichloropropanoate and monochloroacetate biodegradation[16,17] have demonstrated that the initial attack on the molecule is catalysed by dehalogenase enzymes in what appears to be a hydrolytic removal of halogen from the C-2 position yielding the corresponding hydroxy acid. Many dehalogenases can be present in single species of bacteria, and these can have different substrate and enantiomer specificities.[18] At the start of the ZENECA programme, several dehalogenases had been reported that were specific for the (*S*)-enantiomer of CPA, with no activity being shown against the (*R*)-enantiomer.[19,20] However, no dehalogenase had been reported that was specific for the (*R*)-enantiomer of CPA. From an industrial perspective, if such a novel enzyme could be isolated and its effectiveness and commercial value demonstrated, this would give rise to a sound intellectual property position.

10.4.2 IDENTIFICATION OF AN (*R*)-2-CHLOROPROPANOIC ACID DEHALOGENASE

The approach adopted to find a specific (*R*)-CPA dehalogenase was to screen bacteria isolated from soil sites contaminated with chloro acids, such as areas where agrochemicals were manufactured or applied. Although many strains were found that could grow on (*R*)-CPA but not (*S*)-CPA, suggesting the presence of an (*R*)-CPA-specific dehalogenase, in practice these invariably turned out to be deficient in the metabolism of the (*R*)-lactic acid isomer (the expected product of CPA dehalogenation), rather than having a CPA dehalogenase specificity. This probably indicates the selective disadvantage to an organism of being unable to dehalogenate an isomer of a toxic compound such as CPA. Furthermore, it illustrates the deficiency of microbial selection as a basis for identifying new enzymes. However, several racemic CPA-utilising strains displayed differential growth rates on (*R*)- and (*S*)-CPA, which suggested that these strains might contain dehalogenases of different specificity. Dehalogenases in these strains were readily separated by polyacrylamide gel electrophoresis coupled with an enzyme activity stain[21] and, on examination, these were found to contain multiple dehalogenases of different specificities, including the required (*R*)-CPA specificity (Table 1). Thus, *Pseudomonas putida* AJ1, renamed NCIMB 12018, contained two specific dehalogenases, one specific for (*R*)-CPA and the other for (*S*)-CPA.

Table 1 Specificity of dehalogenases present in CPA-degrading microorganisms isolated by ZENECA

Organism	CPA Enantiomer specificity	
	Dehalogenase 1	Dehalogenase 2
P. putida NCIMB 12018	(*R*)-CPA	(*S*)-CPA
P. fluorescens NCIMB 12159	(*R*)-CPA	(*S*)-CPA
NCIMB 12158	(*R*)-CPA	(*R*/*S*)-CPA
NCIMB 12160	(*R*)-CPA	(*R*/*S*)-CPA
NCIMB 12161	(*R*)-CPA	(*R*/*S*)-CPA

10.4.3 PROPERTIES OF (*R*)-2-CHLOROPROPANOIC ACID DEHALOGENASE AND *PSEUDOMONAS PUTIDA* NCIMB 12018

P. putida NCIMB 12018 was found to be a robust organism able to grow in the presence of relatively high concentrations (>100 mM) of CPA in minimal growth media. The strain also demonstrated rapid growth on simple, cheap carbon substrates such as glucose. As a *Pseudomonas* species it would be amenable to genetic manipulation by either mutation or molecular approaches. No toxicological problems were found with the strain, so it provided a good basis for development of a commercialisable fermentation process for production of the enzyme or whole-cell microbial biocatalyst.

The (*R*)-CPA dehalogenase was readily separated from the (*S*)-CPA dehalogenase by ion-exchange or molecular sieve chromatography, allowing examination of its catalytic properties, which was crucial to determining whether it could form the basis of a bio-transformation process. The enzyme was a tetramer of a molecular weight of 135 000 Da. Temperature and pH optima were in line with other dehalogenases and, most important, the enzyme stability was good under laboratory conditions at 30 °C and pH 7–8. At alkaline pH CPA rapidly racemises, so it was important to use a dehalogenase that retained high specific activity at near neutral pH; this was the case. The product of dehalogenation was (*S*)-lactic acid, which indicated a reaction with inversion of configuration. Activity with (*S*)-CPA as potential substrate was undetectable even in the presence of >0.2 M (*S*)-CPA. No significant inhibition of the enzyme was found in the presence of (*S*)-lactic acid, (*S*)-CPA or chloride ions (the products of the reaction) at concentrations in excess of 0.5 M. The values of K_{cat} ($15\,s^{-1}$) and, most importantly, K_m (1 mM) were acceptable in terms of achieving a fast and complete removal of (*R*)-CPA from the reaction, and thus high enantiomeric purity of the (*S*)-CPA product. A low K_m was an essential prerequisite for this, so that activity remained high at (*R*)-CPA concentrations of less than $1\,g\,l^{-1}$ which are typical of the end of a resolution bio-transformation.

In summary, both the host microbial strain and the basic properties of the enzyme indicated that, in principle, the enzyme and organism could be made in bulk, and that a high-purity (*S*)-CPA product could be made in an intensive reaction, with acceptable plant and capital implications for a multi-thousand tonne per annum facility. The realisation of this potential, however, required a substantial and multi-functional development programme.

10.4.4 DOWNSTREAM PROCESSING REQUIREMENTS

Separation problems can often arise with bio-transformation reactions owing to the presence of proteins and other biological materials which can interfere with solvent extraction and crystallisation processes. Thus, in addition to the fermentation and bio-transformation stages of the proposed (*S*)-CPA process, the principle of separating pure (*S*)-CPA from the reaction mixture also needed to be established. Although the product (*S*)-CPA is unstable at high pH, it has good stability at low pH in its un-ionised form. Therefore, at low pH, solvent extraction was applicable, and a carefully chosen solvent and operating conditions could separate (*S*)-CPA from the other major organic component, lactic acid. At low pH, proteinaceous material precipitated in a filterable form which was readily removed from the product stream. This offered flexibility in the nature of the biocatalyst, and implied that a whole cell process, where significant amounts of contaminating biological material would be present, was a viable option.

It was confirmed that all stages of a required process were feasible and that, provided appropriate plant volumetric and biocatalyst productivities could be achieved and the overall yield target met, a cost-effective process was attainable.

10.4.5 PROCESS DEVELOPMENT OVERVIEW

The key to success in bio-transformation technology is the process of translation of an embryonic laboratory procedure into a cost-effective, reliable and robust plant-scale operation in an acceptable time-scale, and with a resource appropriate to the opportunity. Many options needed to be considered for the (*S*)-CPA process, including isolated enzyme versus whole microbial cells, immobilised versus free cells or enzyme and batch versus continuous reaction for fermentation, bio-transformation and downstream processing stages. Making correct choices at an early stage in the development process for a bio-transformation is critical to economic success, and the learning required has become a key part of the intellectual property of ZENECA and other successful biotechnology businesses.

For simplicity and speed of scale-up, whole-cell systems offer several advantages over isolated enzymes. This is because several process steps, needed to prepare a cell free enzyme preparation, can be eliminated. *P. putida* NCIMB 12018 could not be used directly in a process as a whole-cell system owing to the presence of an (*S*)-CPA dehalogenase in the strain, which had first to be removed or inactivated. In the initial development work this was done by chemical mutagenesis, which proved effective at completely eliminating the unwanted dehalogenase activity, and resulted in strain NCIMB 12018-23. No other enzyme present in this strain interfered with the (*R*)-CPA dehalogenase or attacked the (*S*)-CPA product. (This can often be an issue with whole-cell systems; an isolated enzyme system is often required to overcome the problem.) Therefore, for the (*S*)-CPA system, a whole-cell approach was envisaged as being the most cost effective, with the shortest development time-scale, and consequently this approach was followed.

Early fermentation work concentrated on batch growth, although it was intended eventually to establish a continuous fermentation process and so derive the productivity benefits that are possible with this technology, which is well established in ZENECA. The dehalogenase enzyme was not constitutively expressed and substrate induction was required. Both isomers of CPA would induce (*R*)-CPA dehalogenase activity in *P. putida* NCIMB 12018-23. However,

(*S*)-CPA was relatively expensive to use in this way, and provided significant selection pressure on the mutant organism to revert to its former wild-type state. (*R*)-CPA, also expensive, was rapidly metabolised and had to be used at a concentration in excess of 50 mM to achieve good induction at target cell dry weights of >30 g l^{-1}. At this level, the resultant concentration of free chloride ion following dehalogenation causes corrosion problems with stainless steel, so specialist fermentation equipment would have been required. Racemic CPA was also an effective inducer but, like (*S*)-CPA, provided pressure on the culture to select for mutant revertants to wild type, and hence limited the options for fermentation on a longer time-scale.

Although a strain constitutive for (*R*)-CPA dehalogenase could have been constructed, it was found that some chlorinated acids, such as dichloropropanoic acid (DCP) and trichloroacetic acid (TCA), were not dechlorinated, but would act as powerful dehalogenase inducers at concentrations well below 1 mM. Furthermore, compounds such as TCA did not give selection pressure for strains reverting to (*S*)-CPA dehalogenase production. A low level of TCA or DCP, included in a sugar-based growth medium, allowed high fermentation culture dry weight of *P. putida* NCIMB 12018-23 with high dehalogenase activity. Over extended fermentation times typical of continuous fermentation mode, however, use of DCP resulted in a dramatic and unexpected total loss of the cells ability to make dehalogenase enzyme (Figure 4), possibly owing to the cell adapting to prevent uptake of the toxic and unmetabolisible DCP.

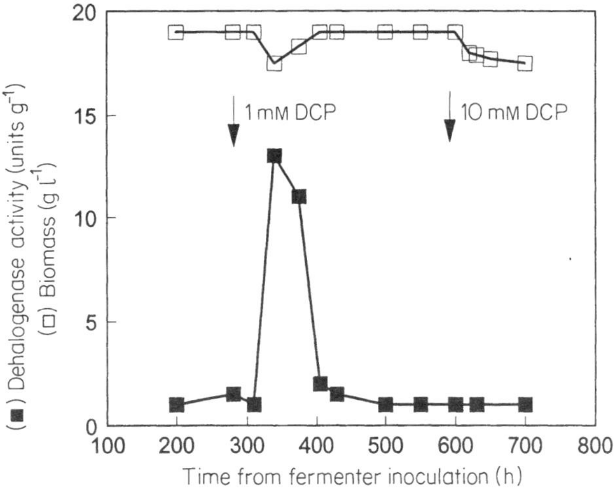

Figure 4 Profile of a continuous fermentation with the (*S*)-CPA production strain using the non-metabolised dichloropropanoate as dehalogenase inducer. The continuous fermentation was established in the steady state prior to introduction of dichloropropanoate (DCA) at 1 mM concentration at 300 h. This was increased to 10 mM at 590 h

The ability of *Pseudomonas* species to display novel methods of regulatory control over dehalogenases has been reported previously,[22] and this may be another example. The effect was observed with all non-metabolisable chlorinated acids tested, which precluded use of this type of enzyme induction system for continuous fermentation operation.

To overcome these problems, an alternative inducer, monochloroacetic acid, was used. This compound is slowly dechlorinated and metabolised by *P. putida* NCIMB 12018 strains, and could induce dehalogenase activity at an acceptable concentration (<50 mM) from a metallurgical perspective. Furthermore, it also provided the organism with an additional potentially utilisable carbon source, glycolate, arising from dechlorination. As an achiral compound, monochloroacetic acid was fully dechlorinated by the (*R*)-CPA dehalogenase and provided no selection pressure for mutant reversion; these properties combined allowed for long-term continuous fermentation (Figure 5).

The cells of *P. putida* NCIMB 12018-23 could be used directly in a bio-transformation process. In the wet form, however, the stability of the intact cells as a bio-catalyst was poor, and a method for converting microbial cells into a stable biocatalyst was required. This is a situation where immobilisation may be applied. However, technology for drying cells and retaining high enzyme activity and viability had been researched by ZENECA, and a proprietary cell drying process was developed and applied to the dehalogenase-containing microorganism. This gave >90% retention of the dehalogenase enzyme activity and allowed for storage of a solid biocatalyst for over 12 months without significant loss of

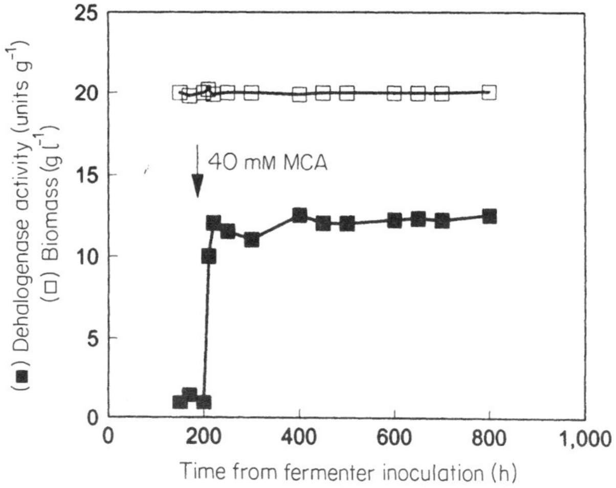

Figure 5 Profile of a continuous fermentation with the (*S*)-CPA production strain using monochloroacetate as dehalogenase inducer. The continuous fermentation was established in the steady state prior to introduction of monochloroacetate (MCA) at 40 mM concentration at 190 h

activity. The adopted strategy was, therefore, to uncouple fermentation and bio-transformation reactions and site them independently.

In order to achieve an (*S*)-CPA product of high enantiomeric purity, a fed-batch rather than continuous bio-transformation process was developed. Optimisation of the process required understanding the effect of, and then controlling, several variables, including temperature, pH and alkali addition, standing (*R*)-CPA concentration, feedstock quality, mixing regime, oxygen tension and buffering capacity of the reaction medium. The importance of these parameters is illustrated by the effect of pH on the reaction; the activity and stability of the biocatalyst are affected differently, and subtle control of pH is required (Figure 6).

By this approach, a robust, intensive process was developed which could be used on a tonne scale in routine operation. The (*S*)-CPA product was of high enantiomer purity (90–99% *ee*) which could be tailored to meet the market need by adjustment of the bio-transformation cycle time; for example, the pharmaceutical opportunity demands a higher enantiomeric purity product, but can tolerate higher cost than would be the case for material for an agrochemical outlet, thus allowing longer cycle times to be used. A downstream processing route based on solvent extraction gave a high recovery yield with no racemisation of (*S*)-CPA , and this element of the process could be operated continuously utilising established ZENECA technology.

The final process was established 18 months after invention of the concept, and utilised the fermentation and biocatalyst preparation capacity of ZENECA BioProducts on Teesside, UK, and the bio-transformation reaction and downstream processing assets of ZENECA Fine Chemicals at Huddersfield, UK. The overall process, shown diagrammatically in Figure 7, has a capacity for multi-thousand tonne production.

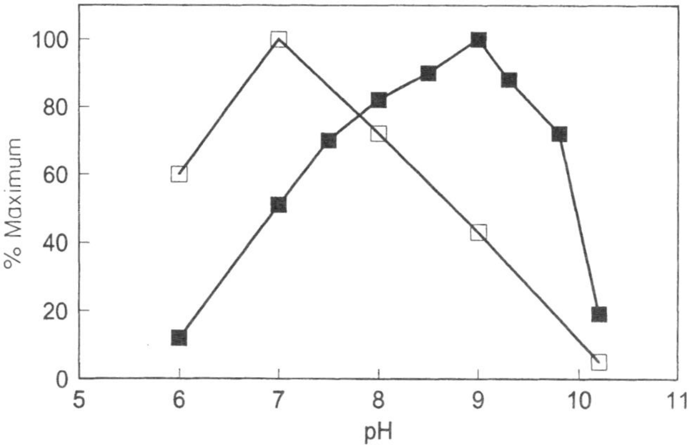

Figure 6 Effect of pH on the activity and stability of the (*S*)-CPA biocatalyst. (■) Dehalogenase activity; (□) dehalogenase stability. The relative stability was determined by measuring the half-life of the biocatalyst held at the appropriate pH and 30 °C

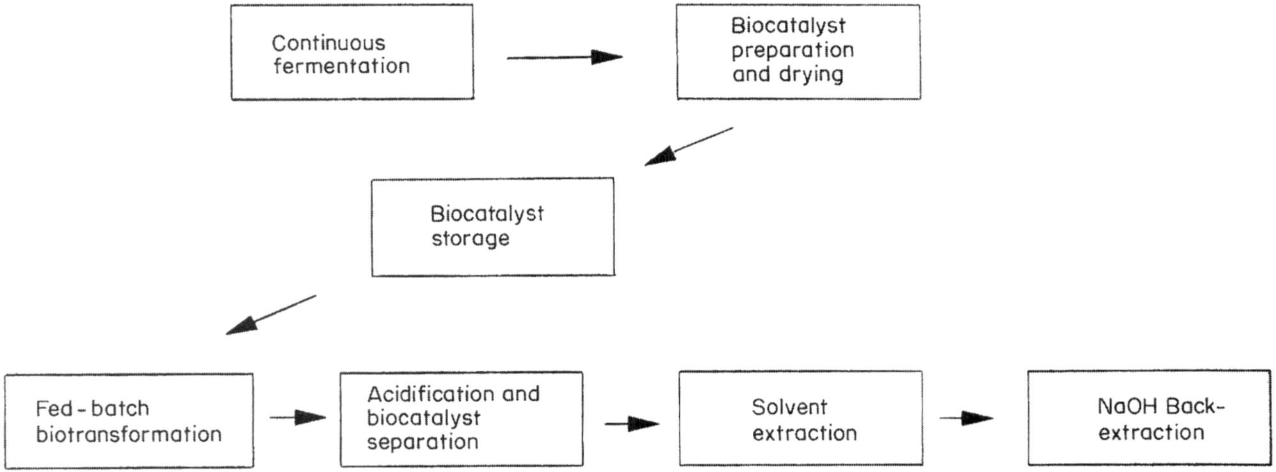

Figure 7 Flowsheet for the ZENECA (*S*)-2-CPA process

10.4.6 DEVELOPMENTS THROUGH MOLECULAR BIOLOGY

Although the initial *Pseudomonas putida* NCIMB 12018 biocatalyst was derived by mutagenesis, it was recognised that an approach which utilised molecular biology techniques offered the potential for achieving step changes in performance, albeit over a longer time-scale, as little was known of the genetics of this particular system. Using sequence information derived from a sample of purified (*R*)-CPA dehalogenase, the Had-D gene encoding the enzyme was identified, cloned and expressed in a strain of *Escherichia coli*.[23] Manipulation utilising a variety of proprietary ZENECA plasmid systems and constructs led to development of strains of both *E. coli* and *P. putida* which produced elevated levels of active (*R*)-CPA dehalogenase in minimal growth media over many generations. By utilising molecular biology and with a good understanding of *Pseudomonas* physiology, a *P. putida* derivative and strain-selective growth environment were established which enabled the use of continuous fermentation culture for many hundreds of hours, maintaining a high level of (*R*)-CPA dehalogenase enzyme expression. The performance of this strain in biotransformation and downstream processing was comparable with that of the mutant, *P. putida* NCIMB 12018-23, but with much higher biocatalyst efficiency and productivity. This strain and its derivatives have been in production at ZENECA since late 1991.

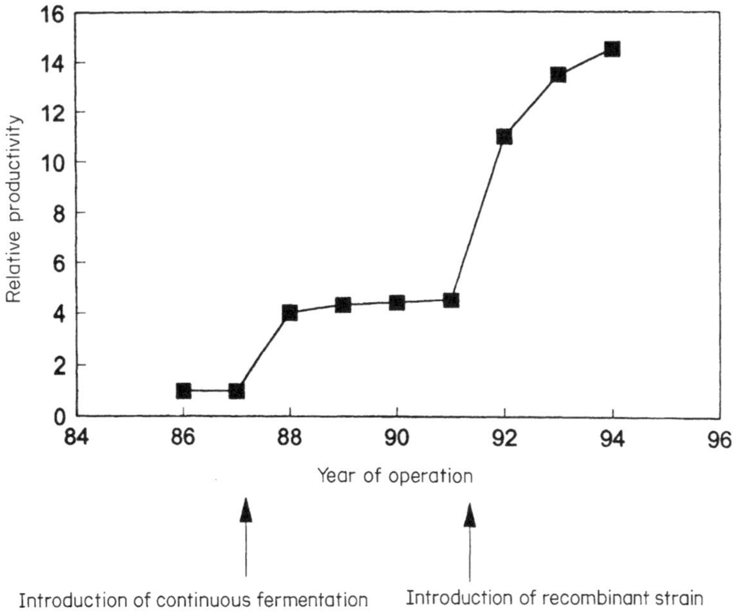

Figure 8 Development in fermenter productivity for dehalogenase enzyme, 1986–94, indicating the impact of continuous operation of a recombinant strain

The molecular studies of the (*R*)-CPA dehalogenase have also served to confirm the novelty of this particular enzyme. No significant sequence homology exists between this enzyme and the (*S*)-CPA dehalogenase also present in *P. putida* NCIMB 12018,[24] or with other dehalogenases that have been sequenced.[25] This seems to rule out a common evolutionary heritage, an aspect of academic interest on which more light may be shed as data on the active sites of these enzymes is accumulated.

The impact of molecular biology and continuous fermentation on this process can be judged by examination of the overall fermenter productivity, measured in units of dehalogenase enzyme made per hour of fermenter occupation. The 20-fold increase (Figure 8) from an already relatively intensive process with a high natural level of dehalogenase within *P. putida* NCIMB 12018 has made a significant impact upon production economics, and has resulted in even lower cost, commodity herbicides being cost-effectively switched to single enantiomer forms.

10.4.7 DEHALOGENATION AS A GENERAL ROUTE TO CHIRAL MOLECULES

The (*R*)-CPA dehalogenase was examined for its ability to generate products from a wide range of racemic and non-chiral halogenated compounds. Substrate specificity was very limited (Table 2) and only compounds with a carbon chain length of two to four, and with an acid at C-1 and a single chlorine or bromine substitution at C-2, were effective substrates. Additional halogen substitution on C-3 was also accepted by the dehalogenase. Where relevant, the enantiomeric purity of products was ascertained and without exception, the dehalogenase demonstrated single-enantiomer specificity.

Hence the (*R*)-CPA dehalogenase, whilst able to provide for the production of very high enantiomeric purity products, was limited in the range of substrates that could be utilised. The system does provide, *inter alia*, for high-purity (*S*)-2-bromopropanoic acid and the other acids shown in Figure 9, which have several potential applications in drug development programmes.

Table 2 Substrate range of the (*S*)-CPA dehalogenase from *P. putida* NCIMB 12018

2-Chloropropanoate
2-Bromopropanoate
Dibromoacetate
2,3-Dibromopropanoate
Iodoacetate
Bromomalonate
Chloroacetate
2,3-Dichloropropanoate
2-Bromo,3-hydroxypropanoate
Dichloroacetate

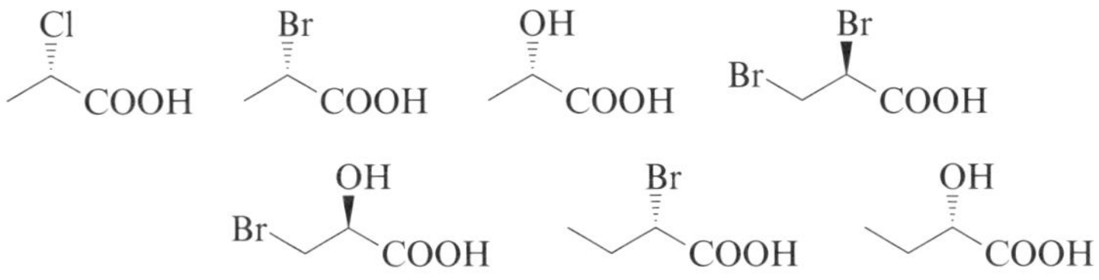

Figure 9 Chiral products derived by use of the ZENECA (*S*)-CPA dehalogenase technology

When 2,3-dibromopropanoic acid is used as substrate, the product is (*S*)-2,3-dibromopropanoate, an unexpected stereochemical outcome.[26] Studies with the (*S*)-CPA dehalogenase from *P. putida* NCIMB 12018 established a similar substrate profile for this complementary enzyme.

Hence, although dehalogenases can clearly be used to very good effect to produce single-isomer products on a very large commercial scale, there is no evidence to show the broad specificity and utility that have become characteristic of esterase resolution systems. Consequently, current dehalogenases are unlikely to find wide utility. However, they will find application where individual chemicals in development command sufficient value to justify a specific application programme. Exceptions to this situation will arise when the chloro- and bromopropanoic derivatives provide appropriate chiral building blocks.

10.5 CONCLUSIONS

The ZENECA development of a bio-transformation process for (*S*)-CPA has clearly demonstrated that these types of enzyme reactions can be scaled up and operated economically on commercial plant on a >1000 tpa scale. In order to realise the full economic benefits that biotechnology can bring to chiral chemistry, a holistic view of all elements of the process must be taken. In developing new technologies of this sort, options in one aspect of the process, such as the genetic make-up of the microbial strain, can provide ways to address problems in another distant aspect, e.g. separation problems downstream of the bio-transformation. The essence of success in this, as in so many applications of bio-technology, is a multi-disciplinary technical and commercial project team, focused on a common goal.

That the development time-scale for the (*S*)-CPA process was relatively long is, perhaps, not surprising for a novel technology. Subsequent bio-transformations developed by ZENECA for chiral molecules have benefited from the (*S*)-CPA learning experience, and the development time-scale for new bio-transformations has been reduced dramatically. Time-scale has now become possibly the most critical aspect in meeting the needs of the pharmaceutical industry, and what used to be a major constraint to the exploitation of biotechnology in this area has now largely been overcome. The key determinant of development timescale now is generally not identification of an appropriate enzyme, but rather the speed of scale-up reflected in the ability to mobilise a cohesive team covering molecular biology, microbiology, enzyme and fermentation technology, chemistry and engineering, and direct it at the key issues while drawing on experience to make early choices from process options. The importance of integrated process development is the key lesson for future bio-transformation developments.

10.6 ACKNOWLEDGEMENTS

The development of a successful (*S*)-CPA process would not have been possible without the strong support of ZENECA business and corporate management, who recognised the long-term value of this aspect of biotechnology. Most important, however, it is the R&D staff of ZENECA BioProducts, ZENECA Fine Chemicals, the ZENECA Fine Chemicals Manufacturing Organisation and ZENECA Pharmaceuticals whose combined expertise made this bio-transformation concept a reality.

10.7 REFERENCES

1. Blacker, A. J., ZENECA FCMO, unpublished information.
2. Worthing, C. R. (ed.) *The Pesticide Manual*, 7th edn, British Crop Protection Council, 1983.
3. ZENECA Agrochemicals, unpublished information.
4. Smeets, J. W. H., and Kieboom, A. P. G., *Recl. Trav. Chim. Pays-Bas*, **111**, 490 (1992).
5. Rhône Poulenc, Eur. *Pat*., EP 266258, 1988.
6. Daicel Chem. Ind., *Eur Pat.*, EP 190 770, 1986.
7. Cambou, B., and Klibanov, A. M., *Appl. Biochem. Biotechnol.*, **9**, 255 (1984).
8. Stauffer, *US Pat.* 4 613 690, 1986.
9. Hansen, H., Ladner, W., Rettenmaier, H., and Zipperer, B. (BASF), *Ger. Pat.*, DE 4 328 231, 1993.
10. Banko, G., Buchner, M., Estermann, R., and Mayerhofer, H. (Chemie-Linz), *Ger. Pat.*, DE 4 117 255, 1991.
11. Bodnar, J., Gubicza, L., and Szabo, L. P., *J. Mol. Catal.*, **61**, 353 (1990).
12. Baioru, K. K., *Jpn.Pat*., JP 5 049 498, 1991
13. Babin, J. E., and Whiteker, G. T. (Union Carbide), *Eur. Pat.*, EP 600 020, 1993.
14. Duhamel, L., Angibaud, P., Desmurs, J. R., and Valnot, J. Y., *Synlett*, **11**, 807 (1991).
15. Taylor, S.C. (ZENECA), *US Pat.*, 4 758 518, 1988.
16. Slater, H., Lovatt, D., Weightman, A. J., Senior, E., and Bull, A. T., *J. Gen. Microbiol.*, **114**, 125 (1979).
17. Berry, E. K. M., Allison, N., Skinner, A. J., and Cooper, R. A., *J. Gen. Microbiol.*, **110**, 39 (1979).
18. Weightman, A. J., Weightman, A. L., and Slater, J. H., *J. Gen. Microbiol.*, **128**, 1755 (1982).
19. Little, M., and Williams, P. A., *Eur. J. Biochem.*, **21**, 99 (1971).

20. Motosugi, K., Esaki, N., and Soda, K., *Agric. Biol. Chem.*, **46**, 837 (1982).
21. Weightman, A., and Slater, J. H., *J. Gen. Microbiol.*, **121**, 187 (1980).
22. Weightman, A., Weightman, A.L., and Slater, J. H., *Appl. Environ. Microbiol.*, **49**, 1494 (1985).
23. Barth, P., Bolton, L., and Thomson, J. C., *J. Bacteriol.*, **174**, 2612 (1992).
24. Jones, D. H. A., Barth, P., Byrom, D. and Thomas, C. M., *J. Gen. Microbiol.*, **138**, 675 (1992).
25. Schneider, B., Muller, R., Frank, R., and Lingens, F., *J. Bacteriol.*, **173**, 1530 (1991).
26. Crosby, J., ZENECA Pharmaceuticals, unpublished information.

11 Development of a Full-scale Process for a Chiral Pyrrolidine

H.-J. FEDERSEL
Astra Production Chemicals AB, Sodertalje, Sweden

11.1 Introduction . . . 225
11.2 Background and Initiation of the Development Project . . . 227
11.3 Route Identification and Evaluation . . . 229
11.4 Experimental Phase: Comparative Studies and Initial Method Development as a Basis for Route Selection . . . 233
11.4.1 Route A: Proline → 1-Ethylproline Ethyl Ester → 1-Ethylprolinamide (**5**) . . . 233
11.4.2 Route B: Proline → Proline Ethyl Ester → Prolinamide → 1-Ethylprolinamide (**5**) . . . 234
11.4.3 Route C: Proline → Proline Ethyl Ester → 1-Ethylproline Ethyl Ester → 1-Ethylprolinamide (**5**) . . . 235
11.4.4 Route D: Proline → 1-Ethylproline → 1-Ethylproline Ethyl Ester → 1-Ethylprolinamide (**5**) . . . 235
11.4.5 Final Step in Routes A–D: 1-Ethylprolinamide (**5**) → (*S*)-2-Aminomethyl-1-ethylpyrrolidine (**2**) . . . 236
11.4.6 Rational Route Selection . . . 236
11.5 Development and Scale-up Phase: Streamlining and Optimising the Process . . . 237
11.6 The Commercial Manufacturing Process . . . 240
11.7 Summary and Conclusions . . . 240
11.8 Acknowledgements . . . 242
11.9 References . . . 242

11.1 INTRODUCTION

The first observation that enantiomers of chiral molecules can exhibit biologically different properties dates back well over a century with the report in 1886 by Piutti,[1] who found (*R*)-asparagine to be sweet and the (*S*)-form to display an insipid character, but it is only in the last 30 years that this issue has become of major industrial significance, especially in the pharmaceutical area.[2] The most important reason for this is probably the gradual accumulation of data indicating that molecular stereoisomerism has a profound impact on the nature of the

Chirality in Industry II. Edited by A. N. Collins, G. N. Sheldrake and J. Crosby

response in living systems; the development of stereospecific analytical techniques such as chiral chromatography, which not only allow a very precise determination of absolute stereochemical purity in isolated compounds but also offer the ability to monitor the fate of stereoisomeric molecular species *in vivo*, is another strongly contributing factor. An illustrative example is the different therapeutic profiles of dextropropoxyphene; the (2*R*,3*S*)-isomer is used as an analgesic, and its enantiomer levopropoxyphene (2*S*,3*R*) exhibits antitussive activity.[3]

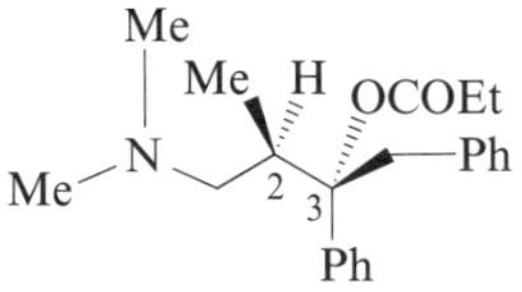

Dextropropoxyphene (2*R*,3*S*)

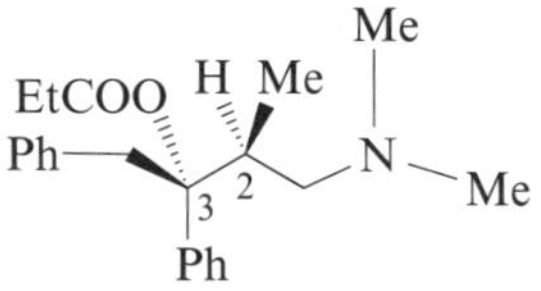

Levopropoxyphene (2*S*,3*R*)

Besides differing in a 'positive' manner in their pharmacological profiles, as in the previous case, often one of the enantiomers shows a toxic side-effect of some kind, or a lower potency, making its presence undesirable or, at best, without any appreciable value. This situation can be illustrated by dopa (3,4-dihydroxyphenylalanine) and penicillamine; the (*S*)- and (*R*)-enantiomers of the former display an anti-Parkinson effect and granulocytopenia (a severe eye disease), respectively,[4] whereas those of the latter are antiarthritic (*S*) and mutagenic (*R*).[5]

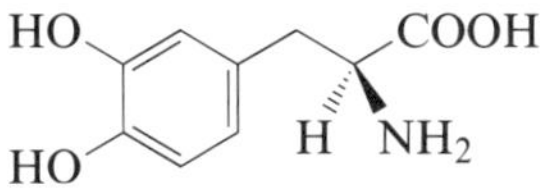

(*S*): anti-Parkinson

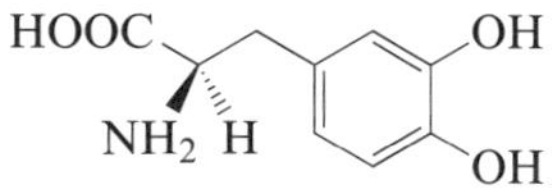

(*R*): causes granulocytopenia

Dopa

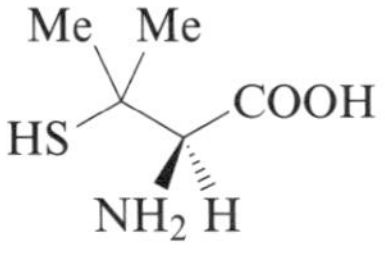

(*S*): antiarthritic

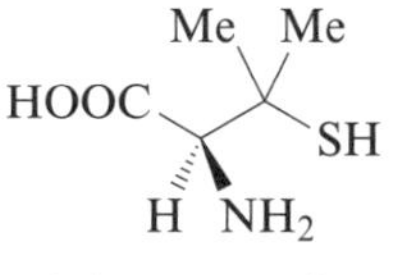

(*R*): mutagenic

Penicillamine

Given this background, the current trend to focus on a single stereoisomer, not only in the development of new drugs[6] but also to an increasing extent in the agrochemicals area,[7] is inevitable. The need to supply kilogram quantities of these materials to conduct extended toxicological and clinical (or field) trials throughout the entire R&D programme, has put a strong emphasis on the scale-up of single-enantiomer materials.[8] It should be pointed out, however, that racemates are not ruled out *a priori*, since instances of beneficial synergistic medicinal effects are known.[9] This is a view taken by representatives of both 'sides,' industry[10] and the various regulatory authorities[11] (see Chapter 2 for a more detailed discussion). The unbiased assessment of the three key parameters *quality*, *safety*, and *efficacy* should be adopted to enable a scientifically sound decision to be taken in favour of the use of either a single enantiomer or a racemate.[12]

11.2 BACKGROUND AND INITIATION OF THE DEVELOPMENT PROJECT

In the late 1960s, it was found that the racemic compound sulpiride exhibited pronounced antipsychotic activity.[13] Further studies revealed that a whole family of compounds belonging to the so-called benzamide group displayed an interesting pharmacological profile with regard to activities in the central nervous system (CNS). Screening and evaluation eventually led Astra to choose a number of

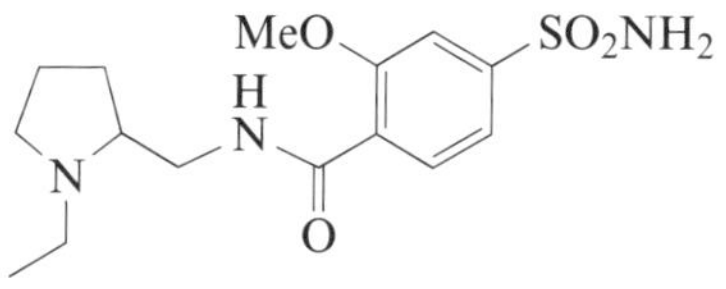

Sulpiride

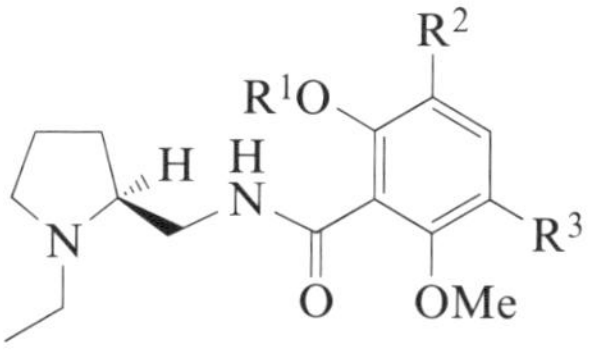

	R^1	R^2	R^3
Remoxipride:	Me	Br	H
Raclopride:	H	Cl	Cl

potent and particularly promising molecules, such as remoxipride and raclopride, as candidate drugs. The prime target was to develop these dopamine D_2-receptor antagonists for the treatment of schizophrenia.[14]

This class of products is composed of a variably polyfunctionalised benzoic acid moiety attached via an amide linkage to a common structural element comprising a chiral N^1-substituted pyrrolidinamine unit. A fairly broad range of different substituents at the pyrrolidine nitrogen has been investigated, but it seems that an ethyl group confers a particularly beneficial combination of properties with regard to key pharmacological parameters, such as efficient receptor binding.

The structural subunits **1** and **2** are easily identifiable as the penultimate build-

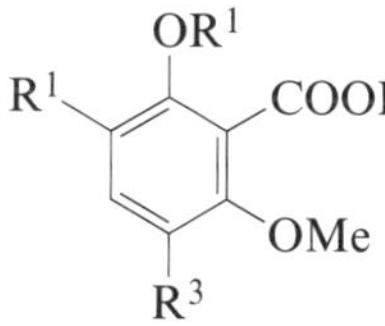

(1)

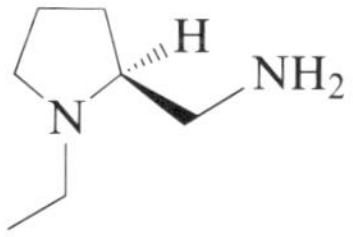

(2)

ing blocks to be combined in a condensation reaction to afford the desired amide products. This convergent synthetic strategy offers the advantage of allowing both moieties, the benzoic acid **1** and the amine **2**, to be assembled in separate reaction sequences.

The aromatic portion can be generated in a fairly straightforward manner using various 'classical' substitution reactions on readily available, inexpensive starting materials, such as 2,6-dimethoxybenzoic acid.[15] However, synthesis of compound **2** is considerably more complex owing to the need to obtain it enantiomerically pure, and to the many routes which can be envisaged for the synthesis. Initially, the pilot-scale demand for this pivotal intermediate was satisfied by a 'classical' resolution of the racemic material (**3**), available on the market in bulk quantities and accessible in a number of ways: (i) from *N*-ethyl-2-pyrrolidinone via catalytic hydrogenation of the 2-nitromethylene derivative,[16] (ii) from 2-substituted tetrahydrofurans, e.g. the chloromethyl compound, via ring opening and reaction with an appropriate alkyl amine,[17] and (iii) via ring contraction of 3-

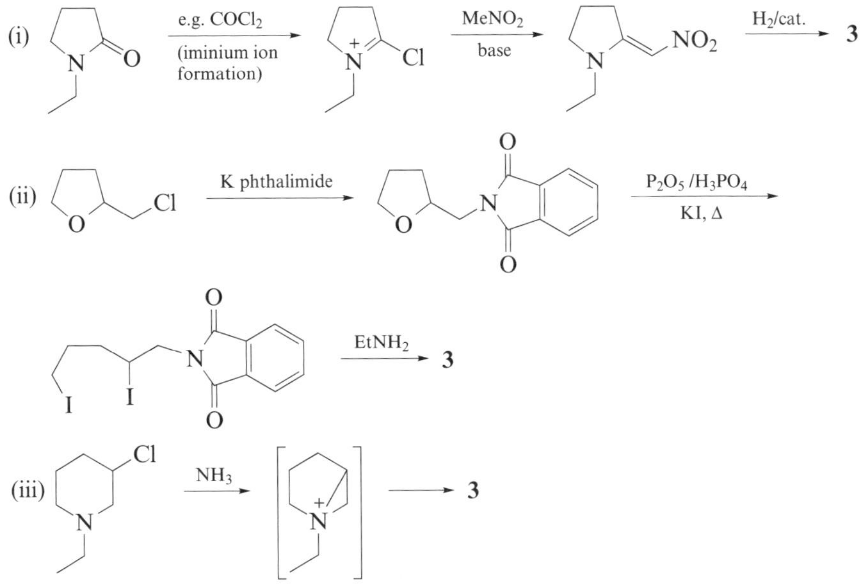

Scheme 1 *Non*-stereospecific syntheses generating racemic pyrrolidinamine (**3**)

halopiperidines upon treatment with ammonia (or an appropriate amine, e.g. benzylamine)[18] (Scheme 1).

On treating racemic amine **3** with expensive, unnatural D-(–)-tartaric acid, a diastereoisomeric 1:2 tartrate salt precipitated which, after neutralisation with base, liberated the desired (*S*)-enantiomer of **2** in *ca* 90% *ee* and an overall yield of *ca* 25% (i.e. 50% of theory). An economic improvement was achieved by par-

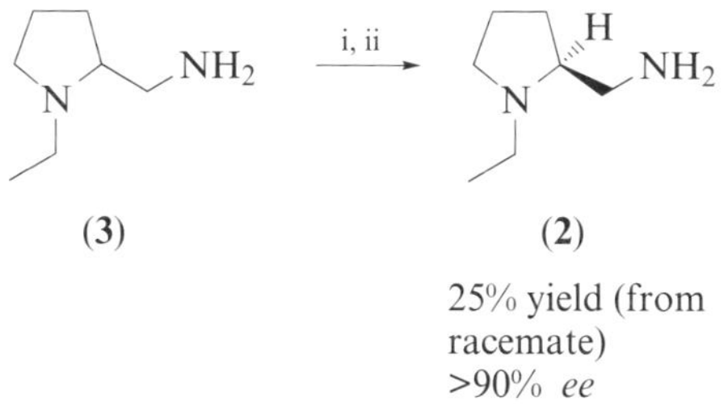

Scheme 2 Resolution of racemic pyrrolidinamine (**3**) to its (*S*)-enantiomer (**2**) using D- and L-tartaric acid. (i) (a) L-(+)-tartaric acid (*ca* 1.7 equiv.); (b) OH^- [liberation of enriched (*S*)-amine]. (ii) (a) D-(–)-tartaric acid (*ca* 1.8 equiv.); (b) recrystallisation ($EtOH/H_2O$); (c) OH^- (liberation of free base)

tially removing the (*R*)- isomer by crystallisation with natural L-(+)-tartaric acid, leaving an (*S*)-enriched mother liquor of the free base (after basification), which was then treated with D-tartaric acid, (Scheme 2). The yield and stereochemical purity were not affected, but the cost situation was improved, since the consumption of expensive D-tartaric acid was reduced considerably.

Evaluation of the potential for the procedure outlined in Scheme 2 revealed a number of weaknesses, besides the possibility of being involved in an interference situation with existing patents.[19] The main identified drawbacks were thus:

- the lack of an obvious method for racemising the unwanted (*R*)-enantiomer remaining in the mother liquor, which results in a 50% loss of expensive starting material;
- low overall yield of the required (*S*)-amine which, despite efforts, could not be improved significantly beyond 50% of theory;
- large usage of expensive D-tartaric acid;
- the need to operate an in-process recovery loop to recycle the D-tartaric acid;
- stability problems in the crystallisation leading to an uncontrolled, scale- and time-dependent precipitation of the tartrate salt of the undesired enantiomer, which was difficult to monitor. In the interval between crystallisation and completion of the product isolation by centrifugation (time consuming on a large scale), deterioration of the optical purity occurred; all efforts to improve the situation, such as changing the solvent composition or modifying the temperature (cooling profile; end-point value), were unsuccessful.

Given this background, the decision was taken to initiate a programme to identify alternative routes to **2**, the outcome of which is now presented[20].

11.3 ROUTE IDENTIFICATION AND EVALUATION

A particularly attractive option, which entirely eliminates the need to conduct a resolution, would be to use a building block with the required configuration from the ever-growing pool of chiral compounds, which could then be transformed into the target molecule with careful control of the stereochemistry. In this case, natural (*S*)-proline (**4**) is the obvious starting material. While its availability on a large scale in high stereochemical purity [bulk quantities of (**4**) at > 99% *ee*] and reasonable price is of paramount importance, the close structural relationship to the final product is also advantageous in order to minimise the number of large-scale synthetic operations.

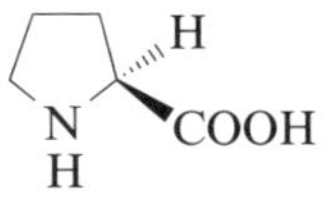

(*S*)-Proline (**4**)

(*S*)-Proline displays a number of structurally important elements relevant in the present case, notably the pyrrolidine nucleus and a C_1-unit (carboxy group) attached to the stereogenic centre (C-2) bearing the correct configuration with respect to the target molecule. In fact, any sequence which incorporates a formal transformation of the carboxy function into an aminomethyl substituent, and the coupling of an ethyl group on to the ring nitrogen, while keeping the stereochemistry intact, would have satisfied the primary objectives for a new route. The real choice, however, between the different options was to a large extent directed by secondary, process-oriented parameters such as yields, ease of operation, product quality and reaction conditions. Patentability would also increase the value of a given methodology in that an exclusive entry to a specific product is guaranteed for some time. This is especially so for entities such as the pyrrolidine **2**, which is useful not only for the purposes described here, but equally in the synthesis of a broad range of analogues.

A literature survey revealed that several sequences where the stereochemistry is retained have indeed been patented, two starting from (*S*)-proline[21,22] and others starting with either (*S*)-prolinol or (*S*)-glutamic acid.[21] These are compared in Scheme 3.

Their common feature is a linear sequence which, in the case of proline, requires four separate transformations, whilst the other approaches require up to eight steps. More specifically, the proline routes comprise either (i) an esterification, amide formation, pyrrolidine *N*-acetylation and a final reduction of both carbonyl functionalities[21] or (ii) an initial *N*-acetylation followed by a reduction to *N*-ethylprolinol, chlorination of the alcohol and finally an ammonia substitution.[22] The level of detail given in these patented methods is rather superficial with respect to crucial process information. In the former sequence, there are no yield data, stereochemical purities being reported only as optical rotations, and the latter quotes a 35% overall yield of a crude material (**2**) used 'as is' without specifying any purity.

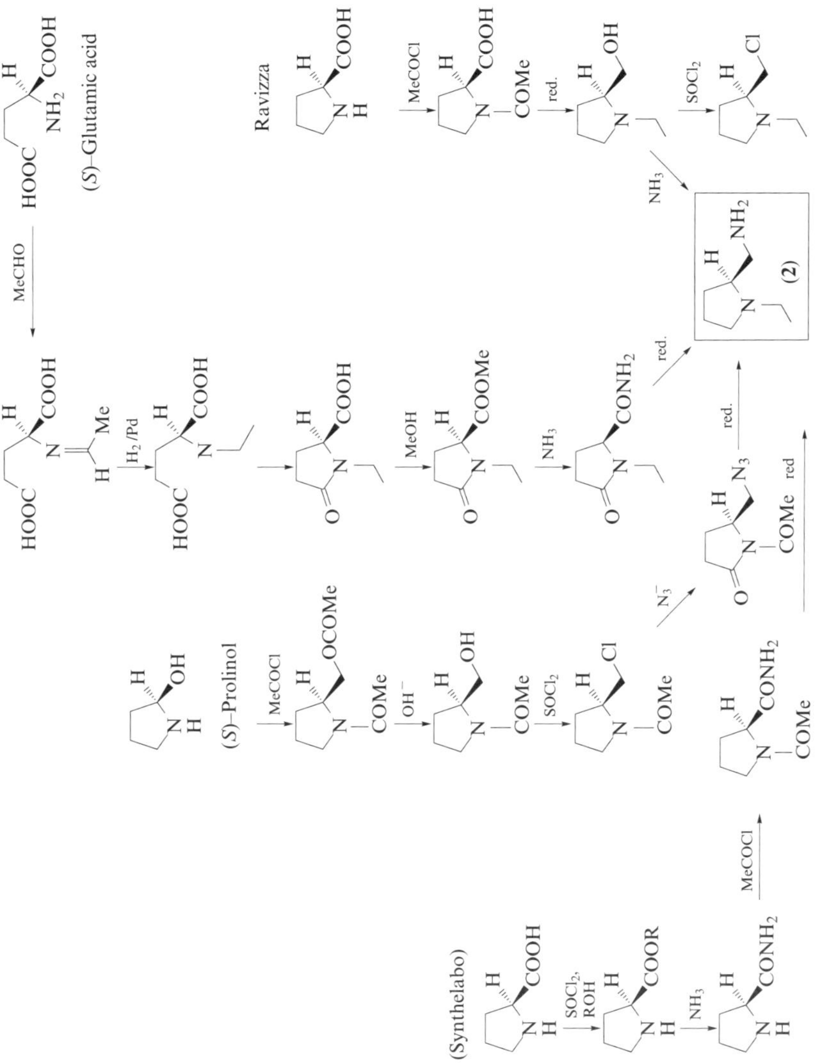

Scheme 3 Patented procedures to pyrrolidinamine **(2)** from stereochemically defined building blocks

Based on what was known in this area at the time, the four routes shown in Scheme 4 were designed as potential pathways for transforming (*S*)-proline in a

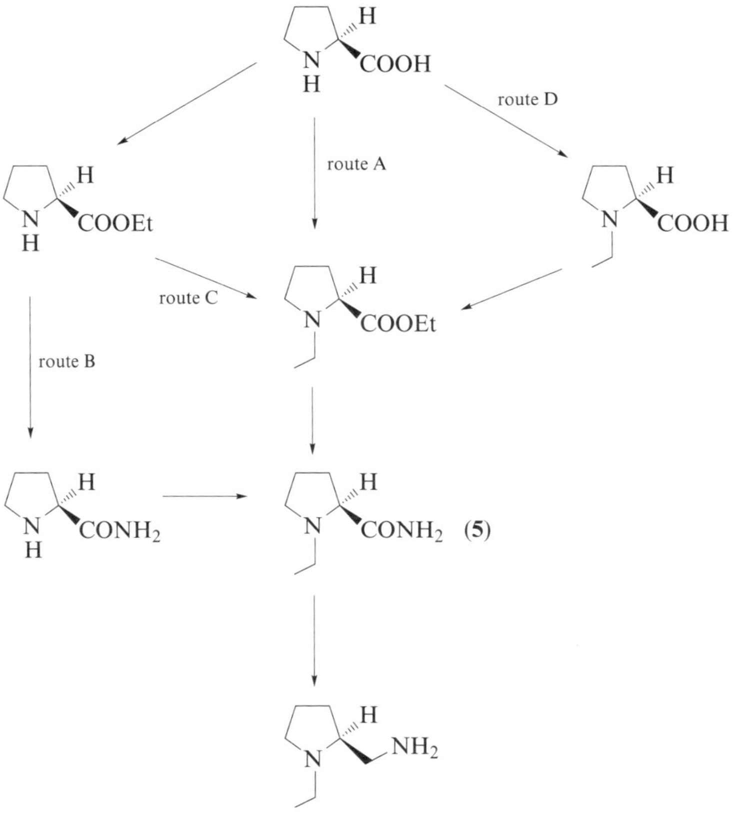

(2)

Scheme 4 Short and stereoconservative Astra patented[23] syntheses of pyrrolidinamine (**2**) starting from (*S*)-proline

concise and straightforward manner into the desired pyrrolidinamine (**2**).[23]

A quick examination of the different routes reveals that each sequence is composed of a set of fairly standard reaction types, an esterification (or *O*-alkylation), an *N*-alkylation, an ester-to-amide transformation and an amide reduction. Although differing in the order of the required steps, all four routes involve a common penultimate intermediate, (*S*)-1-ethylprolinamide (**5**), which clearly becomes a key compound. As will be shown below, it is the fine-tuning of the individual steps which guarantees both a high overall yield and more or less complete control over stereochemistry, eventually ensuring that the desired product can be produced economically and in good quality.

11.4 EXPERIMENTAL PHASE: COMPARATIVE STUDIES AND INITIAL METHOD DEVELOPMENT AS A BASIS FOR ROUTE SELECTION

The first undertaking was to assess the strengths and weaknesses of each sequence outlined in Scheme 4, attempting to identify quickly the most promising in order to direct all efforts towards its further development. The synthetic approaches are detailed individually up to amide **5**, followed by a separate discussion of the common final reductive step to target amine **2**. The results obtained will be presented in a fairly concise format.[24] Factors which received the highest priority during this screening phase were the overall isolated chemical yield, the technical feasibility of a successful scale-up and the stereochemical quality of the product. Of these, the last was given particular attention because the optical purity had to be retained at a very high level to avoid the need for any additional resolution. Monitoring of optical quality was, if possible, performed at each intermediate stage using mainly chiral gas chromatography (on a Chirasil-Val column) and, occasionally, NMR spectroscopy.

11.4.1 ROUTE A: PROLINE → 1-ETHYLPROLINE ETHYL ESTER → 1-ETHYLPROLINAMIDE (5)

Here the first step is a novel and very attractive *O,N*-dialkylation which converts the starting proline into the *N*-alkylated ester in a one-pot procedure. A number of different conditions were investigated systematically, such as (i) type of solvent; (ii) reaction temperature; (iii) bromo- vs iodoethane; (v) time of addition of halide; (v) stoichiometric ratio of ethyl halide/proline; (vi) presence and type of base; and (vii) overall reaction time. The best yields obtained were in the range of 65–70%, and afforded a crude material with a purity of about 90% (by GC). The major by-product (6–8%) was the diketopiperazine **6**, formed in a cyclodimerisation of the intermediate proline ester, a common reaction of α-amino acids (Scheme 5).

In common with the sluggishness frequently encountered when attempting to conduct ester-to-amide conversions, this transformation required rather forcing conditions (prolonged reaction times at elevated temperatures), which meant 1–2 days at about 50 °C when running in a methanolic ammonia (fortunately without causing any appreciable racemisation). Possibilities for improving this process have been investigated in a comparative study[25] where different catalysts were evaluated according to their abilities to effect an acceleration in the formation of amide **5** from the corresponding ethyl ester. Cyanide ion (added in *ca* 10 mol% as sodium cyanide) was significantly better than alternative, well known catalysts such as imidazole, DMAP, 2-hydroxypyridine and potassium iodide. Applying this cyanide method to the current reaction (45 °C/40 h), the crude target compound was obtained in 96% yield at a purity of 98%.

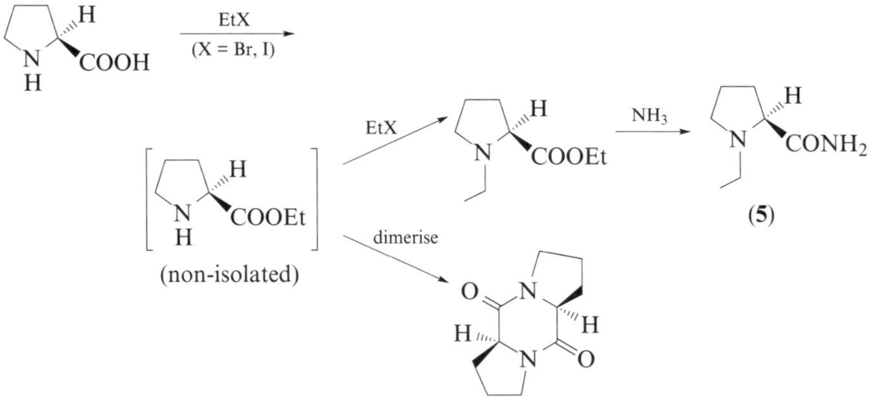

Scheme 5 Conversion of (*S*)-proline to 1-ethylprolinamide according to route A and formation of piperazine by-product (**6**)

The immediate appeal of route A is its brevity (overall only three steps vs four in routes B–D; see Scheme 4). The main drawbacks are the requirement to handle a high-boiling point, water-miscible solvent such as DMF or DMSO, and the fact that the isolated yields in the first step are barely acceptable. Also, the use of cyanide in the amide-forming step would give rise to safety and hazard concerns and call for special handling precautions.

11.4.2 ROUTE B: PROLINE → PROLINE ETHYL ESTER → PROLINAMIDE →1-ETHYLPROLINAMIDE (5)

A standard one-pot procedure was applied to convert the starting material into its ethyl ester (as the hydrochloride salt) via the *in situ* formation of the acid chloride, a reaction which proceeded smoothly and in almost quantitative yield.[26] The next transformation was conducted on the free base of the ester using methanolic ammonia at an elevated temperature and in a sealed reaction vessel, which rapidly caused a transesterification to the methyl ester, analogous to the comparable step in route A,[25] before giving the amide in an excellent yield of ca 90%.[27] Finally, the *N*-alkylation was performed in a heterogeneous system at 30 °C over 24 h with ethanol as solvent and a slight excess of solid potassium carbonate as the main basic component in conjunction with catalytic amounts of triethylamine. Using bromoethane as the alkyl source resulted in yields of about 65%. The sequence is outlined in Scheme 6.

Overall, this sequence was judged, from both the chemistry and the process operability viewpoints, to possess good potential for scale-up. The reactions seemed to be robust and yields were high in spite of the early stage of development.

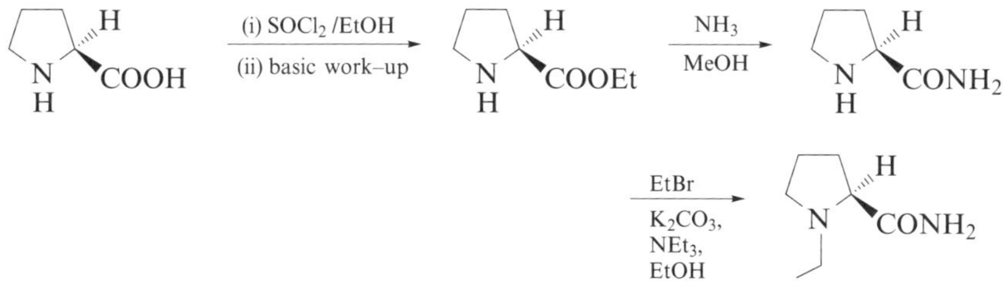

(5)

Scheme 6 Conversion of proline to the amide **5** according to route B

11.4.3. ROUTE C: PROLINE → PROLINE ETHYL ESTER → 1-ETHYLPROLINE ETHYL ESTER → 1-ETHYLPROLINAMIDE (5)

This route is closely related to route A, with the exception that the first two steps are carried out separately and not in a one-pot fashion. This stepwise sequence had a yield benefit since each reaction could be performed with isolated product outputs of well above 90%. With respect to the final amide formation, the same limitations prevail as were outlined previously, meaning that a cyanide-catalysed process would probably be required. For the entire reaction sequence see Scheme 7.

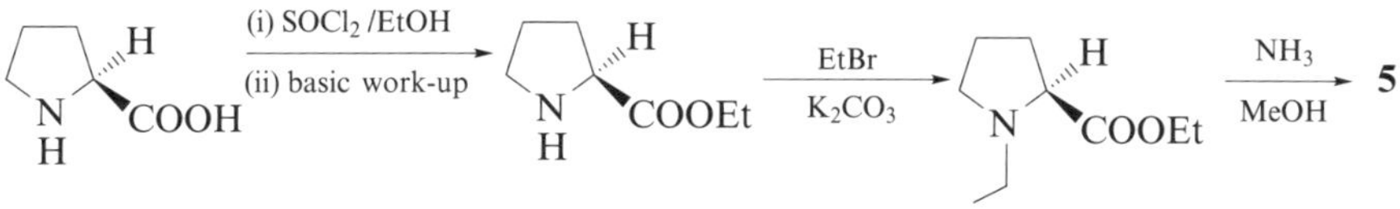

Scheme 7 Formation of 1-ethylprolinamide (**5**) from proline using route C.

11.4.4 ROUTE D: PROLINE → 1-ETHYLPROLINE → 1-ETHYL-PROLINE ETHYL ESTER → 1-ETHYLPROLINAMIDE (5)

This inverted version of route C, with respect to the first two steps, required a selective *N*-alkylation, which could be effected using acetaldehyde under reductive conditions (e.g. $NaBH_4$, H_2/Pd); however, the yields could barely be increased beyond 50%.[28] Also, attempts to conduct this transformation in a direct fashion with bromoethane did not offer any appreciable yield advantages. The subsequent esterification and amide formation were carried out as described previously, and Scheme 8 displays the synthesis step by step.

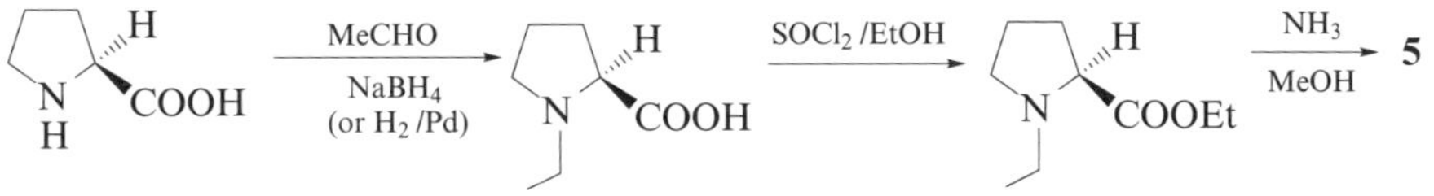

Scheme 8 Route D as a means to generate 1-ethylprolinamide (**5**)

11.4.5 FINAL STEP IN ROUTES A–D: 1-ETHYLPROLINAMIDE (5) → (*S*)-2-AMINOMETHYL-1-ETHYLPYRROLIDINE (2)

Several methods exist for the reduction of an amide to an amine, and a number of them were evaluated. The more common methods, such as using $LiAlH_4$ and $NaBH_4$, readily gave the desired product in yields varying from 60 to 70%. Using sodium bis(2-methoxyethoxy)aluminium dihydride (available as a 70% solution in toluene under the commercial names Red-Al or Vitride), the yield could be increased well above 80%. Apart from the significantly better yields, two further advantages were a better reproducibility in combination with a slightly higher optical quality of the isolated amine (**2**), and the ease of use of the latter as a solution on the large scale compared with the other reagents which would be handled as solids. Scheme 9 summarises the chemistry.

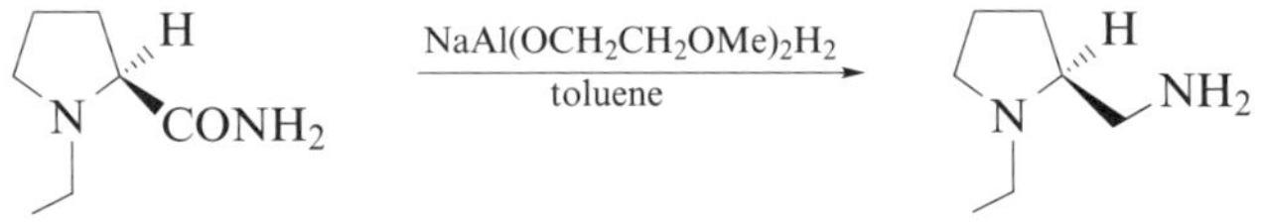

(**5**) (**2**)

Scheme 9 Reduction of amide **5** to target amine **2**

11.4.6 RATIONAL ROUTE SELECTION

The results obtained from the approaches described above were subjected to detailed side-by-side comparison, in addition to scrutiny against the overall project goals. Certain key parameters, as mentioned earlier, were rated as being more critical than others and had more impact on the final choice. Our ultimate choice of route B can be rationalised as follows (most important factors first):

- a high probability that a substantial increase in the chemical yield would be obtained, particularly in comparison with routes A and D;
- the ester, being the stereochemically most sensitive intermediate, can be transformed into the amide under milder conditions before attaching the *N*-substituent, which substantially decreases the risk of an uncontrolled loss of optical purity—a feature unique to route B;

- no requirement to use solvents such as DMF or DMSO (cf. route A), nor any need for a cyanide catalyst in the amide formation (cf. routes A, C and D);
- a technically straightforward process not requiring any extreme reaction conditions, performable with standard equipment using common unit operations;
- no obvious scaleability problems are envisaged;
- bulk availability of all components (starting materials, reagents, solvents);
- with proper precautions, the handling of the required chemicals and process streams is not liable to cause any severe environmental impact, nor would any major safety hazards (e.g. extreme exothermicities) be expected for the reactions and work-up procedures involved;
- a highly competitive production cost.

11.5 DEVELOPMENT AND SCALE-UP PHASE: STREAMLINING AND OPTIMISING THE PROCESS

Efforts were now focused on systematic process development involving both laboratory-scale experiments and scaled-up trial runs in the pilot plant and the production facility. A first step towards full-scale manufacture was taken fairly early using a method based on the preliminary laboratory procedure from the initial selection phase, i.e. route B. The primary purpose of this experimental manufacture was to gather process data regarding technical and operational feasibility, yields, product quality, impurity profile and cycle times, and to gain some experience for the process operators. The outcomes of this first pilot batch, and two subsequent repeats, all conducted more or less consecutively over a limited period (only 4 months), are summarised in Table 1.

A number of conclusions were drawn from the results: (i) the two larger runs proceeded in a fairly consistent manner, both with regard to overall yield and the performance in each single step, (ii) a severe yield penalty was incurred in the larger scale esterifications compared with the smaller reaction, whereas both the *N*-alkylation and the reduction worked significantly better (batch Nos II and III vs I); and (iii) there was a noticeable decrease in stereochemical purity on going

Table 1. Compilation of key data from first pilot batches performed according to route B

Batch No.	Relative scale/ proline amount kg (mol)	Stepwise yield, corrected for assay (%)				Overall yield (%)	Stereochemical purity [% *S* (GC)]
		Esterification[a]	Amide formation	*N*-Alkylation	Reduction		
I	8.6(74.8)	90.2	~97	61.1	79.5	42.5	99.1
II	100(868.6)	70.6	97.5	88.5	91.8	55.9	96.3
III	100(868.6)	76.1	95.2	83.0	88.0	52.9	97.8

[a] Synthesis of the methyl ester.

to the larger scale, albeit to a level which still matched the quality requirement of ≥95% (*S*)-isomer. These findings prompted a careful examination of the manufacturing process, focusing on the areas where shortcomings were observed compared with the laboratory method.

The drastic decrease in yield in the ester step was of particular concern, since it was felt to be technically rather simple with an expected product conversion close to quantitative. On examining the reaction and work-up to the free ester base, it became evident that the problem was related to the stability of the product; a neat sample at ambient temperature (*ca* 25 °C) showed a *ca* 4.6% loss over 24 h, with formation of the piperazinedione **6** through an ester dimerisation.

Fortunately, the situation could be greatly improved by merely switching to the ethyl ester, which is almost twice as stable. The importance of this effect was demonstrated in a dramatic fashion when performing a subsequent first full production size batch in the 4000 l equipment. Thus, when running the ester step with 300 kg (2606 mol) of proline according to the pilot method used earlier and only replacing methanol by ethanol, the gain in yield was in the range 20–25%. The remainder of the sequence was conducted with only very minor modifications (e.g. relative amounts of solvents) with individual steps showing an acceptable to high degree of reproducibility (step 2, 95%; step 3, 75%; step 4, 90%) considering the fairly early stage of process development, resulting in an overall yield of *ca* 62% and an excellent stereochemical purity of *ca* 98% (*S*)-isomer.

Once the successful scale-up had been demonstrated, full attention was given to a stepwise optimisation with the goal of obtaining a streamlined commercial process. The immediate targets in this respect were to minimise the input of chemicals (building blocks, reagents and solvents) in order both to reduce environmental impact and to ensure a high process capacity.

In the step 1 esterification all of the conditions were challenged, eventually leading to the selection of a number of parameters which were examined more closely using a factorial design.[29] Substantial changes were made to the literature procedure: the excess of $SOCl_2$ was reduced (from 1.77 to 1.05 equiv.), as was the temperature (from 60 to 40 °C), time (from 4 to 1 h) and solvent usage (from 6 to 2 kg ethanol per kg proline) without a yield penalty. Interestingly, it was possible to trace the initiation of the formation of a by-product contaminating the target amine (**2**) [its *N-methyl* analogue (**7**)] to this first step. In the latter part of the synthesis it is transformed into the undesired methyl derivative of the final drug substance. The mechanism can be explained by the generation of diethyl sulphite from ethanol and thionyl chloride, a side reaction which occurs even if $SOCl_2$ is added on an equimolar basis relative to proline. This high-boiling impurity (lit.[30] b.p. 157 °C/768 mmHg) undergoes transesterification with the abundantly available MeOH to dimethyl sulphite, a potent methylating agent which results in the formation of 1-methylprolinamide (**8**), as depicted in Scheme 10.

Under the reaction conditions, the ester hydrochloride is formed as the primary product and is subsequently neutralised to the free base, preferably using

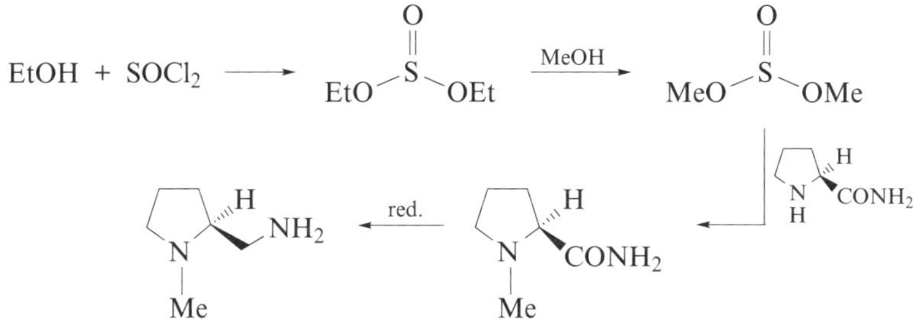

Scheme 10 Formation of *N*-methylpyrrolidinamine (**7**) due to a by-product, diethyl sulphite, from the ester step

aqueous ammonia. Since the free base of the ester is somewhat volatile (b.p. 82–83 °C/22 mmHg), a low-boiling extraction solvent, dichloromethane, had to be used to prevent losses during concentration.

The amide-forming step is the stereochemically most sensitive part of the sequence, requiring a method which minimises racemisation. It was found that reaction temperature is important, 50 °C being established as the upper limit, at which point a 24 h reaction time was required. Although a lower temperature reduces the risk of racemisation, it would extend the reaction time unacceptably. Attempts to replace methanol with ethanol as solvent, and eliminate the trans-esterification, resulted in a drastically decreased reaction rate (50% conversion after 39 h).

The first scale-up batches in the pilot plant revealed the presence of a new impurity at the amide stage, which had not been observed in the laboratory. The cause was identified as dichloromethane, which remains in small amounts in the oily proline ester from the previous stage. This solvent residue, cf. the diethyl sulphite, acts as an alkylating agent under the rather forcing conditions of step 2 forming the novel pyrroloimidazolone **9**. The proposed mechanism for its formation is shown in Scheme 11.[31] This side reaction was easy to avoid by ensuring the complete removal of dichloromethane in a co-evaporation with ethanol during work-up.

Also, in the third step, an experimental plan based on factorial design was deployed in order to monitor the influence of a number of parameters: amounts of potassium carbonate and bromoethane, in addition to the charging time for the latter, the need to add triethylamine and sodium iodide individually and the reaction temperature and time. The results from this study were used to set values for parameters such as a reaction temperature (30 °C), reaction time (20 h) and bromoethane charging time 0.5 h. An additional simplex optimisation was performed in order to determine the relative amounts of potassium carbonate and bromoethane, which turned out to be in the range 2.5–3.0 equiv. at a reaction concentration of 5.5 kg EtOH per kg prolinamide. If, however, the concentration was increased to 2.4 kg kg^{-1}, an increase in yield of *ca* 5% was obtained, whilst at

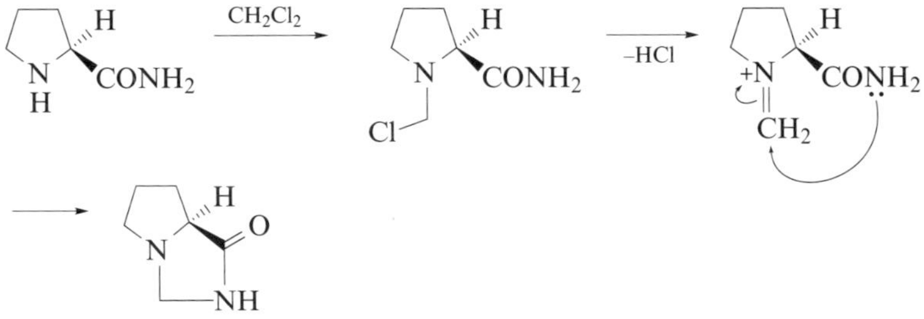

(9)

Scheme 11 Proposed generation of pyrrolo[1,2-c]imidazolone (**9**) from prolinamide and dichloromethane.

the same time allowing the excess of potassium carbonate and bromoethane to be substantially reduced (to 1.8 and 2.2 equiv., respectively).

More detailed investigation of the reduction step revealed that both the temperature and time for the reaction were important with regard to the stereochemical quality of the product. Thus, the addition of the reducing agent had to be performed at less than 20 °C, whereafter the reaction mixture was cautiously heated to 40 °C. This procedure guarantees as short a reaction time as possible (in the range 1–2 h). Finally, during work-up, it is essential to keep the amount of water to a minimum owing to the extremely high solubility of the product amine in aqueous systems.

11.6 THE COMMERCIAL MANUFACTURING PROCESS

The implementation of the improvements and modifications outlined above in successive full-scale trial batches resulted in a procedure which was highly optimised and capable of delivering the required quantity of the target amine **2** with the required quality. Scheme 12 summarises the commercial production in a sequential flow chart indicating unit operations and key process parameters.

11.7 SUMMARY AND CONCLUSIONS

Driven by the need to establish an economically viable process for the synthesis of a chiral aminomethylpyrrolidine compound in high enantiomeric purity, four closely related and patentable routes were identified, all starting from (*S*)-proline. Careful comparative studies and evaluations led to the selection of the process consisting of the four sequential steps esterification, amide formation, *N*-alkylation and reduction as offering the best potential for further development.

Chemical and technical optimisations, coupled with scale-up trials, furnished

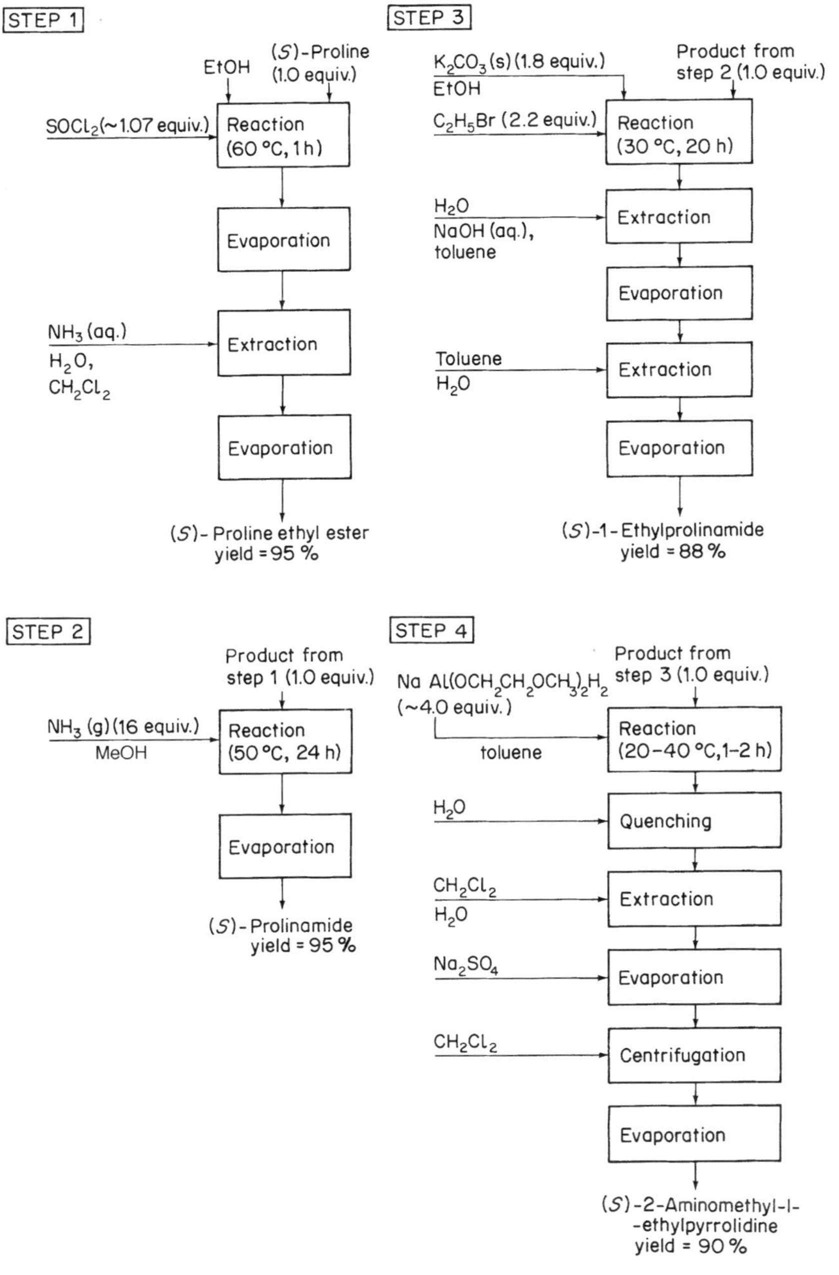

Scheme 12 Process flow chart of four-step synthesis from (*S*)-proline to (*S*)-2-aminomethyl-1-ethylpyrrolidine (**2**)

the final commercial production method for a 4000 l reactor, in which 300 kg (*ca* 2600 mol) of (*S*)-proline can be converted into *ca* 240 kg of the target compound, representing an overall yield of 71%.

It is hoped that this account of an authentic development case has emphasised some of the more important elements of the development of a synthesis from its initially rather rudimentary state in the early laboratory phase to a fully controlled and optimised commercial manufacturing procedure. The chiral nature of the product does not by itself infer any unique feature to the methodology, but may add constraints, e.g. avoiding racemisation. It therefore underscores the importance of following a rational approach and exerting the highest possible degree of control over the different stages and operations.

11.8 ACKNOWLEDGEMENTS

The work reported in this chapter reflects the skill, commitment and endeavour of a combined task force and project team whose foremost members were my close collaborators Erik Könberg and Lars Lilljequist of the Chemical Process Development Laboratory and Dr Thomas Högberg, Dr Sten Rämsby and Peter Ström, all at that time belonging to the Medicinal Chemistry Department of the CNS research unit within Astra.

11.9 REFERENCES

1. Piutti, A., *C. R. Acad Sci.*, **103**, 134 (1886).
2. Selected reading: (a) Ariëns, E. J., *Eur. J. Clin. Pharmacol.*, **26**, 663 (1984); (b) Jamali, F., Mehvar, R., and Pasutto, F. M., *J. Pharm. Sci.*, **78**, 695 (1989); (c) Crossley, R., *Tetrahedron*, **48**, 8155 (1992); (d) Stinson, S. C., *Chem. Eng. News*, **70**, (39), 46 (1992); (e) Testa, B., Carrupt, P.-A., and Gal, J., *Chirality*, **5**, 105 (1993); (f) Waldeck, B., *Chirality*, **5**, 350 (1993); (g) Batra, S., Seth, M., and Bhaduri, A. P., *Prog. Drug Res.*, **41**, 191 (1993).
3. Drayer, D. E., *Clin. Pharm. Ther.*, **40**, 125 (1986).
4. Cotzias, G. C., Papavasiliou, P. S., and Gellene, R., *N. Engl. J. Med.*, **280**(7), 337 (1969).
5. Glatt, H., and Oesch, F., *Biochem. Pharmacol.*, **34**, 3725 (1985).
6. Millership, J. S., and Fitzpatrick, A., *Chirality*, **5**, 573 (1993).
7. Ramos Tombo, G. M., and Bellus, D., *Angew. Chem., Int. Ed. Engl.*, **30**, 1193(1991).
8. (a) Sheldon, R., *Chem. Ind. (London)*, 212 (1990); (b) Crosby, J., *Tetrahedron*, **47**, 4789 (1991); (c) Collins, A. N., Sheldrake, G. N., and Crosby, J. (Eds), *Chirality in Industry. The Commercial Manufacture and Applications of Optically Active Compounds*, Wiley, Chichester (1992); (d) Nugent, W. A., RajanBabu, T. V., and Burk, M. J., *Science*, **259**, 479 (1993); (e) Sheldon, R., *Chirotechnology. Industrial Synthesis of Optically Active Compounds*, Marcel Dekker, New York (1993); (f) Federsel, H.-J., *Chem. Unserer Zeit*, **27**(2), 78 (1993); (g) Federsel, H.-J., *CHEMTECH*, **23**(12), 24 (1993); (h) Kotha, S., *Tetrahedron*, **50**, 3639 (1994); (i) Federsel, H.-J., *Endeavour*, **18**(4), 164 (1994).
9. Testa, B., and Trager, W. F., *Chirality*, **2**, 129 (1990).
10. Cayen, M. N., *Chirality*, **3**, 94 (1991).
11. (a) Swedish Medical Products Agency, *Reg. Affairs J.*, **2**(1), 2 (1991); (b) Madden, S.,

Reg. Affairs J., **2**(1), 20 (1991). (c) Allen, M. E., *Reg. Affairs J.*, **2**(2), 93 (1991); (d) Shindo, H., and Caldwell, J., *Chirality*, **3**, 91 (1991); (e) Hutt, A. J., *Chirality*, **3**, 161 (1991); (f) Rauws, A. G., and Groen, K., *Chirality*, **6**, 72 (1994), and references cited therein.

12. For a summary of recommendations, see Gross, M., Cartwright, A., Campbell, B., Bolton, R., Holmes, K., Kirkland, K., Salmonson, T., and Robert, J.-L., *Drug Inf. J.*, **27**, 453 (1993).
13. Borenstein, P., Champion, C., Cujo, P., and Olivenstein, C., *Ann. Méd-Psychol.*, **126**, 90 (1968).
14. (a) Högberg, T., Rämsby, S., Ögren, S.-O., and Norinder, U., *Acta Pharm. Suec.*, **24**, 289 (1987); (b) Högberg, T., *Drugs Future*, **16**, 333 (1991), and references cited therein; (c) Eriksson, L., *Prog. Neuro-Psychopharmacol. Biol. Psychiatry*, **18**, 619 (1994). (d) Conley, R., Nguyen, J. A., Hain, R., and Tamminga, C., *Psychopharmacol. Bull.*, **30**, 75 (1994).
15. Florvall L., and Ögren, S.-O., *J. Med. Chem.*, **25**, 1280 (1982).
16. (a) Kamiya T., and Hashimoto, M. Société d'Etudes Scientifiques et Industrielles de l'Ile-de-France), *Ger. Offen.*, 1 941 536, 1970; *Chem. Abstr.*, **72**, 100497x (1970); *Ger. offen.*, 1 966 195, 1972; *Chem. Abstr.*, **77**, 48223s (1972); (b) Oto, K., Ichikawa, E., Tamazawa, K., and Takahashi, K., (Yamanouchi Pharmaceutical Co.) *Ger. Offen.*, 2 635 587, 1977; *Chem. Abstr.*, **87**, 5801t (1977); (c) Valenta, V., and Protiva, M., *Collect. Czech. Chem. Commun.*, **52**, 2095 (1987).
17. (a) Fukuzawa, S., Imamura, T., Kakehi, M., and Watanabe, M., (Fujisawa Pharmaceutical Co.), *S. Afr. Pat.*, 69 00 983, 1969; *Chem. Abstr.*, **72**, 90273e (1970); (b) Rossier, J. P., *Swiss Pat.*, 544 754, 1974; *Chem. Abstr.*, **80**, 82643a (1974); (c) *Fr. Demande*, 2 229 693, 1974; *Chem. Abstr.*, **83**, 43181r (1975).
18. (a) Société d'Etudes Scientifiques et Industrielles de l'Ile-de-France, *Fr. Demande*, 1 580 165, 1969; *Chem. Abstr.*, **72**, 132509u (1970); (b) Rossier, J. P., *Swiss Pat.*, 544 753, 1974; *Chem. Abstr.*, **80**, 82640x (1974); (c) *Fr. Demande*, 2 229 692, 1974; *Chem. Abstr.*, **83**, 43180q (1975); (d) Podesva, C., Scott, W. T., and Navratil, M. (Delmar Chemicals), *Can. Pat.*, 1 013 750, 1977; *Chem. Abstr.*, **88**, 22611u (1978); (e) Reitsema, R. H., *J. Am. Chem. Soc.*, **71**, 2041 (1949).
19. Bulteau, G. (Société d'Etudes Scientifiques et Industrielles de l'Ile-de-France), *S. Afr. Pat.*, 68 02 593, 1968; *Chem. Abstr.*, **71**, 30354b (1969).
20. For a brief account of the development work, see Federsel, H.-J., *Celgene Chiral Rep.*, **1**(1), 1 (1993).
21. Kaplan, J. P., Najer, H., and Obitz, D. C. L. (Synthelabo), *Ger. Offen*, 2 735 036, 1978; *Chem. Abstr.*, **88**, 152414t (1978).
22. Mauri, F. (Ravizza SpA), *Ger. Offen.*, 2 903 891, 1979; *Chem. Abstr.*, **91**, 211259h (1979).
23. Federsel, H.-J., Högberg, T., Rämsby, S., and Ström, P. (Astra Läkemedel AB), *PCT Int. Pat. Appl.*, WO 87 07 271, 1987; *Chem. Abstr.*, **109**, 22839b (1988); in accepted form as *US Pat.*, 5 300 660, 1994.
24. For a full paper discussing these reactions in greater detail, see Högberg, T., Rämsby, S., and Ström, P., *Acta Chem. Scand.*, **43**, 660 (1989).
25. Högberg, T., Ström, P., Ebner, M., and Rämsby, S., *J. Org. Chem.*, **52**, 2033 (1987).
26. (a) Deimer, K.-H., in *Methoden der Organischen Chemie (Houben–Weyl)*, 4th edn, Georg Thieme, Stuttgart, 1974, Vol. XV/1, pp 315ff; (b) for a recent corresponding synthetic description of the analogous *methyl* ester, see Elliott, R. L., Kopecka, H., Lin, N.-H., He, Y., and Garvey, D. S., *Synthesis*, 772 (1995).
27. In close agreement with Flouret, G., Morgan, R., Gendrich, R., Wilber, J., and Seibel, M., *J. Med. Chem.*, **16**, 1137 (1973).
28. Methodology essentially according to Möhrle, H., and Sieker, K., *Arch. Pharm. (Weinheim)*, **309**, 380 (1976).
29. For a description of this methodology, see for example Carlson, R., *Design and*

Optimisation in Organic Synthesis, Elsevier, Amsterdam, 1992.
30. Weast, R. C. (Ed.), *Handbook of Chemistry and Physics*, 60th edn, CRC Press, Boca Raton, FL, 1979, p. C-508.
31. Federsel, H.-J., Könberg, E., Lilljequist, L., and Swahn, B.-M., *J. Org. Chem.*, **55**, 2254 (1990).

12 Development of a Multi-stage Chemical and Biological Process for an Optically Active Intermediate for an Anti-glaucoma Drug

A. J. BLACKER
ZENECA Process Technology Department, Huddersfield, UK

and

R. A. HOLT
ZENECA Life Science Molecules, Billingham, UK

12.1 Introduction 245
12.2 Development of the Multi-stage Chemical and Biological Process 248
12.2.1 Synthesis of Methyl (*R*)-3-hydroxybutyrate 248
12.2.2 Formation of Methyl (*S*)-3-(2-thienylthio)butyrate 249
12.2.3 Hydrolysis of Methyl (*S*)-3-(2-thienylthio)butyrate 250
12.2.4 Synthesis of the Ketosulphide and Ketosulphone 250
12.2.5 Development of an Asymmetric Bioreduction of the Ketosulphone 253
12.3 Conclusion 260
12.4 Acknowledgements 260
12.5 References 260

12.1 INTRODUCTION

Glaucoma is a disease of the eye characterised by increased intraocular pressure which results in defects in the field of vision. If left untreated, the disease can cause irreversible damage to the optic nerve, eventually leading to blindness. The most common form of the disease is open-angle or chronic glaucoma, in which the trabecula network which drains aqueous humour from the eye is obstructed whilst inflow of aqueous humour is unrestricted.

There are a number of drug treatments for glaucoma which act by opening the trabecula network, including pilocarpine and physostigmine. Adrenaline has a similar mode of action but also reduces aqueous humour formation, and β-blockers, which have also been used successfully in the treatment of glaucoma,

Chirality in Industry II. Edited by A.N. Collins, G. N. Sheldrake and J. Crosby

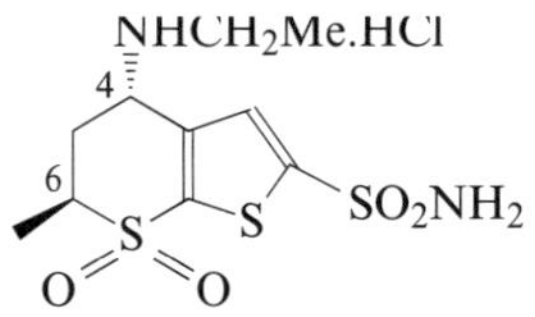

Figure 1 Structure of MK-0507 (trade name Trusopt)

also act by reducing aqueous humour formation. Unfortunately, all of these treatments have side effects which, in certain instances, can preclude their use. A further class of compounds useful in reducing intraocular pressure are the carbonic anhydrase inhibitors (CAIs). These act by reducing the levels of hydrogencarbonate in the aqueous humour, which has the effect of reducing the inflow of water into the eye. Until recently, CAIs were available only as orally administered drugs and, consequently, suffered from a number of systemic side effects. Recently, however, Merck have discovered a new range of CAIs which, by virtue of being water soluble, can be applied topically, which avoids most of the side effects associated with systemic administration.[1] One such compound, designated MK-0507, has recently been launched by Merck under the trade name Trusopt (Figure 1).

MK-0507 contains two chiral centres with defined absolute configuration and Merck have recently published a synthetic route (Scheme 1).[2] Incorporation of the chiral methyl group at C-6 was accomplished by tosylation of methyl 3-(*R*)-hydroxybutyrate followed by S_N2 displacement with 2-(lithiomercapto)thiophene. A major consideration for Merck in developing a manufacturing route to MK-0507 was how to obtain the desired stereochemistry at C-4. Following a full analysis of the options, they opted for a route which introduced the acetamide through a Ritter reaction (Scheme 1). This Ritter reaction is unusual in that it proceeds predominantly with retention of configuration when the hydroxysulphone reactant has the (4*S*,6*S*)-configuration (a full discussion of the proposed mechanism is given in refs 2 and 3).

The stereoselectivity of the Ritter reaction required that the hydroxysulphone be synthesised in the (4*S*,6*S*)-configuration. Unfortunately, reduction of the ketosulphone or ketosulphide by borane–THF or lithium aluminium hydride, respectively, resulted in the formation of predominantly the (4*R*,6*S*)-configuration (>95%). Although the hydroxysulphide could be epimerised using sulphuric acid, the 4*S*:4*R* ratio obtained was only 76:24.[2] The most elegant solution to this problem is the asymmetric reduction of the keto group, and ZENECA's approach was to identify a system capable of achieving this with the desired selectivity. Since this is a redox reaction, a whole-cell bio-transformation system was employed to avoid the need for enzyme purification and allow the *in vivo* recycling of the nicotinamide cofactor involved in the reduction. Consideration of the water solubilities of the ketosulphide and ketosulphone suggested the latter as the preferred substrate for the biological reaction. The manufacturing

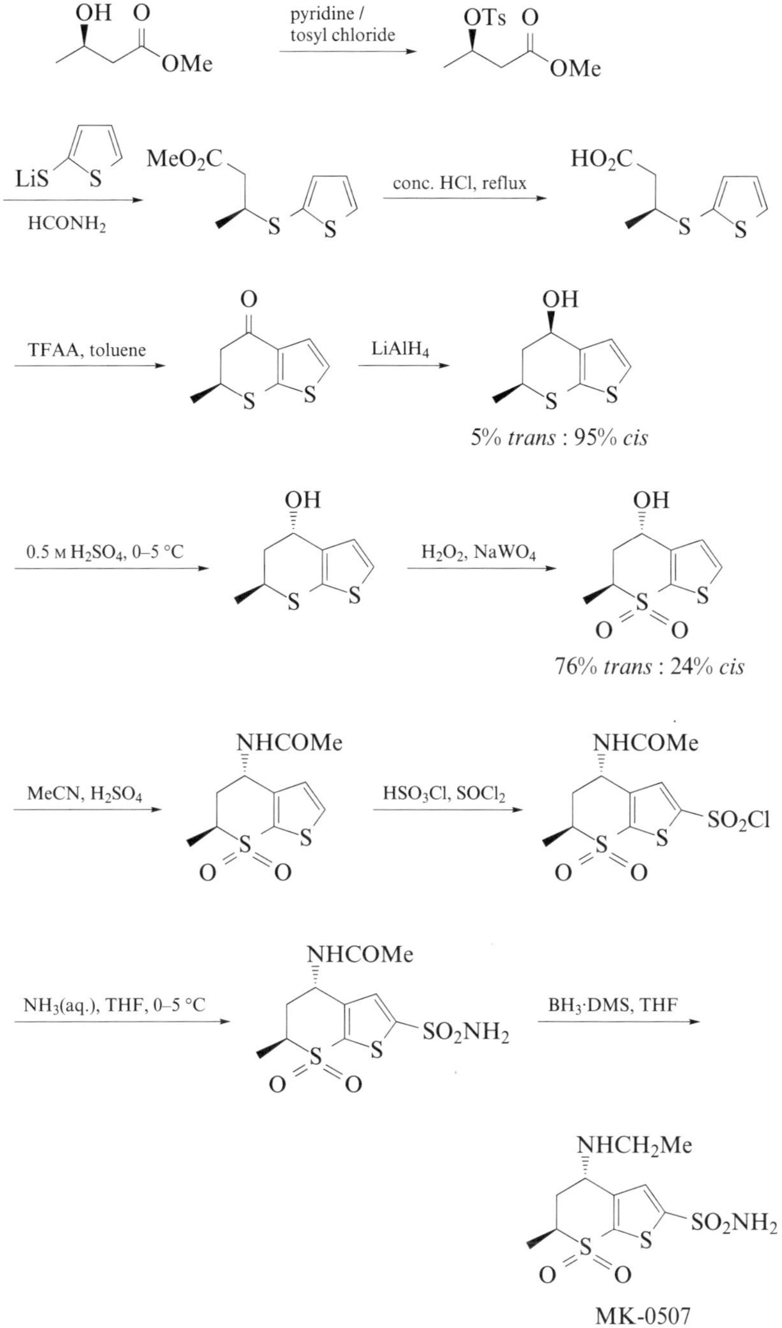

Scheme 1 After ref. 2

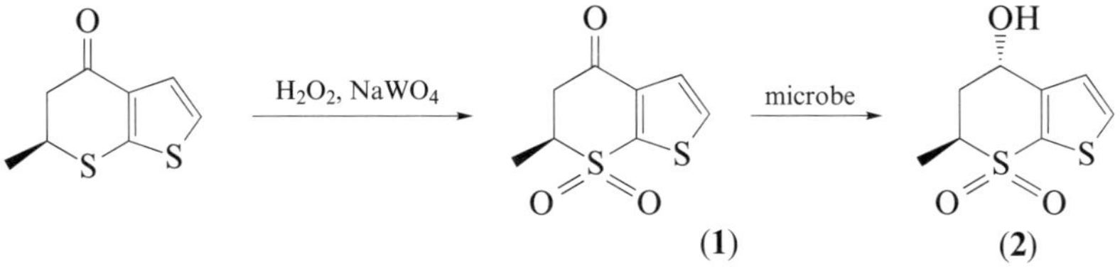

Scheme 2

route employed by ZENECA for the synthesis of the hydroxysulphone (**2**) is shown in Scheme 2, and a more detailed description of the process and its development is given below.

12.2 DEVELOPMENT OF THE MULTI-STAGE CHEMICAL AND BIOLOGICAL PROCESS

12.2.1 SYNTHESIS OF METHYL (*R*)-3-HYDROXYBUTYRATE

There are currently two commercial sources of (*R*)-3-hydroxybutyrate esters. The first is by depolymerisation of the microbially produced polymer poly[(*R*)-3-hydroxybutyrate] and the second via asymmetric hydrogenation of acetoacetate esters using Group VIII transition metal catalysts with optically active ligands. These processes are described in more detail below.

Biopol™ is a biodegradable plastic made by ZENECA. The name covers the generic family of polyhydroxyalkanoates, which are natural polymers produced by some microorganisms as storage compounds when carbon is abundant but growth is not possible owing to the limited supply of a nutrient, such as nitrogen. The polymers are intracellular inclusion bodies made from glucose and carboxylic acids. Polymer granules are isolated from the fermented microorganism, first by cell lysis, and then from the cellular debris by washing. These granules may, if necessary, be further purified before plasticising. Copolymers of (*R*)-3-hydroxybutyrate and (*R*)-3-hydroxyvalerate are preferred for the manufacture of Biopol used in biodegradable plastics, whereas it is the homopolymer that is used by ZENECA to produce methyl (*R*)-3-hydroxybutyrate for MK-0507.

The homopolymer has a molecular weight that may be varied depending on cellular growth conditions, but generally lies between 400 and 600 kDa. It is an optically active polyester that is insoluble in water and moderately soluble in chlorinated hydrocarbons. One of the problems in producing the monomer by saponification of the polymer is the insolubility of the latter [despite the high aqueous solubility of (*R*)-3-hydroxybutyric acid], and another is the formation of significant amounts of the by-product methylcrotonate. Processes have been developed for hydrolyses and transesterifications in chlorinated solvents,[4] but for manufacturing these suffer from problems of low reaction rates, complex work-ups and undesirable solvents.

An improved process operated by ZENECA to produce methyl (*R*)-3-hydroxybutyrate involves transesterification of the polymer in hot, acidic, methanolic medium. In this manner the insoluble polymer may be transesterified with good conversions (Scheme 3). After work-up, the desired product is recovered by fractional distillation under reduced pressure, which gives the methyl (*R*)-3-hydroxybutyrate as a clear, colourless liquid, with a chemical purity of >98% and of >99.5% *ee*. The process has been operated routinely on a multi-tonne scale.

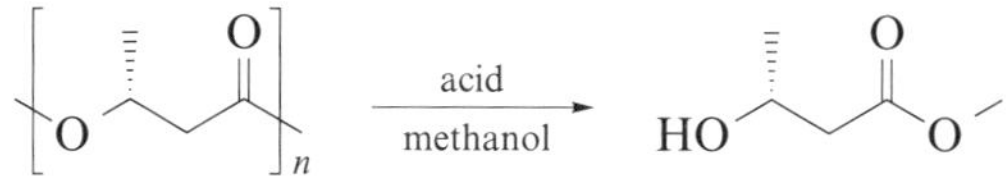

Scheme 3

Whilst methyl (*R*)-3-hydroxybutyrate is used for the manufacture of an intermediate to MK-0507, it may also be used as a starting material for syntheses of other optically active molecules. Much of the chemistry of the 3-hydroxybutyrates has been elegantly elaborated by Seebach's group;[4] examples are its conversion to (*R*)-butane-1,3-diol and (*R*)-3-hydroxybutyrolactone.[5] Should it be required, a simple method for inverting the stereochemistry is given in ref. 6.

Optically active hydroxybutyrates may also be made by asymmetric hydrogenation of acetoacetate esters using catalysts that employ rhodium or ruthenium complexed to optically active phosphine-based ligands. An example of such a catalyst is cycloocta-1,4-dienyl-2,2′-bis(diphenylphosphino)-1,1′-binaphthylruthenium, Ru(COD)(BINAP) (Scheme 4). This catalyst was invented by Noyori and co-workers, developed by Takasago Co. (Japan),[7] and is used to produce both (*R*)-and (*S*)-enantiomers of 3-hydroxybutyrate esters, albeit with a lower optical purity than the biologically derived material.

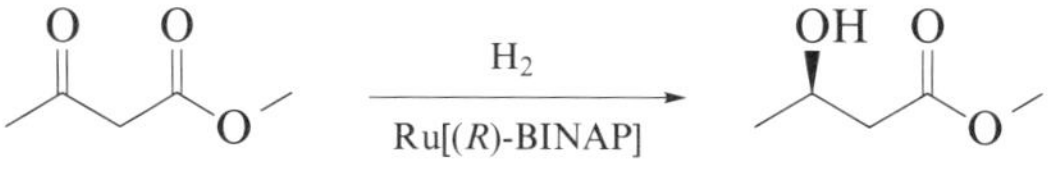

Scheme 4

12.2.2 FORMATION OF METHYL (*S*)-3-(2-THIENYLTHIO)BUTYRATE

Starting from methyl 3-(*R*)-hydroxybutyrate, the process developed by Merck involves activation of the 3-hydroxy group by tosylation in pyridine.[2] After isolating this product by crystallisation from an aqueous drown-out, the sulphonyl ester is treated with 2-lithiomercaptothiophene generated *in situ* by reaction of thiophene, butyllithium and sulphur. The optical activity of the product methyl

(*S*)-3-(2-thienylthio)butyrate indicates a clean inversion of stereochemistry. The reaction is complex, and requires careful control of stoichiometries and other parameters to minimise the formation of by-products.

12.2.3 HYDROLYSIS OF METHYL (*S*)-3-(2-THIENYLTHIO)BUTYRATE

The conditions needed to effect this reaction are fairly harsh: 6M HCl at reflux for several hours. Unfortunately, this procedure favours an undesirable rearrangement of the product (*S*)-3-(2-thienylthio)butyric acid to give the regioisomeric product (*S*)-3-(3-thienylthio)butyric acid. A rearrangement of this type has been reported to occur in an analogous system which, according to one hypothesis, passes through an unstable *ipso* compound (Scheme 5).[8] The regioisomer is virtually inseparable from the desired product and may be carried through the subsequent reactions to the final product, so it constitutes an important and undesirable impurity. Careful process optimisation has minimised formation of the regioisomer.

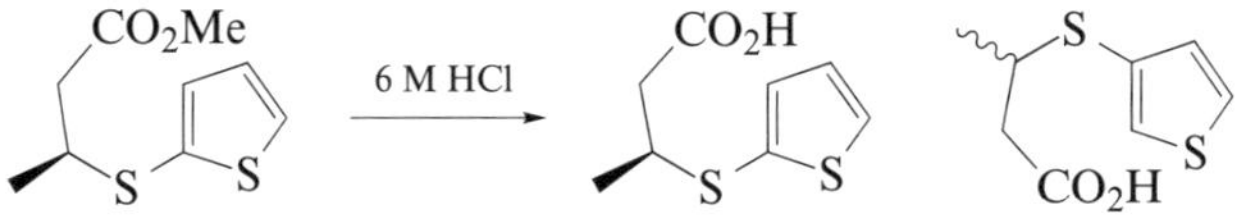

Major product Regioisomeric impurity

Scheme 5

12.2.4 SYNTHESIS OF THE KETOSULPHIDE AND KETOSULPHONE

The cyclisation procedure employed by ZENECA is essentially that developed by Merck, using trifluoroacetic anhydride, which was shown to be superior to standard Friedel–Crafts cyclisations.[2,9,10] It appears that use of trihaloacetic anhydrides has a further, elegant advantage over the Friedel–Crafts acylation, but before describing this, it is of interest to note another method of oxidation introduced by the Merck chemists.

The ketosulphide, in a toluene solution is oxidised to ketosulphone using hydrogen peroxide with a sodium tungstate catalyst. The reaction is highly exothermic and is, therefore, operated by controlled addition of the peroxide. The tungstate catalyst is highly active, and the peroxide is consumed as added, preventing an undesirable accumulation. Moreover, the sulphide is oxidised rapidly to sulphone, and the intermediate sulphoxide is barely observed. The use of toluene is advantageous over mixed solvents since it reduces costs and simplifies the work-up, which is carried out by separating the biphasic reaction mixture and drowning the aqueous phase into a solution of sodium hydrogensulphite. After washing and concentrating the toluene solution, the product can be isolated by crystallisation from mixtures of toluene and methanol.

In an attempt to improve this process still further, it was noted that the trifluoroacetic acid generated in the cyclisation reaction could, potentially, be used to generate trifluoroperacetic acid, an oxidation catalyst for the formation of ketosulphone. Indeed, it was demonstrated that the reaction mixture at the end of the cyclisation reaction could be used directly for the oxidation simply by addition of hydrogen peroxide (Scheme 6).[11] Initial experiments which gave ketosulphone of

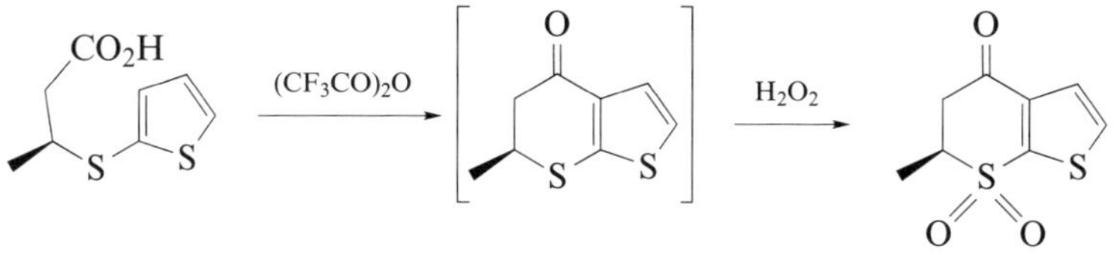

Scheme 6

high enantiomeric excess were carried out in the laboratory by relatively fast additions of hydrogen peroxide. However, as with the tungstate-catalysed oxidation, the heat of reaction was considerable, and it was necessary to slow the rate of hydrogen peroxide addition on a larger scale. Unfortunately, in simulating this in the laboratory we observed that the enantiomeric excess of the ketosulphone dropped dramatically from 98% to about 60%. On further investigation it was found that using slow addition of hydrogen peroxide allowed the build-up of the intermediate ketosulphoxide (Figure 2). Moreover, we observed that both *cis*- and *trans*-(6*S*)-methyl sulphoxides were formed and consumed at different rates. Acid-catalysed racemisation of sulphoxides has been reported,[12] and one of the mechanisms proposed is by sulphenolisation and consequent loss of the chiral centre at C-6 (Scheme 7). Although it might have been possible to increase the rate of peroxide addition on the plant to reduce racemisation, this would have required a far greater cooling capacity than was available, hence the choice of the sodium tungstate-catalysed oxidation.

Although it was recognised at an early stage that the ketosulphone was base sensitive, an unexpected observation was made during process development. In the course of the work-up of the oxidation reaction, an aqueous sodium hydrogensulphite wash is used to destroy residual peroxides. The separation of both phases is not always clean, and this can result in a small carry-over of hydrogensulphite solution into the organic phase. After atmospheric azeotropic drying of the organic solution, and upon cooling, the ketosulphone crystallises and precipitation of insoluble sodium hydrogensulphite is observed. When this mixture is redissolved in a polar solvent, such as boiling methanol, decomposition of the

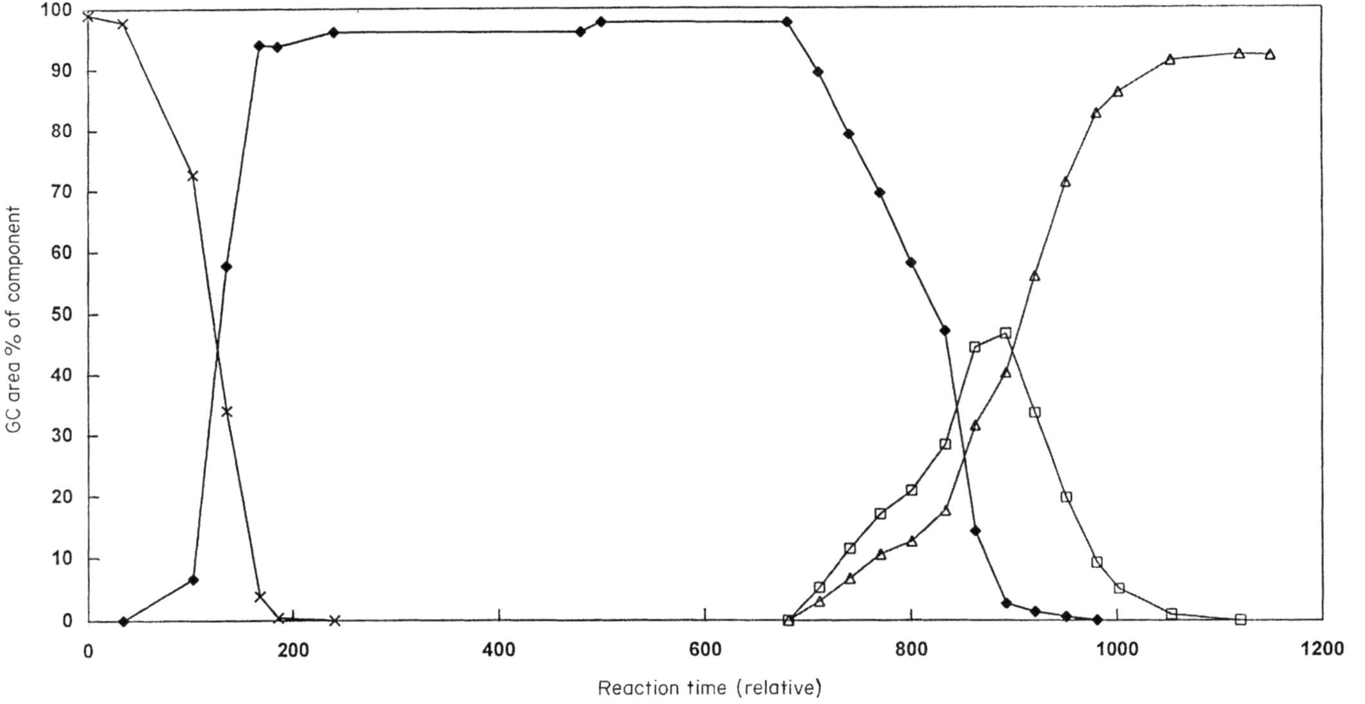

Figure 2 Reaction profile for telescoped cyclisation–oxidation reaction (◆) ketosulphide; (□) ketosulphoxide, (△) ketosulphone; (×)(*S*)-3-(2-thienylthio)butyric acid

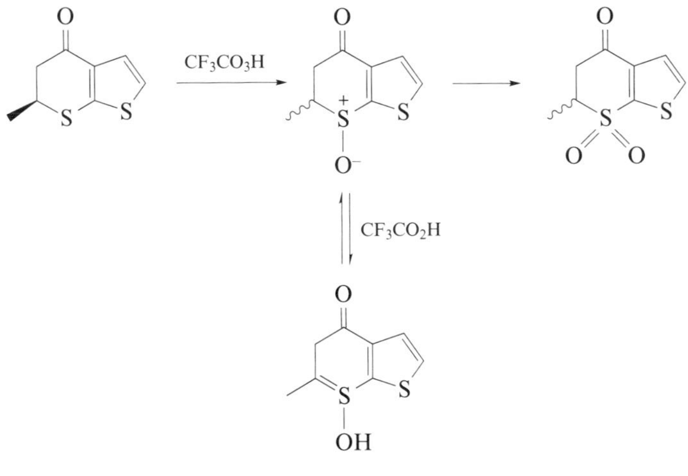

Scheme 7

sodium hydrogensulphite to sulphur dioxide and sodium hydroxide occurs. The latter reagent is then able to slowly racemise the ketosulphone. Clearly, this problem may be easily overcome by efficient washing of the product with water prior to the solvent drying step, but the episode serves to illustrate the importance of a thorough understanding of the process during the development stage.

12.2.5 DEVELOPMENT OF AN ASYMMETRIC BIOREDUCTION OF THE KETOSULPHONE

The use of microorganisms to carry out asymmetric reductions of ketones is well documented and many reactions have been described in the literature.[13] Most of these reactions have been carried out with strains of the yeast *Saccharomyces cerevisiae* and, although with certain compounds good stereoselectivity has been observed, this is by no means always the case. ZENECA's approach to the reduction of the ketosulphone began with a screen to identify potential microorganisms.

Initial studies were carried out using a range of carefully selected microorganisms (bacteria, yeasts and fungi) which were chosen on the basis of previous experience. Each microorganism was grown under the appropriate conditions, recovered by centrifugation and resuspended in a phosphate buffer medium at pH 7.0. To the cell suspension was added ketosulphone plus an appropriate carbon source to supply the microorganisms with reducing power. After incubation, the reaction mixtures were analysed for the presence of hydroxysulphone (**2**). As the hydroxysulphone is diastereoisomeric, it was possible to determine the ratio of *trans*-to *cis*-hydroxysulphone using conventional, non-chiral HPLC. From the results in Tables 1–3, it can be seen that a range of selectivities were obtained depending on the microorganism used.

Table 1 Reduction of ketosulphone (**2**) by bacteria

Organism	Conversion (%)	*trans*-Alcohol (%)
Lactobacillus salivarius	96.9	69.6
L. buchneri	36	33.9
L. fermentum	84.3	54.2
L. brevis	76.6	20.9
L. plantarum	6.1	91.3
Streptococcus faecium	4.9	3.9
Streptomyces lividans	61	26
Arthrobacter petroleophagus	99.5	42.3
Gluconobacter oxydans	86.0	10.2
Corynebacterium species	99.6	27
C. paurometabolum	97.0	20
Mycobacterium M156	100	40
M. smegmatis	99.4	53.6
Pseudomonas putida UV4	0	–
Ps. putida	20	12
Ps. oleovarans	48	74
Ps. species LC2	41	77
Ps. aeruginosa	80	17.4
Proteus vulgaris X19	24	13
P. mirabilis	26	13
Xanthomonas campestris	69	34
Alcaligenes eutrophus	98.4	43.5
Clostridium tyrobutyricum	91.8	40.0
C. tyrobutyricum	24.3	5.9
C. pasteurianum	60.2	36.4
Nocardia species	84.2	34.9

Table 2 Reduction of ketosulphone (**1**) by yeast

Organism	Conversion (%)	*trans*-Alcohol (%)
Candida chalmersii	37.3	42.1
C. diddensiae	66.7	52.2
C. lipolytica	95	35
C. utilis	32.9	51.4
C. oleophilia	1.2	9.9
Geotrichum candidum	99.3	15
G. candidum	95.7	31.6
Hansenula anomala	99.8	51.2
Pichia farinosa	13	45.2
P. farinosa	12.3	51.6
P. haplophila	97.7	68.2
P. pastoris	15.1	41
P. trehalophilia	98.8	36.6
Saccharomyces cerevisiae	21.4	15.4
S. cerevisiae	24	15.0
S. cerevisiae	8	35

(*continued*)

Table 2 (*continued*)

Organism	Conversion (%)	*trans*-Alcohol (%)
S. cerevisiae	40.4	40.2
Zygosaccharomyces rouxii	11.1	45
Torulaspora hansenii	10.8	50.3
Pichia capsulata (*Torulopsis molischiana*)	99.8	10.3

Table 3 Reduction of ketosulphone by fungi

Organism	Conversion (%)	*trans*-Alcohol(%)
Aspergillus flavus	5	68
Beauveria bassiana	46	21
B. bassiana	41	53
B. brogniartii	49	26
Fusarium graminearum	73.5	56.9
Neurospora crassa	87	71
Penicillium species	16.9	39.5

A number of organisms reduced the ketosulphone in almost quantitative yield, but did not simultaneously demonstrate the required stereoselectivity under the chosen reaction conditions. One microorganism which was notable for the selectivity of the reduction was *Lactobacillus plantarum*, but the extent of reduction was low. Given the need for both high stereoselectivity and high yield, further investigations and development work were required. Initially, studies were restricted to *L. plantarum* in an attempt to understand better how variations in the growth and bio-transformation conditions would affect the yield and stereoselectivity of the reaction.

A reaction was carried out which was sampled at intervals with measurement of key variables (Table 4). It is interesting that the glucose provided was almost entirely consumed during the first 2 h of the reaction, being converted nearly quantitatively into racemic lactic acid, the normal product of *L. plantarum* fermentation. Coincident with the formation of lactic acid was a rapid fall in pH.

Table 4 Time course of *Lactobacillus plantarum*-catalysed reduction of ketosulphone (**1**)

Time (h)	pH	Glucose ($g\,l^{-1}$)	Lactic acid ($g\,l^{-1}$)	Ketosulphone reduction (%)	*trans*-Diastereo-isomers (%)
0	7.0	10	0	0	–
2	5.0	0.1	9	13	91
8	5.3	nd[a]	9	18	90
20	5.3	nd[a]	9	27	89

[a]Not detectable.

Table 5 Effect of glucose concentration on the extent and selectivity of ketosulphone reduction by *Lactobacillus plantarum*

Glucose concentration (gl^{-1})	Time (h)	pH	Ketosulphone reduction (%)	*trans*-Diastereo-isomers (%)
0	2	6.9	1.2	61
	4	6.8	1.4	69
	20	6.5	2.4	55
10	2	4.6	5	88
	4	4.8	6	91
	20	5.0	12	89
50	2	3.6	25	90
	4	3.5	36	92
	20	3.6	36	92

The highest rate of ketosulphone reduction was seen during the first 2 h of the reaction, which suggested that glucose exhaustion, or a fall in pH, could be responsible for the slowing of the reaction, and these two variables were therefore examined.

In order to study the effect of glucose concentration, a batch of *L. plantarum* was grown, resuspended in buffer (pH 7.0) and supplemented with different amounts of glucose. Table 5 shows that glucose is necessary for any significant reduction of ketosulphone to occur; this is not surprising, as glucose is required to recycle the reductant, NAD(P)H, which is present in the cell in only catalytic amounts. In the absence of added glucose the pH of the reaction medium stayed close to its initial value, as little lactic acid was formed, but the hydroxysulphone which was formed was rich in the *cis* diastereoisomer.

In order to study the effect of pH on the reduction of ketosulphone by *L. plantarum,* a series of experiments was carried out in which the reaction pH was maintained at a predetermined value by the automatic addition of sodium hydroxide to neutralise the lactic acid formed during metabolism of glucose. Two important observations were made during these experiments: neutral pH was favoured over acidic pH as far as yield was concerned, but the opposite was true for the formation of the *trans* diastereoisomer (Table 6). There were two possible explanations: first, *L. plantarum* contains more than one enzyme system capable of reducing ketosulphone (these systems having opposite stereoselectivities and different sensitivities to pH); second, the ketosulphone or hydroxysulphone may be undergoing racemisation under the reaction conditions. Since the ketosulphone used in these experiments has a 6*S*:6*R* ratio of greater than 99:1, it was possible to test the racemisation hypothesis by measuring the ratio of the four diastereoisomers of the hydroxysulphone produced by *L. plantarum.* It was apparent from these measurements that racemisation at the C-6 position was occurring (Figure 3). Since previous experiments had indicated a strong influence of pH on the stereoselectivity of the reduction reaction, the effect of pH on the

Table 6 Effect of pH on the stereochemical outcome of ketosulphone reduction by *Lactobacillus plantarum*

Reaction pH	Time (h)	Ketosulphone reduction (%)	*trans*-Diastereo-isomers (%)
4.0	2	23	87
	22	46	81
5.0	1	28	82
	19	60	72
7.2	2	79	44
	15	80	45

optical purity of the ketosulphone was examined. Ketosulphone was suspended in aqueous buffers of various pH values and stirred; the mixtures were extracted at intervals and the optical purity was measured by HPLC with a chiral stationary phase. The extreme optical lability of the ketosulphone can be seen clearly in Figure 4. Racemisation appears to occur through a reverse Michael reaction following loss of an acidic hydrogen at C-5 (Scheme 8); this has been demonstrated by deuterium exchange NMR experiments.[2]

The requirement for an acidic pH to avoid ketosulphone racemisation was clearly incompatible with the requirements of *L. plantarum*, which only gave acceptable yields when the pH was maintained near neutral. Consequently, it was

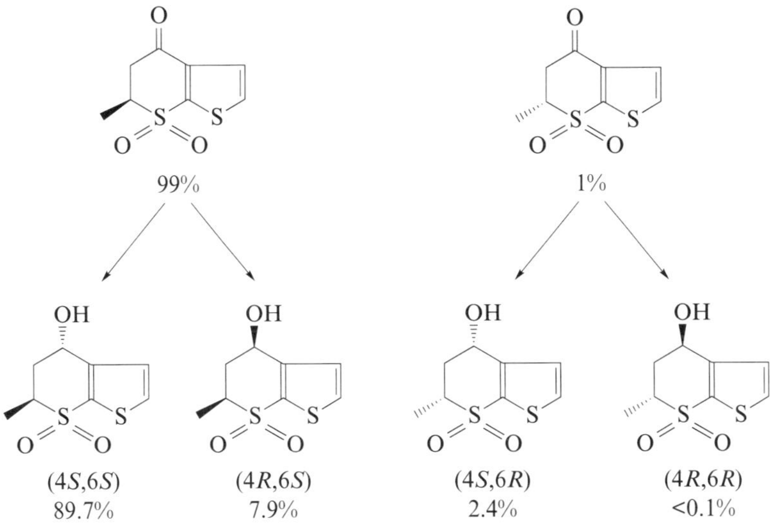

Figure 3 Ratio of hydroxysulphone diastereoisomers obtained from an *L. plantarum*-catalysed reaction. Initial pH 7.0; no pH control

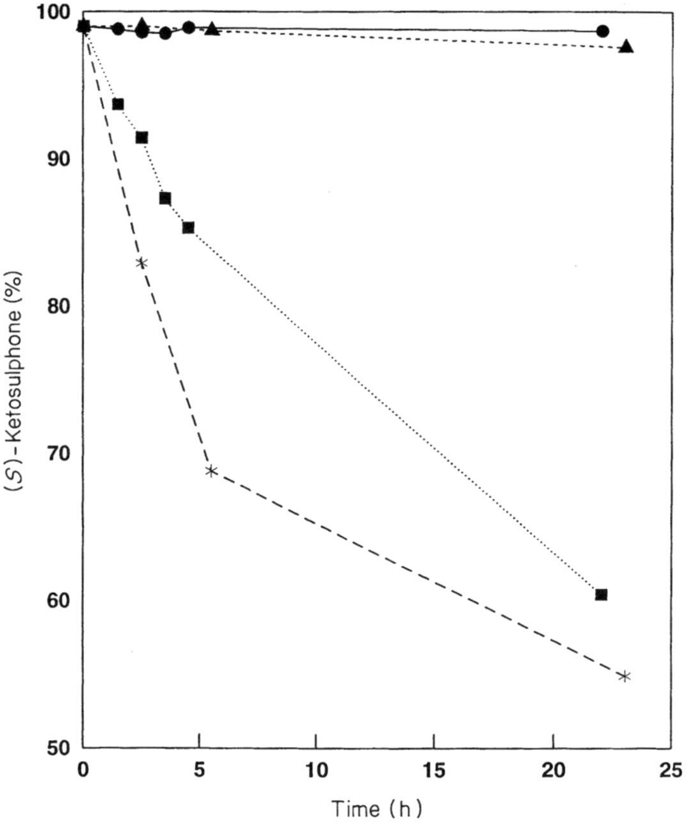

Figure 4 Optical stability of ketosulphone over a range of pH values: (●) 4.3; (▲) 5.0; (■) 6.5; (×) 9.0

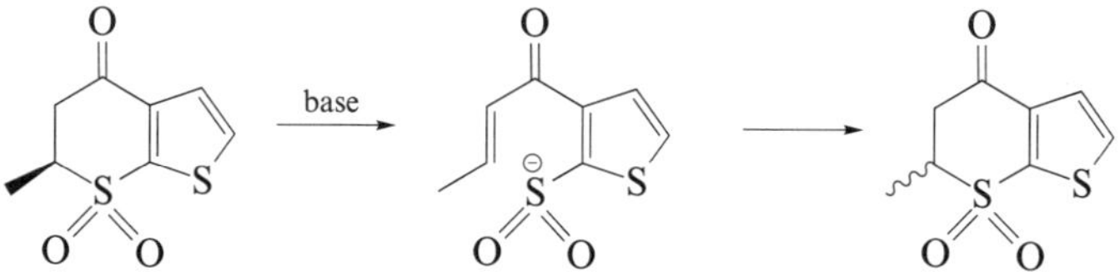

Scheme 8

necessary to identify an alternative microorganism demonstrating the desired selectivity coupled with the ability to tolerate pH values below 5.0.

From the results of the initial screening work, the fungus *Neurospora crassa* appeared worthy of closer investigation, having given a yield of 87% with a

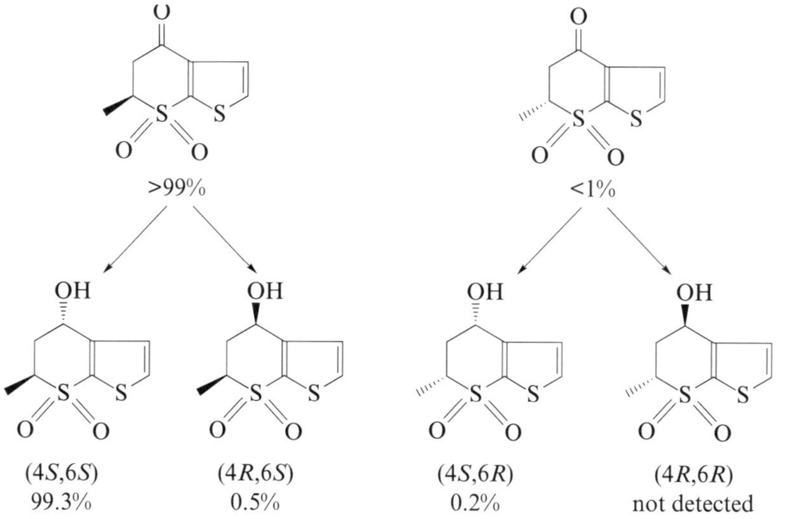

Figure 5 Ratio of hydroxysulphone diastereoisomers obtained from an *N. crassa*-catalysed reaction. Reaction controlled at pH 4.3

trans:*cis* hydroxysulphone ratio of 71:29. One of the advantages of *N. crassa* is its ability to grow at low pH values, which avoids the problem of ketosulphone racemisation. When tested under these conditions *N. crassa* provided hydroxysulphone of very high optical purity (Figure 5). On the basis of these results and our previous experience of developing processes based on the submerged culture of filamentous fungi, *N. crassa* was chosen as the prime candidate for the rapid development of a full-scale process. Optimisation of growth and bio-transformation variables such as culture medium, temperature and, most important, pH were carried out to define the optimum growth and bio-transformation conditions for application at plant scale.

The final process involves the growth of *N. crassa* in a fully controlled fermenter. At a specific stage in the growth the bio-transformation is initiated by the slow addition of ketosulphone to the fermenter. The rate of ketosulphone addition is matched to the rate of reduction to hydroxysulphone so that ketosulphone does not accumulate in the reactor, thus minimising the possibility of ketosulphone racemisation. On completion of the bio-transformation, the fungus is removed and the bio-transformation broth is extracted with organic solvent. Finally, the hydroxysulphone is recovered by crystallisation from the solvent phase. On a plant scale this process provides hydroxysulphone in > 80% isolated yield (based on ketosulphone) with high chemical purity (> 99%) and high optical purity [> 98% of the (4*S*,6*S*)-hydroxysulphone and < 0.5% of (6*R*)-diastereoisomers].

The highly stereospecific dehydrogenase responsible for the reduction of the ketosulphone is found in the cytoplasmic fraction of *N. crassa* cell extracts and is NADPH-specific. The enzyme has been purified using a two-stage procedure involving ion-exchange and affinity chromatography and the enzyme's *N*-terminus has been sequenced (38 amino acid residues) as a first step towards isolating and cloning the gene (work now being actively pursued by ZENECA).

12.3 CONCLUSION

This process demonstrates that complex, optically active compounds can be efficiently synthesised on a plant scale using bio-transformations. The application of bio-transformation technology was particularly vital to the success of this project, both chiral centres being derived through biological reactions. Of note was the availability of enantiopure methyl (*R*)-3-hydroxybutyrate from the bacterial polymer; although this may be considered a natural product, the chirality of this polymer is derived through asymmetric reduction of a ketone [acetoacetyl-CoA to (*R*)-3-hydroxybutyryl-CoA to poly (*R*)-3-hydroxybutyrate]. The novel,[14] highly stereoselective reduction of the ketosulphone was also vital to the success of the project, and marks a step forward in the development of industrial bio-transformations from resolutions (such as those catalysed by hydrolase enzymes) towards more complex, cofactor-requiring, synthetic reactions. Equally important was the development of processes that would avoid racemisation of the chiral centres, particularly at C-6, during the oxidation of the ketosulphide and during the biological reduction of the ketone to form the hydroxysulphone.

12.4 ACKNOWLEDGEMENTS

The authors thank all the members of the ZENECA project team, especially Dr D. J. Moody. Special thanks are also due to Dr E. J. J. Grabowski, Dr D. J. Mathre and Dr D. Matsumoto of Merck for useful discussions.

12.5 REFERENCES

1. (a) Baldwin, J. J., Ponticello, G. S., and Christy, M. E. (Merck) *Eur. Pat.*, EP 0 296 879; (b) Ponticello, G. S., Freedman, M. B., Habecker, C. N., Lyle, P. A., Schwam, H., Varga, S. L., Christy, M. E., Randall, W. C., and Baldwin, J. J., *J. Med. Chem.*, **30**, 591 (1987); (c) Baldwin, J. J., Ponticello, G. S., Anderson, P. S., Christy, M. E., Murcko, M. A., Randall, W. C., Schwam, H., Sugrue, M. F., Springer, J. P., Gautheron, P., Grove, J., Mallorga, P., Viader, M.-P., McKeever, B. M., and Navia, M.-A., *J. Med. Chem.*, **32**, 2510 (1989).
2. Blacklock, T. J., Sohar, P., Butcher, J. W., Lamanec, T., and Grabowski, E. J. J., *J. Org. Chem.*, **58**, 1672 (1993).

3. Sohar, P., Mathre, D. J., and Blacklock, T. J., *Eur. Pat.*, 0 617 037.
4. (a) Seebach, D. Z., and Zuger, M., *Helv. Chim. Acta*, **65**, 495 (1982); (b) Seebach, D., Beck, A. K., Breitschuh, R., and Job, K., *Org. Synth.*, **71**, 39 (1993); (c) *Eur. Pat*. EP 0 043 620 and EP 0 320 046.
5. (a) Fräter, G., *Helv. Chim. Acta*, **62**, 2825 (1979); (b) Seebach, D., and Zuger, M. F., *Tetrahedron Lett.*, **25**, 2747 (1985) (c) Kramer, A., and Pfanler, H., *Helv. Chim. Acta,* **65**, 293 (1982); (d) Seebach, D., Roggo, S., and Zimmermann, J., in *Stereochemistry of Organic and Bioorganic Transformations* (ed. W. Bartmann, and K. B., Sharpless), Proceedings of the Seventh Workshop Conference Hoechst, Verlag Chemie, Weinheim, 1986, and New York, 1987.
6. (a) Cainelli, G., Manescalchi, F., Martelli, G., Panunzio, M., and Plessi, L., *Tetrahedron Lett.*, **26**, 3369 (1985); (b) Griesbeck, A., and Seebach, D., *Helv. Chim. Acta*, **70** , 1320 (1987); (c) Breitschuh, R., and Seebach, D., *Chimia*, **44**, 216 (1990).
7. (a) Noyori, R., and Kitamura, M., in *Modern Synthetic Methods* (ed. R. Scheffold), Springer, Heidelberg, 1989, Vol. 5, p. 115; (b) Noyori, R., *Asymmetric Catalysis in Organic Synthesis*, Wiley, New York, 1994; (c) Akutagawa, S., in *Chirality in Industry* (ed. A. N. Collins, G. N. Sheldrake, and J, Crosby), Wiley, Chichester, 1992, p. 325; (d) Kitamura, M., Tokunaga, M., Ohkuma, T., and Noyori, R., *Org. Synth.*, **71**, 1 (1990).
8. (a) Gronowitz, S., and Pinchas, M., *Acta Chem. Scand.*, **16**, 155 (1962); (b) De Sales, J., Greenhouse, R., and Muchowski, J. M., *J. Org. Chem.*, **47**, 3668 (1982).
9. Shinkai, I., *J. Heterocycl. Chem.*, **29**, 627 (1992).
10. Jones, T. K., Mohan, J. J., Xavier, L. C., Blacklock, T. J., Mathre, D. J., Sohar, P., Turner-Jones, E. T., Reamer, R. A., Roberts, F. E., and Grabowski, E. J. J., *J. Org. Chem.*, **56**, 763 (1991).
11. Merck ZENECA, *PCT Int. Pat. Appl.*, 60/002 890, 1995.
12. House, H. O., *Modern Synthetic Reactions*, Benjamin, New York, 1972, p. 726.
13. (a) Sih, C. J., and Chen, C.-S., *Angew. Chem.*, *Int. Ed. Engl.*, **23**, 570 (1984); (b) Servi, S., *Synthesis*, 1 (1990).
14. Holt, R. A., and Rigby, S. R., *PCT Int. Pat. Appl.*, 93/01 776.

13 Synthesis and Application of Chiral Liquid Crystals

D. PAULUTH and A. E. F. WÄCHTLER
H. Merck K GaA, Darmstadt, Germany

13.1 Introduction 263
13.2 Properties and Applications of Liquid Crystals 264
13.2.1 Properties of Thermotropic Liquid Crystals 264
13.2.2 The Cholesteric Phase and the TN Display 266
13.2.3 The Chiral Smectic C Phase and the FLC Display 267
13.2.4 Properties of Thermochromic Liquid Crystals 272
13.3 Requirements for Commercial Liquid Crystals 273
13.3.1 Nematic Mixtures 273
13.3.2 Dopants for Nematic Mixtures 275
13.3.3 Ferroelectric Mixtures 277
13.3.4 Ferroelectric Dopants 277
13.3.5 Thermochromic Liquid Crystals 279
13.4 Synthesis of Chiral Liquid Crystals 280
13.4.1 Chiral Building Blocks for Dopants Used in Nematic Mixtures 280
13.4.2 Chiral Dopants for Nematic Mixtures 280
13.4.3 Chiral Building Blocks for Ferroelectric Dopants 282
13.4.4 Ferroelectric Dopants 283
13.4.5 Thermochromic Liquid Crystals 283
13.4.6 Optical Purity 283
13.5 Trends and Outlook 284
13.6 Acknowledgements 285
13.7 Notes and References 285

13.1 INTRODUCTION

Liquid crystalline behaviour was first described by the Austrian botanist Friedrich Reinitzer more than 100 years ago.[1] He found that on heating optically active cholesteryl benzoate there was no direct transition from the crystalline state to the liquid state. In a well defined temperature range above the melting point, a hitherto unknown, strangely coloured, turbid liquid phase existed,

Chirality in Industry II. Edited by A. N. Collins, G. N. Sheldrake and J. Crosby

which changed into a clear and colourless ordinary liquid phase on raising the temperature above the 'clearing point.' Since that time, similar temperature-dependent (thermotropic) liquid crystalline behaviour has been found for numerous anisotropic molecules, which are either rod- or disc-shaped. This chapter will deal only with rod-like molecules, because these have been of greater industrial importance to date.

13.2 PROPERTIES AND APPLICATIONS OF LIQUID CRYSTALS

13.2.1 PROPERTIES OF THERMOTROPIC LIQUID CRYSTALS

In the crystalline state, molecules are held together by strong intermolecular forces. They are located at defined positions in the crystal lattice, giving rise to a long-range positional and orientational order. In contrast to the highly ordered arrangement in the crystalline state, there is neither positional nor orientational order in the liquid state; the molecules move randomly and there is no net intermolecular alignment. Positional and orientational order are destroyed by heating an ordinary substance from the crystalline state to the liquid state. For 'thermotropic' liquid crystalline compounds this is a stepwise process. At the melting point the positional order is lost and, at the 'clearing point,' the orientational order vanishes.

There are different types of liquid crystalline phases (mesophases), classified according to their structural features, which reflect their degree of order. We shall confine our discussion to the nematic (N), the smectic A (S_A) and the smectic C (S_C) phases. The least ordered is the nematic phase. In a similar manner to the liquid phase, the molecules in the nematic phase move almost randomly, but the long axes of the rod-like molecules tend to be oriented in a certain direction defined by the director **n** (physically there is no difference between $+\mathbf{n}$ and $-\mathbf{n}$). The order in the 'nematic phase' is described by the order parameter S, with $S=1$ in the ideal case with parallel alignment of all molecules and $S=0$ in the non-ordered isotropic phase.[2] Typical values for the order parameter S are in the range between 0.3 and 0.7 at temperatures not too close to the clearing point (Figure 1).

The translational freedom of molecules in the nematic phase is restricted in the more ordered 'smectic phases.' For example, in the S_A phase the molecules are arranged in layers with their long axes perpendicular to the planes of the layer or, in other words, with the director parallel to the normal of the layers. Within a given layer the parallel-aligned molecules move randomly, and there is a statistical exchange of molecules between adjacent layers. The structure of the S_C phase is closely related to the S_A phase, the only difference being the tilted director (tilt angle ϑ) with respect to the normal of the layer's surface (Figure 2).

On decreasing the temperature, the phases are observed in the following sequence: liquid phase, [clearing point], N, S_A, S_C, S_B, S_I, S_F, higher ordered smectic phases, [melting point], crystalline phase. In practice, only a selection of phas-

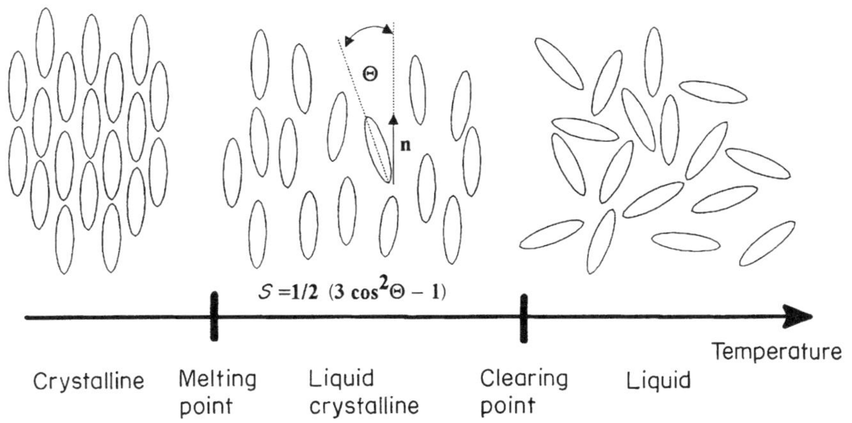

Figure 1 Molecular order in the crystalline, liquid crystalline and liquid states. The order parameter S is defined by the angle Θ between the molecular long axis and the director **n**

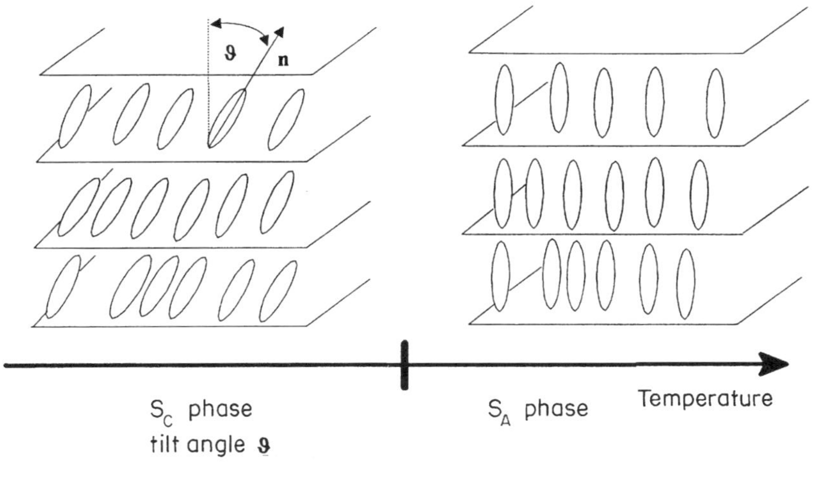

Figure 2 Schematic drawing of the S_C and S_A phases. In the S_C phase the director is tilted by an angle ϑ relative to the normal of the layers

es is found for a given liquid crystalline system, but the phase sequence will not be inverted. While it is difficult to supercool below the clearing point, it is possible to find liquid crystalline phases at temperatures below the melting point (metastable monotropic liquid crystalline phases).

As a consequence of both orientational order and short-range positional order, physical properties in the liquid crystalline phases must be described as vectors, although they can be described as scalars in the isotropic liquid phase.

This means that physical properties in the direction parallel to the director, or perpendicular to the director of the liquid crystalline system, are different.

For technical applications, the most important anisotropic parameters are the refractive indices (n) and the dielectric constants (ε). The birefringence (Δn) of a liquid crystal is the difference between the ordinary refractive index n_0 (light propagation perpendicular to the director) and the extraordinary refractive index n_e (light propagation parallel to the director):

$$\Delta n = n_e - n_o > 0 \tag{1}$$

In a similar way, the dielectric anisotropy ($\Delta\varepsilon$) is described by the dielectic vectors parallel ($\varepsilon_{\parallel}$) and perpendicular ($\varepsilon_{\perp}$) to the director according to the equation

$$\Delta\varepsilon = \varepsilon_{\parallel} - \varepsilon_{\perp} \tag{2}$$

Depending on the molecular structure, the dielectric anisotropy can be positive (molecular dipole parallel to the long axis of the molecule) or negative (polar substituent perpendicular to the long axis of the molecule).

13.2.2 THE CHOLESTERIC PHASE AND THE TN DISPLAY

A special variety of the nematic phase is the 'cholesteric phase' (Ch, chiral nematic phase), in which the liquid crystalline phase is made up of optically active molecules (e.g. cholesteryl benzoate) or a nematic phase is doped with optically active molecules. Looking at one hypothetical layer of the cholesteric phase (Figure 3), all molecules are aligned parallel to the director lying in the plane of the layer. In a second adjacent layer (parallel to the first) the same situation exists, but the director has changed its direction within the plane of the layer. From one layer to the next the director rotates gradually about an axis which is identical with the normal of the layers.

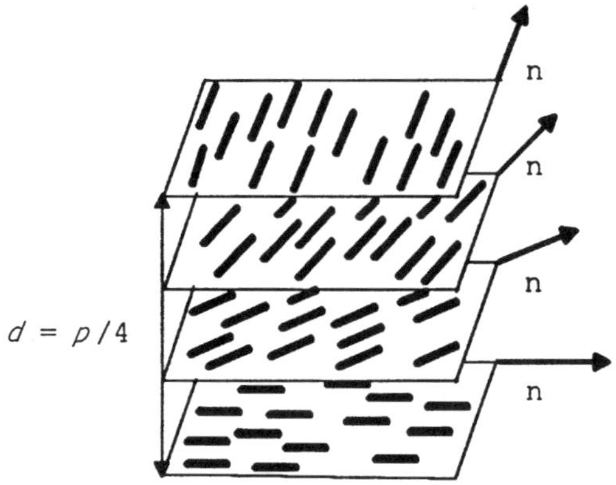

Figure 3 Schematic drawing of the Ch phase showing the precession of the director around the normal of the layers of the Ch phase

The pitch, p, is the distance for one full revolution of the director, and its magnitude depends on the structure of the optically active molecules. If the cholesteric phase is induced by an optically active dopant, the pitch varies linearly with the concentration (c). This property of the dopant is termed the 'helical twisting power' (HTP):[2]

$$\text{HTP} = (pc)^{-1} \quad (3)$$

From a consideration of its structural features, the cholesteric phase can be interpreted as a helically twisted nematic phase. The interaction of polarised light with a helically twisted nematic phase controlled by an electrical field is the basic principle of the most widely used liquid crystal electrooptic effect in display technology,[3] and constitutes the most important technical application of liquid crystals. The plane of linearly polarised light passing through a helically twisted nematic phase of thickness $d = p/4$ along the helical axis follows the rotation of the director. Consequently, the planes of polarisation of the entering and the leaving light are perpendicular to each other. A device with a dielectrically positive liquid crystal sandwiched between two glass plates which are coated with alignment layers to induce the director to be parallel to the surface (homogeneous alignment), and with the direction of the two alignments twisted by 90°, transmits light if it is viewed between crossed polarisers (Figure 4a). As nematic liquid crystals with positive dielectric anisotropy orient their molecular long axes parallel to an electric field, the arrangement of the liquid crystals in this device can be changed by applying a voltage across two transparent electrodes on the glass plates. In this case the director of the liquid crystals switches to a position perpendicular to the glass plates (homeotropic alignment) and the entering polarised light does not change the direction of its plane of polarisation by passing through the liquid crystal parallel to the director (Figure 4b).

On switching off the electrical field, the original helically twisted arrangement of the liquid crystal is restored by elastic forces resulting from the interaction of the liquid crystal with the surfaces of the alignment layers. The twisted nematic arrangement in this type of electrooptic cell (TN cell) is structurally equivalent to a cholesteric phase.

13.2.3 THE CHIRAL SMECTIC C PHASE AND THE FLC DISPLAY

The structure of the S_A phase is the same for optically active and for achiral molecules but, by analogy with the structural features of the nematic and cholesteric phase, there is a chiral smectic C (S_C*) phase, which differs from the S_C phase only in its helical structure. From one layer to the next the tilted director rotates gradually around an axis perpendicular to the layer, thus forming a helix with a pitch p (Figure 5).

Symmetry considerations led Meyer *et al*[4]. to the conclusion that smectic C liquid crystals composed of optically active molecules must have a spontaneous

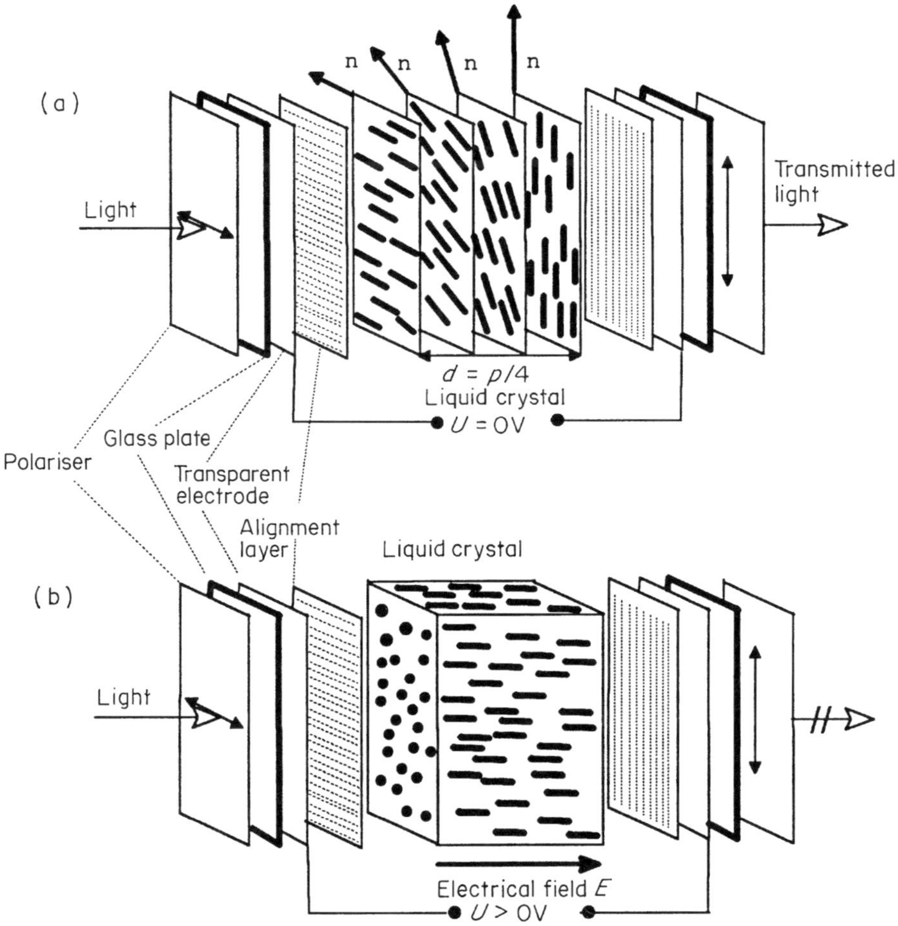

Figure 4 The TN cell with crossed polarisers is the most widely used display. (a) The bright 'off' state with no voltage applied; (b) the dark 'on' state

electrical polarisation characteristic of a ferroelectric state. This was proved by *p*-decyloxybenzylidene *p*′-amino-2-methylbutylcinnamate (DOBAMBC), a material designed and synthesised to exhibit the above-mentioned properties (Figure 6).

Any macroscopic physical property of a material must include all symmetry elements given by the structure of the material; thus ferroelectricity is compatible only with an S_C* phase and not with an S_C phase. Considering the local symmetry of an S_C layer averaged over time, we find C_{2h} symmetry with a twofold rotation axis C_2, a mirror plane σ_h and a centre of inversion i (Figure 7). These symmetry elements do not allow the existence of a macroscopic vectorial physical property such as the spontaneous polarisation P (p_x, p_y, p_z).

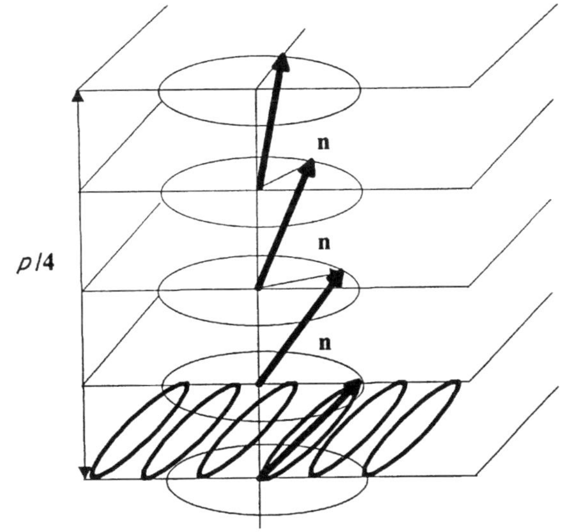

Figure 5 Schematic drawing of the S_C* phase showing the precession of the director around the normal of the layers of the S_C* phase

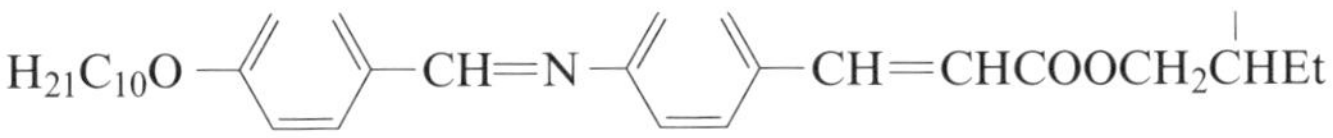

$K76\ S_I^*(63)\ S_C^*\ 95\ S_A\ ^{117}I$ $\quad P_S{-}3\,nC\,cm^{-2}$

Figure 6 Chemical structure, phase range and spontaneous polarisation of the first reported ferroelectric S_C* compound[4,18]

$$\text{point group } C_{2h}: \quad \mathbf{P}(p_x, p_y, p_z) \xrightarrow{C_2} \mathbf{P}'(-p_x, p_y, -p_z) \qquad (4)$$

$$\mathbf{P}(p_x, p_y, p_z) \xrightarrow{\sigma_h} \mathbf{P}'(p_x, -p_y, p_z) \qquad (5)$$

$$\mathbf{P}(p_x, p_y, p_z) \xrightarrow{i} \mathbf{P}'(-p_x, -p_y, -p_z) \qquad (6)$$

$$\Longrightarrow p_x = p_y = p_z = 0$$

Figure 7 Allowed symmetry operations for the S_C phase (point group C_{2h})

In an S_C* layer composed of optically active molecules, there is no mirror plane and no centre of inversion because these symmetry elements are incompatible with optically active molecules. The remaining twofold C_2 axis allows the existence of a permanent dipole moment parallel to this axis (parallel to the y-coordinate) and perpendicular to the director **n** (Figure 8), because the compo-

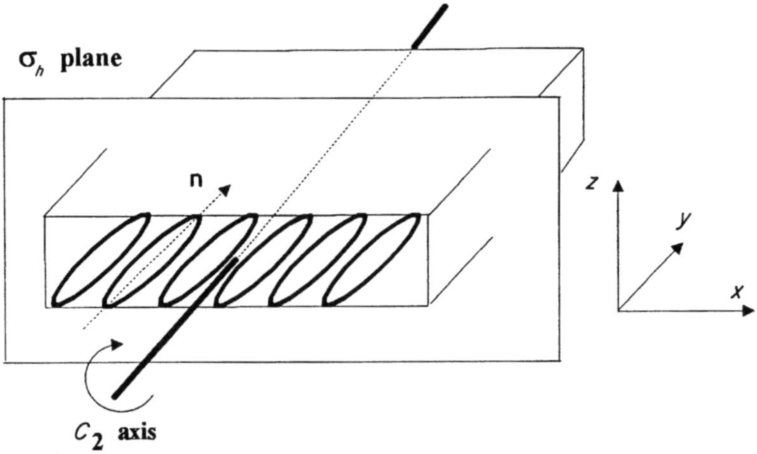

Figure 8 Schematic drawing showing the tilted molecules within one S_C layer. The molecules are lying in the σ_h plane with the C_2 axis being perpendicular to the σ_h plane

nent p_y of the spontaneous polarisation **P** remains unchanged upon rotation around C2:

$$\text{point group } C_2\text{: } \mathbf{P}\,(p_x, p_y, p_z) \xrightarrow{C_2} \mathbf{P}'\,(-p_x, p_y, -p_z) \qquad (7)$$
$$\Rightarrow p_x = p_z = 0;\ p_y \text{ may be different from } 0$$

As a consequence of the helically twisted arrangement of the layers in the S_C* phase, the direction of the polarisation rotates with the helix and the polarisation

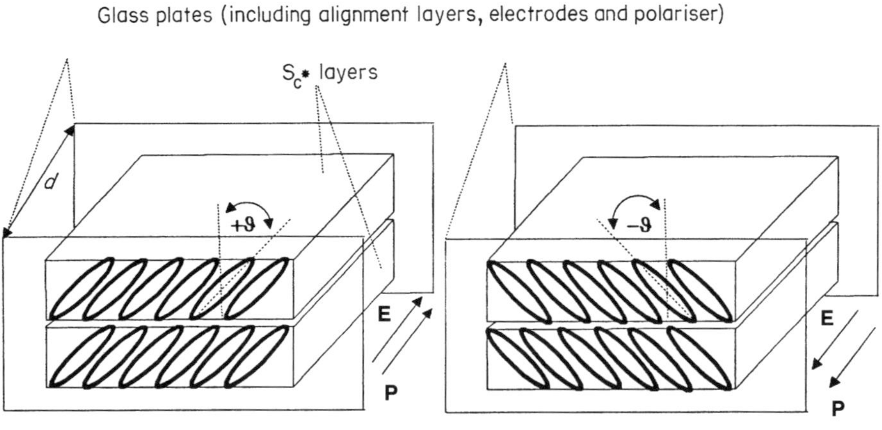

Figure 9 Operating principle of the SSFLC cell. **E**, electric field; **P**, spontaneous polarisation

is averaged to zero in the bulk. By unwinding the helix by surface action, Clark and Lagerwall invented a device showing a bistable electrooptic effect.[5] In this surface-stabilised ferroelectric liquid crystal (SSFLC) cell (Figure 9), the smectic layers are arranged perpendicular to the cell glass plates. By interaction with the surface, the rod-like molecules are constrained to a position parallel to the glass plates, thus preventing the S_C* phase from forming a helical structure.

By applying an electric field **E**, the spontaneous polarisation **P** is adjusted parallel to **E**. Reversing the direction of **E** causes a reversal of **P** by rotation of the molecules on a cone around the layer's normal (Figure 10). If the electric field is switched off, the positions of the molecules are retained. In this way, an SSFLC cell can be switched between two stable states by changing the polarity of an electrical pulse. Given the condition that the tilt angle ϑ is 22.5° and

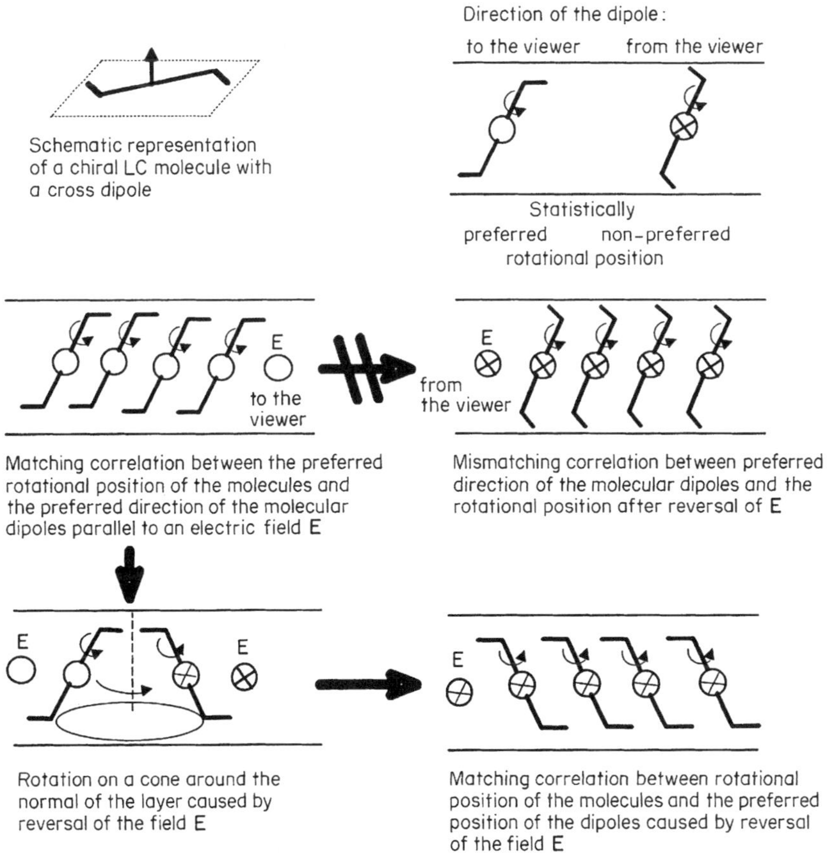

Figure 10 Schematic drawing showing the switching of a chiral material in the S_C* phase on the molecular level

$$d\Delta n = \lambda/2 \tag{8}$$

(where d = cell gap, Δn = birefringence, and λ = wavelength of the light), one of these two states transmits light and the other absorbs light when observed through crossed polarisers.

In order to understand the switching process on the molecular level, we have to look at the rotational motions of the molecules. In the isotropic, nematic and S_A phases the molecules rotate freely around their molecular long axis. Statistically (averaged over time) even molecules with C_1 symmetry can be regarded as having an axial symmetry around their molecular long axes. In the tilted S_C phase the situation is different. The rotation around the molecular long axes is unbalanced, giving rise to statistically preferred or non-preferred rotational positions of the molecules. This means that during one revolution a molecule remains for a longer time in the preferred position than in the other positions (Figure 10).

In an electric field **E**, molecules align themselves in such a way that, in the statistically preferred position, their dipoles are parallel to **E**. A reversal of the electric field causes a mismatch between the preferred direction of the dipoles and the preferred rotational positions of the molecules. Excluding an intramolecular rotation of the molecular dipole by rigidly attaching the dipole to the rest of the molecule, this mismatch situation can be overcome only by rotation of the molecules on a cone around the normal of the smectic layer (Figure 10).

Compared with the TN cell, this special switching mode of the SSFLC cell is much faster because, in the switching process, the extent of reorganisation of the S_C phase is smaller than that of the liquid crystalline phase in the TN cell. Additionally, the optical performance of the SSFLC cell depends less on the viewing angle. The prominent feature of the SSFLC display is bistability, offering data storage without power consumption. A problem arising from bistability is the difficulty of realising states between black and white (grey levels). Drawbacks of the SSFLC displays are higher power consumption compared with other advanced LC displays and the technological problems associated with the large-scale manufacture of displays with a very small cell gap of 2–3 μm (which is necessary in order to unwind the helix of the S_C* phase).

13.2.4 PROPERTIES OF THERMOCHROMIC LIQUID CRYSTALS

If circularly polarised light enters a cholesteric phase, and the wavelength of the light is of the order of the pitch, left-handed circularly polarised light can follow a right-handed helical path and the right-handed circularly polarised light is reflected. The corresponding effect is found for a left-hand helix.[6] As a consequence, a cholesteric phase can reflect light of an appropriate wavelength depending on pitch and handedness of the helix selectively, giving rise to highly iridescent colours.[7] The pitch of the chiral nematic phase usually shows a weak temperature dependence.[7] This results in a shift of the reflected colour from the blue to the red limit of the visible spectrum with decreasing temperature.[8] Such thermochromic liquid crystals find applications in temperature-indicating devices, inks and other areas.

13.3 REQUIREMENTS FOR COMMERCIAL LIQUID CRYSTAL MIXTURES

13.3.1 NEMATIC MIXTURES

Liquid crystals are organic molecules consisting of a rod-like rigid core and two terminal substituents (R^1 and R^2). Twenty-five years of industrial research have shown that six membered carbo- and heterocycles (A) are the most efficient building blocks for construction of the rigid core. These building blocks are either linked by a single bond, or by groups such as ethylene or ester (Z). The terminal substituents are generally unbranched alkyl and alkoxy groups or polar substituents, such as nitrile or a fluorine atom (Figure 11).

Displays for different applications require different liquid crystal properties. The common requirements are long-term stability, wide phase range and low viscosity, and important parameters are clearing point, dielectric and optical anisotropy and ohmic resistance. Not all of these requirements can be fulfilled by a single compound, and mixtures of, typically, 10–20 compounds are required (Figure 12).

For a uniform optical performance of a display based on the TN effect, it is necessary to have a uniform chirality of the helical twist of the liquid crystal.

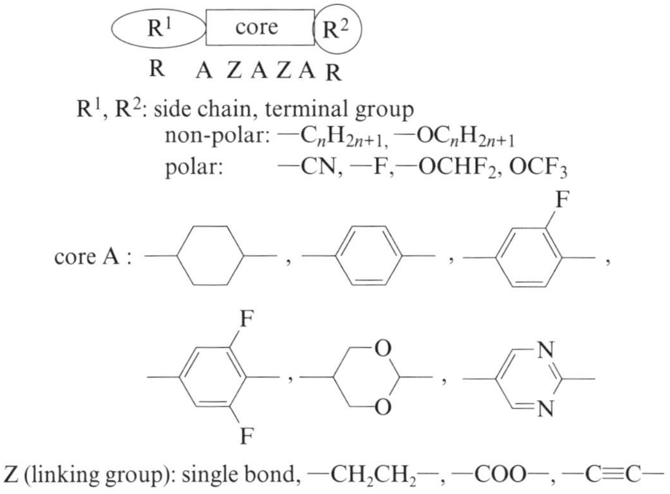

Figure 11 General formula of a rod-like liquid crystal. Ring systems for construction of the core and linking groups are limited here to those of commercial importance

Acronym	Structure
PCH-3	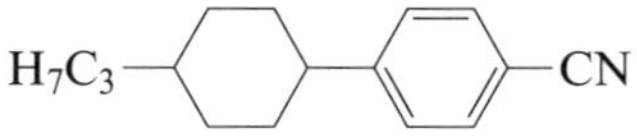 K 45 N 46 I
ME-2N.F	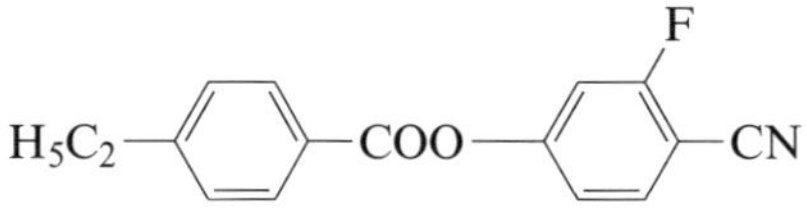 K 78 N (11) I
BCH-5	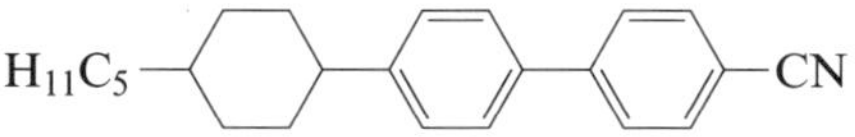 K 96 N 219 I
CBC-53	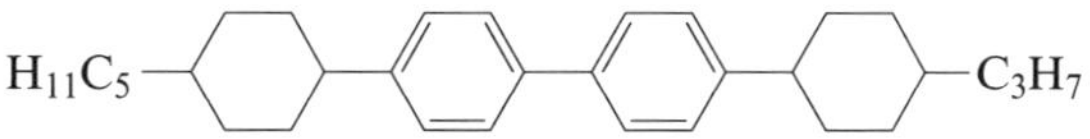 K 64 S_B 237 S_A 260 N 317 I
CCP-3OCF$_3$	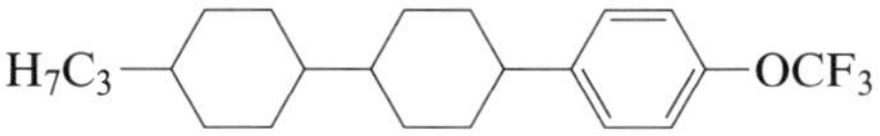 K70 N 155 I
ECCP-5F.F	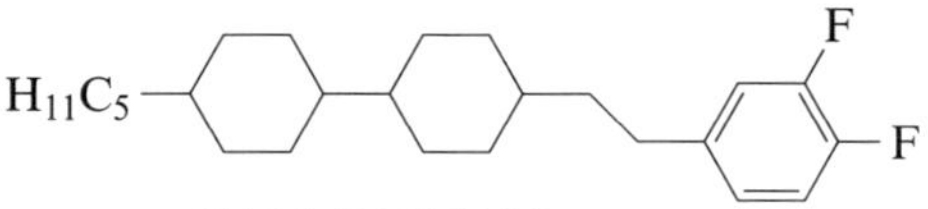 K47 S_B74 N 118 I
CCP-3OCF$_2$.F.F	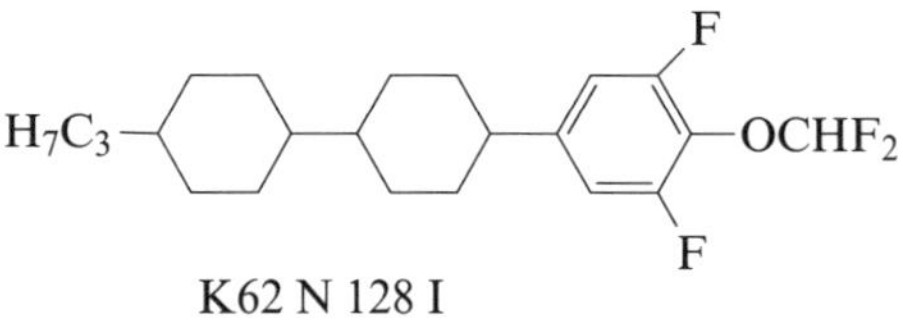 K62 N 128 I
CUP-3F.F	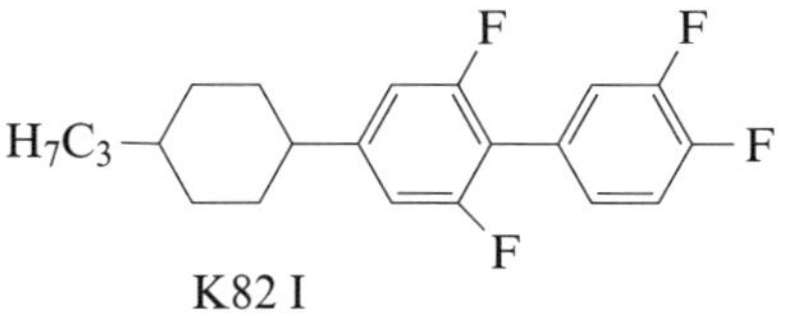 K82 I

Figure 12 The Merck Group produces more than 100 different compounds for application in nematic mixtures. The compounds shown here demonstrate different core structures and linking and terminal groups. Compounds with a cyano group are used in TN and STN mixtures. Compounds with fluorine atoms and SFM groups are mainly used in TFT mixtures

While the twist is generated by the interaction between the alignment layers of the cell, the uniform handedness of the twist is achieved by the addition of a chiral dopant. This dopant should not affect either the phase range or the viscosity of the mixture. For practical reasons, the concentration should not be significantly lower than 0.1%. Additionally, the pitch of the mixture has to fit the cell gap of the display. A consequence of these factors is that the HTP of the dopants should be of the order of 10 μm^{-1}.

Further development of the TN cell has resulted in the more advanced 'super twisted nematic' (STN) cell,[9a] requiring shorter pitch values for the mixture. Fine tuning of the pitch is achieved more readily by the use of a dopant with a larger HTP rather than increasing the dopant concentration.[9b]

Another approach to liquid crystal cells for high information content displays is 'thin film transistor' (TFT) technology. The TFT display differs from the conventional TN display only by integration of thin-film transistors into the transparent electrodes. This technology requires liquid crystals with superior stability and a larger resistivity than conventional TN technology. Generally, liquid crystals made from building blocks such as heterocycles and ester groups are less stable than those built up from six-membered carbocycles linked by single bonds or ethylene groups. All building blocks with a high affinity for ions result in materials with low resistivities. In particular, structural elements such as cyano groups must be avoided in materials for TFT displays. In this case polarity is introduced by fluorine atoms or multiply fluorinated groups such as OCF_3 or $OCHF_2$ ['super fluorinated materials', (SFM)].[10]

13.3.2 DOPANTS FOR NEMATIC MIXTURES

Initially, readily available cholesterol esters such as cholesteryl nonanoate (CN) were used as chiral dopants. The poor stability of these compounds to ultraviolet light[8] resulted in the search for more robust dopants. An additional target was to develop dopants with a less complex structure than steroids.

In the 1970s, liquid crystal core structures were mainly of the ester type, and an obvious approach to these was to link typical rod-like building blocks with commercially available chiral alcohols by an ester group. This resulted in the development of S-811, which requires (*S*)-octan-2-ol, and CB-15 with opposite handedness. As a consequence of the high price and the limited availability of (*S*)-octan-2-ol, an in-house synthesis was developed according to the method of Kenyon.[11] As this approach was based on resolution, (*R*)-octan-2-ol also became available, and R-811, a dopant having opposite handedness, but with all other properties being the same, was synthesised.

The introduction of the STN cell resulted in the need for a dopant with higher HTP, for example (*R*)- and (S)-1011.[9b] These new ester type dopants are based on (*R*)- and (*S*)-1-phenylethane-1,2-diol. The rapidly growing importance of TFT-TN displays in the last five years has created a need for a dopant with HTP comparable with that of S-811 but with improved stability and resistivity. This was

Acronym	Structure	Phase range	HTP (μm^{-1}) ZLI-1132	HTP (μm^{-1}) MLC-6012
CN	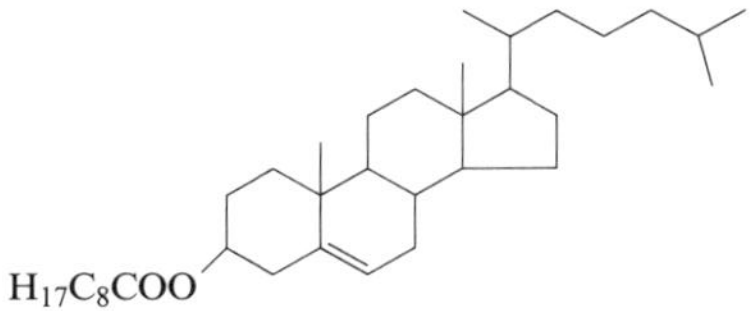	K 26 Ch 41 I	–4.4	
S-811	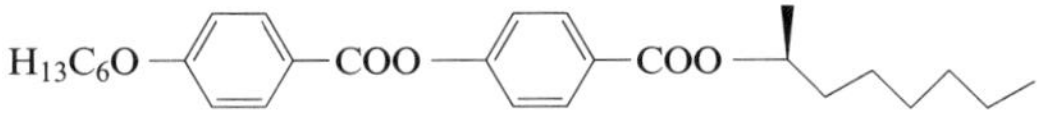	K 48 I	–13.3	–14.0
R-811	Enantiomer to S-811		+13.3	+14.0
C-15	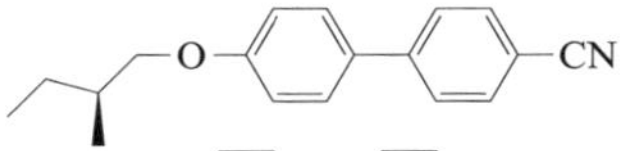	K 154 I	–1.6	
CB-15	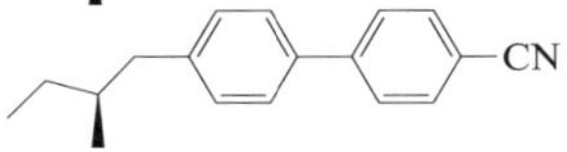	K 4 Ch (–30) I	+9.6	
S-1011	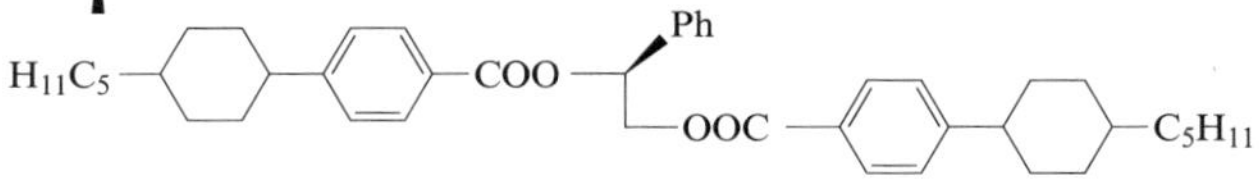	K 134 I	–39.5	
R-1011	Enantiomer of S-1011		+39.5	
S-2011	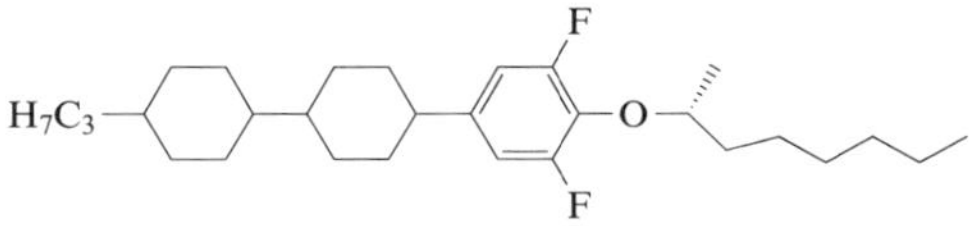	K 19 S_B 45 Ch 72 I		–13.6
R-2011	Enantiomer of S-2011			+13.6

Figure 13 Commercial dopants for nematic mixtures. HTP values are calculated (equation 3) from pitch values determined in a Grandjean wedge[2] in liquid crystal mixtures ZLI-1132 and MLC-6012 at 20 °C

achieved by combining core structures developed for SFM materials with enantiopure octan-2-ol via an ether linkage (Figure 13).

13.3.3 FERROELECTRIC MIXTURES

Much research effort on non-chiral S_C and chiral S_C* materials[12] in the last 10 years has resulted in several commercial ferroelectric mixtures.[13] Whereas the nematic materials are mainly based on cyclohexylphenyl and bicyclohexylphenyl core structures, with a short alkyl chain on the cyclohexane ring and a polar substituent at the phenyl ring, S_C materials are based on biphenyl and analogous heterocyclic core structures, with two long-chain alkyl or alkoxy groups (Figure 14). The S_C phase is converted into an S_C* phase by the addition of a chiral dopant. This dopant must not affect the phase range (S_C or S_C* respectively) of the base mixture, or interfere with the unwinding of the S_C* helix in the SSFLC cell by a large HTP. High values of the spontaneous polarisation are required to switch the mixture.[14]

13.3.4 FERROELECTRIC DOPANTS

The first optically active ferroelectric S_C* compound was synthesised in 1974 (Figure 6),[4] and industrial research on S_C* materials began ten years later. It became obvious that the optically active materials are of much greater importance for smectic formulations than for nematic mixtures. In addition to the more complex properties of the latter, the fact that they are used in concentrations up to 10% contributes to their economic importance. Consequently, building blocks

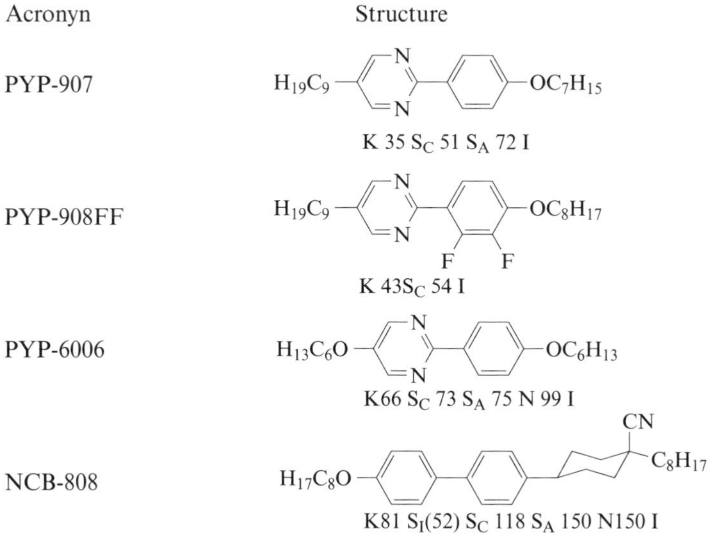

Figure 14 Overview of compounds used in commercial FLC mixtures[13]

from the chiral pool, and derivatives thereof, were screened intensively. Much work was devoted to terpenes and terpene derivatives,[15] amino acid derivatives[16] and lactic acid derivatives.[17] A general disadvantage of terpene-derived compounds is that these materials offer very low values for the spontaneous polarisation. During the systematic variation of the substituents it was found that only compounds with a strong dipole at the chiral centre result in large values for the spontaneous polarisation. For a given structure, spontaneous polarisation increases when the chiral centre is moved closer to the core.[18] Thus, high values of spontaneous polarisation are found for materials with the chiral centre and the cross-dipole coupled rigidly to the core. This was effected by using optically active α-chloro acid esters, ethers of lactic acid esters and ethers of optically active cyanohydrins. Later, optically active 2-fluoroalkanols[19a] were identified as allowing the synthesis of dopants with a wide phase range and high spontaneous polarisation[19b] (Figure 15).

As in the case of the nematic dopants, narrowing of the phase range of the base mixture can be avoided by using the same core structure for dopants and

Code No.	Structure	HTP (μm^{-1})	P_s ($nC\,cm^{-2}$)
IS-2175	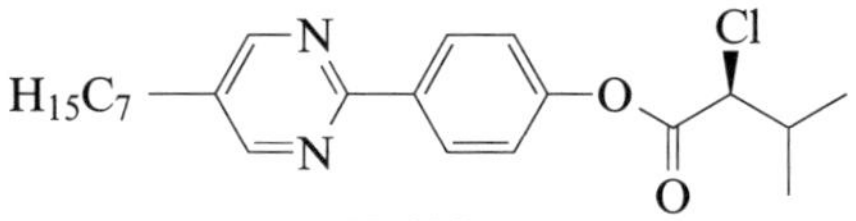 K 64 I	+0.3	14.3
IS-2874	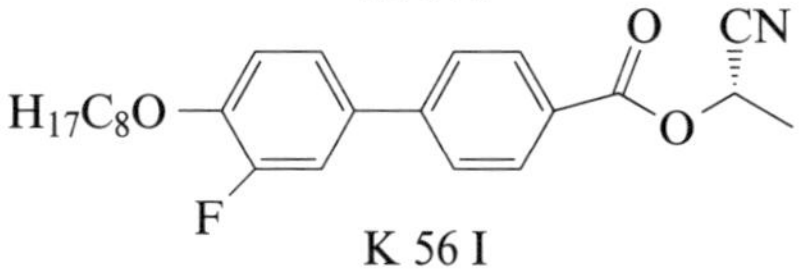 K 56 I	+2.1	31.3
IS-2875	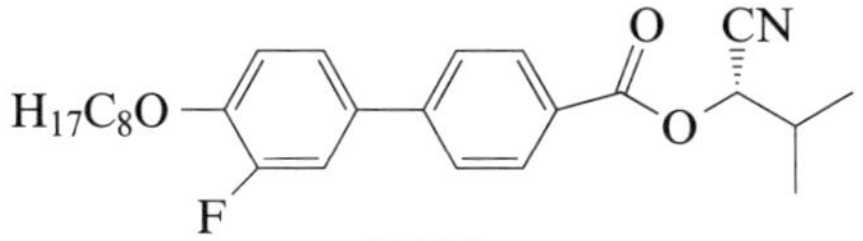 K49 I	+5.9	36.5
IS-3512	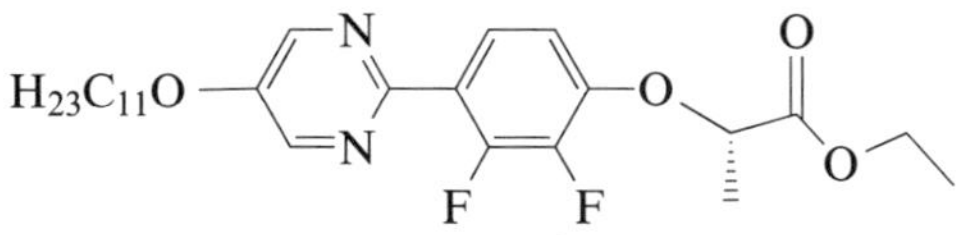 K38 I	+0.2	15.3
IS-4006	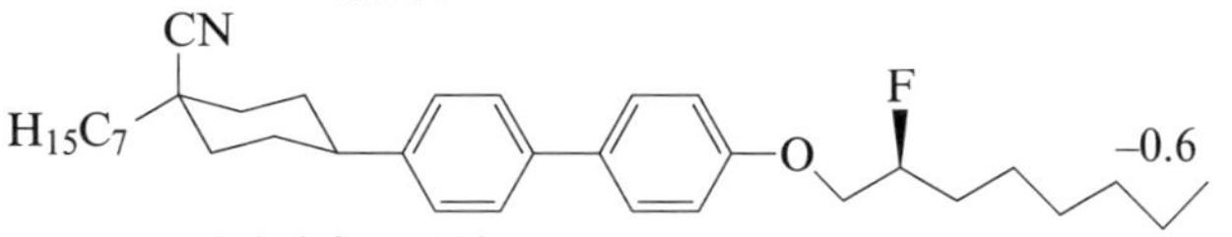 K110 Sm_A 159 I	−0.6	−19.5

Figure 15 Dopants developed by Merck for FLC mixtures[13]

base materials. Ideally, dopants for ferroelectric mixtures should combine a broad S_C* phase and a high value for spontaneous polarisation, but this combination of properties is still difficult to realise in practice.

13.3.5 THERMOCHROMIC LIQUID CRYSTALS

Thermochromic liquid crystals find many applications, mainly in the areas of thermometry, medical thermography, non-destructive testing, radiation sensing and colours/inks.[7] Common requirements for thermochromic liquid crystals are stability and low melting points, and commercial compounds with these features have been developed using benzene and biphenyl core structures. Their properties can be tuned by variation of the length of the non-chiral alkyl and alkoxy side chains (Figure 16). The compounds are commercially available in different formulations including mixtures of thermochromic compounds, micro-encapsulated thermochromic materials and combinations of micro-capsules with various inks and surface coatings.

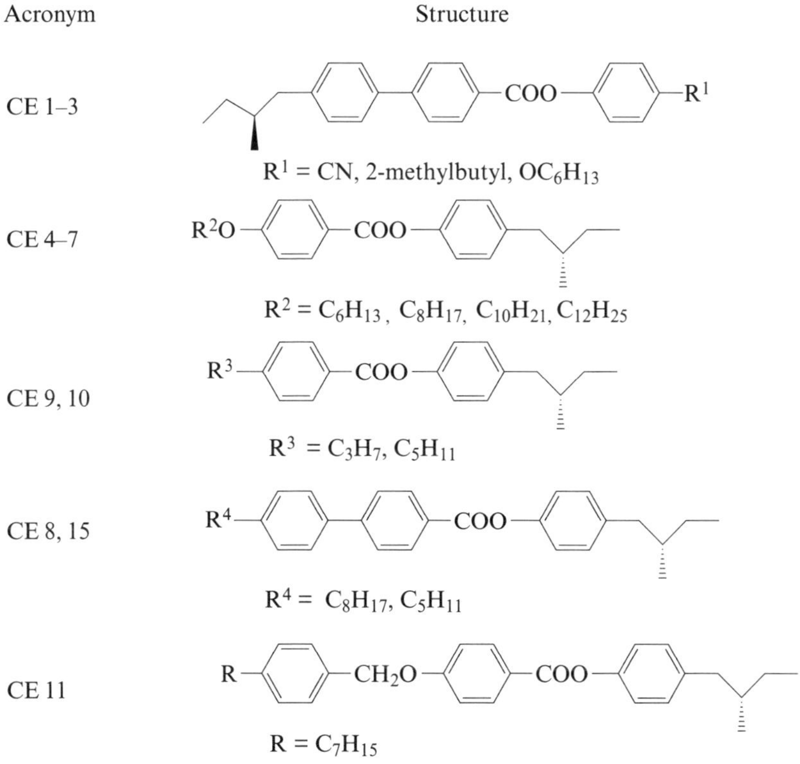

Figure 16 Commercially available thermochromic liquid crystals from Merck

13.4 SYNTHESIS OF CHIRAL LIQUID CRYSTALS

13.4.1 CHIRAL BUILDING BLOCKS FOR DOPANTS USED IN NEMATIC MIXTURES

The world market for nematic liquid crystal mixtures in 1994 was of the order of 16.5 tons.[20] Bearing in mind the typical dopant concentration of 0.1%, it is clear that the dopants are produced on a small scale. As noted previously, the required chiral building blocks are cholesterol, (*S*)-2-methylbutanol, (*R*)- and (*S*)-octan-2-ol and (*R*)- and (*S*)-1-phenylethane-1,2-diol. Cholesterol is found in animal tissue and (*S*)-2-methylbutanol is a by-product formed in alcoholic fermentation. Both of these naturally occurring optically active alcohols have been available in commercial quantities for a long time. As already noted, (*R*)- and (*S*)-octan-2-ol are prepared by classical resolution, which is effected with brucine via the octyl hydrogenphthalate.[11] Although this method is convenient, great care has to be taken in handling the highly toxic brucine which, owing to its high price, must be recycled.

Many other approaches have now been developed, mainly based on enzymatic and microbiological methods,[21] to prepare the necessary enantiopure secondary alcohols. Today, both enantiomers of octan-2-ol and 1-phenylethane-1,2-diol are readily available in larger quantities, making small-scale in-house production unattractive.

13.4.2 CHIRAL DOPANTS FOR NEMATIC MIXTURES

The nematic dopants can be classified according to the linking group between the liquid crystal core structure and the chiral building block into esters, alkylated aromatics and ethers. Starting from optically active alcohols, the esterification processes can be carried out under standard conditions, since the chiral centre is not involved in the reaction. As an example, the synthesis of R-1011 is shown in Figure 17.[22]

Direct attachment of a branched chiral side chain to an aromatic core can be achieved by metal-catalysed cross-coupling of an aryl Grignard reagent and an

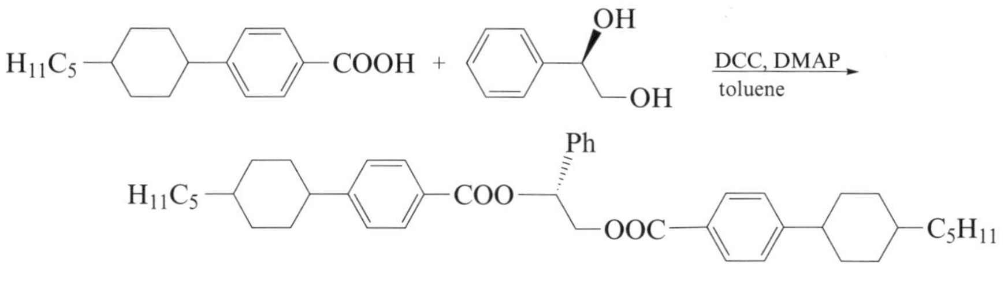

Figure 17 Synthesis of R-1011

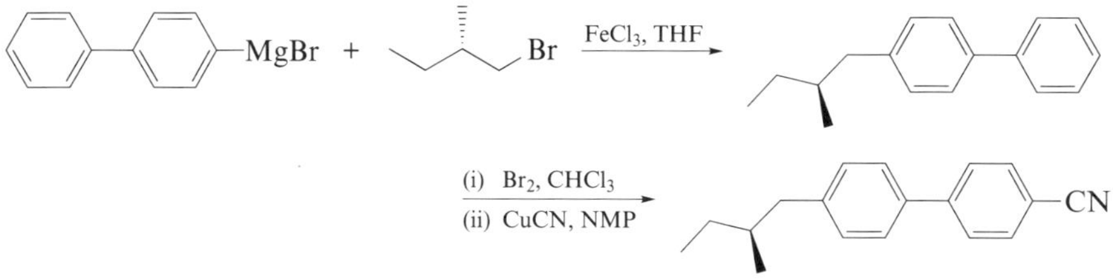

Figure 18 Synthesis of CB-15

alkyl bromide, exemplified in the synthesis of CB-15[23] (Figure 18). The ethers R- and S-2011 are prepared from 4-[*trans*-4-(*trans*-4-propylcyclohexyl)cyclohexyl]-2,6-difluorophenol via the Mitsunobu reaction with inversion of the chiral centre[24] of the corresponding enantiopure octan-2-ol (Figure 19).[25] For safety reasons, diisopropyl azodicarboxylate (DIAD) is used instead of the explosive diethyl azodicarboxylate (DEAD).[26]

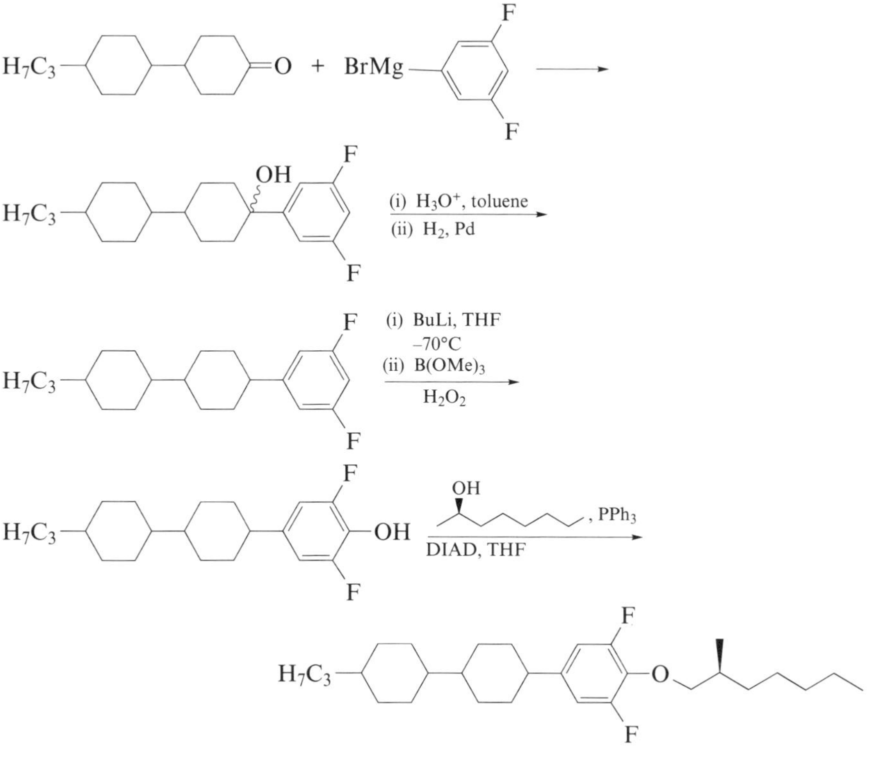

Figure 19 Synthesis of R-2011

13.4.3 CHIRAL BUILDING BLOCKS FOR FERROELECTRIC DOPANTS

The first-generation dopant for commercial ferroelectric mixtures was a derivative of the optically active α-chloroacid derived from L-valine by diazotisation in the presence of HCl[27] (Figure 20). The α-chloro acids were succeeded by derivatives of optically active α-hydroxy acids. The latest generation of dopants are derivatives of optically active 2-fluoroalkan-2-ols which are readily available from commercial optically active 1,2-epoxyalkanes, (see also Volume I, Chapter 7).

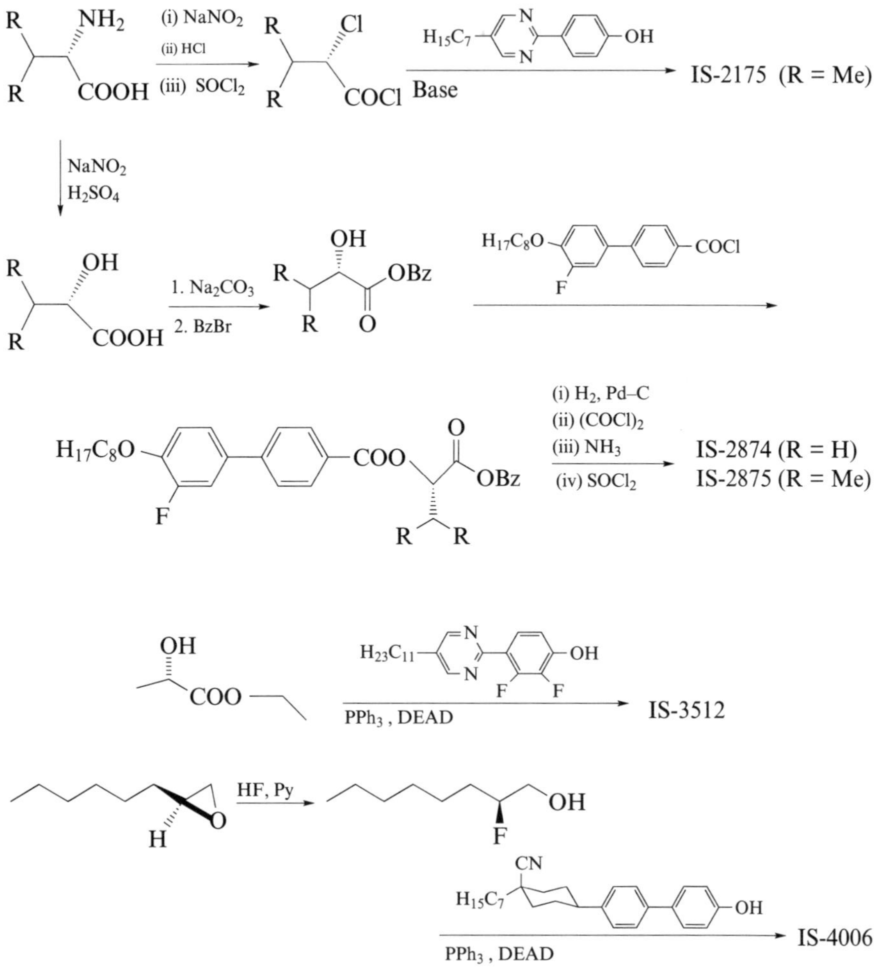

Figure 20 Synthesis of dopants for S_C mixtures from commercial enantiopure building blocks. The synthesis of the liquid crystal intermediates is described in the references

13.4.4 FERROELECTRIC DOPANTS

As in the case of the nematic dopants, the ferroelectric dopants can be classified according to the linking group into esters and ethers. The ester IS-2175 is made by esterification of enantiopure (*S*)-2-chloro-3-methylbutanoic acid chloride with 4-(5-heptyl-1,3-pyrimid-2-yl)phenol under basic conditions (Figure 20). The syntheses of the cyanohydrin esters IS-2874 and IS-2875[28] are more difficult. To make IS-2875, L-valine is transformed, with retention of configuration, into the corresponding optically active α-hydroxy acid, which, as the sodium salt, is treated with benzyl bromide to give the benzylic ester. The hydroxy group of this intermediate is esterified with 3′-fluoro-4′-octyloxybiphenyl-4-carboxylic acid chloride. The benzylic ester is cleaved by hydrogenolysis, and the resulting acid is treated successively with oxalic acid chloride, ammonia and thionyl chloride to produce IS-2875 (Figure 20). Starting from lactic acid, the dopant IS-2874 is made using an analogous reaction sequence.

The ether-type compounds IS-3512 and IS-4006[29] are prepared via the Mitsunobu reaction of 4-(5-undecanyl-1,3-pyrimid-2-yl)-2,3-difluorophenol with ethyl (*S*)-lactate and reaction of 4′-(4-cyano-4-*trans*-heptylcyclohexyl)biphenyl-4-ol with (*S*)-2-fluorooctanol, respectively. The latter compound is available from (*R*)-1,2-epoxyoctane by treatment with hydrogen fluoride–pyridine complex[19a] (Figure 20).

13.4.5 THERMOCHROMIC LIQUID CRYSTALS

With a demand of the order of 1 tpa, chiral compounds for thermochromics are produced on a larger scale than chiral dopants for display application. The products described here compete with inexpensive cholesterol esters and, for economic reasons, they are all based on two key optically active intermediates derived from cheap (*S*)-2-methylbutan-1-ol. These intermediates are combined with building blocks from the production of thermotropic liquid crystals.

The first key intermediate is (*S*)-4-(2-methylbutyl)-biphenyl-4′-carboxylic acid prepared from CB-15 by hydrolysis of the nitrile group under basic conditions. Treatment with thionyl chloride yields the corresponding acid chloride, which is then esterified with 4-alkylphenols to give the thermochromic liquid crystals CE 1 to 3[23] (Figure 21).

The second key intermediate is (*S*)-4-(2-methylbutyl)phenol, which is synthesised by a metal-catalysed cross-coupling reaction between 4-methoxyphenylmagnesium bromide and (*S*)-1-bromo-2-methylbutane. The methoxy group is then cleaved with hydrogen bromide to yield (*S*)-4-(2-methylbutyl)phenol. Esterification of this phenol yields the thermochromic liquid crystals CE 4 to 15.[23]

13.4.6 OPTICAL PURITY

The optical purities of the chiral building blocks used in the synthesis of the dopants are listed in Table 1. As noted above, the dopants are used to induce a

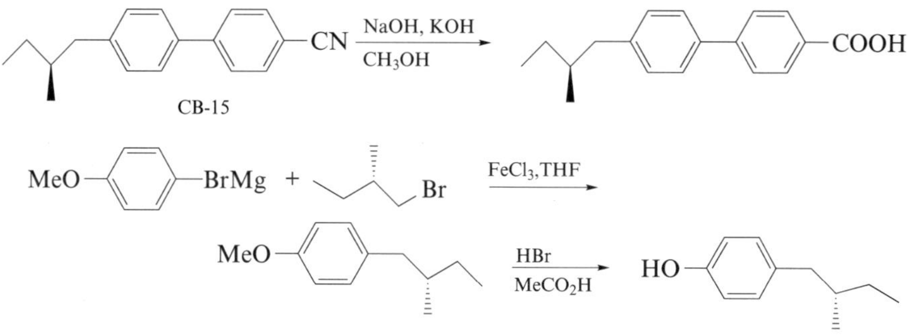

Figure 21 Synthesis of the key chiral intermediates for thermochromic liquid crystal

Table 1 The *ee* values of chiral building blocks used in the synthesis of dopants

Compound	*ee*(%)	Ref.
(*S*)-2-Methylbutanol	80	12
(*R*)-Octan-2-ol	> 99	30
(*S*)-Octan-2-ol	> 99	30
(*S*)-Valine	> 99	30
(*S*)-2-Chloro-3-methylbutanoic acid	98	27
Ethyl (*S*)-lactate	> 98	30

specific pitch in nematic mixtures or a specific spontaneous polarisation in ferroelectric mixtures. Consequently, the HTP and the spontaneous polarisation are used in the characterisation of the dopants.

13.5 TRENDS AND OUTLOOK

The world market for liquid crystal displays is expected to grow from US$5 billion in 1994 to $12 billion in 2000, with the largest growth expected in the area of TFT displays.[31] The corresponding market for liquid crystal mixtures will grow in this period from 16.5 tons ($89 million) to 27.3 tons ($179 million).[20] The market is demanding larger displays with improved optical performance, and reduced power consumption for portable devices. These targets will be approached both by display development by the use of new driver chips and also by the development of new TFT modes[32] and technologies.[33] Large improvements of materials performance are necessary, requiring mainly the development of new polar materials with low viscosity and improved electrical purity. Another target is the development of systems prepared from prepolymers/polymers and liquid crystals in which conventional orientation layers are replaced by molecular networks. The development of new TFT modes might also require dopants with properties fine-tuned appropriately.

The future of ferroelectric liquid crystals will depend heavily on the introduction and success of the first commercially available SSFLC display in 1995.[34]

13.6 ACKNOWLEDGEMENTS

The authors would like to dedicate this chapter to Professor Dr H. C. Michael Hanack on the occasion of his 65th birthday. The authors are grateful to Dr M. Bremer, Dr D. Coates, Dr J. Krause and Mr V. Reiffenrath for helpful discussions and Ms B. Nipkow for preparing the manuscript.

13.7 NOTES AND REFERENCES

1. Reinitzer, F., *Monatsh. Chem.*, 9, 421 (1888).
2. Vertogen, G., and de Jeu, W. H., *Thermotropic Liquid Crystals, Fundamentals*, Springer, Berlin, 1988.
3. Schadt, M., and Helfrich, W., *Appl. Phys. Lett.*, **18**, 127 (1971).
4. Meyer, R. B., Liebert, L., Strzelecki, L., and Keller, P., *J. Phys. Lett.*, **36**, 4 (1975).
5. (a) Clark, N. A., and Lagerwall, S. T., *Appl. Phys. Lett.*, **36**, 899 (1980); (b) Lagerwall, S.T., and Dahl., I., *Mol. Cryst., Liq. Cryst.* **14**, 151 (1984).
6. Bouligand, Y., *J. Phys.*, **34**, 603 (1973).
7. McDonnell, D. G., in *Thermotropic Liquid Crystals*, (ed. G. W. Gray), Wiley, Chichester, 1987, Vol. 22; Sage, I., in *Liquid Crystals Applications and Uses* (ed. B. Bahadur), World Scientific, Singapore, 1992, Vol. 3.
8. Parsley, M., in *Licritherm* (ed. I. C. Sage and K. G. Archer), BDH Poole, 2nd edn, Sage, I., in *Ullmann's Encyclopedia of Industrial Chemistry*, VCH, Weinheim, 1990, Vol. A15.
9. (a) Waters, C. M., and Raynes, E. P. (Ministry of Defence), *Br. Pat.* 2 123 163; (b) Heppke, G., Lötzsch, D., and Oestreicher, F., *Z. Naturforsch., Teil A.*, **41**, 1214 (1986); Hochgesang, R., Plach, H. J., and Reiffenrath, V., *Produktinformation*, Merck, Darmstadt, 1983; Heppke, G., and Oestreicher, F. (Merck), *Eur. Pat.*, EP 0 168 043 1990.
10. Eidenschink, R., and Pohl, L. (Merck), *Eur. Pat.*, EP 0 051 738, 1984; Kurmeier, H. A., Scheuble, B., Poetsch, E., and Finkenzeller, U. (Merck), *Eur. Pat.*, EP 0 334 911 1, 1993; Reiffenrath, V., Kurmeier, H. A., Poetsch, E., Plach, H. J., and Finkenzeller, U. (Merck) *PCT Int. Pat Appl.*, WO 91/03450, 1991.
11. Kenyon, J., *Org. Synth.*, *Coll. Vol.*, **1**, 418.
12. Goodby, J. W., Blinc, R., Clark, N. A., Lagerwall, S. T., Osipov, M. A., Pikin, S. A., Sakurai, T., Yoshino, K., and Zeks, B., *Ferroelectric Liquid Crystals, Principles, Properties and Applications*, Gordon and Breach, New York, 1991.
13. Merck has terminated all FLC activities and has sold its patent portfolio to Hoechst (February 95). Correspondingly, Hoechst has transferred all patent rights for nematic liquid crystals to Merck.
14. Finkenzeller, U., Pausch, A. E., Poetsch, E., and Suermann, J., *Kontakte* (*Darmstadt*), **2**, 3 (1993).
15. Coates, D., Gray, G. W., Lacey, D., Young, D. J. S., Toyne, K. J., and Bone, M. F. (Ministry of Defence), *PCT Int. Pat. Appl.*, WO 86/04328, 1986.

16. Bahr, C., and Heppke, G. *Mol. Cryst. Liq. Cryst. Lett.*, **4**, 31 (1986); Demus, D., Zaschke, H., Weissflog, W., Mohr, K., Köhler, S., and Worm, K. (VEB Werk), *Ger. Pat. Appl.*, DE 362 796 4, 1985; Eidenschink, R., Escher, C., Geelhaar, T., Hittich, R., Kurmeier, H. A., Pauluth, D., and Wächtler, A.E.F (Merck), *Eur. Pat.* EP 0 257 049, 1990.
17. Sage, I., Jenner, J., Kurmeier, H. A., Pauluth, D., Escher, K., and Poetsch, E. (Ministry of Defence), *Eur. Pat.*, EP 0 263 843, 1986; Geelhaar, T., Kurmeier, H. A., and Wächtler, A.E.F., paper presented at the 12th ILCC, Freiburg, 1988.
18. Walba, D. M., Slater, S. C., Thurmes, W. N., Clark, N. A., Handschy, M. A., and Supon, F., *J. Am. Chem. Soc.*, **108**, 5210 (1986).
19. (a) Nohira, H., Kamei, M., Nakamura, S., Yoshinaga, K., and Kai, M. (Canon KK), *Jpn. Pat.*, JP 62 093 248, 1985; (b) Pausch, A.E., Geelhaar, T., Poetsch, E., and Finkenzeller, U., in *Liquid Crystal Materials, Devices and Applications* (ed. P. S. Drzaic and E. Uci), *Proc. SPIE*, 1665, **128** (1992).
20. Mentley, D. E., and Castellano, J. A., *Liquid Crystal Display Manufacturing, Materials and Equipment*, Stanford Resources, San Jose, CA, 1994.
21. Kawashima, M., (Kankyo Kagaku Sentaa KK)., *Jpn. Pat.*, JP 05 317 090, 1992; Wong, C. H., Wang, Y. F., Hennen, W. I, and Babiak, K. (Searle), *Eur. Pat.*, EP 0 357 009, 1989; Nishio, T., Seto, K., Ohashi, M., Terao, Y., Tsuji, K., and Achinami, K. (Sapporo Breweries), *Jpn. Pat.* JP 02 142 495, 1988; Murakami, N., and Hara, S., (Idemitsu Kosan), *Jpn. Pat.*, JP 01 257 484; Barton, P., and Page, M. I. (ICI), *Int. Pat.*, 93/02207, 1993; Ogura, M., Shiraishi, T., Takahashi, H., and Hasegawa, J. (Kanegafuchi), *Eur. Pat.*, EP 0 317 998, 1988; Suzuki, G., and Murakami, N. (Idemitsu Kosan), *Jpn. Pat.*, JP 63 042 700, 1986.
22. Heppke, G., and Oestreicher, F., (Merck), *Eur. Pat.*, EP 0 168 043, 1990.
23. Gray, G. W., and McDonnell, D. C. (Ministry of Defence), *Br. Pat.*, 1 592 161, 1977.
24. Bittner, S., and Assaf, Y. *Chem. Ind.* (London), 281 (1975); Mitsunobu, O., *Synthesis*, 1 (1981).
25. Pauluth, D., and Plach, H. J., (Merck) *Ger. Pat. Appl.*, DE 4 322 905, 1993.
26. Camp, D., and Jenkins, I. D., *J. Org. Chem.*, **54**, 3045 (1989).
27. Fu, S.-C., Birnbaum, S. M., and Greenstein, J. P., *J. Am. Chem. Soc.*, **76**, 6054 (1954); Koppenhöfer, B., personal communication.
28. Coates, D., Greenfield, S., and Sage, I. C. (Ministry of Defence), *US Pat.*, 5 230 830, 1992.
29. Wächtler, A. E. F., Geelhaar, T., Hittich, R., Bartmann, E., Krause, J., and Poetsch, E., (Merck), *PCT Int. Pat Appl.* WO 90/13 611, 1990.
30. *Chemika–BioChemika Catalogue*, Fluka, Buchs, 1991, pp. 1044, 1375.
31. Trish, C., Flat Panel Display Industry Overview in *Technology and Application Trends*, (ed. J. A. Castellano), Stanford Resources., San Jose, CA, 1993: worldwide LCD market opportunity past present and future, flat panel display industry overview.
32. Shimada, S. (Sharp KK), *Eur. Pat. Appl.*, EP 0 626 607, 1994.
33. Kiefer, R., Weber, B., Windscheid, F., and Baur, G., presentation at the 12th International Display Research Conference, Hiroshima, Japan, October 1992: in-plane switching of nematic liquid crystals.
34. The start of the production of 15 inch SSFLC displays at 5000 per month by Canon in the spring of 1995 was announced by, amongst others, Nikkon Kogyo on 19 November 1994 and by Handontai Sangyo on 21 December 1994.

14 Bio-transformation in the Production of L-Carnitine

Th. P. ZIMMERMANN, K. T. ROBINS,* J. WERLEN and F. W. J. M. M. HOEKS
LONZA AG, Visp, Switzerland

In memory of Hans G. Kulla

14.1 Introduction 287
14.2 Strain Isolation 288
14.3 Butyrobetaine/L-Carnitine Metabolism 288
14.4 Transport 292
14.5 Genetic Background 293
14.6 The Bio-transformation Strain 294
14.7 The Recombinant Bio-transformation Strain 296
14.8 Process Design 298
14.9 Kinetic Analysis 300
14.10 Raw Materials Aspects 303
14.11 Conclusion 304
14.12 References 304

14.1 INTRODUCTION

L-Carnitine, (*R*)-3-hydroxy-4-(trimethylammonio)butanoate, belongs to a group of vitamin-like nutrients. Also known as vitamin B_T, it is synthesised in the human liver from lysine and methionine and functions in the transport of activated fatty acids over the mitochondrial inner membrane in eukaryotes. L-carnitine plays an essential role in the β-oxidation of long-chain fatty acids by mitochondria; skeletal and cardiac muscle cells rely upon this mechanism as a source of metabolic energy. Insufficient L-carnitine is produced in infants,[1] adolescents and adults under certain physiological conditions, and a dietary deficiency of lysine or methionine can lead to an L-carnitine deficiency. This condition is associated with muscle weakness, fatigue and elevated levels of triglycerides in the blood. The role of L-carnitine in infant, health, sport and geriatric nutrition is receiving increasing interest, and applications in the medical field have been reviewed and discussed recently.[2] L-Carnitine is considered to be an essential nutrient for

*K. T. R. was not directly involved with the invention covered by the patent applications.

Chirality in Industry II. Edited by A. N. Collins, G. N. Sheldrake and J. Crosby

infants because endogenous biosynthesis is not adequate to meet infant needs.[1] Infants fed unsupplemented soya-based formulae receive virtually no dietary L-carnitine until other foods are introduced. It has been demonstrated that supplementation of the diet of healthy people can improve physical performance,[3] and the beneficial effect of L-carnitine for the heart has led to its use in geriatric nutrition. The use of L-carnitine also has a large commercial potential in the animal feed sector; improved growth rates and improved feed efficiencies have been reported for various species.[4,5]

The metabolically active form of carnitine is L-carnitine. D-Carnitine competes with L-carnitine for active transport into the eukaryotic cell,[1] and it has been shown in fish cultures that D-carnitine causes slower growth.[6] For this reason, the pure L-enantiomer is required for use in the medical, nutritional and feed applications of carnitine.[7] LONZA has developed a new bio-transformation route from the achiral compound 4-butyrobetaine, which gives L-carnitine in 99.5% yield and 100% *ee*. Other possibilities for the bio-technological production of L-carnitine[8] which involve resolution of racemic carnitine and its precursors have an inherent yield disadvantage and afford less than 100% *ee*. As at least 50% of this yield loss results from complete degradation of the D-isomer, the recycling of D-carnitine via chemical racemisation should, in principle, offset the yield loss, but a chemical resolution process developed at LONZA turned out to be uncompetitive.[7]

14.2 STRAIN ISOLATION

Betaines of the type shown in Figure 1 occur widely in nature. One of their functions is as osmoregulatory substances in plants and bacteria. We therefore expected that a microorganism capable of enantioselective synthesis of L-carnitine from 4-butyrobetaine or crotonobetaine could be isolated from soil samples using a selective medium. Strains isolated in this way were tested for their ability to use 4-butyrobetaine, crotonobetaine and L-carnitine as the sole carbon, nitrogen and energy source under aerobic conditions. An important selection criterion was the inability of the strains to utilise D-carnitine.

One of these strains, HK4, was chosen for further work. This strain was phenotypically characterised (Table 1) but, despite extensive studies, it could not be assigned to one genus. It was found to be related taxonomically to the common soil microorganisms *Agrobacterium* and *Rhizobium*, a close relationship to *Rhizobium meliloti* being established using the UPGMA method[9] and by the high genetic homology that was found between the two microorganisms. The strain was assumed to belong to a new, as yet undefined, genus.[10,11,12]

14.3 BUTYROBETAINE/L-CARNITINE METABOLISM

There are several known pathways for the metabolism of L-carnitine in microorganisms, and an excellent review was given by Jung *et al.*[8] It appeared that the

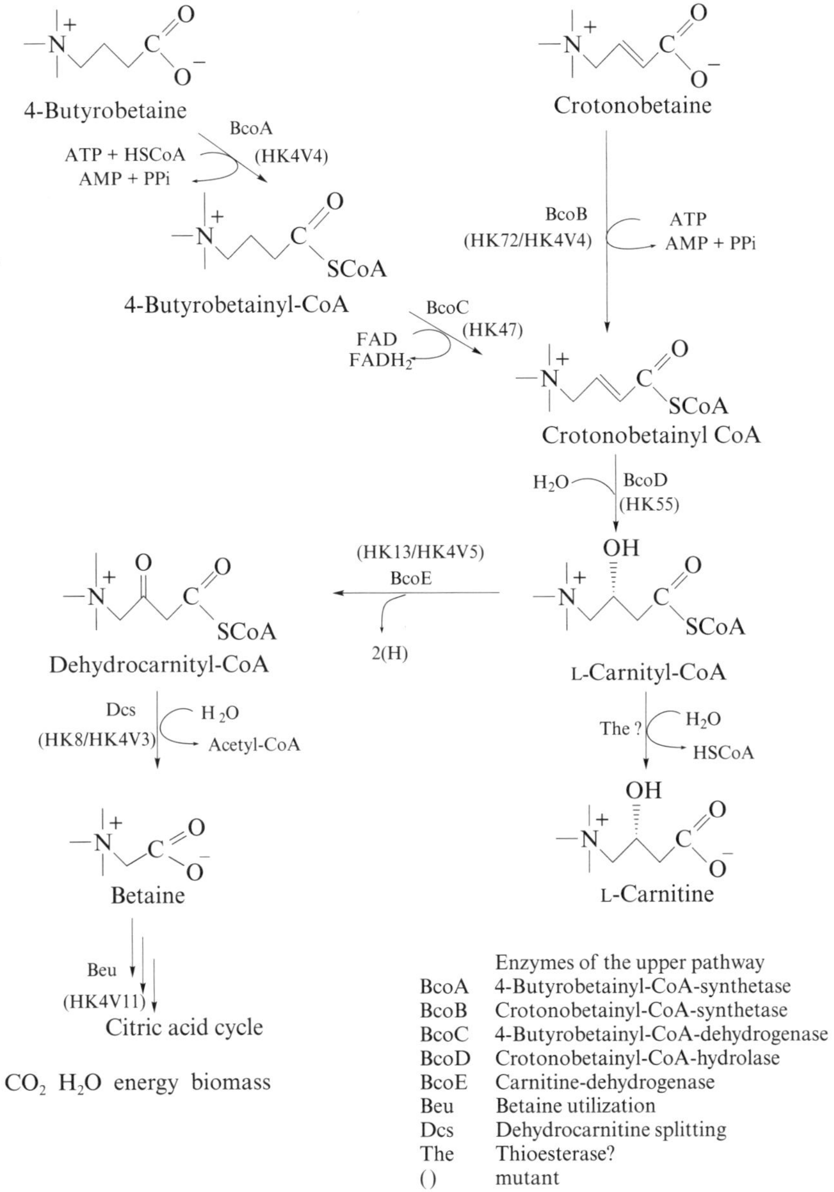

Figure 1 Catabolic pathway of betaine derivatives in the strain HK4. The first steps of the degradation of 4-butyrobetaine and/or crotonobetaine via L-carnityl-CoA to dehydrocarnityl-CoA are under coordinated regulation and described as the upper pathway of the catabolism. The further mineralisation of dehydrocarnityl-CoA is independent of the upper pathway on both a genetic and a biochemical level and is therefore termed the lower pathway of the catabolism. For each step frame shift or transposon insertion mutants were obtained as indicated in the scheme (compare Figure 2 and Table 2)

Table 1 Taxonomic description of the strain HK4 and composition of the nutrient medium[a]

Form of the cell:			
Shape	Rods partly pleomorphic		
Length	1–2 μm		
Width	0.5–0.8 μm		
Mobility	+		
Flagellata	Peritric		
Gram reaction	–		
Spores	–		
Formation of poly-β-hydroxybutyrate	–		
Oxidase	+		
Catalase	+		
Growth:			
Anaerobic	–		
30 °C	+		
37 °C	–		
MacConkey agar	+		
SS agar	–		
Cetrimide agar	–		
Utilisation of substrate:			
Acetate	–		
Citrate	–		
Malonate	–		
Glycine	–		
Norleucine	–		
Xylose	+		
Fructose	+		
Glucose	+		
Autotrophic growth with H_2	–		
3-Ketolactose	–		
Growth on:			
4-Butyrobetaine	(–)	+	(–)
Crotonobetaine	(–)	+	(–)
L-Carnitine	(–)	+	(–)
Betaine	(+)	+	(+)
Choline	(+)	+	(+)
Dimethylglycine	(+)	+	(+)
Sarcosine	(+)	+	(+)
L-Glutamate and 4-butyrobetaine	(+/–)	+/–	(+)
L-Glutamate and crotonobetaine	(+/–)	+/–	(+)
L-glutamate and L-carnitine	(+/–)	+/–	(+)

Formation of acids (OF-test) from:	
Glucose aerobic	–
Anaerobic	–
Fructose aerobic	–
ASS glucose	+
Xylose	+
Trehalose	+
Ethanol	–
Gas formation from glucose	–
ONPG	+
Arginine dehydrolase	–
Lysine decarboxylase	–
Phenylalanine deaminase	–
Ornithine decarboxylase	–
H_2S	–
Voges–Proskauer	–
Indole	–
Nitrite from nitrate	+
Denitrification	+
Formation of lavan	–
Lecithinase	–
Urease	+
Decomposition of:	
Starch	–
Gelatin	–
Casein	–
Tyrosine	–
Tween 80	–
DNA	+
Aesculin	+
Formation of pigment:	
Not diffusing	–
Diffusing	–
Fluorescing	–

(*continued*)

Table 1 (*continued*)

Composition of the nutrient medium:			
L-Glutamate	2 g	*Trace elements solution*:	
Betaine	2 g	$ZnSO_4 \cdot 7H_2O$	100mg
Crotonobetaine	2 g	$MnCl_2 \cdot 4H_2O$	30 mg
Buffer solution	100 ml	H_3BO_3	300 mg
Mg–Ca–Fe solution	25 ml	$CoCl_2 \cdot 6H_2O$	200 mg
Trace elements solution	1 ml	$CuCl_2 \cdot 2H_2O$	10 mg
Vitamin solution	1 ml	$NiCl_2 \cdot 6H_2O$	22 mg
With water to	1ℓ	$NaMoO_4 \cdot 2H_2O$	30 mg
Buffer solution:		With water to	1ℓ
Na_2SO_4	1 g		
$Na_2HPO_4 \cdot 2H_2O$	25.08 g	*Vitamin solution*:	
KH_2PO_4	10 g	Pyridoxal·HCl	10 mg
NaCl	30 g	Riboflavin	5 mg
With water to	1ℓ	Nicotinamide	5 mg
		Thiamine·HCl	5 mg
		Biotin	2 mg
Mg–Ca–Fe solution:		Sodium pantothenate	5 mg
$MgCl_2 \cdot 6H_2O$	16 g	*p*-Aminobenzoic acid	5 mg
$CaCl_2 \cdot 2H_2O$	0.58 g	Folic acid	2 mg
$FeCl_3 \cdot 6H_2O$	0.032g	Vitamin B_{12}	5 mg
With water to	1 ℓ	With water to	1ℓ

[a] The strain is deposited at the German Culture Collection under the registration number DSM2938. The phenotypical changes of the mutant strains HK13 (DSM2903) and HK1349 (DSM3944) are given in parentheses.

conversion of 4-butyrobetaine into L-carnitine in HK4 did not follow one of the known pathways.[10] Small, but measurable, amounts of L-carnitine were formed and excreted into the medium during growth on crotonobetaine and 4-butyrobetaine.

A lag phase was observed when the cells were grown on succinate or glucose and then transferred to betaine derivatives as the carbon source. In contrast, no lag phase occurred when the cells were grown on one betaine derivative and then transferred to a medium containing a different betaine derivative, which indicated a close coupling of the metabolism of 4-butyrobetaine, crotonobetaine and L-carnitine, and that the degradation of these substrates occurred under coordinated induction. The biochemical evidence showed that the metabolism of the betaine derivatives was analogous to the β-oxidation of fatty acids (Figure1). The activities of each catabolic step could be measured *in vitro* with a direct enzymatic assay or by product formation, and the requirement for coenzyme A, FAD and ATP was also demonstrated. The enzymes of the pathway were enriched by chromatography and gel filtration for the biochemical assays, but not purified because of their instability (Table 2). 4-Butyrobetainyl-CoA is a key intermediate of this pathway and could be isolated from *in vitro* assays using ion-exchange chromatography (its structure being confirmed by ^1H NMR spectroscopy). The CoA-synthetase and the hydrolase appear to form an enzymatic complex that

Table 2 Biochemical, phenotypical and genotypical features of HK4 and selected mutants[a]

Strain	Phenotype				Activity						DNA-fragment Complementing	Mutated gene
	4-bb	cb	L-c	b	BcoT	BcoB	BcoA	BcoC	BcoD	BcoE		
HK4	+	+	+	+	+	+	+	+	+	+	nt.	Wild-type
HK13	–	–	–	+	+	+	+	+	+	–	1	*bcoE*
HK1349	–	–	–	+	+	+	+	+	+	–	1	*bcoE*
HK47	–	+	+	+	+	+	+	–	+	+	1	*bcoC*
HK55	–	–	+	+	+/–	+	+	+	–	+	1	*bcoD*
HK72	+	–	+	+	+	–	+	+	+	+	1	*bcoB*
HK4V4*	–	–	+	+	–	–	–	+	–	+	1	*bcoA/B*
HK4V3*	–	–	–	+	+	+	+	+	+	+	2	*dcs*
HK4V5*	–	–	–	+	–	–	–	–	–	–	1	*bcoE*
HK4V11*	–	–	–	–	+	+	+	+	+	+	3	*beu*

[a]The phenotype characterises the use of betaine derivatives as growth substrates, e.g. 4-bb = 4-butyrobetaine, cb = crotonobetaine, L-c = L-carnitine, b = betaine. The listed enzyme activities correspond to the catalytic steps indicated in Figure 1. The gene loci were identified by the ability of cosmid gene bank clones to restore a wild-type function to the mutant (Figure 2). Asterisks indicate transposon insertion mutants

can be easily separated from the 4-butyrobetainyl-CoA-dehydrogenase. Both CoA-synthetase activities have been assigned to the same protein, but they can be distinguished from one another by mutation (Table 2). Mutants for each metabolic step were obtained using the frame shift mutagen Acridin ICR191[13] followed by conventional counter-selection.

Further mutants were obtained by Tn5 transposon insertion mutagenesis[14,15] (Table 2), and more detailed information about the butyrobetaine-L-carnitine metabolism in HK4 was obtained using these strains. The strain HK13, a mutant of HK4 with an irreversibly deactivated carnitine dehydrogenase, can transform 4-butyrobetaine and/or crotonobetaine to L-carnitine, the latter being excreted quantitatively into the medium, possibly by an 'overflow' mechanism. This efficient L-carnitine producer was used in ^{18}O-incorporation studies which confirmed that the origin of the O-atom in L-carnitine is water, and not molecular oxygen.[10]

14.4 TRANSPORT

A specific transport system is responsible for the uptake and excretion of betaines by the HK4 strain. It is dependent on a specific periplasmic binding protein, needs energy in the form of ATP and is also induced by the betaine derivatives.[16–20] This 4-butyrobetaine transport system was studied using the mutant *Agrobacterium/Rhizobium* HK47 (DSM2938), which had lost the ability to form crotonobetaine from 4-butyrobetaine. We found that the transport system obeyed Michaelis–Menten kinetics;[16] the saturation constant was very low (0.5 μM), which should, in theory, allow the complete conversion of 4-butyrobetaine. However, the maximum transport velocity for 4-butyrobetaine measured in the transport assay was relatively low at 0.0055 mol C mol^{-1} h^{-1},[16] indicating that the transport system should be rate-limiting in the production of L-carnitine. In practice this was not the case, which implies that another transport phenomenon for 4-butyrobetaine must be involved. Passive diffusion is the most likely, considering the high 4-butyrobetaine concentrations required for fast production.

14.5 GENETIC BACKGROUND

The DNA regions coding for the butyrobetaine-L-carnitine pathway in the bacterial genome were identified by complementation and hybridisation experiments.[21] An HK4 cosmid clone gene bank was prepared and screened for the enzymes of the metabolic pathway. This was achieved by conjugative transfer of the cosmid clones bearing the HK4 DNA fragments into HK4 mutants (Table 2). The appropriate clones that complemented the specific mutation in the HK4 mutants were then isolated. The identity of these DNA fragments was further confirmed by Southern blot hybridisation of the isolated clones with transposon Tn5-labelled and cloned DNA fragments of the transposon insertion mutants.

Four different DNA fragments (Figure 2, fragments1–4) implicated in the complete mineralisation of 4-butyrobetaine or crotonobetaine via L-carnitine and betaine were identified. The first enzymes of the upper pathway leading to dehydrocarnityl-CoA, i.e. 4-butyrobetaine-CoA- and crotonobetaine-CoA-synthetase, 4-butyrobetaine-CoA-dehydrogenase, crotonobetaine-CoA-hydrolase and carnitine-dehydrogenase are coded in one transcription unit as an oper-

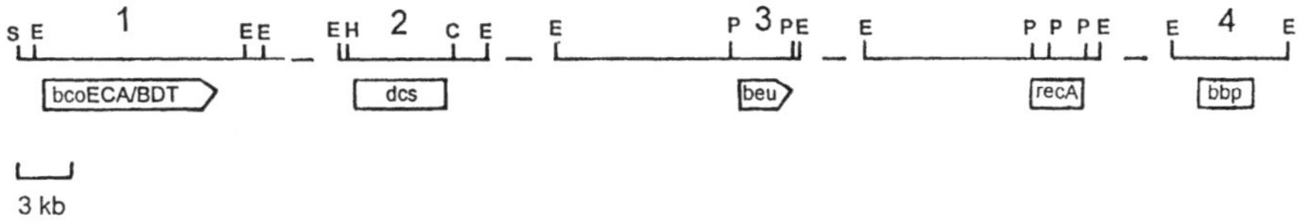

Figure 2 DNA fragments from an HK4 cosmid gene bank encoding the identified gene loci of 4-butyrobetaine catabolism. The DNA from HK4 was extracted, cut into fragments with the restriction enzyme EcoR1 and ligated into the similarly cut and dephosphorylated mobilisable cosmid vector pVK100.[33] *E. coli* S17-1[34] was transfected by lambda phages after *in vitro* packaging of the recombinant cosmids.[22] The cosmid vector contains genes coding for tetracycline and neomycin resistances and therefore cosmid carrying cells were easy to select for by their ability to grow in the presence of selectable amounts of the antibiotics (25 μg ml^{-1}). The gene library contained statistically about 5500 individual clones. S17-1 can mobilise pVK100 hybrids because of its chromosomally integrated R-plasmid coding for transfer functions. The specific cosmid clones bearing the desired genes of the upper pathway of the 4-butyrobetaine catabolism could be screened for after conjugation between the gene bank *E. coli* donors and the receptor HK mutant strains. The restoration of the genetic defects of the different HK mutants by a cosmid clone led to the isolation and identification of the desired gene loci. After subcloning of smaller DNA fragments of the identified gene loci and further *in vivo* complementations of HK mutants the genes of the upper pathway could be localised on the cloned DNA more precisely (Figure 3)

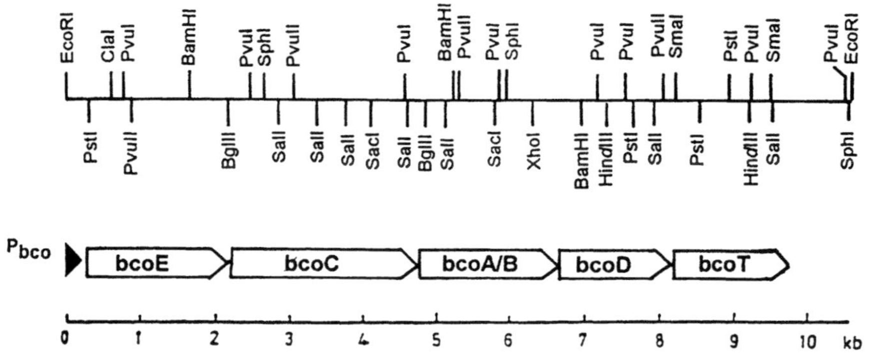

Figure 3 Restriction map of the DNA fragment encoding the catabolic enzymes of the upper pathway and a transport protein. This DNA fragment could be isolated from an HK4 gene bank clone and the different genes were identified via *in vivo* complementation of HK mutants with subfragments of the DNA

on: *bcoECA/BD* (Figures 2 and 3). In view of the fact that an active 4-butyrobetaine uptake can no longer be measured in the transposon insertion mutants (and that it is known that transposon insertion has a polar effect on the transcription of the following genes in a gene cluster), it was assumed that at least one transport component (*bcoT*) is also coded for on the DNA region following the gene cluster for the catalytic protein.

It was also observed that the transport could be restored in the Tn insertion mutants by the same gene bank clone that complemented the catalytic mutations of the upper pathway. The fact that 4-butyrobetaine, crotonobetaine and L-carnitine induce both the transport and the catalytic activities of the enzymes of the upper pathway indicates that the associated genes are on one operon. The catalytic activities responsible for the splitting of dehydrocarnitine (Dcs) and the utilisation of betaine (Beu) that follow in the further degradation of dehydrocarnityl-CoA (the lower pathway of 4-butyrobetaine catabolism) are coded by separate DNA regions of the HK4 genome, and mutations in these catalytic steps were complemented by other cosmid clones (Table 2). A further gene locus *bbp*, responsible for the butyrobetaine binding protein (Figure 2), was also located on the HK4 genome by Southern blot hybridisation with mixed DNA oligomers of the *N*-terminus of the purified and sequenced Bbp [17].

14.6 THE BIO-TRANSFORMATION STRAIN

As already mentioned, the production strain for the fermentative production of L-carnitine was obtained by irreversible inactivation of the carnitine dehydrogenase in HK4 by frame shift mutation.[10–12] This strain, HK13 (Table 2, Figure 1) converts 4-butyrobetaine and crotonobetaine quantitatively into L-carnitine with an enantiomeric purity of 100% and excretes it into the medium. Cofactor regeneration and educt/product transport by the cell during the formation of L-carnitine require large amounts of energy, and occur only in growing cells. Cells in the maintenance state, which is characterised by high metabolic activities at very low growth rates, are ideal for the L-carnitine production, giving nearly complete conversion of the educt. The growth of HK13 is inhibited by product concentrations higher than 3%, although resistant production mutants that tolerate much higher product concentrations (>7%) were isolated from a strain development programme. The improved tolerance to high concentrations of L-carnitine is associated with a reduced uptake affinity for the educt, 4-butyrobetaine, and other modified traits in the HK strain. This can be clearly seen in Table 1 and

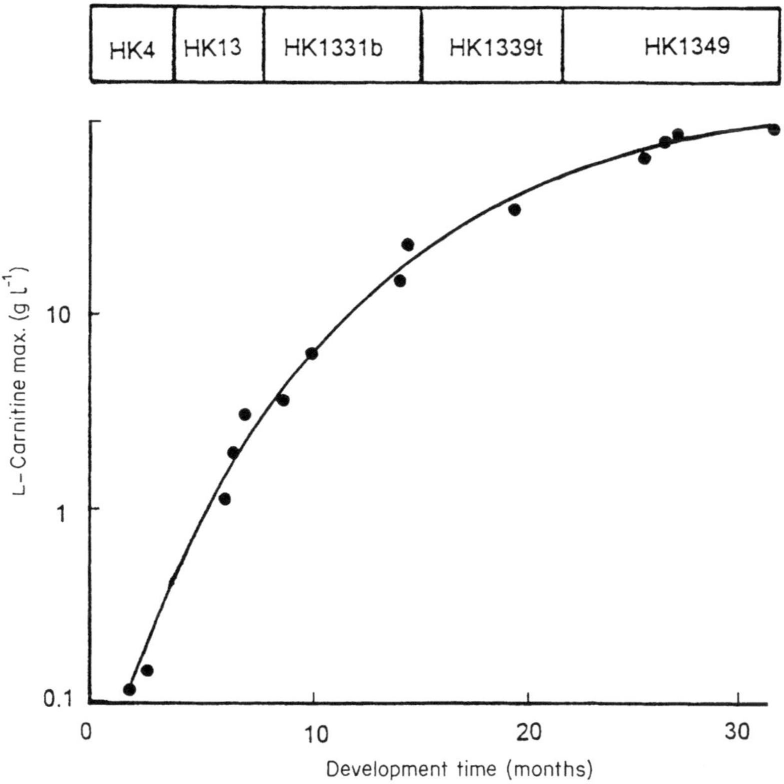

Figure 4 Production strain improvement. It took *ca* 2.5 years to develop the system ready for scale-up.[10] The programme started with the isolation of a microorganism, HK4, from a soil sample. A strain improvement programme, which aimed to isolate mutants that were resistant to high concentrations of L-carnitine, but still retained a high affinity for 4-butyrobetaine uptake, resulted ultimately in the isolation of the production strain, HK1349.

Figure 4. In the production strain, HK1349, isolated from the programme, a balance between an acceptable tolerance to high concentrations of L-carnitine and sufficient uptake affinity for 4-butyrobetaine was achieved.

14.7 THE RECOMBINANT BIO-TRANSFORMATION STRAIN

A recombinant production strain was obtained by cloning the upper pathway gene operon bearing the genes responsible for the conversion of 4-butyrobetaine to L-carnitine (Figures 3 and 5). This strain displays an increased productivity with respect to the mutant HK1349 used in the current production process (Table 3).

The advantages of the recombinant strain are the reduced volume and reduced time-scale needed for the production of L-carnitine.[21] The butyrobetaine-L-carnitine operon *bcoE'CA/BDT* with the deactivated carnitine dehydrogenase gene was isolated from a HK1349 gene bank and cloned into different broad host range vectors with varying copy numbers. The operon was still under the control of the natural promotor (P_{bco}), and the increased gene dosage led to increased

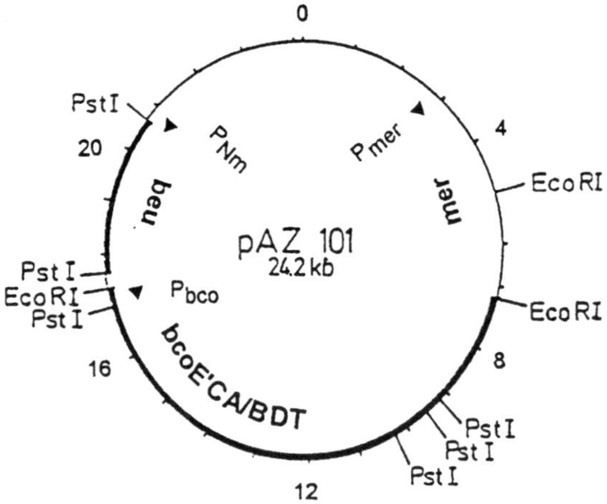

Figure 5 The recombinant hybrid plasmid pAZ101. The mobilisable broad host range vector pME285[35] was used for the construction of a recombinant hybrid plasmid containing the upper pathway operon of the 4-butyrobetaine catabolism from HK1349 instead of the operon from HK4, because of the carnitine dehydrogenase mutation (bcoE′) within the cloned 10.6 kb EcoR1 DNA fragment. Furthermore, the betaine utilisation encoding gene locus *beu* was used to create a selectable gene marker for this plasmid in the *beu* negative deletion mutant HK1349.4. Therefore, a 3.0 kb PstI HK4 DNA fragment was cloned into the newly created PstI site of the vector under the control of the promotor P_{Nm}. Another selectable marker that could be used in the genetic experiments is presented by the vectors *mer* locus transmitting mercury resistance to the plasmids host

Table 3 Bio-transformations with HK1349 and its recombinant descendants[a]

Strain	Medium MM + glycerol	Specific activity (mmol $l^{-1}h^{-1}$ OD^{-1})
HK1349	+ Betaine	0.24
HK1349/pVK100q	+ Betaine	0.36
HK1349/pAZ7	+ Betaine	0.49
HK1349	+ L-Glutamate	0.14
HK1349.4	+ L-Glutamate	0.11
HK1349.4/pLO41	+ L-Glutamate	0.29
HK1349.4	+ Ammonia	0.12
HK1349.4/pVK1011	+ Betaine	0.54
HK1349.4/pAZ7::beu	+ Betaine	0.47
HK1349.4/pAZ101	+ Betaine	1.08

[a] The bio-transformations for the comparison and estimation of L-carnitine productivities with different HK strains were carried out in shake flasks. The nitrogen-free mineral salt medium MM[36] contained 0.4% (w/v) glycerol as carbon source, 0.2% (w/v) 4-butyrobetaine as the educt for the bio-transformation and 0.2 % (w/v) betaine was added as the nitrogen source; 0.2% (w/v) L-glutamate was used for the betaine-negative strains as the nitrogen source and additional carbon source. The formation of L-carnitine from 4-butyrobetaine was measured by the DTNB (5,5′-dithiobis-2-nitrobenzoate) method described by Bergmeyer[37]

productivity in the biocatalyst. Plasmid stabilisation using an antibiotic selection was not attempted for the recombinant bio-transformation process, as the use of antibiotics, e.g. tetracycline, is known to lead to the loss of the plasmid through segregation and also to growth disturbances.[22,23] Antibiotics are, in any case, not desirable in a large-scale fermentation, so an antibiotic-free selection was developed. The gene *beu,* which is necessary for the utilisation of betaine, was completely deleted from the bacterial chromosome and transferred to a plasmid under the control of the neomycin phosphotransferase promotor P_{Nm} (Figure 5). This selectable plasmid also contains the genes encoding for the upper 4-butyrobetaine–L-carnitine pathway (*bco*). Plasmid stabilisation was now achieved by growth on betaine as the sole nitrogen and/or carbon source.[22]

Another problem that had to be overcome with the recombinant production strain was the tendency of the recombinant plasmid to integrate into the chromosome, owing to homologous recombination between the chromosomal and plasmid encoded operon of the catalytic proteins (BcoE′CA/BDT). One way to reduce this tendency is to introduce a *recA* mutation. The *recA* gene plays an important role in the bacterial repair mechanism of naturally occurring mutations which are otherwise frequently fatal.[24] However, the use of a *recA* minus strain is desirable from a biosafety point of view, because it dramatically reduces the survival rate of the strain outside of the fermenter environment.[24] Another solution to the problem of recombination in the HK strain would be the complete deletion of the homologous DNA from the bacterial chromosome. Yet another possible solution, cloning of the *bco* operon into another host, was not considered because the transport system and the cofactor background in the HK strain are important elements of the L-carnitine production system.

14.8 PROCESS DESIGN

The economics of the bio-technological production of fine chemicals are influenced not only by the bio-process but also by the downstream processing and product isolation. Despite this fact, bioprocess technology often concentrates on the bio-process, with the risk that much effort is put into the production of an aqueous product solution which may be unsuitable for downstream processing. This can result in the loss of product during isolation, or poor product quality which can only be improved by numerous costly unit operations. Consequently, the development and optimisation of a bio-process and the subsequent downstream processing (in addition to the upstream processing) should be done in an integrated way which leads to a cost-optimised production process.

There are several options for the process design of the bio-transformation for the production of L-carnitine, e.g. one-stage or multi-stage chemostat with cell recycling or fed batch. The first reported process for the bio-transformation of

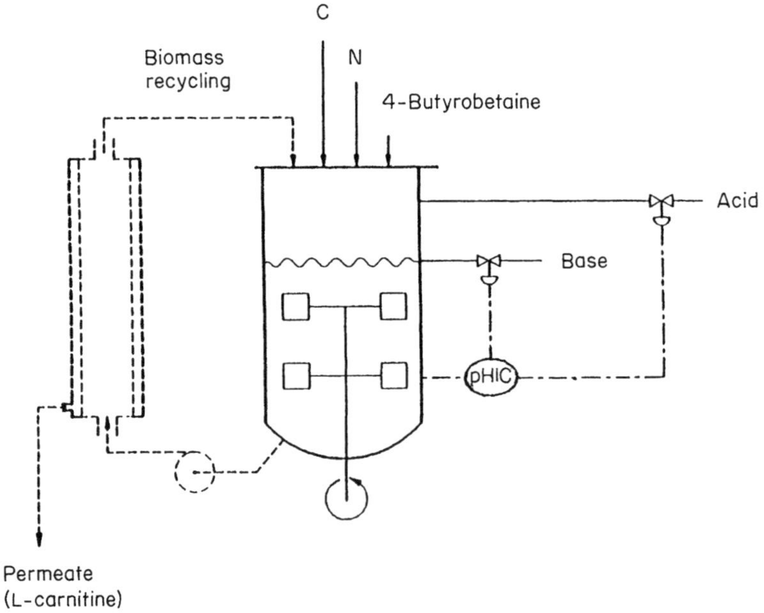

Figure 6 Scheme of the fed-batch bio-transformation process showing the additions of C and N sources, 4-butyrobetaine, acid and base. The dotted line indicates the modifications of the fed-batch system that are necessary for the one-stage continuous process with cell recycling. For the continuous system with cell recycling, the following data have been reported:[25] flow rate (Q) = 32 l h^{-1}; educt concentration at the inlet of bioreactor (c_s,0) = 60 g l^{-1} = 0.41 mol l^{-1}; bioreactor volume (V_{cc}) = 360 l; biomass concentration ($c_{x,cc}$) = 40 g l^{-1} as dry matter; product concentration at the outlet of bioreactor ($c_{p,cc}$) = 60 g l^{-1} = 0.37 mol l^{-1}; educt concentration at the outlet of bioreactor ($c_{s,cc}$) = 4.6 g l^{-1} = 0.032 mol l^{-1}.

chemically synthesised 4-butyrobetaine into L-carnitine consisted of a one-stage chemostat with cell recycling.[10] This process was successfully scaled up to 2000 l,[25] and the phenomenon of maintenance state production was optimally exploited. The process configuration and the cultivation of the strain are shown in Figure 6 and have been described previously.[10,16,25] Betaine was used as the carbon, energy and nitrogen source. The separation of the biomass after the bio-transformation by micro- or ultrafiltration is the first step in the product isolation, which is integrated with the biotransformation in the continuous process (Figure 6). The biomass separation must be reliable and operated aseptically, which is difficult on a large scale, and ceramic cross-flow filters proved to be most suitable for this continuous bio-transformation process.[25]

The downstream processing for L-carnitine is a purification of the aqueous solution after micro- or ultrafiltration. L-Carnitine is highly soluble in water at low, neutral and high pH. Butyrobetaine (educt) and L-carnitine have very similar physio-chemical properties which make their separation from an aqueous solution difficult, so no selective recovery method for L-carnitine is available.

The product isolation consisted of several unit operations which are shown schematically in Figure 7. The continuous bio-transformation process with cell

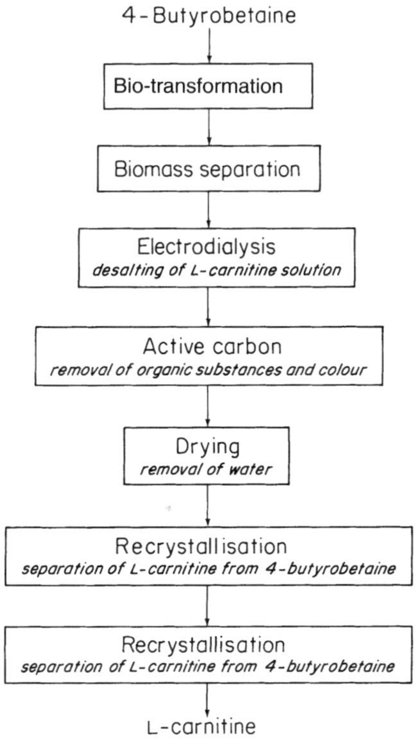

Figure 7 Unit operations for the production of L-carnitine from 4-butyrobetaine

recycling has a high volumetric productivity (130 g l^{-1} d^{-1}) in the steady state.[25] However, the kinetics of the bio-transformation process are such that this is only possible when the concentration of 4-butyrobetaine is relatively high (5 g l^{-1}). This results in a product solution containing a mixture of approximately 92% L-carnitine and 8% 4-butyrobetaine,[10] so two recrystallisations from isobutanol are necessary after the continuous biotransformation process in order to obtain the L-carnitine in the purity required by the market (>99%). These recrystallisations are an expensive part of the work-up of L-carnitine.[25]

A fed-batch system has also been developed (Figure 6) in which the volumetric productivity (30 g $l^{-1}d^{-1}$) was much less than the one-stage chemostat with cell recycling. However, the bioconversion yield was >99%,[25] which simplified the downstream processing.

14.9 KINETIC ANALYSIS

A kinetic analysis of the bio-process was carried out in order to define an optimum design with respect to productivity, downstream processing and costs. For complex biochemical multi-enzyme and multi-component reaction systems with microorganisms, Blackman kinetics is a simple model of the reaction rate as a function of the concentration of a substrate or reaction component.[26–29] In this technique it is assumed that the substrate is rate limiting below a critical concentration (Figure 8). Above the critical concentration other components and/or reaction mechanisms are rate limiting, because the substrate uptake system is saturated. Below the critical concentration, the reaction kinetics approximate to those of a first-order reaction.

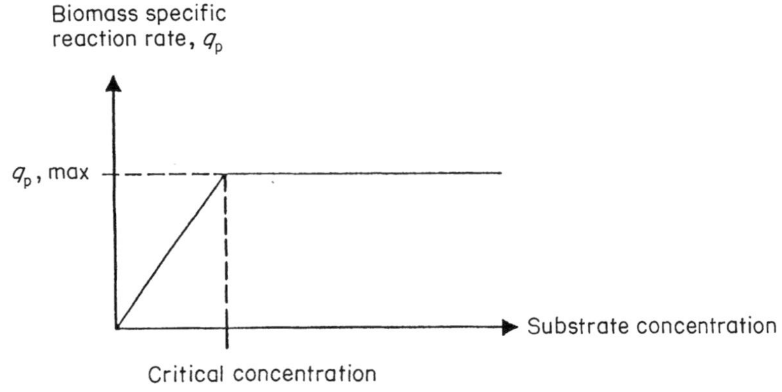

Figure 8 Blackman kinetics: specific product formation rate of whole cells as a function of the substrate concentration

The continuous system with cell recycling (single- or multi-stage) and the fed-batch system were analysed and compared using the kinetic data and concentrations from the continuous process (Figure 6). To derive the volumetric productivity of the fed-batch biotransformation the four periods of the process cycle time must be taken into account (Figure 9).

As can be seen from Table 4, a one-stage continuous bio-process with cell recycling has a high volumetric productivity, but also a much higher residual

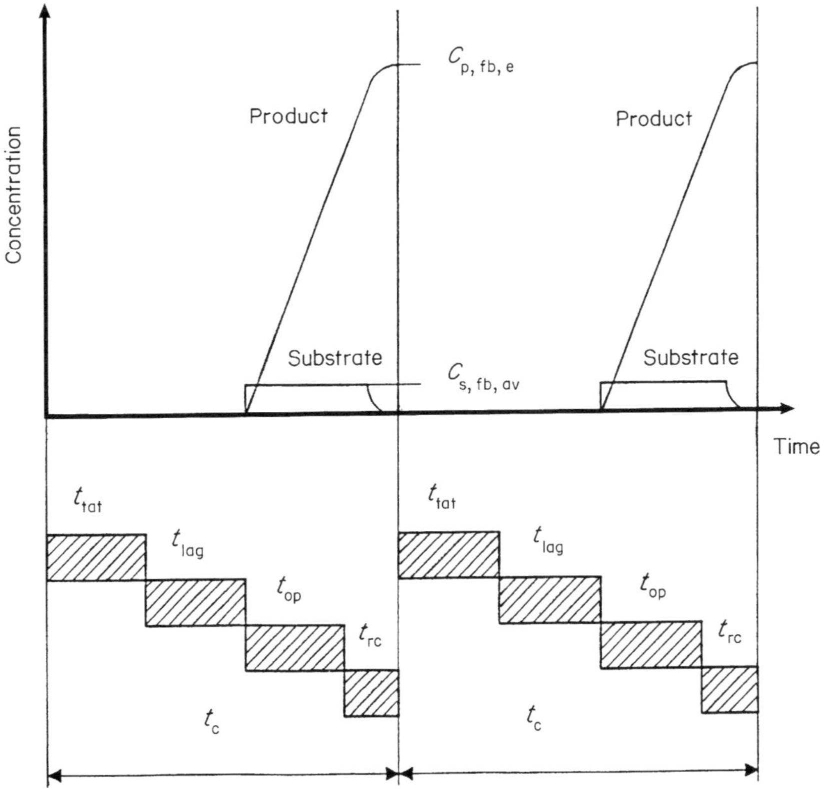

Figure 9 Time periods in the fed-batch bio-transformation of 4-butyrobetaine into L-carnitine showing substrate and product concentrations as a function of time. The duration of both the period with fast production and the period of the residual substrate conversion was calculated:[29] t_{tat} = turnaround time in which the bioreactor is emptied, cleaned, filled, sterilised, etc; it is usually 15–25 h, and here 15 h will be assumed; t_{lag} = time taken after inoculation of the bioreactor for the production to start; growth can take 10–30 h,[30] and here 20 h will be assumed; t_{op} = in this period, the optimum conditions for production are fulfilled,[25] and have t_{op} = 11.5 h; t_{rc} = in this period, when there is no feed, the remainder of the substrate 4-butyrobetaine in the reactor is converted into L-carnitine.[27] t_{rc} = 2.8 h; t_c = total cycle time; c_p, fb, e = final product concentration in the fed-batch process; c_s, fb, av = average educt concentration in the fed-batch process

Table 4 Comparison of continuous processes with cell recycling and a fed-batch process for the bio-transformation of 4-butyrobetaine into L-carnitine[a]

Parameter	One-stage continuous process with cell recycling		Fed-batch process	Two-stage continuous process with cell recycling
Volumetric productivity ($mol\,l^{-1}$)	0.034	0.0019	0.0075	0.0155
Bioconversion yield (as molar % of substrate)	91	99.5	99.5	99.5
Cost estimation for the bio-transformation and the isolation of L-carnitine (arbitrary units)	100	105	60	75

[a] The volumetric productivities and the yields were calculated as in ref. 29.

substrate level, than a fed-batch process. Consequently, the cost of the downstream processing is disproportionally high for the continuous process owing to the necessary separation of L-carnitine and 4-butyrobetaine. In the fed-batch process, 99.5% of 4-butyrobetaine is converted into L-carnitine.[10,25] The continuous process can be adapted so that the conversion of 4-butyrobetaine into L-carnitine is as high as the fed-batch process, but the resulting productivity would

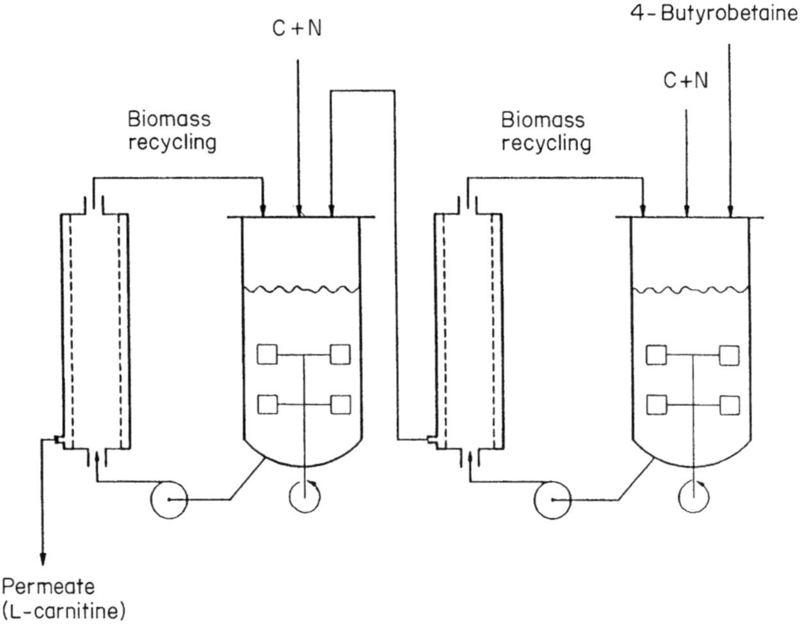

Figure 10 Schematic presentation of a two-stage continuous process with cell recycling for the bio-transformation of 4-butyrobetaine into L-carnitine

be less by a factor of 18. This loss in productivity nullifies the savings in the downstream processing (Table 4).

The low product to substrate ratio of the continuous process can also be improved with a multi-stage reactor system (Figure 10). The volumetric productivity of the multi-stage reactor system is twice as high as that for the fed-batch process (Table 4). Despite this, the system would be unattractive on a production scale because of the technical difficulties associated with a complex plant with two bioreactors in series (and with two cell-recycling systems), which has to be operated in a continuous aseptic manner. The technical difficulties of a one-stage system have been reported before.[25] The two-stage continuous process would undoubtedly require a relatively high investment per unit working volume, as reflected in the estimated production costs (Table 4).

Given that the fed-batch process is economically the most favourable (Table 4), LONZA decided to produce L-carnitine in a multi-purpose 15 m^3 fed-batch plant adapted to the specific needs in the Czech Republic.[31] This process gives a product solution comprising >99% L-carnitine and <1% 4-butyrobetaine. After electrodialysis and active carbon treatment, this solution is dried to a product which meets the required specification. Consequently, the previously described recrystallisations with isobutanol could be omitted, with a large economic advantage.[25] In 1996, the fed-batch process was scaled up to 50 m^3 without altering the yield and quality of the product.

14.10 RAW MATERIALS ASPECTS

Other criteria in process optimisation are the media composition and the quality of the raw materials, and in our case these had important consequences for the downstream processing. Originally, betaine was used as sole carbon and energy source for the microorganism, leading to the production of ammonia, which had to be neutralised with phosphoric acid. The formation of ammonium phosphate led to complications both in the biomass separation and the electrodialysis steps in the downstream processing.

The flux rate through the ceramic filters of the biomass separation system appeared to be twice as high when glycerol was used as carbon and energy source instead of betaine (0.15 compared with 0.075 $m\ h^{-1}\ bar^{-1}$). The reduced flux rate when betaine was used is thought to be caused by the ionic strength of 0.11 that resulted from the ammonium phosphate formation. It has been reported that an ionic strength of this magnitude reduces the flux rate through membranes substantially;[32] ammonium phosphate was not formed when glycerol was used.

The above-mentioned formation of ammonium phosphate when betaine was used as sole carbon source was also disadvantageous in the electrodialysis. There were more salts to be removed from the L-carnitine solution and the duration of the electrodialysis step affected the manufacturing costs of L-carnitine. A longer electrodialysis period implies a higher operating cost and higher product losses. For this reason, glucose or glycerol as carbon and energy sources in the bio-transformation are preferred over betaine. Betaine was retained as a nitrogen source owing to its postulated role in the interrupted catabolic pathway.

Replacement of phosphoric acid with sulphuric acid resulted in a more efficient electrodialysis, as sulphate removal was 30% faster than phosphate removal. The colour of the L-carnitine solution after the bio-transformation process depended on the quality and nature of the raw materials used in the bio-transformation process. Technical-grade betaine, although cheaper, resulted in a coloured product. A white L-carnitine product could be obtained if a pure grade of betaine and glycerol was used as the substrate in the bio-transformation. This made the subsequent active carbon treatment effective, with the extra cost of pure betaine offset by savings in the downstream processing.

14.11 CONCLUSION

The development of the L-carnitine process illustrates the importance of a multi-disciplinary approach. The isolation of the microorganism, strain development and biochemical and genetic studies resulted in a comprehensive understanding of L-carnitine metabolism at a cellular level. This led to strains which could produce high levels of optically pure L-carnitine. The exploitation of the maintenance state of this strain allowed the successful development of the single-stage continuous culture with a cell recycling process. Although this system had a high specific productivity, a drawback was the mixture of product and substrate in the product stream. Costly recrystallisations were required to separate the product from the substrate.

The 'black box' approach (system modelling based on Blackman's kinetics) allowed an evaluation and comparison of other process options, e.g. multi-stage continuous culture with cell recycling and fed-batch processes, with the single-stage continuous culture and cell recycling system. From this analysis LONZA opted for a fed-batch process, in which a lower productivity was compensated for by cheaper and simplified downstream processing and lower investment costs. A careful choice of raw materials for the bio-transformation resulted in a reduced number of unit operations during purification of the product.

14.12 REFERENCES

1. Scholte, H. R., and De Jonge, P. C., in *Carnitin in der Medizin*, (ed.R. Gitzelmann, K. Baerlocher and B. Steinmann), Schattauer, Stuttgart, 1987, p. 22.
2. Ferrari, R., Di Mauro, S., and Sherwood, G., *L-Carnitine and its Role in Medicine from Function to Therapy*, Academic Press, London, 1992.
3. Wagenmakers, A. J. M., *Med. Sport Sci.*, **32**, 110 (1991).
4. Carter, A. L., *Current Concepts in Carnitine Research*, CRC Press, Boca Raton, FL, 1992.

5. Torreele, E., Sluiszen, A. van der, and Verreth, J., *Br. J. Nutr.*, **69**, 289 (1993).
6. Santulli, A., and D'Amelio, V., *J. Fish. Biol.*, **28**, 81 (1986).
7. Voeffray, R., Perlberger, J. C., Tenud, L., and Gosteli, J., *Helv. Chim. Acta*, **70**, 2058 (1987).
8. Jung, H., Jung, K., and Kleber, H. P., *Adv. Biochem. Eng. Biotechnol.*, **50**, 21 (1993).
9. Sneath and Sokal (Eds), *Numerical Taxonomy*, Freeman and San Francisco, 1973, p. 230.
10. Kulla, H., *Chimia*, **45**, 81 (1991).
11. Kulla, H., and Lehky, P., *Eur. Pat.*, 01 58 194, 1985; *US Pat.*, 5 187 093, 1993.
12. Kulla, H., Lehky, P., and Squaratti, A., *Eur. Pat.*, 0 195 944, 1986; *US Pat.*, 4 708 936, 1987.
13. Miller, J. H., *Experiments in Molecular Genetics*, Cold Spring Harbor Laboratory Press, Cold Spring Harbor, NY, 1972.
14. Rella, M., Mercenier, A., and Haas, D., *Gene*, **33**, 293 (1985).
15. Simon, R., O'Connell, M., Labes, M., and Pühler, A., *Methods. Enzymol.*, **118**, 640 (1986).
16. Nobile, S., and Deshusses, J., *J. Bacteriol.*, **168**, 780 (1986).
17. Nobile, S., Baccino, D., Takagi, T., and Deshusses, J., *J. Bacteriol.*, **170**, 5236 (1988).
18. Nobile, S., Baccino, D., and Deshusses, J., *FEBS Lett.*, **233**, 325 (1988).
19. Nobile, S., and Deshusses, J., *Biochimie*, **70**, 1411 (1988).
20. Nobile, S., and Deshusses, J., *J. Chromatogr.*, **449**, 331 (1988).
21. Zimmermann, Th. P., and Werlen, J., PCT Int. *Pat. Appl.*, WO 95/10613, 1995.
22. Zimmermann, Th. P.,Boraschi, C., Burgdorf, K., and Caubere, C., *Eur. Pat. Appl.*, EP 543 344, 1992.
23. Lee, S.W., and Edlin, G., *Gene*, **39**, 173 (1985).
24. Selbitschka, W., Arnold, W., Priefer, U. B., Rottschäfer, Th., Schmidt, M., Simon, R., and Pühler, A., *Mol. Gen. Genet.*, **229**, 86 (1991).
25. Hoeks, F. W. J. M. M., Kulla, H., and Meyer, H. P., *J. Biotechnol.*, **22**, 117 (1992).
26. Atkinson, B., and Mavituna, F., *Biochemical Engineering and Biotechnology Handbook*, Nature Press, New York, 1983.
27. Dabes, J. N., Finn, R. F., and Wilke, C. R., *Biotechnol. Bioeng.*, **15**, 1159 (1973).
28. Roels, J. A., *Energetics and Kinetics in Biotechnology*, Elsevier, Amsterdam, 1983.
29. Hoeks, F. W. J. M. M., Mühle, J., Böhlen, E., and Psenicka, I., *Chem. Eng. J.*, **61**, 53 (1996).
30. Heijnen, J. J., Van Scheltinga, A. H. T., and Straathof, A. J., *J. Biotechnol.*, **22**, 3 (1992).
31. Meyer, H. P., *Chimia*, **47**, 59 (1993).
32. Fane, A. G., Fell, C. J. D., Hodgson, P. H., Leslie, G., and Marshall, K. C., *Filtr. Sep.*, **28**, 332 (1991).
33. Knauf, V. C., and Nester, E. W., *Plasmid,* **8**, 45 (1982).
34. Simon, R., Priefer, U., and Pühler, A., *Bio/Technology*, **1**, 784 (1983).
35. Itoh, Y., and Haas, D., *Gene*, **36**, 27 (1985).
36. Kulla, H. G., Klausener, F., Meyer, U., Lüdeke, B., and Leisinger, Th., *Arch. Microbiol.*, **135**, 1 (1983).
37. Bergmeyer, H. U., *Methoden der Enzymatischen Analyse*, Verlag Chemie, Weinheim, 3rd edn., 1974.

ASYMMETRIC SYNTHESES BY CHEMICAL METHODS

15 The Asymmetric Hydrocyanation of Vinylarenes

A. L. CASALNUOVO
DuPont Agricultural Products, Newark, DE, USA

and

T. V. RAJANBABU
Ohio State University, Columbus, OH, USA

15.1 Introduction and Background 309
15.1.1 Nickel-catalysed Hydrocyanation of Alkenes 310
15.1.2 Asymmetric Hydrocyanation of Norbornene Derivatives . 312
15.1.3 Asymmetric Hydrocyanation of Vinylarenes with Non-carbohydrate Phosphorus Ligands 312
15.2 Asymmetric Hydrocyanation with Carbohydrate-derived Phosphorus Ligands 315
15.2.1 Carbohydrate Ligand Syntheses 315
15.2.2 Hydrocyanation and Analytical Protocol 317
15.2.3 Trials of Asymmetric Hydrocyanation of Vinylarenes . . . 318
15.2.3.1 Steric and Electronic Effects of Phosphorus Substituents 318
15.2.3.2 Site of Phosphorus Substitution 322
15.2.3.3 Effect of Electronic Asymmetry 324
15.2.3.4 Effect of Other Parameters 326
15.2.4 Asymmetric Hydrocyanation Mechanism 327
15.2.5 Laboratory Preparations of Naproxen and Ibuprofen Precursors 331
15.2.6 Asymmetric Hydrocyanation as a Commercial Route to Naproxen 331
15.3 Acknowledgements 332
15.4 Notes and References 332

15.1 INTRODUCTION AND BACKGROUND

The asymmetric, catalytic hydrocyanation reaction is a potentially competitive route to the commercial manufacture of a diverse array of optically active compounds because hydrogen cyanide is a low-cost feedstock, and the cyano group

Chirality in Industry II. Edited by A. N. Collins, G. N. Sheldrake and J. Crosby

can be readily transformed into a variety of other functional groups.[1–3] Thus, chiral nitriles, amines, aldehydes, carboxylic acids, esters and imidates are all potential targets. Even so, the only commercial process based on an asymmetric, catalytic hydrocyanation reaction appears to be in the manufacture of a pyrethroid insecticide.[4] This process utilises a cyclic dipeptide catalyst (**1**) to effect the asymmetric addition of hydrogen cyanide to an aromatic aldehyde.

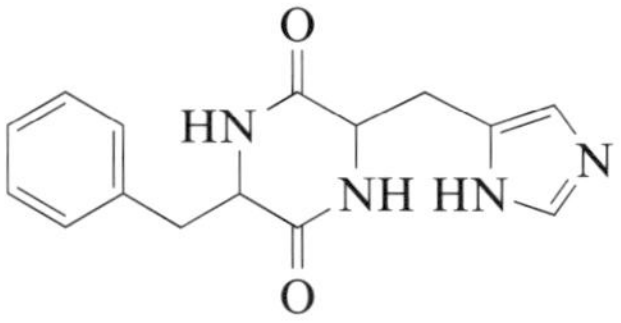

(**1**)

Highly enantioselective hydrocyanations of carbonyl groups have been achieved in the laboratory using both chemical and biochemical catalysts.[2,5,6] In contrast, there are few good examples of either base-catalysed or transition metal-catalysed asymmetric hydrocyanations of alkenes.[7,8] DuPont's discovery and commercialisation of the nickel-catalysed hydrocyanation of alkenes has made the development of an asymmetric version of this reaction a natural research target,[1,3] and towards this end we began a programme that utilised the asymmetric hydrocyanation of vinylarenes as a prototypical reaction.[8] One of the principal, and unanticipated, lessons learned in this exploratory programme was the remarkable impact of subtle changes in the electronic properties of the catalyst on the reaction enantioselectivity. In this chapter we describe the development of this catalyst system and its potential application to the synthesis of the anti-inflammatory agent naproxen.

15.1.1 NICKEL-CATALYSED HYDROCYANATION OF ALKENES

The homogeneous, transition metal-catalysed hydrocyanation of alkenes, generically illustrated in equation 1, has been extensively reported in the literature.[1, 3] Although a comprehensive review of this reaction is beyond the scope of this chapter, a few relevant details are given here as background.

$$R^1R^2C{=}CR^3R^4 + HCN \xrightarrow[\text{[M]}]{} R^1R^2HCC(CN)R^3R^4 \qquad (1)$$

A number of transition metal complexes have been reported to catalyse this reaction, although zero-valent nickel phosphite catalysts, $Ni[P(OR)_3]_x$ ($x = 3, 4$), have generally proved to be the most effective. Using triarylphosphite nickel catalysts, for example, DuPont produces over one billion pounds of adiponitrile from butadiene annually (equation 2).

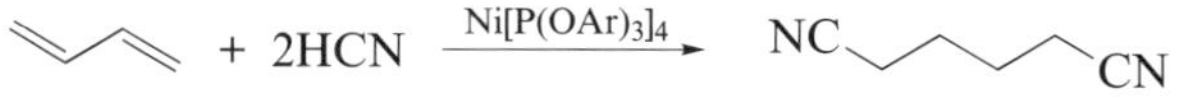

(2)

With some notable exceptions,[1] these catalysts tolerate a broad range of substrates and functional groups. Thus alkynes, dienes, internal and terminal alkenes, and vinylarenes are all suitable, although for certain substrate classes, particularly isolated alkenes, catalytic amounts of Lewis acids (e.g. $ZnCl_2$) are necessary to achieve acceptable catalyst turnover.

The regioselectivity of the hydrocyanation depends on several factors, including the substrate type, ligand and Lewis acid. For isolated terminal alkenes, linear hydrocyanation products usually predominate. Even internal, isolated alkenes will give rise to a preponderance of the linear nitrile, provided alkene isomerisation is possible through a hydride insertion/β-hydride elimination mechanism (Scheme 1). Lewis acids tend to increase the rate of alkene isomerisation and thus promote the formation of linear nitriles from internal alkenes.

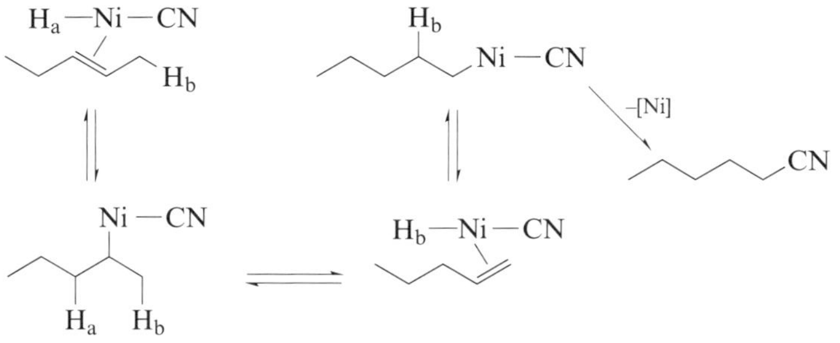

Scheme 1

The hydrocyanation of conjugated dienes occurs via the formation of stable η^3-allyl intermediates.[1,9,10] For example, the η^3-allyl intermediate **2** (equation 3) has been observed by ^{31}P NMR spectroscopy in the $Ni[P(OEt)_3]_4$ and $Ni[P(O\text{-}o\text{-}tolyl)_3]_3$-catalysed hydrocyanation of buta-1,3-diene. Similar allylic interactions have been invoked to explain the predominance of the branched nitrile in the $Ni[(P(O\text{-}o\text{-}tolyl)_3]_3$-catalysed hydrocyanation of styrene (Scheme 2).[1]

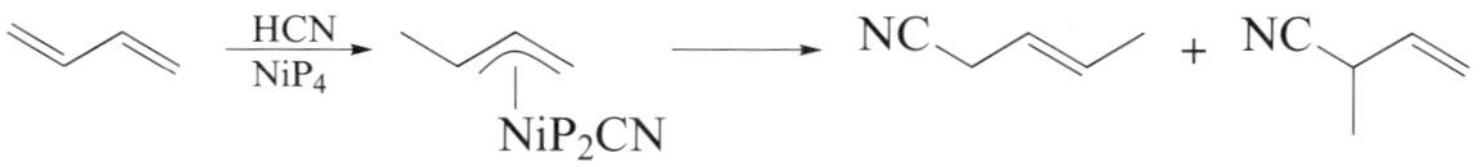

(3)

(**2**)

P=P(*O*-*o*-tolyl)$_3$

One of the primary deactivation pathways of most transition metal hydrocyanation catalysts is metal oxidation by hydrogen cyanide. In the case of zerovalent nickel catalysts, the net redox process is given by equation 4. The overall oxidation rate is thought to be second order with respect to hydrogen cyanide in most cases, whereas the alkene hydrocyanation rate is usually first order. Thus, maintaining a low hydrogen cyanide concentration in the reaction is usually necessary to obtain good catalyst turnover.

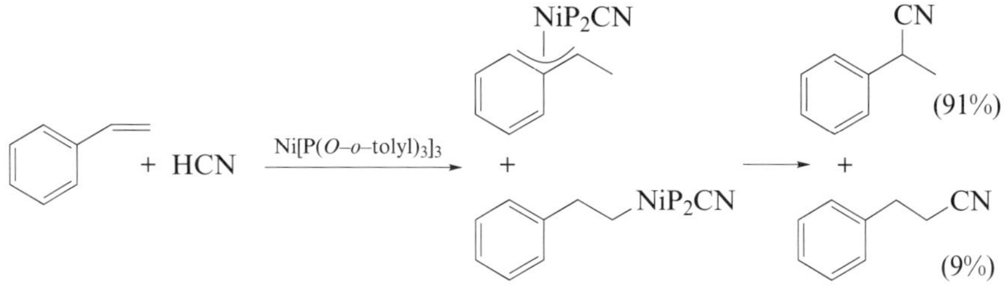

Scheme 2

$$2\,HCN + NiL_n \longrightarrow Ni(CN)_2 + H_2 + n\,L \qquad (4)$$

15.1.2 ASYMMETRIC HYDROCYANATION OF NORBORNENE DERIVATIVES

By comparison with the asymmetric hydrogenation of alkenes, the catalytic, asymmetric hydrocyanation of alkenes is still in its infancy. In fact, there have been only a few published accounts of the catalytic asymmetric hydrocyanation of an alkene since the first report appeared in 1982.[7,8] The earlier work, summarised in Table 1, was focused almost exclusively on the asymmetric hydrocyanation of norbornene or norbornene derivatives (equation 5).

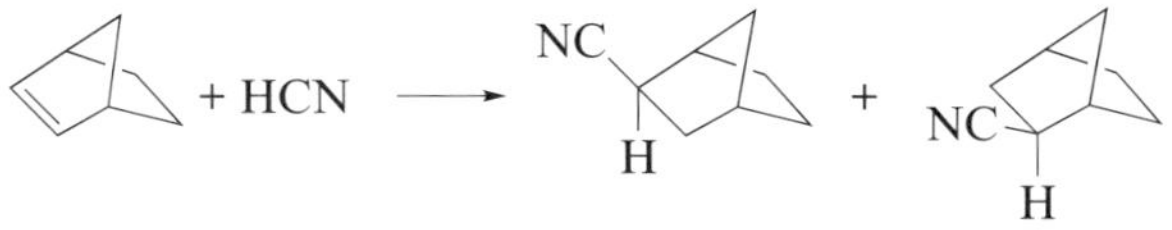

(5)

Although these reactions were completely facially selective (*ca* 100% *exo* nitrile), they met with only modest success in terms of reaction enantioselectivity and yield. For example, the highest *ee* reported for this class of substrate was 38–40% (Table 1; entries 9 and 11) using either Pd(BINAP)$_2$ or the binaphthol-derived phosphite complex represented as Ni(**B**)$_2$.

15.1.3 ASYMMETRIC HYDROCYANATION OF VINYLARENES WITH NON-CARBOHYDRATE PHOSPHORUS LIGANDS

$$ArCH{=}CH_2 + HCN \longrightarrow \underset{\text{2-Arylpropionitrile}}{ArCH(CN)Me} \longrightarrow \underset{\text{2-Arylpropionic acid}}{ArCH(CO_2H)Me} \qquad (6)$$

2-Arylpropionitriles constitute a more interesting target for the asymmetric hydrocyanation reaction because they are precursors to chiral 2-arylpropionic acids; a class of commerically important anti-inflammatory drugs (equation 6).[11–13] In general, the (*S*)-enantiomers of these acids have a greater physiological activity and, in the case of the 6-methoxynaphthalene derivative, naproxen, only

Table 1 Asymmetric hydrocyanation of norbornene and derivatives

Entry	Catalyst[a]	Substrate	Yield (%)	Temperature (°C)	*ee* (%)	Ref.
1	$Ni(DIOP)_2$	Norbornene	20	80	6[b]	7a
2	$Ni(DIOP)_2$	Norbornadiene	78	120	8	7a
3	$Ni(DIOP)_2$	Benzonorbornadiene	85	120	16	7a
4	$Pd(DIOP)_2$	Norbornene	94	80	13	7c
5	$Pd(DIOP)_2$	Norbornadiene	83	120	17	7a
6	$Pd(DIOP)_2$	Benzonorbornadiene	83	120	13	7a
7	$Pd[(-)\text{-}ARSOP)]_2$	Norbornene	<5	120	5[b]	7a
8	$Pd(DIPHIN)_2$	Norbornene	10	120	3[b]	7b
9	$Pd(BINAP)_2$	Norbornene	6	120	40	7c
10	Pd(BPPM)	Norbornene	13	80	25	7c
11	$Ni(B)_2 + BPh_3$	Norbornene	58	100	38	7d
12	$Ni(A)_2$	Norbornene	16	150	10	7d

[a] Ligand abbreviations:

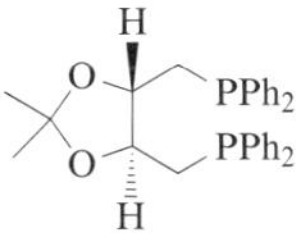

(+)-DIOP

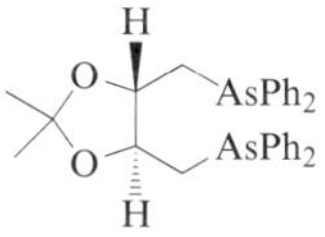

(+)-ARSOP

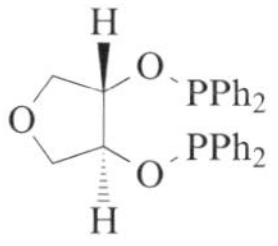

(+)-DIPHIN

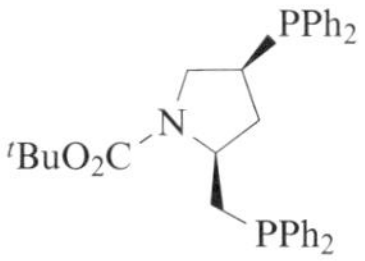

BPPM

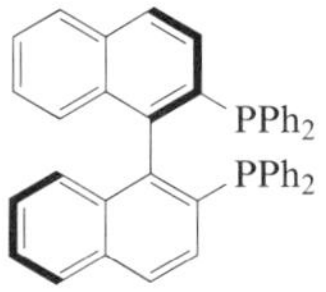

BINAP

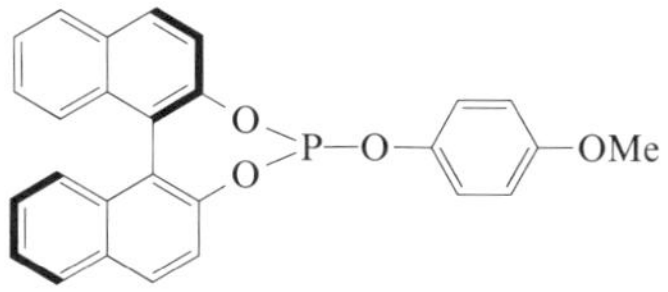

A

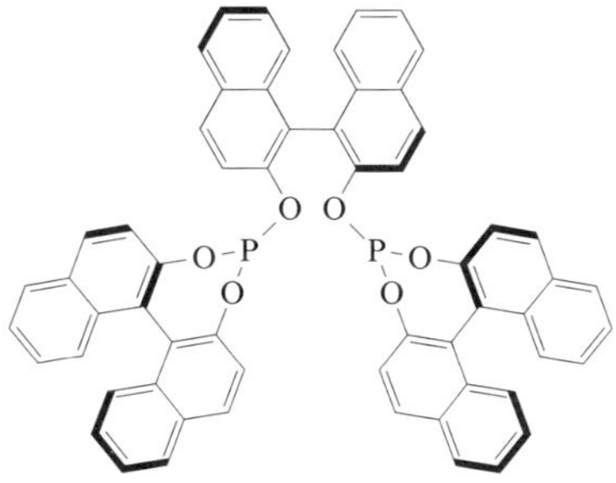

B

[b] *ee* Values have been corrected from the original report (see ref. 7c).

the pure (*S*)-(+)-enantiomer has been licensed for sale (see also Volume I, Chapters 15, and this volume Chapter 3). From a regioselectivity standpoint, the starting prochiral vinylarenes are also attractive substrates for the asymmetric hydrocyanation reaction because an overall Markovnikov addition is expected.

In this regard, Nugent[14] and McKinney first reported the efficient synthesis of racemic ibuprofen and naproxen via the Ni[P(*O*-*p*-tolyl)$_3$]$_4$-catalysed hydrocyanation of the corresponding vinylarene precursors (equations 7 and 8). Styrene derivatives, such as the ibuprofen precursor *p*-isobutylstyrene, suffered from competitive oligomerisation. In the latter case, the addition of Lewis acids decreased the amount of alkene oligimerisation but increased the amount of linear nitrile (equation 8).

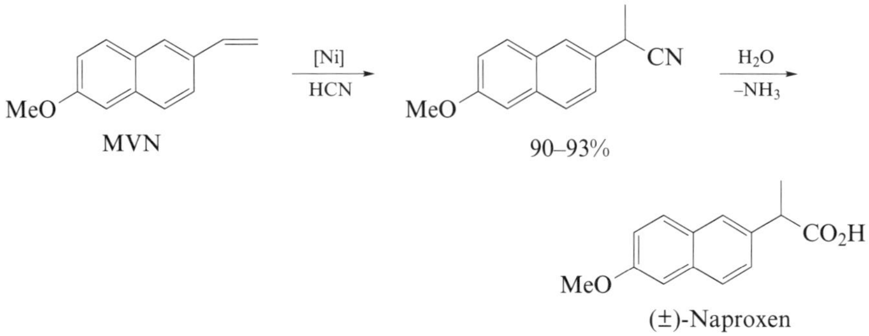

(7)

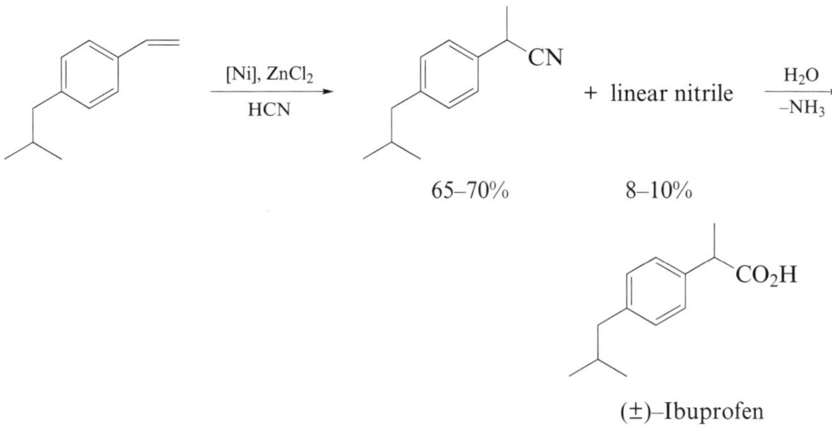

(8)

In what probably constitutes the earliest attempt at the asymmetric hydrocyanation of a vinylarene, Gosser obtained a slight asymmetric induction (*ca* 10% *ee*) in the nickel-DIOP-catalysed hydrocyanation of styrene.[15] Later, Nugent and McKinney used various chiral phosphorus ligands to explore the asymmetric hydrocyanation of 6-methoxy-2-vinylnaphthalene (MVN) and 2-vinylnaphthalene (VN). They obtained modest enantioselectivities, with the highest *ee* (25%, 2-vinylnaphthalene) obtained using the dimethyl tartrate-derived phosphinite ligand **3**.[16]

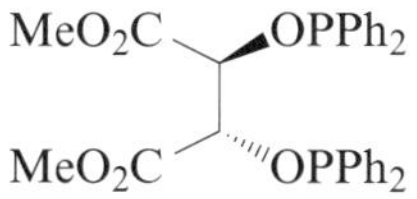

(3)

15.2 ASYMMETRIC HYDROCYANATION WITH CARBOHYDRATE-DERIVED PHOSPHORUS LIGANDS

The catalyst characteristics which control enantioselectivity in asymmetric catalysis can be subtle and complex, particularly in multi-step reaction mechanisms such as the nickel-catalysed hydrocyanation reaction[1] or the rhodium-catalysed asymmetric hydrogenation of alkenes.[17] In such cases, the *a priori* design of suitable chiral ligands is often unproductive and a more empirical approach may be required. Carbohydrates provide many accessible, unique and structurally diverse frameworks for building chiral phosphorus ligands. Indeed, certain glucose-based diphenylphosphinite ligands, such as compound **4**, have been used successfully in the asymmetric hydrogenation of alkenes.[18]

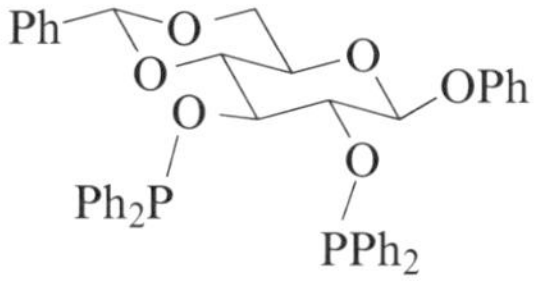

(4)

Furthermore, the existence of well developed, carbohydrate synthetic methodologies allows a modular approach to ligand construction and so simplifies the systematic modification of the ligand. For example, several possible sites for modifying a carbohydrate-derived bidentate phosphorus ligand are illustrated in Figure 1. Variation of these sites using standard synthetic methods leads to a number of potentially useful chiral phosphorus ligands, a few of which are shown Figure 1. With this strategy we have synthesised over 100 carbohydrate-based phosphorus ligands and used them to evaluate systematically the asymmetric hydrocyanation of vinylarenes. Ultimately, this empirical approach to ligand design pointed the way to the discovery of more enantioselective hydrocyanation catalysts and, perhaps more importantly, to some useful insights into electronic effects in asymmetric catalysis.[8]

15.2.1 CARBOHYDRATE LIGAND SYNTHESES[8,18]

In addition to the synthesis of the carbohydrate fragment, the synthesis of the carbohydrate ligands used in these studies generally requires the reaction of

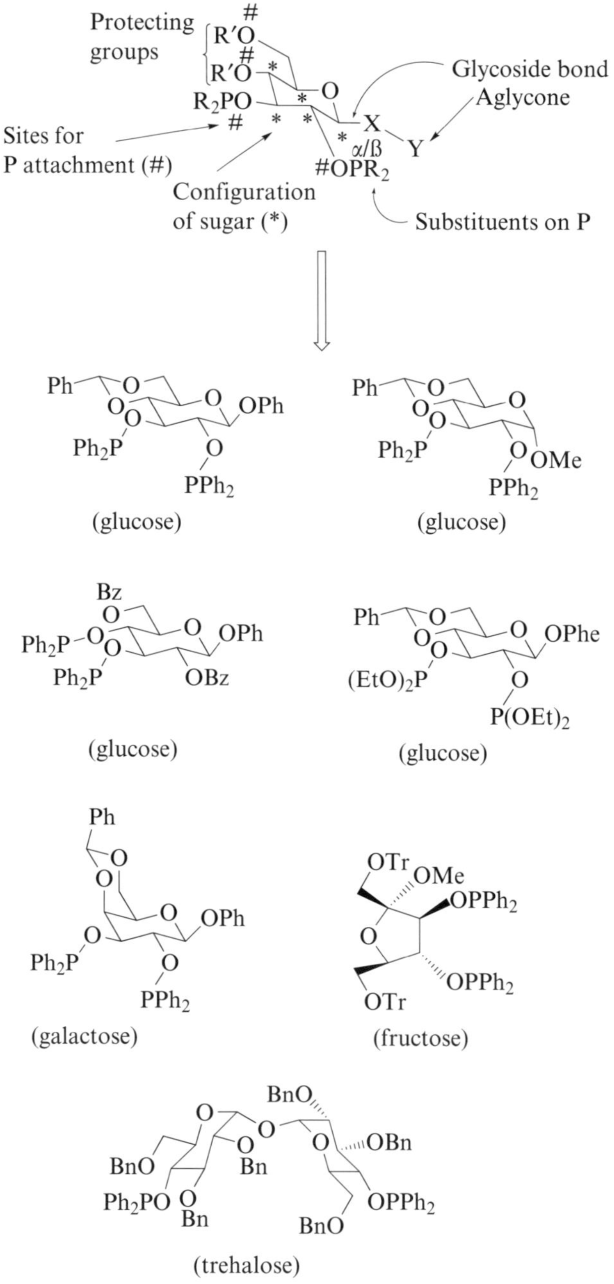

Figure 1 Modifiable sites on a carbohydrate-derived phosphorus ligand and examples of accessible ligands. Bz = benzoyl; Bn = benzyl; Tr = triphenylmethyl. The parent sugars are indicated in parentheses

R_2PCl with the hydroxy groups of a suitably protected carbohydrate in the presence of a base. Commercially unavailable R_2PCl reagents can be prepared by displacement of the chlorine atom of PCl groups with the appropriate Grignard reagents or alcohols. An example of a typical ligand preparation is shown in Scheme 3. The synthesis of diarylchlorophosphines is shown in Scheme 4.

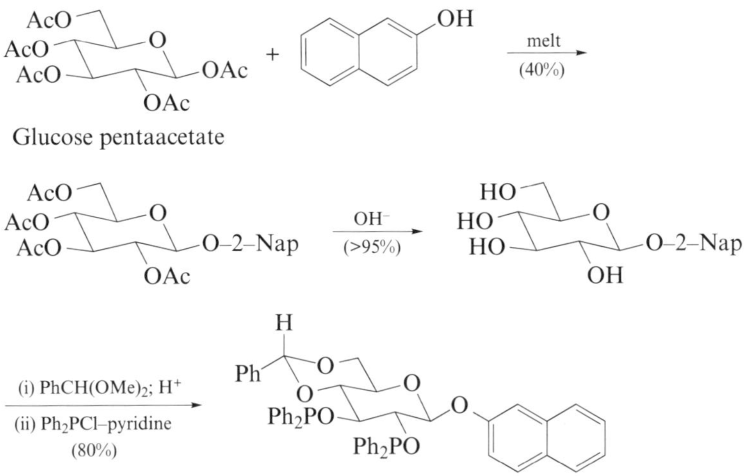

Scheme 3

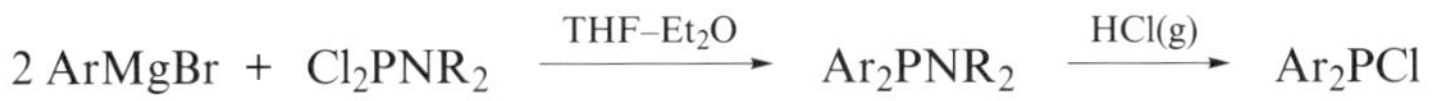

Scheme 4

15.2.2 HYDROCYANATION AND ANALYTICAL PROTOCOL

Hydrocyanation catalysts are conveniently prepared by mixing commercially available $Ni(COD)_2$ (COD = cycloocta-1,5-diene) with solutions of the chiral phosphorus ligand (P) at room temperature under nitrogen. In most cases, the formation of the typical catalyst precursors $NiP_2(COD)$ or NiP_4 occurs rapidly at this temperature and any unreacted $Ni(COD)_2$ is catalytically inactive. For trial reactions, HCN is usually added dropwise to a mixture of the vinylarene and catalyst because this procedure minimises oxidation of the catalyst by HCN. Conventional GC methods can be used to evaluate reaction yields and regioselectivity. Chiral HPLC has proved to be the most useful and reliable method for measuring the enantiomeric purities of the 2-arylpropionitriles.

15.2.3 TRIALS OF THE ASYMMETRIC HYDROCYANATION OF VINYLARENES

15.2.3.1 Steric and Electronic Effects of Phosphorus Substituents

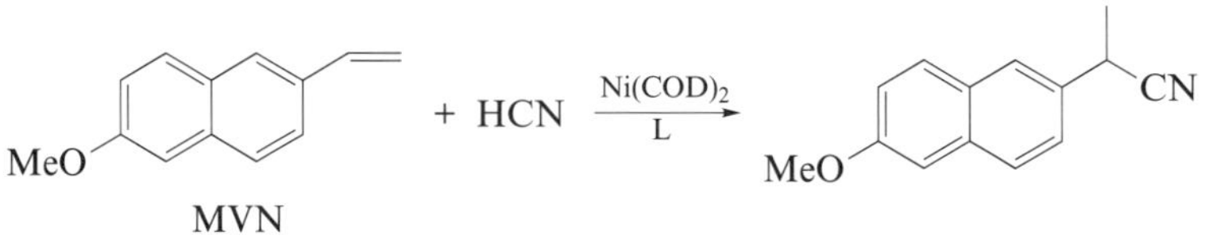

(9)

Trial studies on the asymmetric hydrocyanation of 6-methoxy-2-vinylnaphthalene (MVN, equation 9) and 2-vinylnaphthalene (VN) were initially carried out using various *O*-substituted, glucose-derived phosphite [$P(OR)_3$] and phosphinite [$PR_2(OR')$, R′ = glucose fragment] ligands. For both types of ligands, the results pointed to the important effect of the phosphorus substituents on the reaction enantioselectivity. In the case of the β-phenyl glucophosphite ligands of type **5** [R = $P(OR')_2$], the data suggested that the size and orientation of the phosphorus alkoxy groups were the primary factors influencing enantioselectivity. As shown in Table 2, the *ee*s ranged from 13% (*R*) to 60% (*S*), with the largest enantioselectivities (entries 4 and 6) obtained when chiral diols were used in constructing the phosphite group. A comparison of these results with those obtained with the phosphinite ligands showed that this glucose moiety imposes a transition state arrangement which favours the formation of (*S*)-nitrile; the presence of other chiral centres near phosphorus can either enhance or counteract this effect (compare entry 4 with 5 and entry 6 with 7).

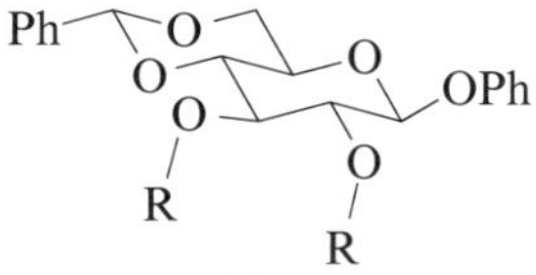

(**5**)

In contrast to the glucophosphite ligands, a similar trial study of the asymmetric hydrocyanation of MVN and VN using *o*-, *m*- or *p*-substituted glucodiarylphosphinite ligands (**5**, R = PAr_2) led to the unexpected discovery that the electronic characteristics of the phosphorus–aryl substituents have a much greater impact on the enantioselectivity than the inherent size of the substituent. Only a few instances of such ligand electronic effects have been reported in the field of asymmetric catalysis, the principal examples being Jacobsen's asymmetric alkene epoxidation reaction and Nishiyama's ketone hydrosilylation reaction.[19] This electronic effect can be seen clearly from the result obtained using the *m*-disubstituted ligands (**5a–d**), as shown in Table 3. Here the *ee*s increased dramatically (from 16% to 78% for MVN) as the electron-withdrawing power of phosphorus–aryl substituents increased (for **5d–5a**, σ_m = –0.07, 0, 0.34, 0.43,

Table 2 Asymmetric hydrocyanation of VN using glucophosphite ligands (**5**)

Entry	R	*ee* %
1	$(EtO)_2P$—	18 (*S*)
2	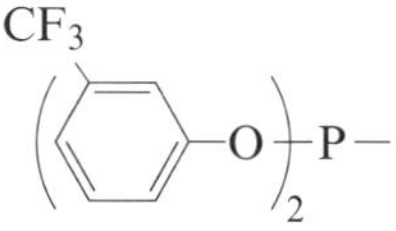	35 (*S*)
3	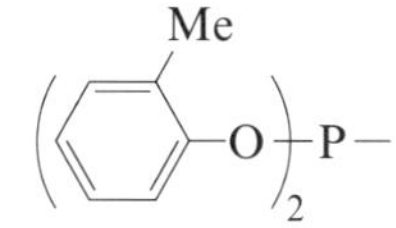	39 (*S*)
4	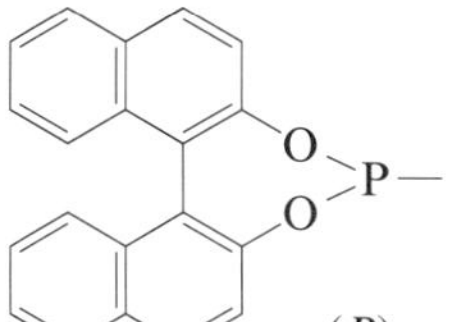 (*R*)	49 (*S*)
5	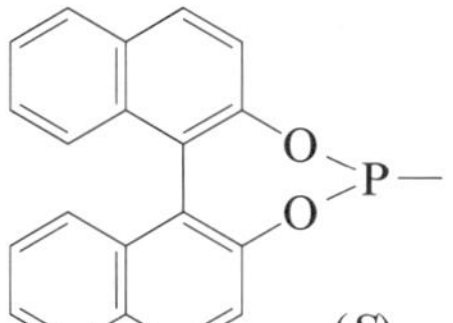 (*S*)	13 (*R*)
6	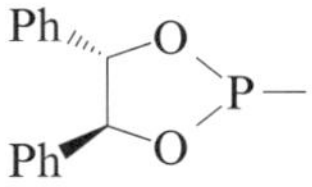 (*S*,*S*)	60 (*S*)
7	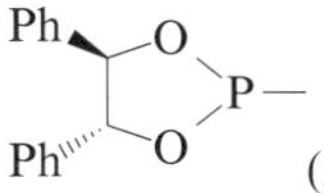 (*R*,*R*)	26 (*S*)

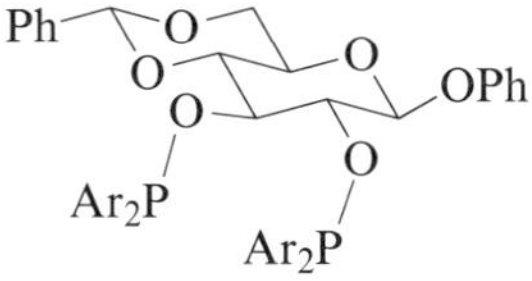

5a–d

Ar =

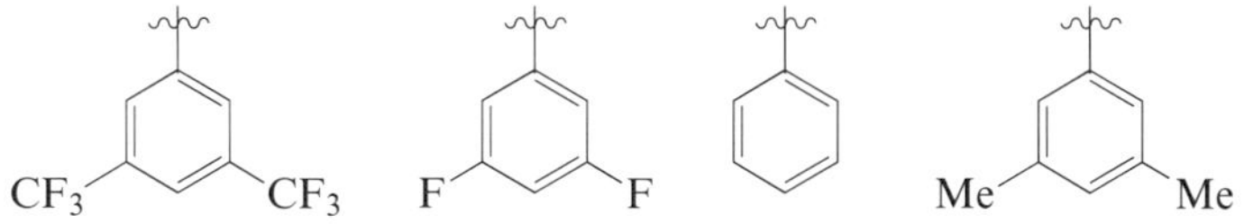

(**a**) (**b**) (**c**) (**d**)

Table 3 Hydrocyanation of MVN and VN using ligands **5a–d**[a]

Substrate	*ee* (%)			
	5a	**5b**	**5c**	**5d**
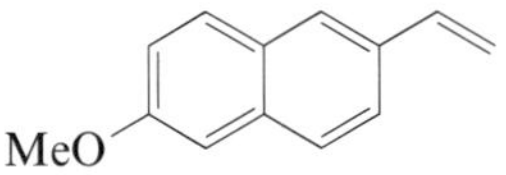	85–91[b]	78[c]	35[d]	16[d]
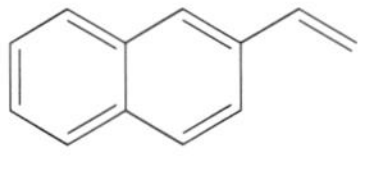	77	75	46[d]	25[d]

[a]0.10–0.20 M alkene, 1.0–5.0 mol% $Ni(COD)_2$/**5** in hexane.
[b]91% at 0°C in heptane.
[c]Toluene.
[d]Benzene.

Table 4 Effect of other aryl substituents on the *ee* for the hydrocyanation of MVN

Aryl substituents in **5**	*ee* (%)[a]
3,4,5-F_3	61
3,5–Cl_2	64
4-CF_3	65
4-F	38
2-Me	0
3,5-$(TMS)_2$	16
2,5-Me	9
4-MeO	17

[a]0.10–0.20 M MVN, 1–6 mol% $Ni(COD)_2$/**5**, hexane

respectively). A study of other phosphorus–aryl substituents (Table 4) also showed that the highest *ee*s were generally obtained with the more electron-withdrawing substituents (although a strict correlation between enantioselectivity and the substituent σ values was not observed, cf. 4-CF_3 *vs* 3,5-Cl_2).

The asymmetric hydrocyanation of other vinylarenes showed a similar response to the electron-withdrawing character of the phosphorus–aryl substituent. For example, the results from the use of ligands **5a**–**5c** in the asymmetric hydrocyanation of a series of *p*-substituted styrene derivatives are shown in Table 5. In every case the highest enantioselectivities were obtained with the most electron-withdrawing trifluoromethyl derivative (**5a**). Even so, the enantioselectivities showed a strong dependence on the substrate (*ee* range: 14–70% for **5a**) and some of the data suggests that the electronic nature of the *substrate* is also a major factor (compare entries 1 and 9).

Studies with other ligand frameworks (see Table 6), both carbohydrate and non-carbohydrate, support the notion that the electronic enhancement of the enantioselectivity is a general feature of the diaryl phosphinite-catalysed asym-

Table 5. Hydrocyanation of 4-substituted styrene derivatives[a]

		ee (%)		
Entry	4-Substituent	**5a**	**5b**	**5c**
1	Me	70	47	1
2	Ph	68	41	8
3	$Me_2C{=}CH$	63	–	–
4	PhO	60	38	7
5	Me_2CHCH_2	56	38	6
6	MeO	52	39	6
7	Cl	40	–	–
8	F	28	15	4
9	CF_3	14	9	1

[a] 0.65 mmol alkene, 0.020 mmol $Ni(COD)_2$, 0.020 mmol ligand, 0.65 mmol HCN in 5 ml of hexane after 24 h.

Table 6 Enantioselectivities (% *ee*) of MVN hydrocyanation using other carbohydrate and non-carbohydrate ligands

		p-Aryl	
Entry	Ligand	Ph	3,5-$(CF_3)_2C_6H_3$
1	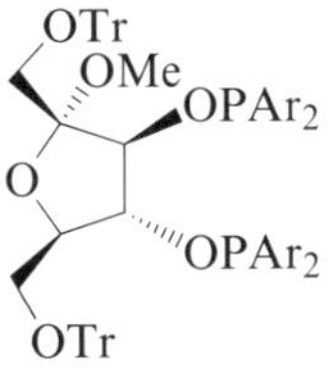	30 (or 43)	56
2	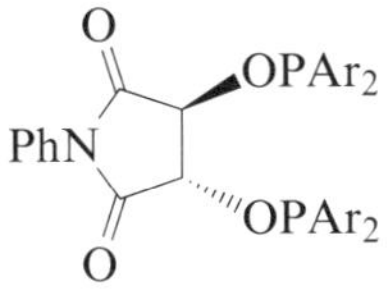	54	72
3	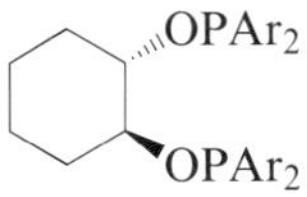	26	35

metric hydrocyanation reaction. However, other subtle catalyst features must be important in determining the overall enantioselectivity because the enantioselectivity does not always correlate directly with substituent σ values and the magnitude of the effect varies considerably from ligand system to ligand system.

15.2.3.2 Site of Phosphorus Substitution

One of the frequently cited drawbacks of using chiral catalysts derived from nature's chiral pool is the unavailability of both catalyst enantiomers. This is a particularly pertinent criticism for many of the unnatural carbohydrate derivatives. From the perspective of making precursors to profen drugs such, as naproxen, it was fortunate that the 2,3-*O*-bis(diarylphosphino)glucopyranoside ligands, such as **5a**, gave the desired (*S*)-nitrile as the predominant enantiomer. However, if the other nitrile enantiomer had been needed, the results from an investigation on the nickel-catalysed asymmetric cross-coupling reaction[20] suggested a possible solution. In a prototypical cross-coupling reaction, shown in Figure 2, the cross-coupled product's sense of chirality changed from (*S*) to (*R*)-when a 3,4-*O*-disubstituted glucophosphinite was used instead of a 2,3-*O*-disubstituted glucophosphinite. In accordance with this observation, the asymmetric hydrocyanation of MVN using 3,4-*O*-disubstituted glucophosphinite ligands gave the (*R*)-nitrile as the major enantiomer, although the magnitude of the *ee* was much less than in the 2,3-system.

Another clue was found when reactions conducted with 2,3-*O*-disubstituted α-methylfructofuranoside ligands, **(6)** also gave an excess of the (*R*)-nitrile. A summary of these results and those obtained from other ligand frameworks is given in Table 7. Inspection of these ligand structures suggests that the local chirality defined by the phosphorus-substituted diol controls the sense of asymmetric induction. In this approximation, illustrated in Figure 3, groups adjacent to the P—O—C—C—O—P linkage are ignored. Thus, the 2,3- and 3,4-*O*-disubstituted

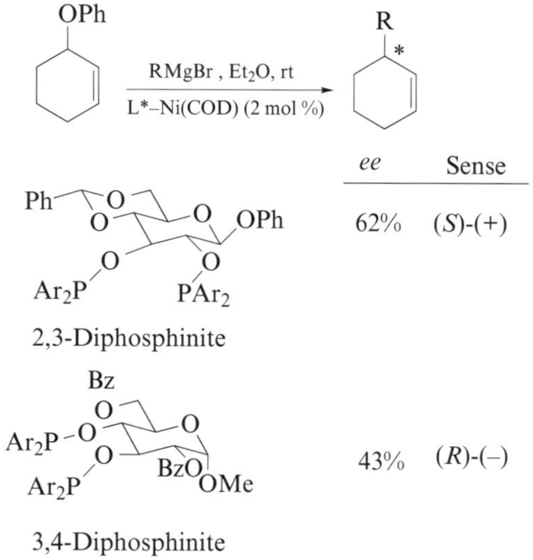

Figure 2 Effect of the phosphorus substitution site on the major product enantiomer for an asymmetric cross-coupling reaction

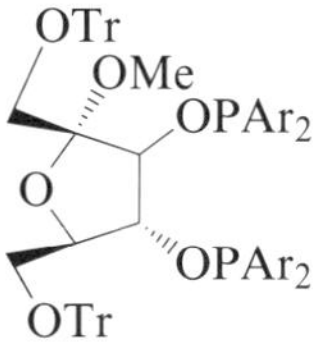

(**6**)

Table 7 Nitrile configuration for MVN hydrocyanation using various ligand systems

Entry	Ligand	*p*-Aryl	*ee*(%)
1	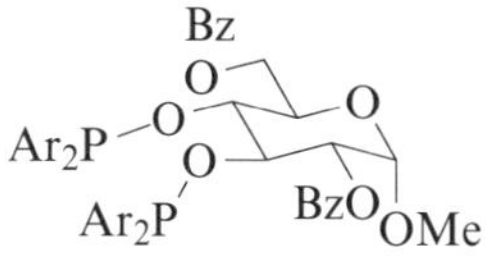	3,5-$(CF_3)_2C_6H_3$	33 (*R*)
2	As above	3,5 $Me_2C_6H_3$	13 (*R*)
3	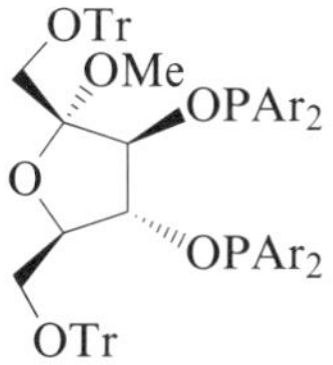	3,5-$(CF_3)_2C_6H_3$	56 (*R*)
4	As above	Ph	30 (*R*)
5	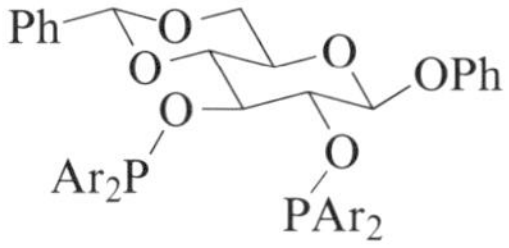	3,5-$(CF_3)_2C_6H_3$	78 (*S*)
6	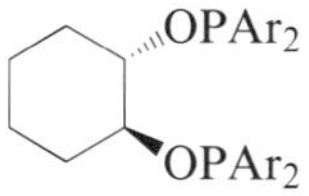	3,5-$(CF_3)_2C_6H_3$	35 (*S*)
7	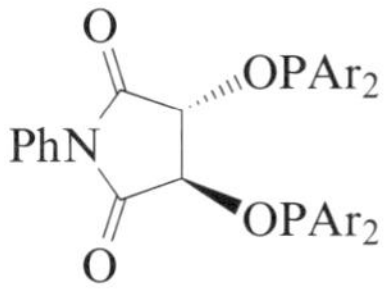	3,5-$(CF_3)_2C_6H_3$	72 (*S*)

glucophosphinites can be viewed as near mirror images or 'quasi-enantiomers,' as can the 2,3-*O*-disubstituted glucophosphinites and 2,3-*O*-disubstituted fructophosphinites.

One explanation of the influence of the diol stereochemistry on the stereochemical outcome of the reaction is that the positions of the phosphorus aryls in

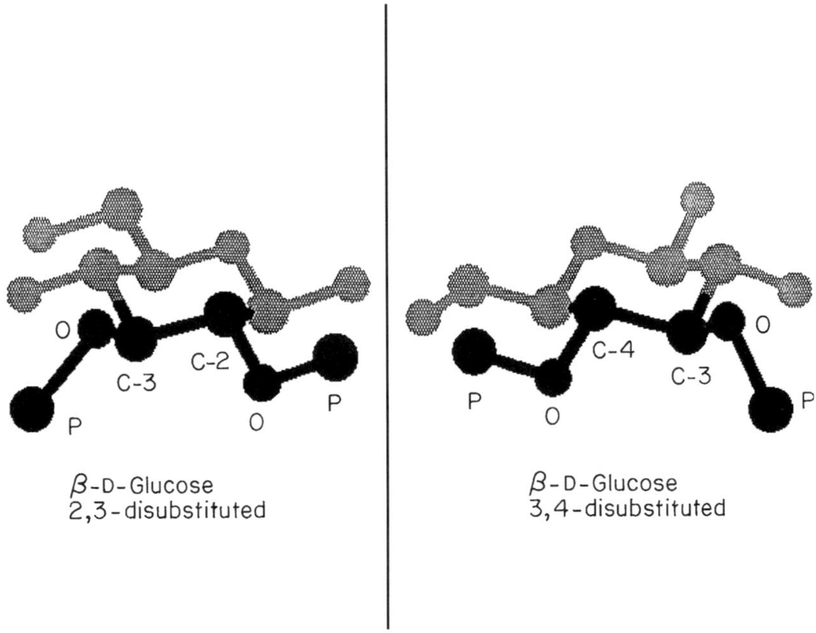

Figure 3 'Quasi-enantiomeric' relationship of 2,3- and 3,4-*O*-disubstituted glucophosphinites

the metal complex define the overall shape of the catalyst's chiral pocket.[21] We presume, therefore, that the approximate inversion of the diol stereochemistry effectively inverts the conformational positions of the phosphorus-aryl groups in the seven-membered chelate ring.

Because these quasi-enantiomers are not true mirror images, the magnitude of the *ee* may change, and further catalyst optimisation may be necessary. For example, the *ee* drops from 85% (*S*) to 56% (*R*) on going from the glucose-based ligand **5a** to the fructose based ligand **6a** [Ar = 3,5-$(CF_3)_2C_6H_3$]. Nevertheless, one should be able to predict which phosphorus-substitution sites are likely to produce a particular product enantiomer, and thus profit from the wide variety of possible carbohydrate precursors. As suggested by the initial cross-coupling experiment, one would also expect this principle to apply to other asymmetric, catalytic reactions. In fact, the use of the 2,3- and 3,4-*O*-disubstituted glucophosphinites ligands in the rhodium-catalysed asymmetric hydrogenation of alkenes allows the synthesis of either the (*R*) or (*S*) product enantiomers in very high *ee*.[22]

15.2.3.3 Effect of Electronic Asymmetry

C_2-Symmetric chiral ligands are often targeted in asymmetric catalysis because they reduce the number of competing diastereoisomeric intermediates in the catalytic cycle and are usually easier to synthesise. Nature, however, is replete with both chiral catalysts and chiral building blocks that lack a C_2 symmetry element. Although, strictly, the carbohydrate frameworks used in these hydrocyanation studies have C_1 symmetry, it can be argued, as we have done with the quasi-enantiomers, that the local chiral environment around the phosphorus atoms is essentially C_2. In principle, further steric or *electronic* differentiation of the two ligating phosphorus centres should provide an interesting test of the importance of C_2 symmetry in these ligand systems. Given the significant ligand electronic effects observed in asymmetric hydrogenation, we were particularly eager to explore the effects of ligand electronic asymmetry in this reaction.

Although there are synthetic obstacles to preparing bidentate diarylphosphinite ligands where the two phosphorus atoms have different aryl substituents, we were successful in preparing a series of such electronically asymmetric ligands using an α-methylfructofuranoside framework (**6′**). The results of using these ligands in the asymmetric hydrocyanation of MVN, summarised in Table 8, demonstrated that the enantioselectivity depended strongly on the ligand electronic asymmetry. In every case, the highest enantioselectivities were obtained when the phosphorus attached to the 4-hydroxy group carried a more electron-withdrawing aryl group than the phosphorus attached to the 3-hydroxy group. For example, the highest enantioselectivity (89%, entry 1) was obtained with 3,5-bis(trifluoromethyl)phenylphosphino groups at the 4-*O*-position and the corresponding diphenylphosphino groups at the 3-*O*-position. Simply

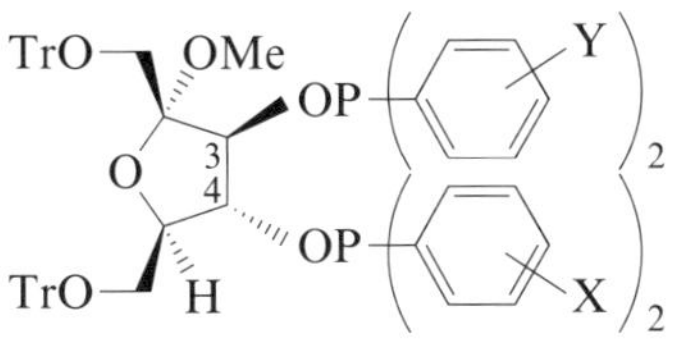

(**6′**)

Table 8 Effect of electronic asymmetry on MVN hydrocyanation using a fructofuranoside framework

Entry	X	Y	*ee* (%) of (*R*)-nitrile
1	3,5-$(CF_3)_2$	H	89
2	3,5-$(CF_3)_2$	4-F	88
3	3,5-$(CF_3)_2$	4-MeO	84
4	3,5-$(CF_3)_2$	3,5-$(Me)_2$	78
5	H	3,5-$(CF_3)_2$	58
6	3,5-$(CF_3)_2$	3,5-$(CF_3)_2$	56
7	3,5-$(CH_3)_2$	3,5-$(Me)_2$	40
8	4-MeO	4-MeO	25
9	3,5-$(CF_3)_2$	3,5-F_2	78
10	3,5-F_2	3,5-$(CF_3)_2$	42
11	3,5-F_2	3,5-F_2	45

exchanging the positions of these groups or having both sites substituted with 3,5-bis(trifluoromethyl)phenylphosphino groups decreases the *ee* from 89% to 58% and 56%, respectively (entries 5 and 6). Similar changes in enantioselectivity were observed with the 3,5-difluorophenyl groups (entries 9–11).

There is also some evidence that electronic asymmetry can enhance enantioselectivity in other ligand systems. Diaryl phosphinites (**7**), derived from the C_2-symmetric tartranil framework, showed a similar, although less dramatic, effect. Thus, the highest *ee*, 77%, was obtained with the mixed phenyl–3,5-bis(trifluoromethyl)phenyl derivatives, whereas the C_2-symmetric phenyl and 3,5-bis(trifluoromethyl)phenyl derivatives gave *ee*s of 54% and 70%, respectively.

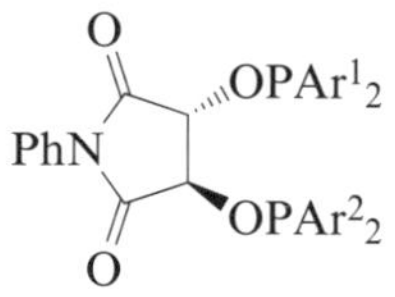

(**7**)

These results clearly demonstrate that the incorporation of electronically different phosphorus chelates can markedly enhance the enantioselectivity obtained with certain ligand arrangements. Thus, the selective manipulation of the electronic properties of the ligand on two different carbohydrate frameworks allows the preparation of either the (*R*)- or (*S*)-enantiomers of 2-(6-methoxy-2-naphthalene)propionitrile in high enantioselectivity [85% (*S*), 89% (*R*) at 25 °C].

15.2.3.4 Effect of Other Parameters

The glycoside bond is one of the most readily modified groups in carbohydrate synthesis. Unfortunately, in the case of the 2,3-*O*-disubstituted glucophosphinites the effect of this group on the enantioselectivity was very modest. For example, varying the glycoside groups in ligands of the general structure **8** (Table 9) led to a net change of only 10% in the *ee* for the hydrocyanation of VN. On the other hand, strong solvent effects were observed for the compound **5**-catalysed hydrocyanation of MVN, with the highest *ee*'s obtained in non-polar solvents. For example, *ee*s of 27%, 65%, 78% and 85% were obtained in MeCN, THF, benzene and hexane, respectively, in the **5a**-catalysed hydrocyanation of MVN. Modest increases in enantioselectivity have been observed at temperatures of 0–10 °C but these reactions have been accompanied by a much lower catalyst turnover.

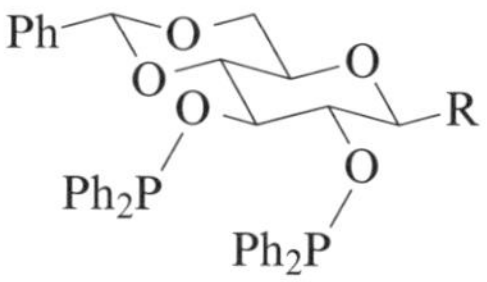

(**8**)

Table 9 Effect of C-1 substituent in β-glucophosphinite (**8**) on the hydrocyanation of VN

Entry	C-1 substituent	*ee* (%) (*S*)
1	OCH_3	45
2	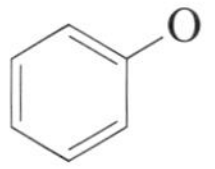	46
3	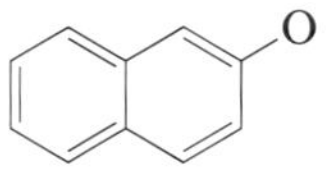	52
4	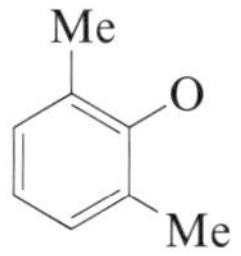	53
5	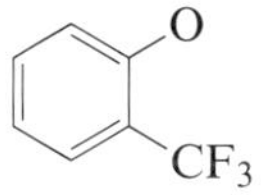	55
6	Ph	49

15.2.4 ASYMMETRIC HYDROCYANATION MECHANISM[8c]

The mechanism proposed for the **5**–Ni(COD)$_2$-catalysed asymmetric hydrocyanation of MVN is shown in Scheme 5. Note that this simplified mechanism is meant to represent all of the diastereoisomers which result from the C_1 symmetry of the ligand. The catalytic cycle is based on a number of experimental observations for the **5**–Ni(COD)$_2$–MVN system and also on previously proposed hydrocyanation mechanisms. Several key observations regarding this mechanism include:

- the preparation of a series of other catalyst precursors showed that the active catalyst contains **5** and Ni in a 1:1 ratio;
- the *ee* is independent of conversion, the addition of other substrates and the addition of enantiomerically enriched nitriles;
- the only directly observed intermediate is the Ni(**5**)(COD) complex **9**, which forms rapidly upon mixing Ni(COD)$_2$ and glucophosphite ligand **5**; this same complex is the catalyst resting state in the absence of or at low concentrations of HCN.
- one other proposed intermediate, Ni(**5a**)(MVN) (**10**), can be independently prepared, and gives the same *ee* as the Ni(**5a**)(COD) catalyst;
- the only completely irreversible step is the final reductive elimination of the 2-arylpropionitrile from intermediate **13**.

Reaction kinetics for the Ni(**5a**)(COD)-catalysed hydrocyanation of MVN in toluene were zero order for both MVN and HCN at concentrations above 0.04 M in each reagent. This saturation behaviour implies a shift of the catalyst resting

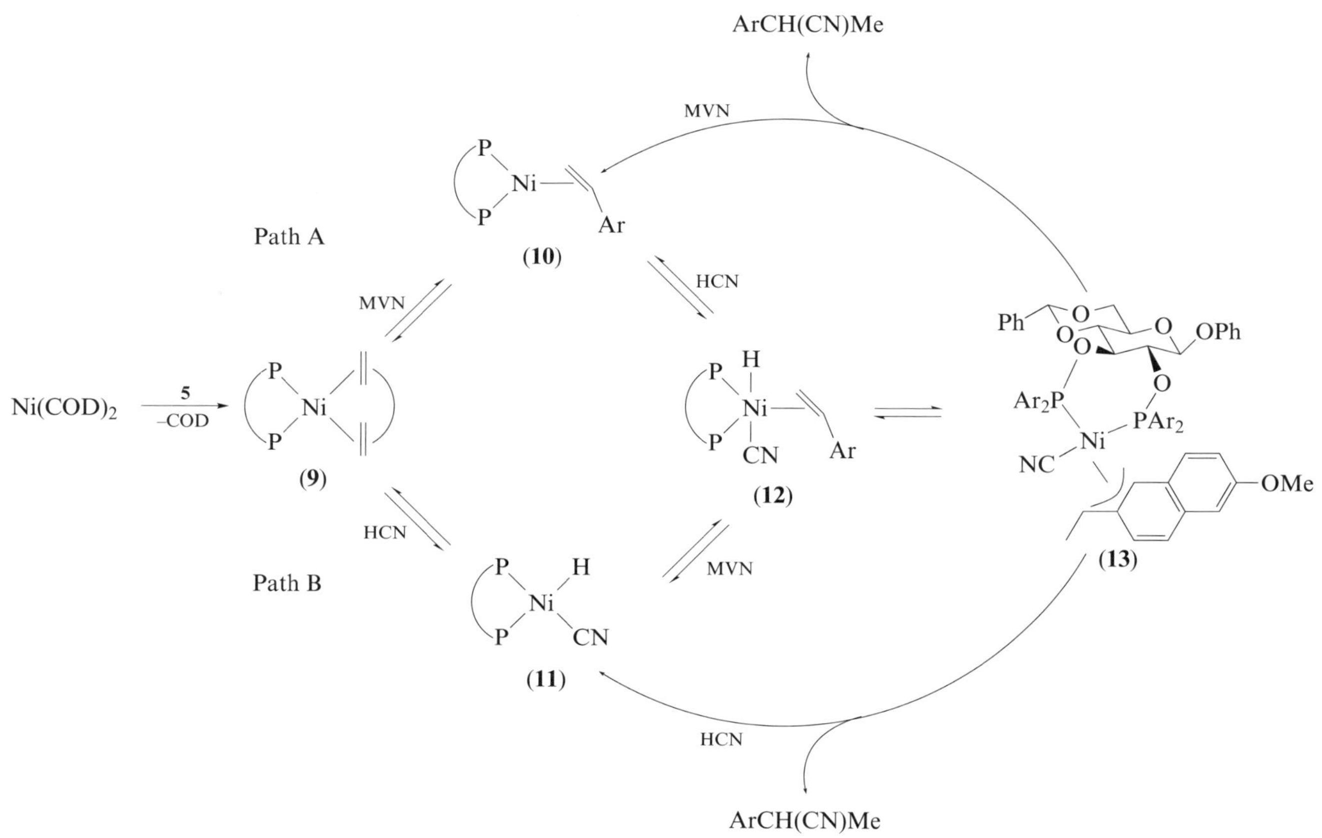

Scheme 5 Proposed mechanism for the asymmetric hydrocyanation of MVN using **5**–Ni(COD)$_2$

state from Ni(**5a**)(COD) (**9**) to either complex **12** or complex **13**. Based on the known stability of allylic-type hydrocyanation intermediates and the complete regioselectivity of this reaction, we believe that complex **13** is the catalyst resting state under most hydrocyanation conditions. At room temperature under these saturation conditions, a maxium activity of about 2000 turnovers h^{-1} (turnover = mol nitrile/mol Ni) was observed. Deactivation of the catalyst by HCN (equation 4) limits the catalyst lifetime to about 700–800 turnovers.

Two fundamental questions regarding this mechanism are, (i) when is the enantioselectivity determined, and (ii) how does the ligand electronic effect bring about an enhancement in the enantioselectivity? Our data suggest that the enantioselectivity is determined either in the formation of the η^3-benzylic intermediate **13**, or in the final reductive elimination of the product nitrile from this intermediate, and not in the initial coordination of MVN to form complex **10** or **12**. This conclusion is based upon the rapid exchange rates of free and complexed MVN in complex **10**, the very low ground-state differentiation of the *re* and *si* faces of MVN in this complex and deuterium labelling studies (equation 10), which suggested that the formation of intermediate **12** was also readily reversible.

$$\underset{\text{MVN-}d_2}{\text{ArCH=CD}_2} + \text{HCN} \rightarrow \text{ArCH(CN)(CD}_x\text{H}_{3-x}) \quad (10)$$

In this latter deuterium-labelling study, terminally labelled MVN, MVN-d_2, was hydrocyanated using $Ni(COD)_2$ and the ligands **5a**–**d**. Analysis of the 2-arylpropionitrile product showed that the extent of deuterium scrambling decreased markedly with the more electron-withdrawing ligands **5a** and **5b** (Table 10). The decrease in deuterium scrambling is believed to be due to an increase in the rate of reductive elimination of nitrile from **13** (k_2, equation 11) relative to the rate of β-hydride elimination from **13** to form **12** (k_{-1}). This explanation is consistent with studies which suggest that less basic phosphorus ligands (i.e. more electron deficient) accelerate the rate of reductive elimination reactions. The electronic effect of the phosphorus aryl substituents on the nickel centre was further

Table 10 Nitrile deuterium content from MVN-d_2 hydrocyanation[a]

	Deuterium content (%)				
Ligand	d_3	d_2	d_1	d_0	Conversion (%)
MVN-d_2		92.3	7.2	0.5	
5a	1	94	5	0	76
5a[b]	2	90	8	0	87
5b	6	87	7	0	50
5c	31	50	15	3	69
5d	29	50	18	3	68

[a] 0.49 mmol MVN d_2, 0.015 mmol $Ni(COD)_2$, 0.015 mmol ligand, 0.50 mmol HCN in 5 ml of hexane.
[b] 0.25 mmol MVN-d_2, 0.18 mmol $Ni(COD)_2$, 0.015 mmol **5a**, 0.26 mmol HCN in 2 ml of hexane.

$$\underset{(\mathbf{12})}{\text{Ni}(\mathbf{5})(\text{MVN})(\text{H})\text{CN}} \underset{k_{-1}}{\overset{k_1}{\rightleftharpoons}} \underset{(\mathbf{13})}{\text{Ni}(\mathbf{5})(\text{CN})(\eta^3\text{–benzyl})} \xrightarrow{k_2} \text{nitrile} + \text{Ni}(\mathbf{5}) \quad (11)$$

demonstrated via IR analysis of a series of *m*- and *p*-substituted nickel carbonyl derivatives Ni(**5**)$(CO)_2$. In this study, a linear relationship was obtained from a plot of the CO stretching frequencies (A_1) vs the Hammett parameters σ_m or σ_p, with an overall increase of 34 cm^{-1} in ν_{CO} as the aryl substituent was varied from 4-CH_3O to 3,5-$(CF_3)_2$ (Figure 4).

If the primary effect of the electron-withdrawing phosphorus–aryl substituents is to accelerate the final reductive elimination, then the enhancement in enantioselectivity may result because one of the diastereoisomers of **13** [i.e. a diastereoisomer that produces the (*S*)-nitrile] is disproportionately affected. The precise factors favouring a particular diastereoisomer are at present unclear, but the results with the α-methylfructofuranoside ligand framework strongly suggest the importance of a *stereoelectronic* component. For example, the effect of electronic asymmetry may reflect the importance of a *trans* relationship between the η^3-aryl fragment and the phosphorus bearing the electron-withdrawing aryl groups in complex **14**.

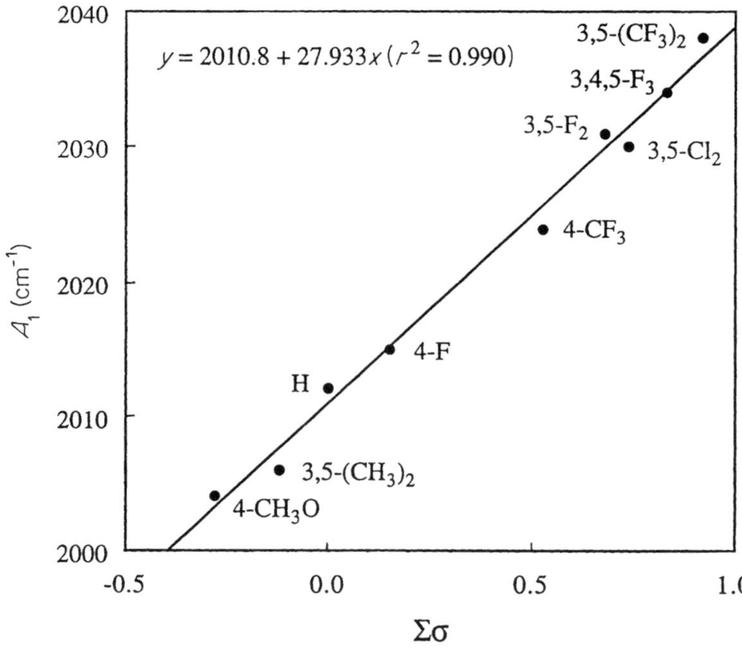

Figure 4 Correlation of σ_m and σ_p with the A_1 stretching frequency in Ni(**5**)$(CO)_2$ complexes

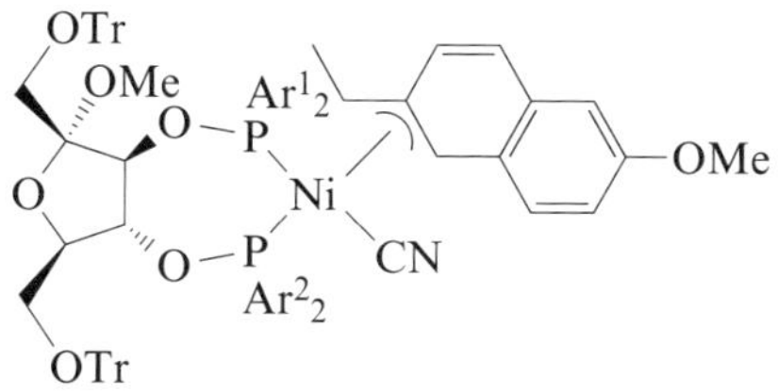

$Ar^1 = Ph$
$Ar^2 = 3,5\text{-}(CF_3)_2C_6H_3$

(**14**)

Although these explanations for the electronic effect are highly speculative, they provide a rationale for the ligand electronic effects observed in another asymmetric reaction: the rhodium-catalysed asymmetric hydrogenation of acetamido acrylates. Using similar diaryl phosphinite ligands, we found that electron-*donating* phosphorus aryl substituents dramatically enhanced the enantioselectivity of the reaction.[22] Here, the hydrogen oxidative addition rate is thought to play a key role in determining the reaction enantioselectivity,[17] and such oxidative addition reactions should be accelerated by more electron-donating phosphorus groups.

15.2.5 LABORATORY PREPARATIONS OF NAPROXEN AND IBUPROFEN PRECURSORS

The nickel catalyst derived from ligand **5a** is a practical laboratory reagent for the asymmetric hydrocyanation of vinylarenes. For example, the asymmetric hydrocyanation of MVN and *p*-isobutylstyrene has been carried out at room temperature on 5 and 10 g scales using 0.01 and 0.02 equiv. of the catalyst $Ni(COD)_2$-(**5a**), respectively. The resulting 2-arylpropionitriles were obtained in 96% (MVN) and 77% (*p*-isobutylstyrene) isolated yields with vinylarene conversions greater than 99%. In both cases, only the branched nitriles were detected and no yield losses due to vinylarene polymerisation were observed. Although the Ni(**5a**)(COD) catalyst is limited to about 700–800 turnovers in the asymmetric hydrocyanation of MVN, greater *ligand* utility can be obtained because deactivation of the catalyst occurs with dissociation of the free ligand (equation 4). Thus, we have obtained 4000–5000 turnovers based on ligand **5a** in the hydrocyanation of MVN without loss of enantioselectivity (84% *ee*, hexane–toluene) by periodically adding more $Ni(COD)_2$ to the reaction mixture.

15.2.6 ASYMMETRIC HYDROCYANATION AS A COMMERCIAL ROUTE TO NAPROXEN

Several potentially practical routes to naproxen have been reported and reviewed.[13] Of the current commercially practised methods, none is believed to utilise asymmetric catalytic methods. Although our work in this area did not focus on process and development issues for naproxen manufacture, several aspects of the Ni(**5a**)(COD)-catalysed hydrocyanation of MVN make it a feasible route for the commercial manufacture of naproxen: the regioselectivity is 100% and few, if any, by-products are produced; high yields, reasonable reaction rates and high levels of ligand utility can be obtained, and the ligand **5a** is readily accessible. Although the nitrile is not produced in >99% *ee*, it is enantiomerically enriched, and can be crystallised to afford optically pure nitrile. The hydrolysis of the nitrile to naproxen has some precedent in the related area of chiral cyanohydrins (see also Volume I, Chapter 14). For example, the hydrolysis of optically active cyanohydrins, such as 4-methoxymandelonitrile, to amino acids can be carried out without racemisation.

Some of the most pressing technical and economic issues concern the cost and handling of $Ni(COD)_2$. Unless ligand **5a** can be readily separated from the reaction mixture and recycled, $Ni(COD)_2$ (or an equivalent zero-valent nickel source) must be fed to the hydrocyanation reaction so that the use of **5a** can be minimised. Unfortunately $Ni(COD)_2$, currently available in only research quantities, is a relatively expensive complex that is also very air unstable and thermally sensistive. Other key issues for this process include the supply and safe handling of HCN, finding low-cost routes to MVN and the expense of incorporating trifluoromethyl groups into the ligand.

15.3 ACKNOWLEDGEMENTS

We gratefully acknowledge the seminal contributions of T. Ayers and T. Warren to the work reported here. We also thank M. Beattie, E. Duvall, G. Halliday, D. Johnson, S. Bernard and K. Messner for their invaluable technical assistance.

15.4 NOTES AND REFERENCES

1. Tolman, C. A., McKinney, R. J., Seidel, W. C., Druliner, J. D., and Stevens, W. R., *Adv. in Catal.*, **33**, 1 (1985).
2. Kruse, C. G., in *Chirality in Industry* (ed. A. N. Collins, G. N. Sheldrake and J. Crosby), Wiley, Chichester, 1992, p. 279.
3. Casalnuovo, A. L., McKinney, R. J., and Tolman, C. A., in *The Encyclopedia of Inorganic Chemistry*, Wiley, New York, 1994.
4. (a) Stoutamire, D. W., and Tieman, C. H. (Shell Oil Co.), *US Pat.*, 4 560 515, 1985; (b) Stoutamire, D. W., and Dong, W. (Shell Oil Co.), *US Pat.*, 4 594 196, 1986; (c) Stoutamire, D. W., and Dong, W. (Du Pont, *US Pat.*, 4 723 027, 1988.
5. Tanaka, K., Mori, A., and Inoue, S., *J. Org. Chem.*, **55**, 181 (1990).
6. North, M., *Synlett*, 807 (1993).
7. (a) Elmes, P. S., and Jackson, W. R., *Aust. J. Chem.*, **35**, 2041 (1982); (b) Jackson, W. R., and Lovel, C. G., *Aust. J. Chem.*, **35**, 2053 (1982); (c) Hodgson, M., Parker, D., Taylor, R. J., and Ferguson, G., *Organometallics*, **7**, 1761 (1988); (d) Baker, M. J., and Pringle, P. G., *J. Chem. Soc., Chem. Commun.*, 1292 (1991).

8. (a) Casalnuovo, A. L., and Rajanbabu, T. V., DuPont *US Pat.*, 5 175 335, 1992; (b) RajanBabu, T. V., and Casalnuovo, A. L., *J. Am. Chem. Soc.*, **114**, 6265 (1992); (c) Casalnuovo, A. L., RajanBabu, T. V., Ayers, T. A., and Warren, T. H., *J. Am. Chem. Soc.*, **116**, 9869 (1994); (d) RajanBabu, T. V., and Casalnuovo, A. L., *Pure Appl. Chem.*, **66**, 1535 (1994).
9. Tolman, C. A., Seidel, W. C., Druliner, J. D., and Domaille, P. J., *Organometallics*, **3**, 33 (1984).
10. Andell, O. S., and Bäckvall, J. E., *Organometallics*, **5**, 2350 (1986).
11. Rieu, J., Boucherle, A., Cousse, H., and Mouzin, G., *Tetrahedron*, **42**, 4095 (1986).
12. Sonawane, H. R., Bellur, N. S., Ahuja, J. R., and Kulkarni, D. G., *Tetrahedron: Asymmetry*, **3**, 163 (1992).
13. Giordano, C., Villa, M., and Panossian, S., in *Chirality in Industry* (ed. A. N. Collins, G. N. Sheldrake and J. Crosby), Wiley, Chichester, 1992, pp. 303–312 .
14. Nugent, W. A., and McKinney, R. J., *J. Org. Chem.*, **50**, 5370 (1985).
15. See contributions of Gosser, L., in ref. 8a.
16. See contributions of McKinney, R., and Nugent, W., in ref. 8a.
17. Landis, C. R., and Halpern, J., *J. Am. Chem. Soc.*, **109**, 1746 (1987).
18. (a) Cullen, W. R., and Sugi, Y., *Tetrahedron Lett.*, 1635 (1978); (b) Selke, R., *React. Kinet. Catal. Lett.*, **10**, 135 (1979); (c) Jackson, R., and Thompson, D. J., *J. Organomet. Chem.*, **159**, C29-C31 (1978); (d) Yamashita, M., Kobayashi, M., Sugiura, M., Tsunekawa, K., Oshikawa, T., Inokawa, S., and Yamamoto, H., *Bull. Chem. Soc. Jpn.*, **59**, 175 (1986); (e) Habus, I., Raza, Z., and Sunjic, V., *J. Mol. Catal.*, **42**, 73 (1987); (f) Selke, R., Facklam, C., Foken, H. and Heller, D., *Tetrahedron: Asymmetry*, **4**, 369 (1993); (g) Selke, R., Schwarze, M., Baudisch, H., Grassert, I., Michalik, M., Oehme, G., and Stoll, N., *J. Mol. Catal.*, **84**, 223 (1993).
19. (a) Jacobsen, E. N., Zhang, W., and Guler, M. L., *J. Am. Chem. Soc.*, **113**, 6703 (1991); (b) Nishiyama, H., Yamaguchi, S., Kondo, M., and Itoh, K., *J. Org. Chem.*, **57**, 4306 (1992). Until recently (see ref. 22), only modest effects were observed in the asymmetric hydrogenation reaction. See (c) Morimoto, T., Chiba, M., and Achiwa, K., *Chem. Pharm. Bull.*, **41**, 1149 (1993); (d) Morimoto, T., Chiba, M. and Achiwa, K., *Chem. Pharm. Bull.*, **40**, 2894 (1992); (e) Hengartner, U., Valentine, D., Jr., Johnson, K. K., Larscheid, M. E., Pigott, F., Scheidl, F., Scott, J. W., Sun, R. C., Townsend, J. M., and Williams, T. H., *J. Org. Chem.*, **44**, 3741 (1979); (f) Werz, U., and Brune, H., *J. Organomet. Chem.*, **365**, 367 (1989).
20. RajanBabu, T. V., and Ayers, T. A., unpublished results. See also ref. 22.
21. (a) Knowles, W. S., *Acc. Chem. Res.*, **16**, 106 (1983); (b) Oliver, J. D., and Riley, D. P., *Organometallics*, **2**, 1032 (1983); (c) Brown, J. M., and Evans, P. L., *Tetrahedron*, **44**, 4905 (1988); (d) Pavlov, V. A., Klabumovskii, E. I., Struchkov, Y. T., Voloboev, A. A., and Yanovsky, A. I., *J. Mol. Catal.*, **44**, 217 (1988); (e) Bogdan, P. L., Irwin, J. J., and Bosnich, B., *Organometallics*, **8**, 1450 (1989); (f) Seebach, D., Plattner, D. A., Beck, A. K., Wang, Y. M., and Hunziker, D., *Helv. Chim. Acta*, **75**, 2171 (1992).
22. RajanBabu, T. V., Ayers, T. A., and Casalnuovo, A. L., *J. Am. Chem. Soc.*, **116**, 4101 (1994).

16 Enantioselective Protonation in Fragrance Synthesis

C. FEHR
Firmenich SA, Geneva, Switzerland

16.1 Introduction 335
16.2 The Appeal of Enantioselective Protonation 336
16.3 (*S*)-α-Damascone 337
16.3.1 The Rose Ketones 337
16.3.2 Design of the Chiral Reagent 339
16.3.3 Interaction of Chiral Reagent, Counter Ion and Ligand . . 339
16.3.4 (*S*)-α-Damascone by Catalytic Enantioselective Protonation 341
16.3.5 Extension to the Syntheses of the Enantiomers of α- and γ-Cyclogeranic Acid, α- and γ-Cyclocitral and γ-Damascone 343
16.4 Enantioselective Synthesis of a New Tetralin Musk Perfume Component 348
16.5 Acknowledgements 350
16.6 Notes and References 350

16.1 INTRODUCTION

The fragrance industry is not subjected to the same regulatory constraints as the pharmaceutical industries with respect to the enantiomeric purity of chiral compounds. However, very often one enantiomer shows superior odour properties compared with the other, justifying its synthetic development. As an example, the odour of (*S*)-α-damascone [(*S*)-**1**] is reminiscent of rose petals, whereas the (*R*)-enantiomer has a more pronounced apple note and is also responsible for an undesired cork odour. The difference in their odour threshold concentrations is also remarkable. Whereas the odour of the (*S*)-enantiomer is still perceived at 1.5 ppb, the (*R*)-antipode requires a 70-fold higher concentration for odour perception.[1] In the field of aromatic musk compounds, it has been established that the extremely strong and characteristic note of the racemic new Firmenich musk (±)-**2**[2] is almost entirely due to the (–)-enantiomer [(–)-**2**], the odour of which is extremely strong, both on the smelling strip and in perfume compositions.[3]

In this chapter, the syntheses of the fragrances (*S*)-(–)-**1** and (*S*,*S*)-(–)-**2**, together with related compounds are described, employing as the key step the promising new methodology of enantioselective protonation.[1,3–11]

Chirality in Industry II. Edited by A.N. Collins, G. N. Sheldrake and J. Crosby

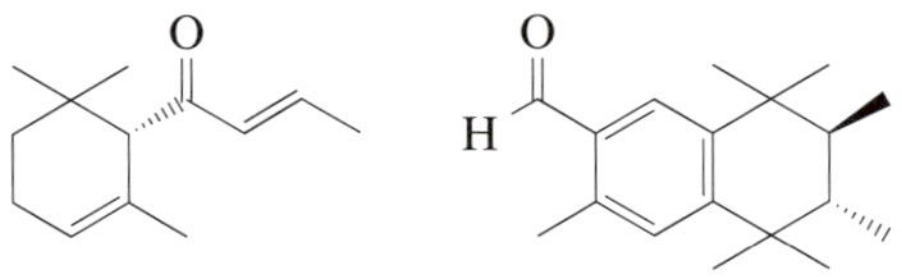

[(S)-(–)-1] [(S,S)-(–)-2]

16.2 THE APPEAL OF ENANTIOSELECTIVE PROTONATION

Conceptually, enantioselective protonation is extremely simple. A prochiral enolate possesses two enantiotopic faces. Using a chiral, non-racemic proton source, the proton transfer to the enolate will be kinetically favoured either from the top face or from below, thus affording an (*R*)- or (*S*)-enantiomerically enriched carbonyl compound (Scheme 1). However, it should be stated immediately that a generally applicable chiral reagent is not yet available. In addition, the outcome of enantioselective protonation depends strongly on the structure and configuration of the enolate, the presence or absence of lithium salts (LiOR, $LiNR_2$, LiX) and the solvent. In spite of the complexity of the reaction, due to aggregated reactive species, application of enantioselective protonation is straightforward and very attractive. The enolate, obtained by deprotonation from the parent carbonyl compound, or directly from a synthetic reaction step, e.g. by addition of an organometallic reagent to a ketene or an enone, is 'quenched' with the chiral proton donor to afford the enantiomerically enriched carbonyl compound. After addition of water, the chiral reagent is regenerated to its protonated state and recovered by simple extraction.

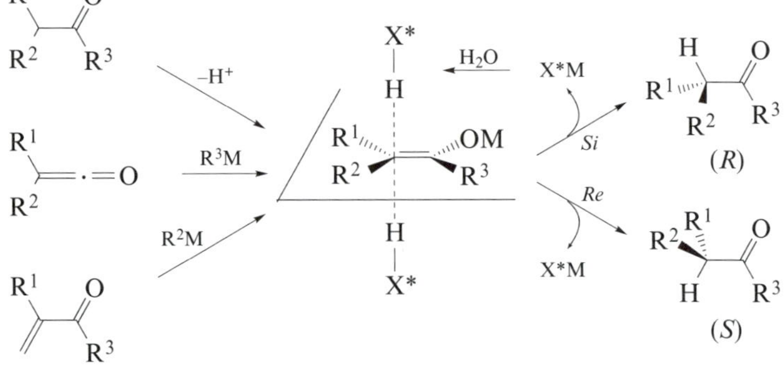

Scheme 1 Enolate generation and enantioselective protonation

16.3 (*S*)-α-DAMASCONE

16.3.1 THE ROSE KETONES

In recent years the rose ketones, namely α-damascone [(±)-**1**], β-damascone (**3**), γ-damascone [(±)-**10**] and β-damascenone (**4**), have become important perfume components owing to their typical fruity–flowery scent and exceptional odour strength.[12] γ-Damascone, which has never been found in nature, has a slightly weaker odour, but nevertheless possesses a very elegant damascone-type note. Despite the similarity of their odours, each rose ketone has an intrinsic value, so it has been essential to develop efficient syntheses of each of these individual fragrance compounds. Although they are already effective at concentrations lower than 0.1%, their annual production is steadily increasing and has grown from 1 tonne in 1978 to 10 tonnes in 1988, of which α-damascone [(±)-**1**] represents the major part.

In the 1980s, the development of a unique reaction scheme for the synthesis of rose ketones (±)-**1**,**3** and **4** starting from the corresponding, readily accessible, esters **5**,**6**, and **7** constituted an important breakthrough. Whereas the normal allyl-Grignard reaction mainly leads to the tertiary alcohols via di-addition, the selectivity for mono-/di-addition could be completely reversed when the Grignard reaction was performed in the presence of LDA.[13] Under these conditions, the ketone formed *in situ* is protected from further reaction by deprotonation to the enolate and, after acid-catalysed isomerisation of the terminal double bond, the three rose ketones (±)-**1**,**3** and **4** are isolated in high yield.

This methodology has been applied to other sterically hindered esters or amides and gives the best results with allylic Grignard reagents, as the product ketones are rapidly deprotonated.[14]

Evidently, for the synthesis of (*S*)-α-damascone [(*S*)-**1**] by enantioselective protonation, the regioisomeric, achiral ketone enolate **8** is required (compare Schemes 2 and 3), which can be obtained by Grignard reaction on the *deprotonated* ester **5**.[15] This reaction, in which the ester enolate represents the electrophilic species, merits comment. Owing to the delocalisation of the negative charge over five centres, the carbonyl group possibly retains part of its electrophilic character. On the other hand, it is also known that ester enolates have only limited stability: elimination of LiOMe would then allow the Grignard reaction to occur via a ketene or ketenoid species.[16] Application of the same reaction conditions to methyl β-cyclogeranate (**6**) constitutes a highly efficient and direct synthesis of γ-damascone [(±)-**10**].[15] The addition of organometallic reagents to ketenes also represents an interesting access to prochiral enolates, as exemplified by the synthesis of α-damascone [(±)-**1**] starting from the readily accessible and stable ketene **14**[17] (see Scheme 5).

The aforementioned Grignard reactions, usually in combination with strong lithium bases, have now been developed into highly efficient industrial procedures.[18] In addition, the prochiral enolate intermediates **8** and **9**, which afford

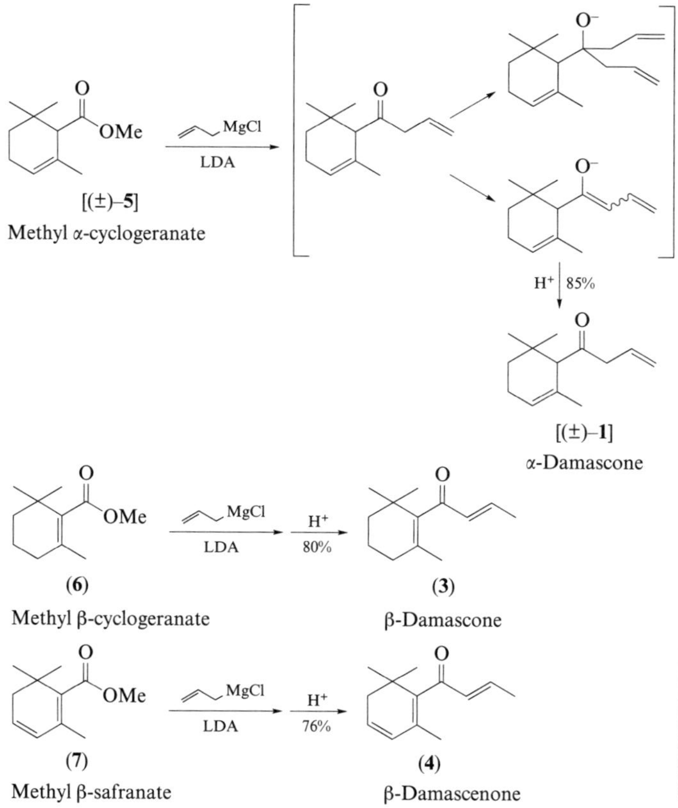

Selectivity ketone/alcohol	
With LDA	Without LDA
98 : 2	28 : 72
90 : 10	15 : 85
96 : 4	15 : 85

Scheme 2

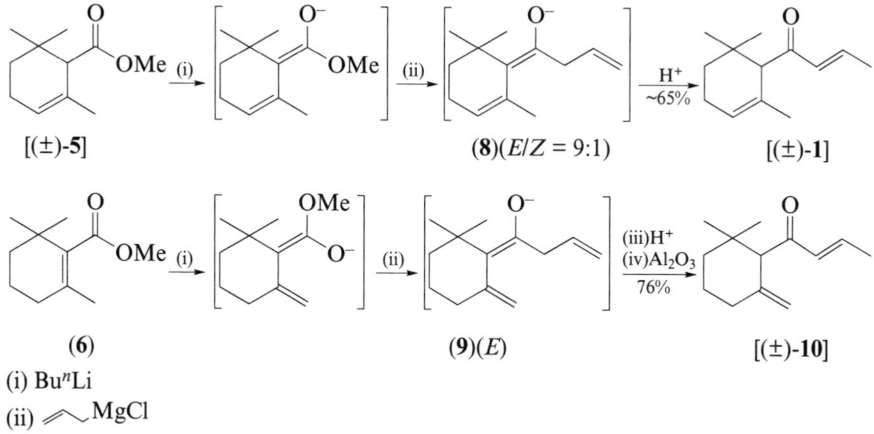

Scheme 3

racemic α- and γ-damascone [(±)-**1** and (±)-**10**] upon quenching with H_2O, offer the possibility of enantioselective protonation by a chiral proton source. Thus, a minimal change of reaction conditions should allow the production of enantio-enriched α- and γ-damascones.

16.3.2 DESIGN OF THE CHIRAL REAGENT

In the search for an efficient and synthetically useful chiral proton source, we were guided by the following criteria (Scheme 4). Ideally, the chiral reagent should be only weakly acidic to allow better transition-state discrimination. It should also contain electron-rich groups with coordination or chelation ability, which would enhance conformational rigidity in the transition state. For obvious reasons, the transferred proton should be located in the proximity of the stereogenic centre (within the 'chiral environment') and, further, the chiral reagent should be readily accessible in both enantiomeric forms and be easily recoverable. These criteria are fulfilled with the ephedrine derivatives **11** and **12**, of which only one enantiomer is illustrated. The imidazolidinone **11** has an N—H bond confined in a cyclic system, and the amino alcohol **12** can attain conformational rigidity through chelation. It should be added that (+)- and (–)-**11** are commercially available (Merck), and that (+)- and (–)-**12** are readily obtained by treatment of (+)- or (–)-ephedrine with acetone by catalytic hydrogenation.[19]

—pK_A 15–20

—Chelation; rigidity 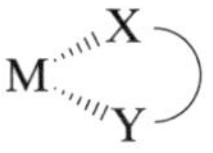

—Proximity H—X〰C*

—Accessibility of both enantiomers

—Recoverability

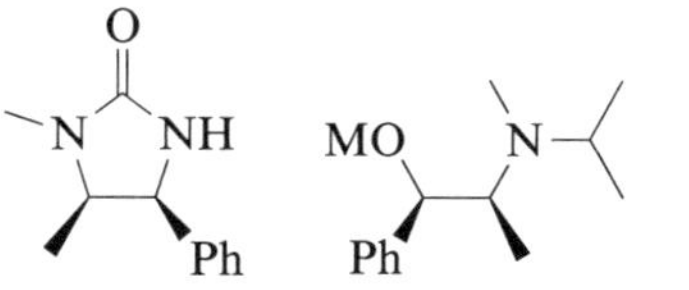

[(+)-**11**] M = H [(–)-**12**] M = Li [(–)-**13**]

Scheme 4 Criteria for chiral reagent

16.3.3 INTERACTION OF CHIRAL REAGENT, COUNTER ION AND LIGAND

In an initial experiment, the ester (±)-**5** was deprotonated with *n*-butyllithium and the resulting ester enolate was treated with allylmagnesium chloride to afford ketone enolate **8** (*E*/*Z* ≈ 9:1). Protonation of enolate **8** with ephedrine derivatives (+)-**11** or (+)-**12** at –50 to –10 °C and subsequent isomerisation of the terminal double bond afforded the desired product (*S*)-**1** with 58% and 70% *ee*, respectively (Scheme 5). The moderate chemical yield (60%) was due to competing nucleophilic attack of *n*-butyllithium on the ester function. Some additional

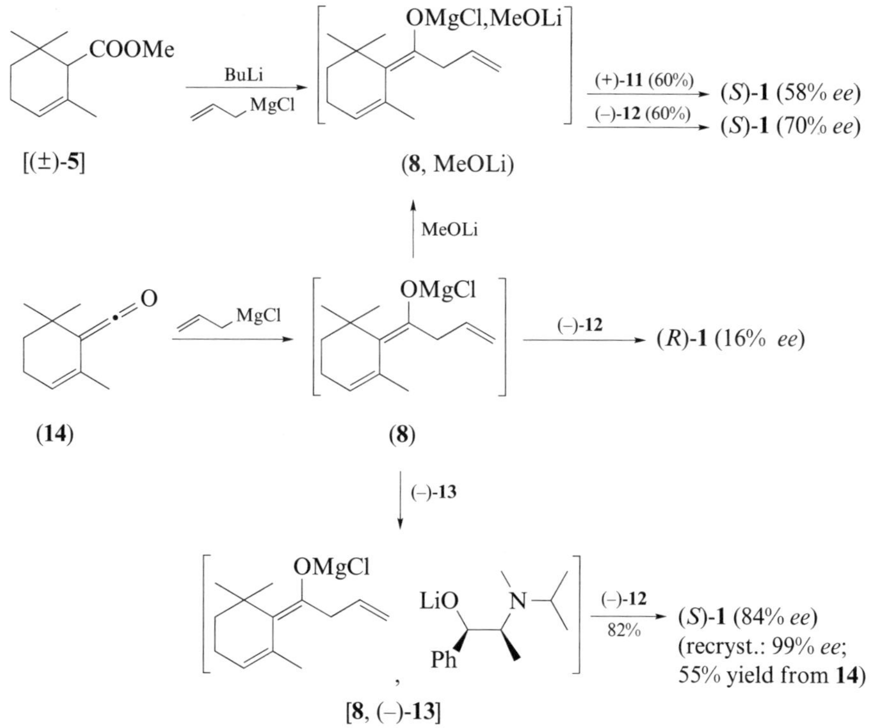

Scheme 5

screening experiments with differently *N*-substituted ephedrine derivatives revealed that inferior inductions were obtained both with less hindered (NHMe, NMe_2, $NHPr^i$) and more hindered [$NMeCH(Et)_2$] proton donors, which highlights the subtle balance between steric factors and ease of chelation. Therefore, the amino alcohol (–)-**12** [or (+)-**12**] was chosen as the standard chiral reagent for further investigations. This novel protonating agent is not only efficient, but also very practical, as it can be recovered almost quantitatively by a simple acid–base extraction.

To understand better the influence of the counter ion and ligand, and also to avoid the use of *n*-butyllithium, we next investigated the reaction of allylmagnesium chloride with ketene **14** (Scheme 5). The THF-solvated (but otherwise ligand-free) enolate **8** (*E*/*Z* ≈ 9:1) thus formed was protonated with amino alcohol (–)-**12**. Much to our surprise, protonation of the ligand- and lithium-free enolate **8** afforded (*R*)-**1** (16% *ee*) as the major enantiomer. On the other hand, addition of 1 equiv. of MeOLi prior to protonation, thus restoring the conditions present when starting from ester (±)-**5**, furnished (*S*)-**1** with 70% *ee*. Next, enolate **8** was treated with the chiral lithium alkoxide (–)-**13** derived from *N*-isopropylephedrine, and protonated with (–)-**12** to afford (*S*)-α-damascone with

84% *ee* in an improved yield of 82%.[6] This product could easily be purified further by crystallisation (99% *ee*; 55% yield from **14**), but for use as a fragrance building block, a 9:1 enantiomeric mixture was judged to be sufficient because of cost considerations.

At this stage of development, the major drawbacks of the (*S*)-α-damascone synthesis were the large amounts of chiral reagent employed [*ca* 3 kg of (–)-**12** for 1 kg of (*S*)-**1**], and the impractically low protonation temperature. In the industrial process which produced the first kilograms of (*S*)-**1**, the lithium alkoxide ligand (–)-**13** was replaced by Bu^tOLi (from Chemetal), and the protonation was effected at 0 °C with only minimal loss of enantioselectivity (79% *ee*).

16.3.4 (*S*)-α-DAMASCONE BY CATALYTIC ENANTIOSELECTIVE PROTONATION

As shown above (Scheme 5), protonation of the complex between magnesium enolate **8** and the lithium alkoxide (–)-**13** affords (*S*)-**1** with 84% *ee*, whereas the ligand- and lithium-free magnesium enolate affords (*R*)-**1** with a low *ee* of 16%. In order to examine the reactivity of the alkoxide-free lithium enolate, the magnesium enolate **8** was quenched with Me_3SiCl. Almost pure (*E*)-silyl enol ether **15** was readily obtained by fractional distillation. Treatment of this compound with methyllithium generated the lithium(*E*)-enolate **16** which, upon addition of (–)*N*-isopropylephedrine [(–)-**12**] at +60 °C, underwent protonation with excellent enantioselectivity (94% *ee*).[10] Apparently, the lithium enolate does not require a lithium alkoxide ligand for selective protonation. Evidently, this reaction is only ligand-free and enolate-rich at low conversion, as 1 equiv. of the lithium alkoxide ligand (–)-**13** is generated during this process.

To test whether the enolate protonation is more enantioselective at low conversion, enolate **16** was treated with only 0.5 equiv. of (–)-**12** (–60 °C, 2 min) and the reaction mixture was quenched with Me_3SiCl (Scheme 6). Much to our surprise, no trace of enol silyl ether **15** was detected. Instead, approximately equal amounts of ketone (*S*)-**17** and enol silyl ether (*S*)-**19** of high enantiomeric purity (94% *ee*) were formed. This unique enantioselective protonation is thus catalytic: the protonated intermediate (*S*)-**17** makes available its most acidic hydrogen (C-2) for the protonation of (–)-**13** or, more likely, of a mixed aggregate of alkoxides **16** and (–)-**13**. This catalytic cycle is based on subtle kinetic differences for the proton transfer reactions between chiral reagents, enolate and non-inducing[20] proton donor. When the quantity of ligand (–)-**12** is lowered below *ca* 0.2 equiv., a marked decrease in reaction rate and selectivity was observed. Finally, for the synthesis of (*S*)-α-damascone [(*S*-**1**)], enolate **16** was treated with 0.3 equivalents of (–)-**12** (–60 °C, 10 min) and the intermediate enolate (*S*)-**18** was quenched with dilute HCl (Scheme 7). After isomerisation of the terminal double bond, (*S*)-**1** was obtained in 86% yield and 93% *ee*.[10] Ongoing experiments indicate that this reaction is not very temperature dependent, and we are confident that this elegant access to (*S*)-α-damascone is also the most efficient one.

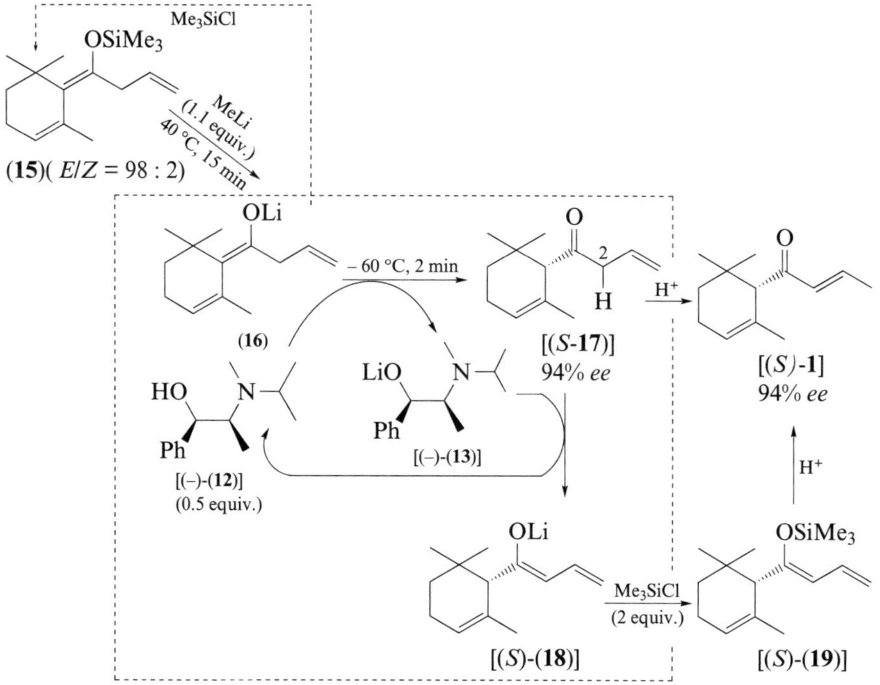

Scheme 6

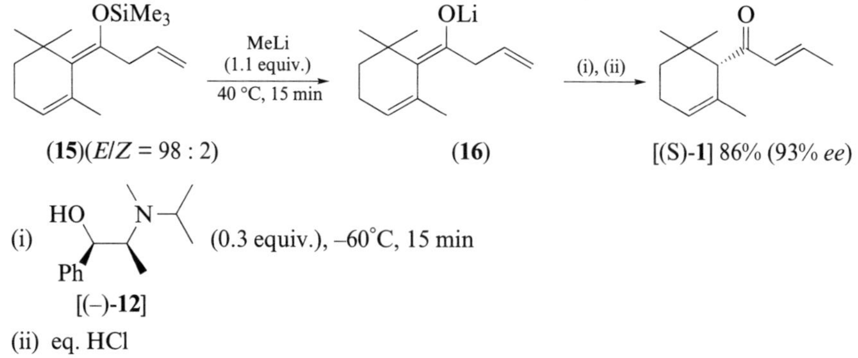

Scheme 7

The aforementioned case is, of course, very substrate specific, as the product ketone (*S*)-**17** is rapidly and exclusively deprotonated at C-2. For example, an analogous autocatalytic reaction is not possible with enolate **20**, as the acidity of the C-2 protons of (*S*)-**21** is substantially weaker. To circumvent this problem, 1-phenylacetone has been used successfully as an external, achiral proton donor (Scheme 8). Thus, protonation of **20** (*E*/*Z* ≈ 97:3), readily obtained from ketene **14** and *n*-butyllithium, with equimolar amounts of (–)-**12** afforded butyl ketone (*S*)-**21** with 96% *ee*, and catalytic enantioselective protonation with 0.2 equiv. of (–)-**12**, followed by the addition of 0.85 equiv. of 1-phenylacetone, also furnished ketone (*S*)-**21** with high enantioselectivity (94% *ee*). With only 0.1 equiv. of (–)-**12** the *ee* decreased to 85%.[10]

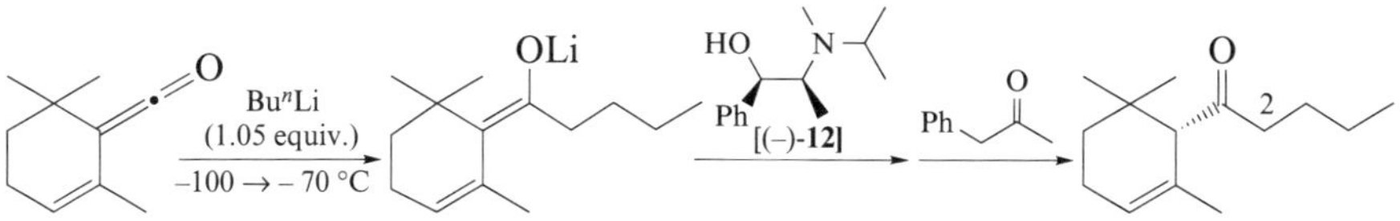

(**14**) (**20**)(*E*/*Z* = 97:3) [(*S*)-**21**]

1 equiv.	–	96% *ee* (90%)
0.2 equiv.	0.85 equiv.	94% *ee* (94%)
0.1 equiv.	0.95 equiv.	85% *ee*

Scheme 8

16.3.5 EXTENSION TO THE SYNTHESES OF THE ENANTIOMERS OF α- AND γ-CYCLOGERANIC ACID, α- AND γ-CYCLOCITRAL AND γ-DAMASCONE

The single enantiomers of α- and γ-cyclogeranic acids and esters represent versatile intermediates in the field of fragrance and medicinal chemistry, and the racemic α- and γ-methyl esters (±)-**5** and (±)-**33** are used as fragrance materials which complement the widely used rose ketones. Applying the aforementioned technology for enolate generation/protonation, methyl α-cyclogeranate [(±)-**5**] was deprotonated with *n*-butyllithium and the resulting ester enolate **24** was protonated with (–)-**12**. Unfortunately, the isolated methyl ester (*S*)-**5** showed a low enantiomeric excess of 36%. This marked difference with the related ketone enolates (see above) may be ascribed to an over-rapid protonation of the enolate (due to the high pK_a of **5**), coupled with an insufficient structural differentiation of the enolate substituents on C-1 (OMe vs OLi) for efficient enantiotopic recognition. For this reason, the phenyl ester (±)-**22** and the phenyl thioester (±)-**23** were chosen as substrates. Indeed, protonation of enolates **25** and **26** with 2 equiv. of (–)-*N*-isopropylephedrine [(–)-**12**] at –100 °C, followed by gradual warming to –10 °C, afforded the corresponding esters with 77% and 99% *ee*, respectively (Scheme 9, Table 1).[7,9] Likewise, almost enantiomerically pure (*R*)-thioester [(*R*)-**23**] was obtained by protonation of enolate **26** with (+)-**12**. Application of the catalytic version on thioester enolate **25** is also possible: use of 0.5 equiv. of (–)-**12** and 1.55 equiv. of 1-phenylacetone gives almost the same result (98% *ee*) as

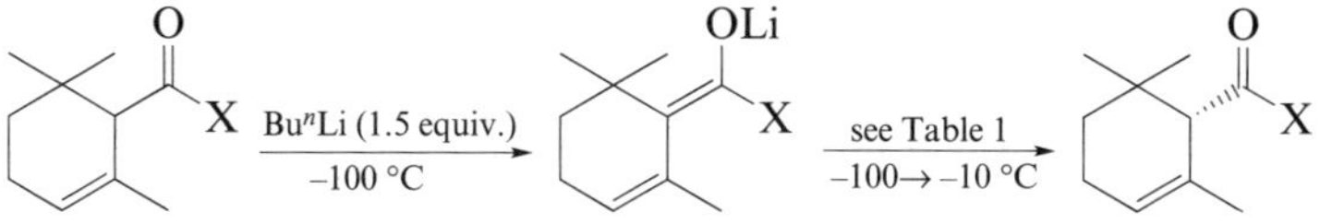

[(±)-**5**] X = OMe
[(±)-**22**] X = OPh
[(±)-**23**] X = SPh

(**24**)(*Z*)
(**25**)(*Z*)
(**26**)(*Z*)

[(*S*)-**5**]
[(*S*)-**22**]
[(*S*)-**23**]

Scheme 9

Table 1 Protonation conditions for Scheme 9

Substrate	Reagents (equiv.)	*ee* %	yield %
(±)-**5**	(–)-**12** (2.0)	36 (*S*)	–
(±)-**22**	(–)-**12** (2.0)	77 (*S*)	–
(±)-**23**	(–)-**12** (2.0)	99 (*S*)	87
(±)-**23**	(+)-**12** (2.0)	99 (*R*)	87
(±)-**23**	(–)-**12** (0.5), Ph 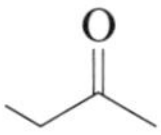 (1.55)	98 (*S*)	81
(±)-**23**	(–)-**12** (0.2), Ph 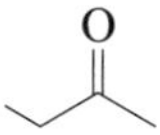 (1.85)	81 (*S*)	

2.0 equiv. of (–)-**12**. However, a substantial loss in selectivity is observed with smaller amounts of (–)-**12** [81% *ee* with 0.2 equiv. of (–)-**12**].

The foregoing protonations are performed at –100 °C because, above – 80 °C, the thioester enolates eliminate LiSPh to afford ketene **14** (see Section 16.3.3). To circumvent this problem, we studied the enantioselective addition of aromatic thiols to ketene **14**[8,9] (Scheme 10). After some experimentation, it was found that slow addition of the chiral amino alcohol (–)-**12** to a mixture of ketene **14** and ArSLi at –60 °C leads to the thioesters (*S*)-**23** and (*S*)-**27** with 95 and 97% *ee*,

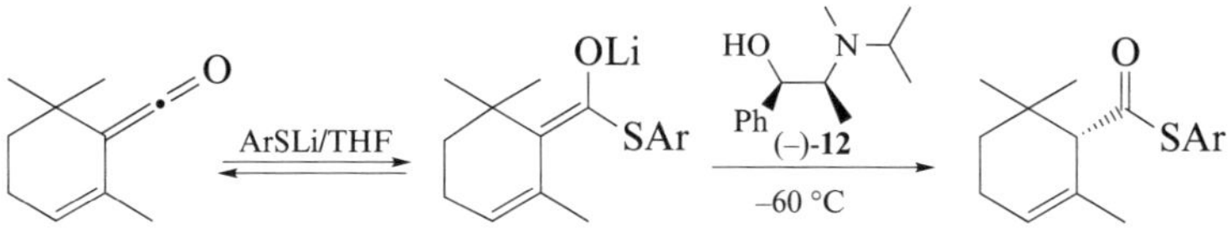

(**14**)

(**26**)

[(S)-**23**] Ar = Ph, 95% *ee* (80% yield)
[(S)-**27**] Ar = 4-ClPh, 97% *ee* (83% yield)

Scheme 10

respectively. Since, under these conditions, the equilibrium between ketene **14** and enolate **26** lies on the side of the former, the reaction rate is controlled by the addition of (–)-**12**. This allows rapid and irreversible *C*-protonation of the incipient enolate. In this manner, the concentration of (–)-**12** in the reaction medium is kept low, thus minimising the risk of proton exchange with the thiolate.

In the overall process, 1 equiv. of thiol is incorporated irreversibly into ketene **14**; however, the lithium ephedrate (–)-**13** is recovered unchanged. We next turned our attention, therefore, to the possibility of a catalytic process, in which the aromatic thiol serves as nucleophile *and* proton source, in the presence of catalytic amounts of lithium-ephedrate (–)-**13**. To allow this catalytic reaction to succeed, conditions had to be found which guaranteed the rapid consumption of the thiol, thus maintaining its concentration lower than that of the chiral lithium alkoxide (–)-**13**. The required rate acceleration could be achieved by working at higher temperatures and higher concentrations. Thus, continuous slow addition of ArSH to a reaction mixture containing 1 equiv. of ketene **14** and 0.05 equiv. of (–)-**13** at –27 °C furnished the thioesters (*S*)-**23** and (*S*)-**27** with excellent *ee*'s (89–90%) (Scheme 11). At lower temperatures the reaction becomes sluggish and at 0 °C competing addition of (–)-**13** to **14** is observed, leading to an accumulation of thiol and a corresponding drop in the *ee* of **23** (*ca* 50% *ee*).[8,9]

In comparison with the catalytic enantioselective enolate protonation, the thiol addition described here shows improved efficiency in terms of the overall selectivity and turnover of chiral reagent. This is undoubtedly due to the fact that the concentrations of achiral proton donor (ArSH) and enolate are very low throughout the course of the reaction.

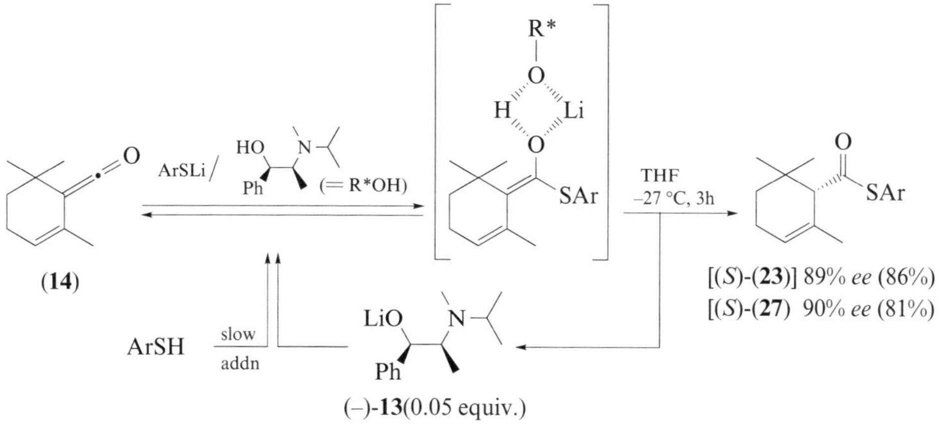

Scheme 11

The thiol esters (*R*)-**23** and (*S*)-**23** have been transformed readily into the corresponding methyl esters (*R*)-**5** and (*S*)-**5** [of which the (*R*)-enantiomer is the more valuable fragrance], the α-cyclocitrals (*R*)-**28** and (*S*)-**28** and the α-damascones (*R*)-**1** and (*S*)-**1**, all without racemisation.

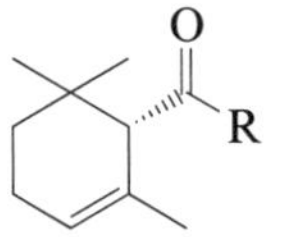

[(*S*)-**5**] R = OMe
[(*S*)-**28**] R = H
[(*S*)-**1**] R = 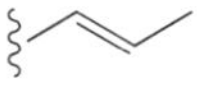

Finally, the most recent extension of enantioselective protonation to the synthesis of the hitherto unknown enantiomers of γ-damascone [(*R*)- and (*S*)-**10**] and the γ-cyclogeranates is presented briefly for the (*S*)-series.[11] Following the synthetic route used for racemic γ-damascone [(±)-**10**] (Scheme 3), the prochiral mixed lithium–magnesium enolate complex **9** is protonated with (–)-**12**. Because this protonation is slow, and to minimise side reactions between enolate **9** and the product ketone such as double-bond isomerisation and racemisation, enolate **9** was added to excess (–)-**12** at 0 °C. Under these conditions and after treatment with Al_2O_3, a 3:1 mixture of (*S*)-**10** and (*R*)-**10** was isolated (Scheme 12). On the other hand, protonation of the magnesium-free lithium enolate **29** depends critically on the presence of the chiral lithium alkoxide (–)-**13**. The best results were obtained when a solution of **29** in THF was added to a 2:1 mixture of (–)-**13** and (–)-**12** (75% *ee*). By comparison, a 1:1 mixture of (–)-**13** and (–)-**12** gives 68% *ee*, and (–)-**12** alone gives 49% *ee*.

(*S*)-γ-Damascone of a higher enantiomeric purity was obtained by an LDA-mediated Grignard reaction on the corresponding (*S*)-γ-thioester (*S*)-**32**, which was prepared by the route illustrated in Scheme 13. Following the procedure for its positional isomer **23**, the α,β-unsaturated thioester **30** was deprotonated with excess LDA, and the resulting enolate **31** was protonated with an excess of (+)-amino alcohol (+)-**12**. It should be stressed that, unlike the case with thiol ester **23**, deprotonation of **30** leads to an enolate of opposite configuration (*E*), and that the protonation takes place with excellent, but *opposite*, selectivity. Here, use of the dextrorotatory reagent (+)-**12** allows the preparation of the (*S*)-thioester (*S*)-**32** with greater than 96% *ee* (Scheme 13). Unfortunately, competing γ-protonation also affords substantial amounts of starting ester **30**, thus rendering the conversion incomplete. The quantities of base and proton donor could be reduced with minimal loss of enantioselectivity, and we were pleased to find that the combination of phenylacetone and 0.5 equiv. of amino alcohol (+)-**12** also

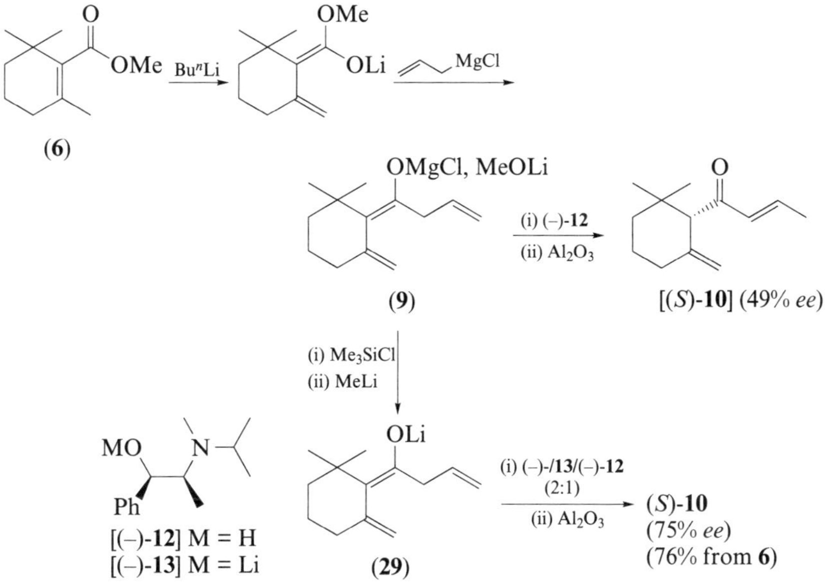

Scheme 12

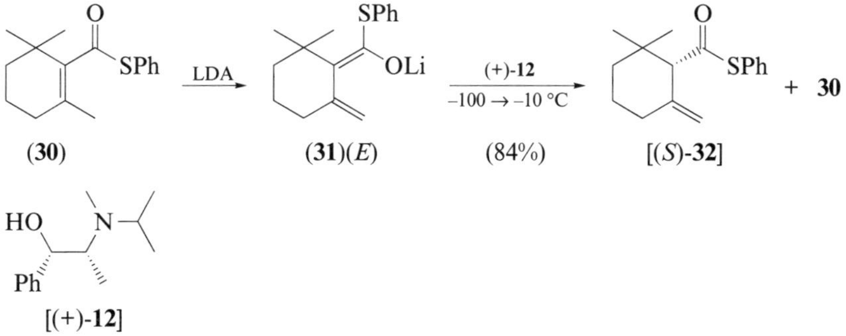

LDA (equiv.)	(+)-**12** (equiv.)	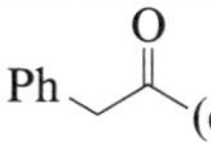 (equiv.)	(*S*)-**32**	**32**:**30**
3.0	4.0	–	≥96% *ee*	53:47
1.5	2.0	–	94% *ee*	55:45
1.5	0.5	1.5	88% *ee*	45:55

Scheme 13

resulted in a high enantiofacial discrimination, affording (*S*)-**32** with 88% *ee*. With the enantiomers of (*S*)-thioester **32** in hand, the enantiomers of methyl γ-cyclogeranate [(*S*)- and (*R*)-**33**], γ-damascone [(*S*)- and (*R*)-**10**] and γ-cyclocitral [(*S*)- and (*R*)-**34**] could thus be readily synthesised in high enantiomeric purity.

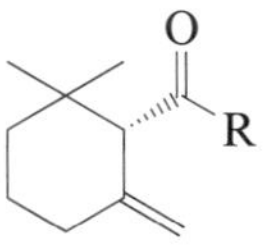

[(*S*)-**33**] R = OMe
[(*S*)-**34**] R = H
[(*S*)-**10**] R =

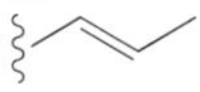

For both γ-damascone (**10**) and methyl γ-cyclogeranate (**33**), the (*S*)-enantiomer exhibits better odour properties, this difference being more pronounced in the latter case; whereas (*S*)-**33** excels with rich aromatic, herbal, floral, damascone-reminiscent notes, (*R*)-**33** has a predominant camphoraceous, cork, cellar note.

16.4 ENANTIOSELECTIVE SYNTHESIS OF A NEW TETRALIN MUSK PERFUME COMPONENT

Several hundred structurally related aromatic musk odourants of different odour strength are known, of which Tonalid [(±)-**35**], first commercialised by Polak Frutal Works, has long been considered the strongest. Its annual consumption is estimated at more than 3000 tonnes. Recently, we have discovered a new aldehyde musk, (±)-**2**, which, according to perfumery performance tests, is approximately ten times stronger than Tonalid.[2]

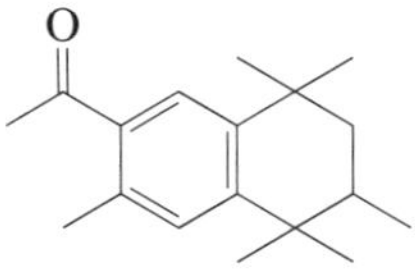

[(±)-**35**]
Tonalid® (*PFW*)
[=Fixolide® (*Givaudan*)]
>3000 tpa

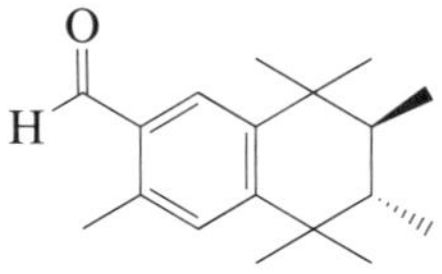

[(±)-**2**]
New *Firmenich* m usk
~10 × stronger than Tonalid®

The synthesis of (±)-**2** is outlined in Scheme 14. Friedel–Crafts alkylation of readily accessible *tert*-butyl ketone (±)-**36** with *o*-xylene furnishes ketone (±)-**37**, the reduction of which affords alcohol (±)-**38** as a 94:6 diastereoisomeric mixture. Under strongly acidic conditions (MsOH, P_2O_5 or H_2SO_4), a Wagner–Meerwein rearrangement generates a tertiary carbocation, which

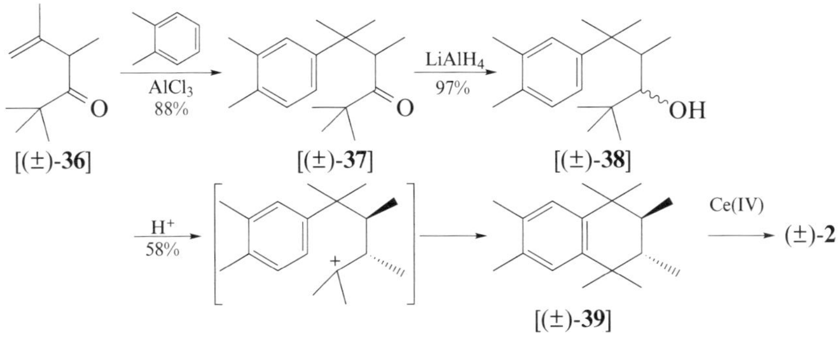

Scheme 14

undergoes cyclisation to furnish, after crystallisation, hydrocarbon (±)-**39** in 58% yield; the mother liquors contain about 15% of the diastereoisomeric *meso*-hydrocarbon. The final step involves Ce(IV)-mediated oxidation to the musk odourant (±)-**2**.

Subsequently, the two enantiomers of **2** were synthesised by a multi-step sequence including a classical resolution step,[21] thus allowing the evaluation of the fragrances of (*R*,*R*)-(+)- and (*S*,*S*)-(–)-**2** (see Introduction). In view of its clear superiority, it was decided to elaborate an enantioselective synthesis of (–)-**2**.[3]

If known enantioselective reactions are examined, it becomes evident that only a few are suited to the construction of sterically crowded chiral centres, and enantioselective protonation is often the method of choice. Our plan was to prepare ketone (*R*)-**36** by enantioselective protonation of either the (*Z*)- or (*E*)-enolate **40**, and to follow the efficient synthetic route shown in Scheme 14. In spite of the risk of racemisation, both in a Friedel–Crafts alkylation with (*R*)-**36** and, even more important, during acid-catalysed cyclisation (Scheme 14; **38** → **39**), this approach was considered the most interesting in view of its simplicity and similarity to the established route.

For the generation of enolate **40**, the deprotonation of both the β,γ-unsaturated enone (±)-**36** and of the readily accessible α,β-unsaturated enone **41** was examined. Whereas deprotonation of (±)-**36** with LDA afforded a 9:1 mixture of (*Z*)-**40** and (*E*)-**40**, chelation-controlled deprotonation of **41** generated (*Z*)-**40** with complete stereocontrol, thus enhancing the chances of an efficient enantioselective protonation (Scheme 15). Indeed, we were gratified to find that protonation of (*Z*)-**40** with (–)-*N*-isopropylephedrine [(–)-**12**] led to the desired ketone (*R*)-**36** with 90% *ee*. The site selectivity for protonation is also satisfactory, as only 31% of starting enone **41** was regenerated. Decreasing the amounts of LDA and (+)-**12** had only a minor effect on enantioselectivity (86% *ee*), but dramatically reversed the site selectivity for α- *vs* γ-protonation. As (*R*)-**36** is much more volatile than its isomer **41**, these compounds are easily separated by distillation. On the other hand, owing to the inertness of **41** to Friedel–Crafts condi-

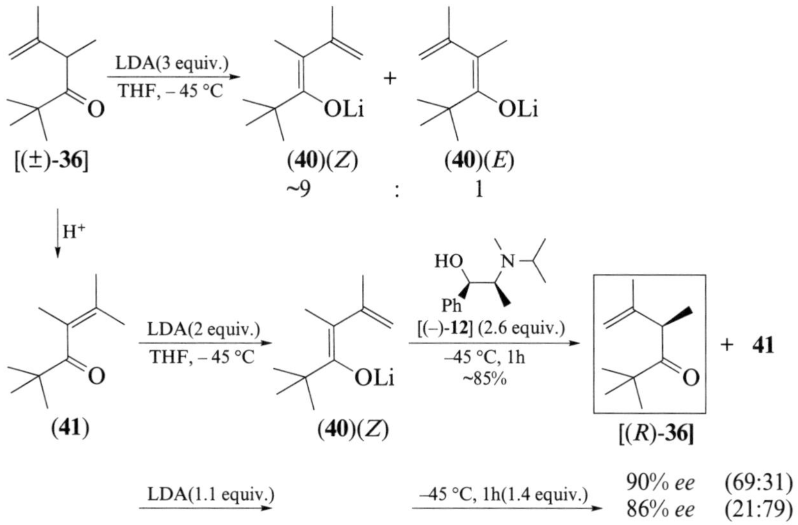

Scheme 15

tions, the mixture (*R*)-**36** and **41** can also be treated with *o*-xylene. To our satisfaction, the synthesis of (–)-**2** according to Scheme 14 could be achieved with minimal racemisation (<10%), and the organoleptic evaluation of a 9:1 mixture of (–)-**2** and (+)-**2** clearly confirmed the superiority of (–)-**2** over the racemic version.[3]

In view of these encouraging results, we are confident that the technique of enantioselective protonation will often be considered for future synthetic applications in industry.

16.5 ACKNOWLEDGEMENTS

The author gratefully acknowledges the postdoctoral contributions of Dr Isabelle Stempf and Dr Nathalie Chaptal-Gradoz and the excellent work of Mr José Galindo.

16.6 REFERENCES AND NOTES

1. Fehr, C., and Galindo, J., *J. Am. Chem. Soc.*, **110**, 6909 (1988); Fehr, C., and Galindo, J. (Firmenich), *Eur. Pat.*, EP 326 869, 1988.
2. Fehr, C., Galindo, J., Haubrichs, R., and Perret, R., *Helv. Chim. Acta*, **72**, 1537 (1989); Fehr, C., and Galindo, J. (Firmenich), *US Pat.*, 5 324 875, 1989; *Eur. Pat.*, EP 405 427, 1989.

3. Fehr, C., Delay, F., Blanc, P.-A., and Chaptal–Gradoz, N. (Firmenich), *Eur. Pat. Appl.*, 1994.
4. Reviews: Fehr, C., *Angew. Chem.*, **108**, 2726 (1996); *Angew. Chem. Int. Ed. Engl.*, **35**, 2566 (1996); Hünig, S., *Houben-Weyl, Methods of Organic Chemistry*, Vol. E 21 d (ed. Helmchen, G., Hoffmann, R.-W., Mulzer, J. and Schaumann, E.), Thieme, Stuttgart, 1995, p. 3851. Duhamel, L., Duhamel, P., Launay, J.-C., and Plaquevent, J.-C., *Bull. Soc. Chim. Fr.* II, 421 (1984) and ref. 5. Recent examples: Vedejs, E., and Lee, N., *J. Am. Chem. Soc.*, **117**, 891 (1995); Vedejs, E., Lee, N., Sakata, S. T., *J. Am. Chem. Soc.*, **116**, 2175 (1994); Ishihara, K., Kaneeda, M., and Yamamoto, H., *J. Am. Chem. Soc.*, **116**, 11179 (1994); Yanagisawa, A., Kuribayashi, T., Kikuchi, T., and Yamamoto, H., *Angew. Chem.*, **106**, 129 (1994); *Angew. Chem., Int. Ed. Engl.*, **33**, 107 (1994); Takeuchi, S., Ohira, A., Miyoshi, N., Mashio, H., and Ohgo, Y., *Tetrahedron: Asymmetry*, **5**, 1763 (1994); Gerlach, U., Haubenreich, T., Hünig, S., and Klaunzer, N., *Chem. Ber.*, **127**, 1989 (1994); Kosugi, H., Hoshino, K., and Uda, H., *Phosphorus Sulfur*, **95–96**, 401 (1994); Yanagisawa, A., Kikuchi, T., Watanabe, T., Kuribayashi, T., and Yamamoto, H. *Synlett*, 372 (1995); Aboulhoda, S. J., Létinois, S., Wilken, J., Reiners, I., Hénin, F., Martens, J., and Muzart, J., *Tetrahedron: Asymmetry*, **6**, 1865 (1995); Fuji, K., Kawabata, T., and Kuroda, A., *J. Org. Chem.*, **60**, 1914 (1995).
5. Fehr, C., *Chimia,* **45**, 253 (1991).
6. Fehr, C., and Guntern, O., *Helv. Chim. Acta*, **75**, 1023 (1992).
7. Fehr, C., Stempf, I., and Galindo, J., *Angew. Chem.*, **105**, 1091 (1993); *Angew. Chem., Int. Ed. Engl.*, **32**, 1042 (1993).
8. Fehr, C., Stempf, I., and Galindo, J., *Angew. Chem.,* **105**, 1093 (1993); *Angew. Chem., Int. Ed. Engl.*, **32**, 1044 (1993).
9. Fehr, C., Stempf, I., and Galindo, J. (Firmenich), *Eur. Pat.*, EP 593 917, 1992.
10. Fehr, C., and Galindo, J., *Angew. Chem.*, **106**, 1967 (1994); *Angew. Chem.*, *Int. Ed. Engl.*, **33**, 1888 (1994).
11. Fehr, C., and Galindo, J., *Helv. Chim. Acta*, **78**, 539 (1995).
12. Kastner, D., *Parfüm. Kosmetik*, **75**, 170 (1994).
13. Fehr, C., and Galindo, J., *Helv. Chim. Acta*, **69**, 228 (1986).
14. Fehr, C., Galindo, J., and Perret, R., *Helv. Chim. Acta*, **70**, 1745 (1987).
15. Fehr, C., and Galindo, J., *J. Org. Chem.*, **53**, 1828 (1988).
16. Häner, R., Laube, T., and Seebach, D., *J. Am. Chem. Soc.*, **107**, 5396 (1985).
17. Naef, F., and Decorzant, R., *Tetrahedron*, **42**, 3245 (1986).
18. *Perfum. Flavorist*, **19**, 59 (1994).
19. Adamski, R. J., and Numajiri, S., *US Pat.*, 3 860 651, 1969; *Chem. Abstr.*, **82**, 170 330t (1975).
20. (*S*)-**17** does not protonate **16** enantioselectively in the absence of (–)-**12**.
21. Fehr, C and Chaptal-Gradoz, N., unpublished work.

17 Thiamphenicol: a Manufacturing Process Involving a Double Inversion of Stereochemistry

L. COPPI, C. GIORDANO, A. LONGONI and S. PANOSSIAN
Zambon Group SpA, Bresso (Milan), Italy

17.1 Introduction . . . 353
 17.1.1 The Drug . . . 353
 17.1.2 The Market . . . 354
17.2. Main Synthetic Approaches . . . 354
 17.2.1 Process 1 (from 4-Methylmercaptobenzaldehyde) . . . 355
 17.2.2 Process 2 (from 4-Methylsulphonylbenzaldehyde) . . . 358
17.3. Conversion of (1*S*,2*S*)-2-Amino-1-[(4-methylthio)-phenyl]propane-1,3-diol into its Enantiomer . . . 358
 17.3.1 Degradation Route . . . 359
 17.3.2 Racemisation Route . . . 359
 17.3.3 Direct Conversion: the Present Zambon Industrial Process . . . 359
17.4. References . . . 362

17.1 INTRODUCTION

17.1.1 THE DRUG

Thiamphenicol[1] is the international non-proprietary name for (1*R*,2*R*)-2-(dichloroacetamido)-1-(4-methylsulphonyl)phenylpropane-1,3-diol. The compound was described for the first time by Sterling Drug[2] in 1951, and is used as an antibiotic for both human and animal health. Thiamphenicol belongs to the family of the so-called 'amphenicols, which includes chloramphenicol[1b] and florfenicol.[1b] It was originally synthesised during a research programme aimed at the identification of compounds with a spectrum of activity similar to that of chloramphenicol, but with a better safety profile.

The configurational requirements for the antibacterial activity of the family are very specific, as shown by the fact that the D-*threo* configuration is essential. Thus, thiamphenicol possesses a fairly wide spectrum of antimicrobial activity against Gram-negative bacteria, whereas its enantiomer is not active. All compounds in this family have the same mechanism of antibacterial activity, which

Chirality in Industry II. Edited by A. N. Collins, G. N. Sheldrake and J. Crosby

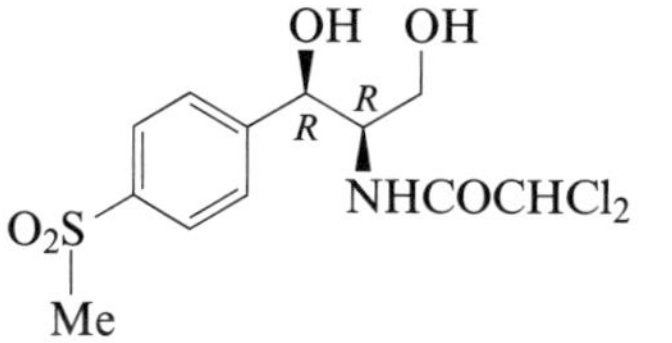

Thiamphenicol

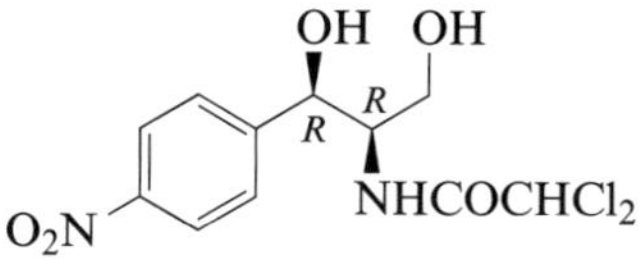

Chloramphenicol

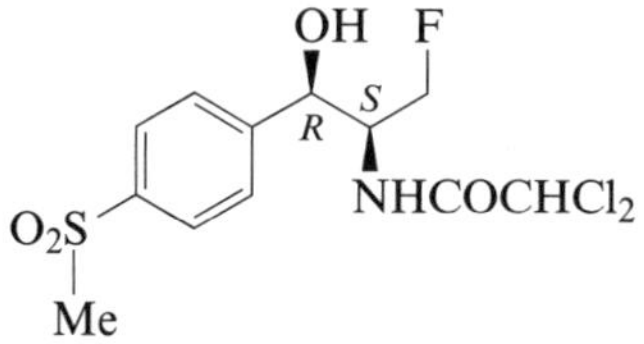

Florfenicol

involves the inhibition of peptidyl transferase following the interaction of the compounds with ribosomal sub-units of sensitive bacteria.

17.1.2 THE MARKET

Thiamphenicol and its derivatives[3] (thiamphenicol glycinate acetylcysteinate, thiamphenicol glycinate hydrochloride, thiamphenicol glycinate 4-hydroxy-isophthalate) are marketed for both human and veterinary applications. The development of the veterinary application of thiamphenicol was undertaken in the early 1960s and the marketing of a different formulation began in Europe and Japan in the second half of the same decade.

The recent prohibition of chloramphenicol from major markets in the treatment of food-producing animals and the positive evaluation of the safety features of thiamphenicol (establishment of MRL according to the EEC Regulation 2377/90) have opened up new opportunities for the product. As a result of the growth of this veterinary application, the worldwide market for thiamphenicol is increasing and is currently estimated to be 100 Mtonne per/year. At present, bulk thiamphenicol is produced by the Zambon Group in Italy and by local producers in China. The average price of the bulk drug is about US$ 150 kg^{-1}.

17.2 MAIN SYNTHETIC APPROACHES

Owing to the presence of two non-equivalent stereogenic carbon centres, (1*R*,2*R*)-thiamphenicol is one of the four possible stereoisomers of 2-(dichloroacetamido)-1-[(4-methylsulphonyl)phenyl]propane-1,3-diol. Therefore, the target of any synthetic approach to thiamphenicol must be effective control of the relative (*threo* vs *erythro*) and absolute (1*R*,2*R* vs 1*S*,2*S*) configurations. In order to satisfy the above stereochemical requirements, all the practical synthetic routes to thiamphenicol follow a similar strategy consisting of the following steps:

(i) preparation of racemic *threo* intermediates;
(ii) classical resolution;
(iii) preparation of thiamphenicol from the 'correct' *threo* resolved intermediate.

To the best of our knowledge, no asymmetric synthesis has yet been found to be commercially viable on a manufacturing scale.

The only two synthetic methods applying the above strategy [(i)–(iii)] which have been shown to be useful for the production of commercial quantities of thiamphenicol are described in Schemes 1 and 2. In both processes, the relative *threo* configuration is determined in the first step, in which 4-methylmercaptobenzaldehyde (Process 1, Scheme 1) or 4-methylsulphonylbenzaldehyde (Process 2, Scheme 2) is condensed with glycine to give a *threo* racemic arylserine, virtually free from the *erythro* epimer. The optical resolution is performed at different stages and with different techniques:[4] entrainment resolution in Process 1 and diastereomeric salt formation in Process 2.

17.2.1 PROCESS 1 (FROM 4-METHYLMERCAPTOBENZALDEHYDE)

Zambon has produced thiamphenicol for several decades using Process 1. Here, 2 mol of 4-methylmercaptobenzaldehyde are condensed with 1 mol of glycine in an aqueous alcoholic medium under basic conditions to provide a heterogeneous mixture of *threo* and *erythro* Schiff's bases **7** and **8**. Prolonged stirring of the reaction mixture allows the equilibration between these compounds, causing the almost complete conversion of **8** into **7**. The *threo* Schiff base obtained (**7**) is hydrolysed with hydrochloric acid to give, in 80% overall yield, *threo*-4-methylthiophenylserine (**1**).

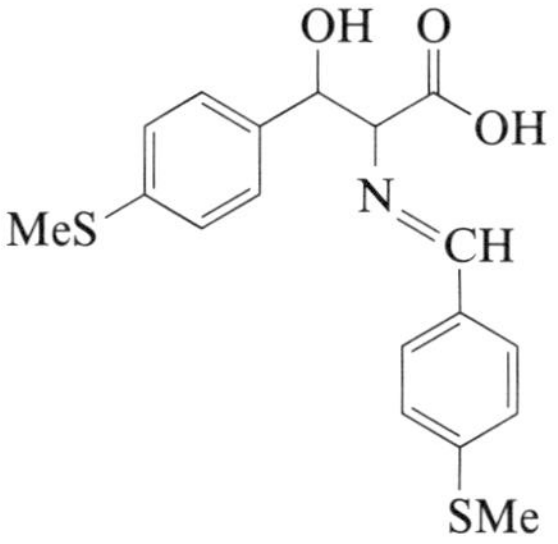

(**7**) *threo* (2*S*, 3*R*) (2*R*, 3*S*)
(**8**) *erythro* (2*R*, 3*R*) (2*S*, 3*S*)

Process 1

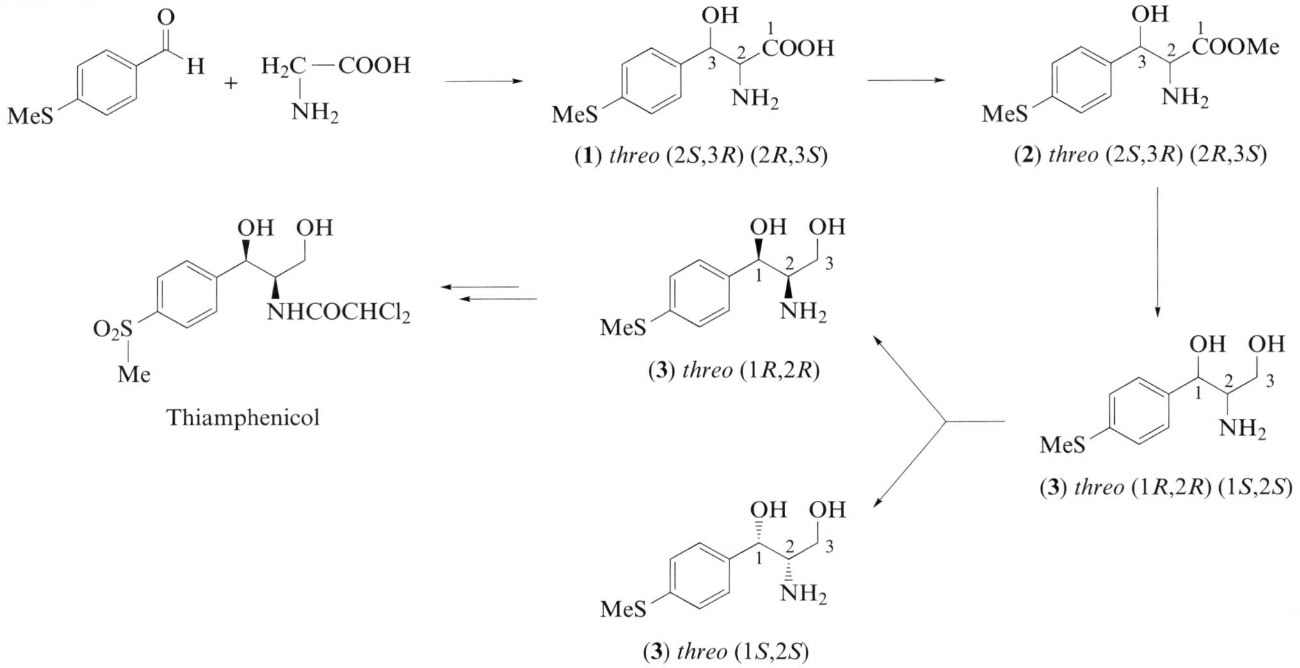

Scheme 1

Process 2

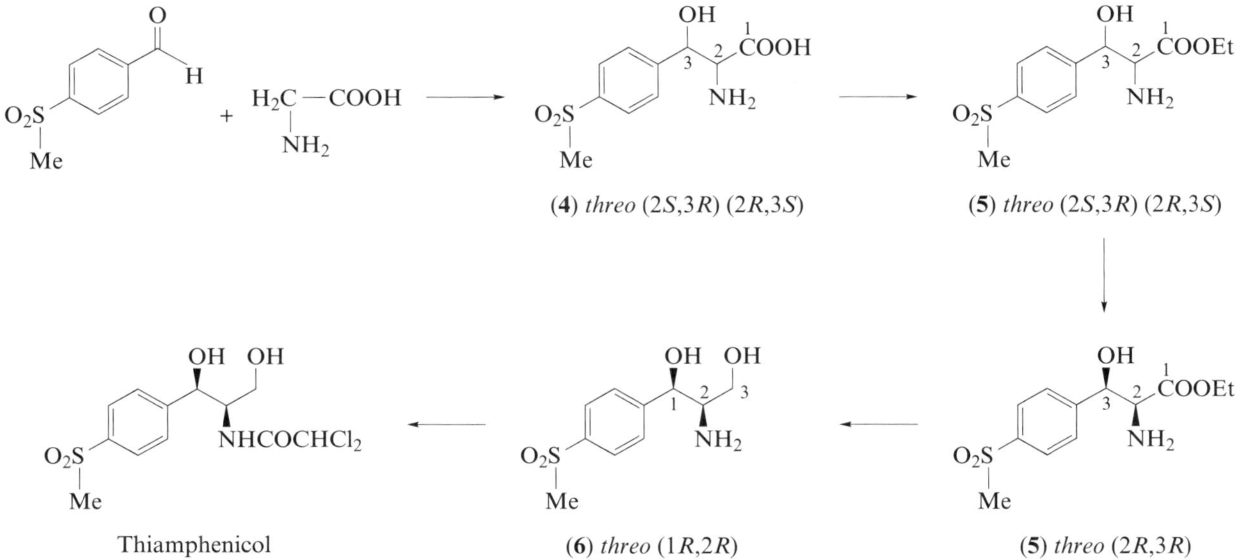

Scheme 2

threo-4-Methylmercaptophenylserine (**1**) is transformed in almost quantitative yield into ester **2** by treatment with thionyl chloride in methanol. The methyl ester so obtained is subsequently reduced to the corresponding *threo*-2-amino-1-[(4-methylthio)pheny]propane-1,3-diol (**3**) by sodium borohydride in the presence of calcium chloride in about 90% yield. The entrainment resolution of the *threo* racemate (**3**),[5] carried out in aqueous acid, provides both enantiomers in enantiomerically pure form. The main features which make this entrainment resolution particularly suitable for industrial application are as follows:

—mild conditions (temperature range 25–50 °C);
—almost quantitative yield (high stability of products under operating conditions);
—high volume productivity;
—an exceedingly high number of cycles (>70);
—the enantiomers are collected as wet free base, which can be used directly in the next step.

The main drawback of this process is that the resolution is perfomed at a late stage and co-produces the 'incorrect' enantiomer, (1*S*,2*S*)-thiomicamine (**3**), whose chemical structure does not allow a straightforward recycle (see Section 17.3). The wet (1*R*,2*R*) enantiomer is converted into thiamphenicol by *N*-acylation followed by sulphur oxidation with hydrogen peroxide in the presence of sodium tungstate.

17.2.2 PROCESS 2 (FROM 4-METHYLSULPHONYLBENZALDEHYDE)

We believe that Process 2 is used by Chinese producers, but only limited information is available. The racemic *threo*-4-methylsulphonylphenylserine (**4**) is converted into its ethyl ester,[5] which is resolved in alcoholic solution in high yield by treatment with the natural (L) isomer of tartaric acid.[6] The sequence to thiamphenicol is completed by converting the enantiomerically pure (2*S*,3*R*)-4-methylsulphonylphenylserine ethyl ester (**5**) into the corresponding aminopropanediol (1*R*,2*R*)-**6** followed by *N*-acylation.

Alternative methods have been described for the resolution of either arylserine[7,8] or aminopropanediol derivatives.[9–12] However, as far as we know, none of these has found industrial application.

17.3 CONVERSION OF (1*S*,2*S*)-2-AMINO-1-[4-METHYLTHIO)PHENYL]PROPANE-1,3-DIOL INTO ITS ENANTIOMER

Historically, Zambon has produced more than 1300 tonnes of thiamphenicol using Process 1 which, in turn, provided approximately 1500 tonnes of the by-product (1*S*,2*S*)-thiomicamine (**3**) (Scheme 1) of high chemical and enantiomeric purity. In the course of efforts to establish how to transform (1*S*,2*S*)-thiomicamine into a source of the desired (1*R*,2*R*)-isomer, three possible strategies were investigated:

—degradation of the molecule to a useful intermediate;
—racemisation (which ultimately requires a resolution);
—direct conversion.

17.3.1 DEGRADATION ROUTE

Compound (1*S*,2*S*)-**3** is transformed into 4-methylthiobenzaldehyde by treatment with an oxidising agent (e.g. sodium hypochlorite) followed by acid hydrolysis.[13]

17.3.2 RACEMISATION ROUTE

Zambon's strategy for racemisation does not involve degradation of the carbon skeleton.[14,15] The preferred approach[15] is shown in Scheme 3, and involves the following steps:

(i) protection of both the amino group and the primary hydroxy group as a 1,3-oxazoline followed by oxidative removal of the benzylic stereogenic centre;
(ii) spontaneous and rapid racemisation at the adjacent stereogenic carbon centre due to the acidity of the hydrogen α to the carbonyl group;
(iii) regeneration of the benzylic stereogenic centre by diastereoselective reduction of the carbonyl group under conditions which favour the *threo* form of the resulting alcohol;
(iv) hydrolytic removal of the protecting group to afford the *threo*-aminodiol.

Using the last strategy, a synthetic procedure has been developed which allows racemisation in four steps with a 50% overall yield.

17.3.3 DIRECT CONVERSION: THE PRESENT ZAMBON INDUSTRIAL PROCESS

The only direct conversion of (1*S*,2*S*)-thiomicamine (**3**) into the (1*R*,2*R*)-enantiomer is, to the best of our knowledge, that investigated and developed to an industrial scale by Zambon, as described below.

An efficient conversion of (1*S*,2*S*)-**3** into thiamphenicol is effected by sequential inversion of configuration at C-2 and C-1, in which each stereogenic centre controls the other and maintains its stereochemical integrity. The process[16] has five stages, starting from (1*S*,2*S*)-**3** (Scheme 4).

Step 1: condensation of (1*S*,2*S*)-**3** with acetone followed by *N*-acetylation provides *trans*-*N*-acetyl-1,3-oxazolidine (**9**) (80% yield).

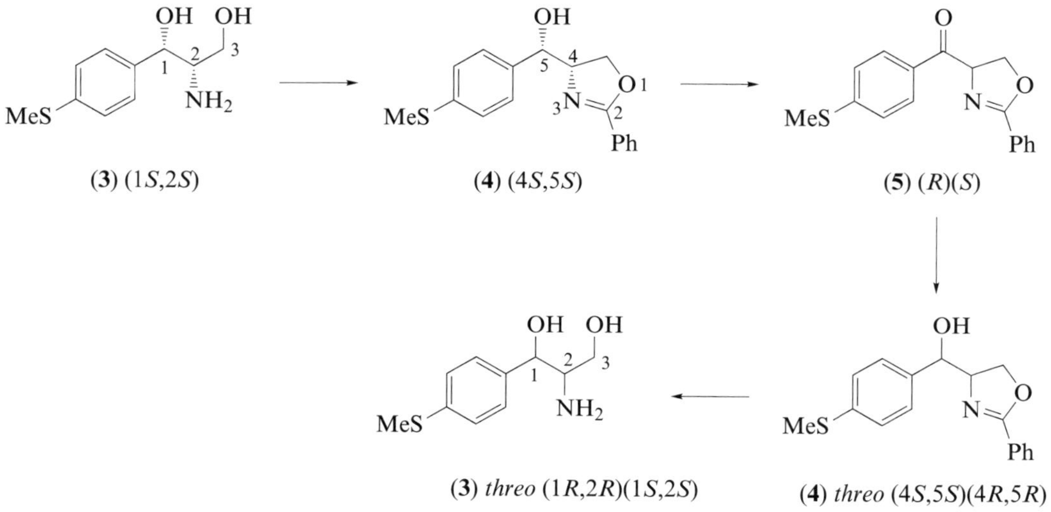

(**3**) (1*S*,2*S*)

(**4**) (4*S*,5*S*)

(**5**) (*R*)(*S*)

(**4**) *threo* (4*S*,5*S*)(4*R*,5*R*)

(**3**) *threo* (1*R*,2*R*)(1*S*,2*S*)

Scheme 3

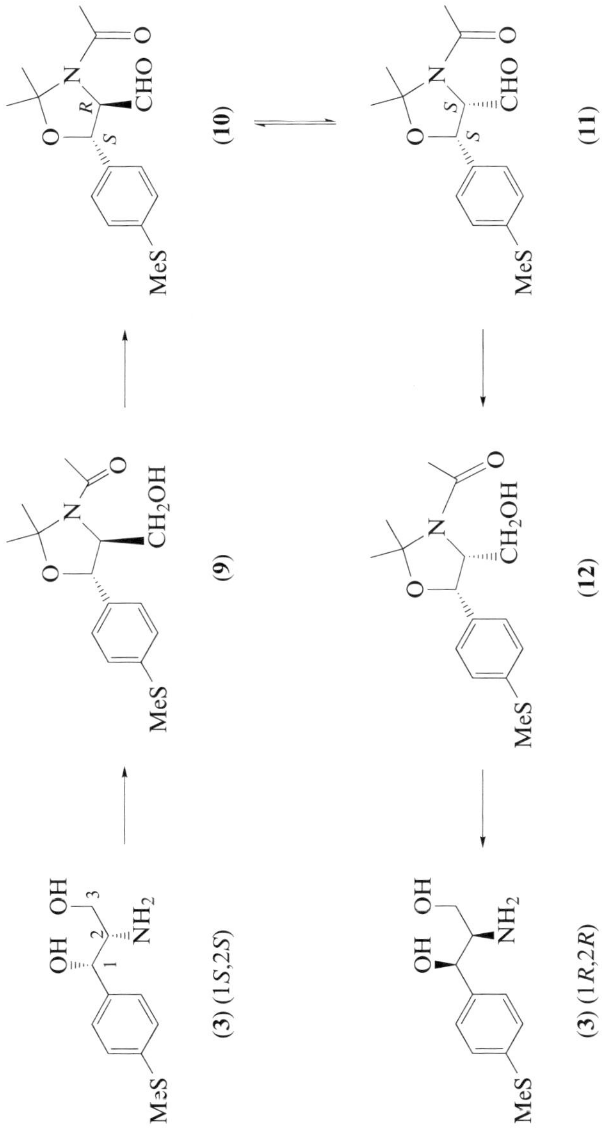

Scheme 4

Step 2: oxidation of the primary alcoholic function at room temperature to aldehyde **10** in 90% yield using DMSO and a Lewis acid.

Step 3: heating neat aldehyde **10** at 35 °C with a catalytic amount of DABCO, causing epimerisation to **11**, which crystallises from the medium, thus allowing virtually complete conversion of **10** into **11**.

Step 4: reduction of aldehyde epimer **11** providing oxazolidine alcohol **12** in almost quantitative yield.

Step 5: *O*-acetylation of alcohol **12**, followed by treatment with methanesulphonic acid and acetic anhydride, providing, after alkaline hydrolysis, crude (1*R*,2*R*)-**3**. Crystallisation then produces pure (1*R*,2*R*)-**3** in 86% yield based on aldehyde **11**.

The above chemical sequence is the result of 4 years of research; two years were needed to complete the scale-up of the process, which became an industrial reality in 1990.

The industrial process, which converts (1*S*,2*S*)-**3** into (1*R*,2*R*)-**3** in 55% yield, does not require any special apparatus or unusual operating conditions, and is carried out in standard stainless-steel and glass-lined reactors. The productivity of each stage is very high; stage 3 in particular is operated with a volumetric productivity of more than 90%, which allows the production of more than 1000 kg of aldehyde **11** in a 2000 l reactor.

This process has allowed Zambon to gain a competitive advantage in the thiampenicol market.

17.4 REFERENCES

1. (a) *The Merck Index*, XI Edn, No. 9230, p. 1445; (b) Elks, J., and Ganellin, C. R., *Dictionary of Drugs*, Chapman and Hall, London, 1990.
2. Sterling Drug, *Br. Pat.*, 745 900, 1951.
3. Kleeman, A., and Engel, J., *Pharmazeutische Wirkstoffe*, Georg Thieme Stuttgart, Vol. 5, 1982.
4. Jacques, J., Collet, A., and Wilen, S. H., *Enantiomers, Racemates and Resolutions*, Wiley, New York, 1981.
5. Parke, Davis and Co., *US Pat.*, 2 767 213, 1952.
6. Sumitomo Chemical, *Ger. Pat.*, 2 938 513, 1969.
7. Zambon Group, *Ital. Pat.*, 1 196 434, 1986.
8. Celgene, *Eur. Pat. Appl.*, EP 507 153, 1991.
9. E Gy.T. Gyogyszervegyeszeti Gyar, *Hung. Pat.* T3611, 1970; *Chem. Abstr.*, **77**, 5165g 1972.
10. Rebstock, M. C., and Bambas, L. L., *J. Am. Chem. Soc.*, **77**, 186 (1955).
11. Du Pont, *US Pat.*, 2 742 500, 1953.
12. Boehringer Mannheim, *Eur. Pat.*, EP 224 902, 1985.
13. Zambon Group, *Ital. Pat.*, 1 227 166, 1988.
14. Zambon Group, *Ital. Pat.*, 1 186 716, 1985.
15. Zambon Group, (a) *Ital. Pat.*, 1 223 563 1987; (b) Giordano, C., Cavicchioli, S., Levi, S., and Villa, M., *Tetrahedron Lett.*, **29**, 5561 (1988).
16. Zambon Group, (a) *Eur. Pat.*, 423 705, 1989; (b) Giordano, C., Cavicchioli, S., Levi, S., and Villa, M., *J. Org. Chem.*, **56**, 114 (1991).

18 Sharpless Asymmetric Epoxidation: Scale-up and Industrial Production

W. P. SHUM and M. J. CANNARSA
ARCO Chemical Company, Newtown Square, PA, USA

18.1 Introduction . . . 363
18.2 Early Development . . . 366
18.3 The Glycidol Process . . . 366
18.4 Process Chemistry . . . 368
18.5 Epoxy Alcohol Recovery Methods . . . 369
18.6 Reduction of the Organic Hydroperoxide . . . 371
18.7 Determination of the Optical Purities of Optically Active Epoxy Alcohols and Their Derivatives . . . 371
18.8 Optical Purification of Optically Active Glycidol Derivatives . . . 373
18.9 Synthetic Applications of Optically Active Epoxy Alcohols . . . 374
18.10 Synthetic Applications of Aqueous Optically Active Glycidol . . . 376
18.11 Conclusion . . . 377
18.12 Acknowledgements . . . 377
18.13 Notes and References . . . 377

18.1 INTRODUCTION

ARCO Chemical Company specialises in the development and commercialisation of olefin epoxidation using alkyl hydroperoxides, a technology which forms the basis of an international business in the manufacture and use of propylene oxide and its derivatives. The company constantly seeks additional applications of its expertise, and an excellent example of this is in the scale-up of the Sharpless asymmetric epoxidation (SAE) route to optically active epoxy alcohols (Figure 1).

Early attempts to prepare enantiopure epoxide intermediates started with optically active compounds available from nature. In this chiral pool approach, D-mannitol has been used for the preparation of optically active glycidol, isopropylideneglycerol and epichlorohydrin. The synthesis of (*S*)-isopropylideneglycerol from D-mannitol was reported as early as 1939[1] and, in the late 1970s, researchers at Merck disclosed in a patent[2] and papers[3,4] the synthesis and utilisation of optically active epichlorohydrin. They synthesised (*R*)- and (*S*)-epichlorohydrin starting with D-mannitol, in 6–8 steps and in 30% overall yield with an *ee* of >97%. The authors described several important chiral intermediates made accessible from a chiral C_3 synthon, and pointed out the general utility

Chirality in Industry II. Edited by A. N. Collins, G. N. Sheldrake and J. Crosby

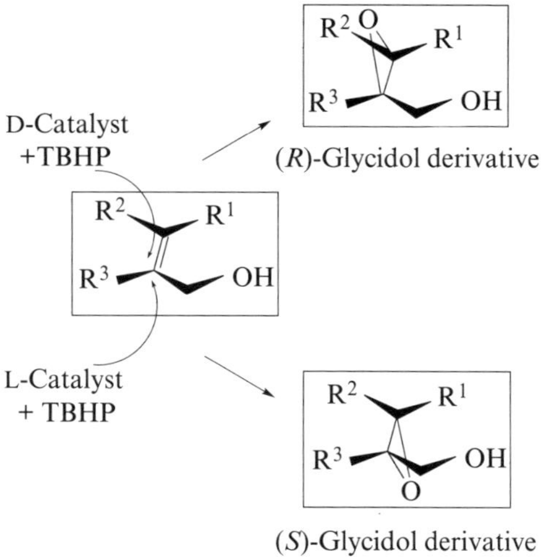

Figure 1 Sharpless epoxidation chemistry

of the C_3 fragment in asymmetric synthesis. In the mid-1980s, important work was also carried out by Whitesides at Harvard University in developing the enzymatic resolution of glycidyl butyrate.[5–6] Porcine pancreatic lipase selectively hydrolysed the (*S*)-ester leaving the unreacted (*R*)-ester with 92% *ee* at 45% isolated yield. There are several other synthetic routes to chiral C_3 synthons based on the use of lipase enzymes and microorganisms.[7–11]

In the early 1980s, we became aware of the work of Sharpless, then at Stanford University, on the asymmetric epoxidation of allylic alcohols using alkyl hydroperoxides and titanium tartrate catalysts.[12] During the same period, single-enantiomer drug development began to generate interest, and we recognised this as an opportunity to extend our experience in epoxidation to the pharmaceutical market. Progress in asymmetric synthesis in academic and industrial laboratories, and also advances in enzymatic resolution, made single-enantiomer drug synthesis economically viable. However, a critical missing link in the utilisation of asymmetric synthesis was the ability to take new methodologies through the appropriate scale-up and engineering studies to develop commercial processes.

Discussions began between ARCO Chemical, Stanford University and Professor Sharpless in 1985, concerning possible licensing and commercial development of the SAE technology. While these discussions were taking place, a better understanding of the potential value of this technology was needed. In the absence of 'in-house' expertise of the pharmaceutical industry, several consultant studies were commissioned to provide an overview of potential markets. The results of these studies were sufficiently encouraging to motivate ARCO Chemical to enter a licensing agreement with Stanford. One attractive aspect of the technology is its broad applicability to a wide range of allylic alcohols. It was recognised that the three-carbon chiral epoxy alcohol building block would be a valuable intermediate for the production of a variety of pharmaceutical compounds (Figure 2).

This chapter does not review the literature pertaining to the discovery or mechanistic studies of the SAE, as these aspects have been covered previously in many excellent reviews, including several by Sharpless and co-workers.[13–14] Here, an overview is provided of the work required to commercialise the process, and

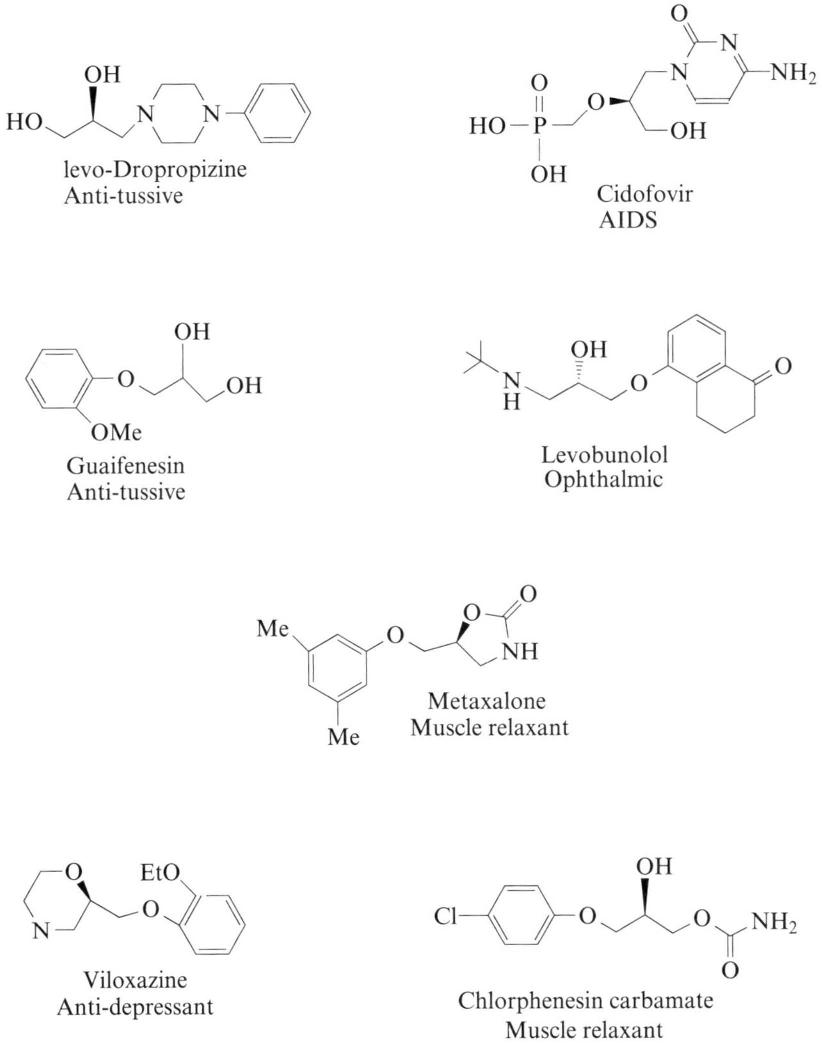

Figure 2 Pharmaceutical compounds accessible from optically active glycidol

also some specific product development work. Also described is some important auxiliary work that was necessary, including the development of safety systems and analytical characterisation. Several important applications of the SAE in the pharmaceutical industry are covered in the final section.

18.2 EARLY DEVELOPMENT

The first paper on SAE, published in 1980,[12] described a process that utilised a stoichiometric quantity of titanium isopropoxide–diethyl tartrate in epoxidations using *tert*-butyl hydroperoxide. It is clear that it would be difficult, if not impossible, to practise this technology on a commercial scale owing to the large quantities of reagents required. For example, based on the original Sharpless papers, production of 100 kg of (*R*)-glycidol would have required 380 kg of titanium isopropoxide and 275 kg of D-(–)-diethyl tartrate. The cost of reagents, difficult product isolation and waste disposal made this technology unworkable.

While the license negotiations were taking place, Hanson and Sharpless published and patented an improved epoxidation process utilising molecular sieves.[15] This process allowed the titanium reagent and the tartrate ligand to be reduced to catalytic amounts. The same 100 kg batch of (*R*)-glycidol could now be prepared using only 20 kg of titanium isopropoxide and 15 kg of D-(–)-diisopropyl tartrate with the addition of 40 kg of relatively inexpensive and easily recycled molecular sieves. This improvement was the first in a series that led, ultimately, to a commercial process. Despite these improvements, the isolation of the water-soluble, unsubstituted (*R*)- and (*S*)-glycidols remained a significant challenge. These particular compounds had been identified from market research as highly desirable chiral starting materials. Several commercial products, including β-adrenergic blocking agents and anti-tussives had already been identified that could use (*R*)-glycidol in their synthesis. Therefore, although we knew that the water-soluble products would be the most difficult to process, especially in the presence of titanium alkoxides, these compounds were chosen for development work.

ARCO Chemical began work to commercialise the Sharpless laboratory-scale process in 1986. Several important process considerations are described below, including reagent concentrations and purity, reaction conditions and product isolation. In addition, critical safety considerations are discussed, including the handling of alkyl hydroperoxides on a large scale, the reduction of residual hydroperoxide and the isolation of a variety of reactive epoxy alcohol products.

18.3 THE GLYCIDOL PROCESS

Optically active glycidol was the first of several epoxy alcohols manufactured by ARCO Chemical in commercial quantities. The generalised process flow diagram for the epoxidation of allyl alcohol to provide optically active glycidol is shown

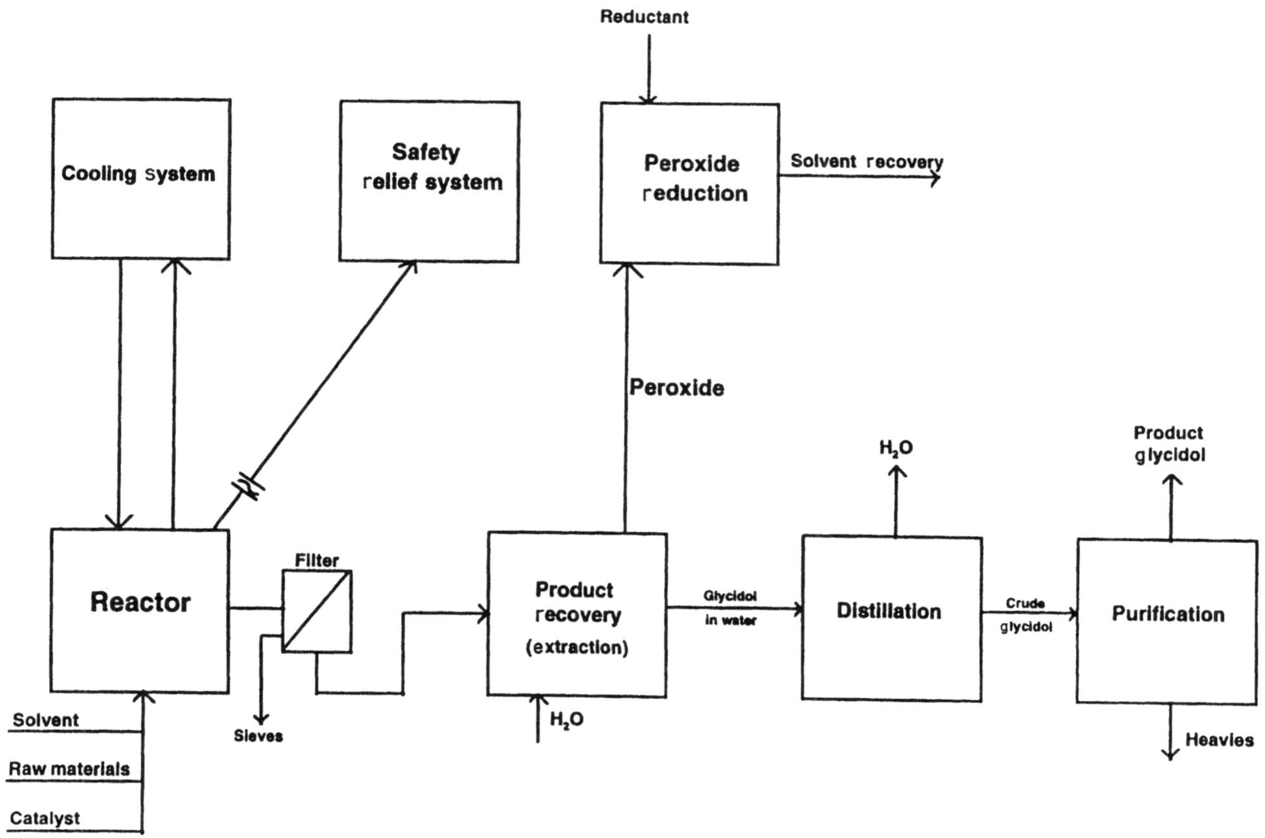

Figure 3 Glycidol process flow diagram

in Figure 3. Allyl alcohol, titanium tartrate catalyst, molecular sieves and solvent are present in the reactor, which is cooled to –20 °C. Cumene hydroperoxide is gradually added to the mixture, and the reaction exotherm is controlled by cooling. At the completion of the epoxidation, the molecular sieves are removed by filtration and the glycidol product is isolated by extraction into an aqueous phase. The deactivated catalyst, 2-phenylpropan-2-ol and residual cumene hydroperoxide remain in the organic solvent, which constitutes a waste stream which is treated with an appropriate reducing agent to remove the hydroperoxide before disposal. The glycidol product in the aqueous extract is concentrated by water removal and, finally, optically active glycidol of 88–90% *ee* is purified by vacuum distillation.

The isolation of optically active glycidol in high yield from the aqueous extract was one challenging aspect of the process development. Glycidol is prone to dimerisation and polymerisation at elevated temperatures, and these processes are sensitive to the pH of the medium.[16] The boiling point of glycidol at atmospheric pressure is reported to be 162 °C with decomposition[17], and so removal of water and high-boiling organic components requires distillation under carefully controlled conditions. In addition, trace alkali metal contamination can promote exothermic polymerisation of glycidol. Many safety systems were built into the distillation units to deal with these issues and, during process development, extensive accelerated rate calorimeter (ARC) evaluations on critical process streams were carried out to ensure safe plant operation.

18.4 PROCESS CHEMISTRY

In this section, catalyst chemistry related to the SAE process development is described. Although the chiral titanium tartrate catalyst formed *in situ* from titanium isopropoxide and the optically active dialkyltartrate ester provides predictable stereoselectivity in the asymmetric epoxidation of allylic alcohols, the optimal yield and enantiomeric purity of the epoxy alcohols are dependent on a number of process variables. In the SAE, a minimum amount of catalyst and solvent is required for the successful production of an epoxy alcohol of acceptable enantiomeric purity. The precipitous decrease in yield and enantiomeric purity of these epoxy alcohols observed in reactions with inadequate catalyst and solvent can be related to the proposed catalyst deactivation mechanism.[18] The Lewis acidic titanium(IV) metal centre is known to catalyse the nucleophilic addition of alcohol to an epoxy alcohol.[19] The diol made from this epoxide-opening reaction can act as a strong catalyst inhibitor. This detrimental effect on the epoxidation catalyst activity has been utilised as a safeguard against a run-away reaction in the scale-up of the SAE, in which propylene glycol is used as the catalyst inhibitor. In view of this chemistry, it is important to operate the batch SAE process under conditions that can suppress the epoxide-opening reactions of the epoxy alcohol product.

The dynamic titanium tartrate catalyst, with vacant sites available for ligand exchange, is also susceptible to hydrolytic degradation; sufficient molecular sieves must be used to adsorb trace water present in the solvent and the organic hydroperoxide. The enantioselectivity of the epoxidation is significantly reduced if trace water is present in the reaction medium. It is noteworthy that the catalyst used for the asymmetric oxidation of sulphides (see Chapter 19) is the SAE catalyst modified by the addition of water.[20] This suggests that the Sharpless titanium tartrate catalyst may undergo structural modification by water. Whether the reversible[21] hydrolytic degradation can be related to an equilibrium between a dimer and its monomer is open for debate in view of the conflicting proposals from Sharpless[22] and Corey.[23] From a practical standpoint, in the chiral glycidol process, which involves the SAE of the relatively inactive allyl alcohol, water must be rigorously eliminated from the reaction medium and cumene hydroperoxide must be used for optimal catalyst performance.

The effect of reactant purity on catalyst productivity and product purity is evaluated prior to production runs. It has been found that trace aldehyde impurities can have an adverse effect on catalyst performance. It is possible that aldehydes are oxidised to carboxylic acids which modify the titanium tartrate catalyst. This rationale is based on the reverse enantioselectivity observed for SAE using the slightly acidified phenethyl hydroperoxide.[24] Apart from the direct structural transformation of the catalyst, the carboxylic acid might catalyse the dehydration of alcohol to form water, which has already been demonstrated to have a detrimental effect on catalyst performance. In this context, the possibility of an acid-catalysed epoxide rearrangement to an aldehyde also has to be considered. Cascading side-reactions of this type would explain the continuous depletion of the epoxy alcohol product in the reaction mixture observed during storage, even at low temperatures.

18.5 EPOXY ALCOHOL RECOVERY METHODS

The inherent instability of the epoxy alcohol in the epoxidation reaction mixture imposes restrictions on the methodology used for product recovery in a production-scale process. The physical properties of the epoxy alcohol product and the organic hydroperoxide reactant must be considered and, as in many other homogeneous catalytic processes, the product recovery and purification steps require considerable process development. This is especially true because of the presence of unreacted organic hydroperoxide in the reaction effluent. The safe handling of the process streams containing hydroperoxide will be discussed in the following section. Product recovery methodologies that have been successfully commercialised can be conveniently categorised according to the physical properties of the epoxy alcohols.

Aqueous extraction works well for a number of water soluble epoxy alcohols such as glycidol (see Section 18.3) and 2- or 3-methylglycidol. In the glycidol case, where cumene hydroperoxide is used for the SAE, the use of this liquid–liquid extraction method can effectively remove the hydroperoxide and the allyl alcohol starting material from the product. The titanium tartrate catalyst is hydrolysed in the aqueous extraction process, and can also be effectively removed from the aqueous effluent containing the glycidol product. This method is suitable for epoxy alcohols such as the 2-substituted glycidols which are susceptible to decomposition at elevated temperatures in the presence of the titanium catalyst. Another advantage of this method is the possibility of direct utilisation of the aqueous glycidol effluent for the synthesis of a variety of synthetically useful glycidol derivatives.

Effective removal of the titanium tartrate catalyst and organic hydroperoxide from the product effluent prior to product isolation is necessary for the recovery of epoxy alcohols that are not water soluble. The titanium tartrate catalyst is hydrolysed, preferably by a caustic wash procedure under carefully controlled conditions to minimise epoxy alcohol loss. After phase separation, the organic hydroperoxide, typically *tert*-butyl hydroperoxide, present in the organic fraction with the epoxy alcohol product can be reduced by a variety of methods to *tert*-butyl alcohol, which can be easily removed by vacuum distillation. The isolated epoxy alcohol can be used without further purification, as in the case of *trans*-3-benzylglycidol for the synthesis of HIV protease inhibitor.[25] For aliphatic epoxy alcohols that are liquids, purification by fractional distillation can be carried out, as exemplified by the synthesis of *trans*-3-propylglycidol.

For solid epoxy alcohols of higher molecular weight, purification by recrystallisation can improve not only the chemical purity, but also the optical purity. This can significantly simplify the product isolation procedures, as illustrated in the synthesis of enantiopure *trans*-3-phenylglycidol.[26] In this process, the titanium tartrate catalyst was hydrolysed, and the residual *tert*-butyl hydroperoxide was removed by an azeotropic distillation method. The crude product that crystallised out from the process effluent was purified to >99% *ee* by recrystallisation.

There are a number of low-boiling epoxy alcohols that can be recovered by an azeotropic distillation approach. Glycidol is known to form azeotropic mixtures with a variety of solvents,[27] and it is possible to recover the glycidol made in the SAE by vacuum distillation with cumene as the azeotropic agent.[28] In one case, the unreacted cumene hydroperoxide was completely reduced prior to distillation. In another case, where cumene hydroperoxide was allowed to remain in the distillation mixture, the addition of excess propylene glycol was found to deactivate the titanium tartrate catalyst and no loss in the optical purity of glycidol was observed in the product recovery step. The azeotropic distillation methodology is an important alternative to the aqueous extraction procedure, especially for medium-chain aliphatic epoxy alcohols that have low solubility in water. The distillate containing the epoxy alcohols can be used without further purification for the synthesis of optically active epoxy alcohol derivatives, e.g. glycidyl arenesulphonates and trityl glycidol.

The recovery of epoxy alcohol would be greatly facilitated if a heterogeneous catalyst could be used for the SAE. A titanium-pillared montmorillonite has been used as a heterogeneous catalyst for the asymmetric epoxidation of allylic alcohols in the presence of an optically active tartrate ester.[29] In another approach, polymer-linked tartrate esters were used with titanium isopropoxide for the SAE of geraniol.[30] However, none of these heterogeneous catalysts gave satisfactory yields or optical purity, and research continues in this area.

18.6 REDUCTION OF THE ORGANIC HYDROPEROXIDE

The safe and complete removal of unreacted organic hydroperoxide is an important development issue in the scale-up of SAE. There are many methods for the effective removal of the hydroperoxide in the reaction effluent. The azeotropic distillation approach and the use of trialkyl phosphite or iron(II) sulphate heptahydrate as reducing agents have been described in the literature.[21] Selection of the best reducing agent for hydroperoxide removal varies with the epoxy alcohol process. In all cases, the reduction step has to be carefully integrated with the product recovery step in the process to ensure satisfactory isolated yield of the epoxy alcohol and compliance with waste disposal regulations. In an ideal situation, where the epoxy alcohol product can be separated from the unreacted hydroperoxide, the subsequent reduction step would be carried out with the more common reducing agents such as aqueous sodium hydrogen sulphite, which cannot be used satisfactorily in the presence of the epoxy alcohol product because of the high acidity of the reaction medium. The high acidity also catalyses dehydration of the alcohols to form olefins and high boiling point species on ageing. Some reduction processes also produce diperoxide (e.g. dicumyl peroxide, di-*tert*-butyl peroxide) by-products which must be taken into consideration.

During process development, the proportions of all of these side reactions had to be accurately determined to address the fate of all streams. In addition, the stability of the reaction effluents containing hydroperoxide at various stages of the process has also been adequately evaluated by a series of ARC tests to ensure safe process design for the epoxidation, product isolation, product purification and waste disposal steps.

18.7 DETERMINATION OF THE OPTICAL PURITIES OF OPTICALLY ACTIVE EPOXY ALCOHOLS AND THEIR DERIVATIVES

Both chemical and optical purity specifications must be addressed in the manufacture of optically active epoxy alcohols. Although the chemical purity of these compounds can be readily determined by methods such as GC, HPLC and NMR spectroscopy, the optical purity determination usually requires more elaborate method development on a case-by-case basis. In the early phase of our process research work, the traditional Mosher ester method[31] had been employed for the determination of *ee* values of optically active epoxy alcohols. This involves the derivatisation of the epoxy alcohols with the enantiopure α-methoxy-α-(trifluoromethyl)phenylacetyl chloride reagent (MTPA chloride) in the presence of triethylamine and 4-(dimethylamino)pyridine (DMAP) catalyst, followed by isolation of the diastereoisomeric Mosher esters of the epoxy alcohols by column chromatography or vacuum distillation. With care, this method is perfectly reliable for *ee* determination by the subsequent NMR analysis of the isolated Mosher esters. However, such a derivatisation method requires a synthetic procedure that is too time consuming to apply to the large number of product samples that must be analysed during process development work, so more convenient analytical methods were investigated

Capillary gas chromatography using liquid derivatives of α-, β- and γ-cyclodextrins as chiral stationary phases emerged as a powerful method for the separation of enantiomers in the late 1980s.[32,33] The application of these commercially available[34] cyclodextrin-based columns to the precise *ee* determination of many epoxy alcohols prepared by the SAE route was successfully accomplished.[35] Gas chromatographic separations of epoxy alcohols were carried out on a 30 m × 0.25 mm i.d. Chiraldex™ permethyl-*O*-hydroxypropyl-α-cyclodextrin A-PH column in the temperature range 40–80 °C. The three short-chain aliphatic epoxy alcohols, namely glycidol, 2-methylglycidol and *trans*-3-methylglycidol, can be resolved without derivatisation. Since these epoxy alcohols are water soluble, their enantiopurity during the course of the epoxidation can be readily followed by direct analysis of an aqueous extract of the reaction mixture. This A-PH column was found to be very resistant to hydrolysis, and it had been used extensively for the determination of enantiopurity of many aqueous glycidol samples in our process development work with no signs of column degradation. There have been several more recent reports[36–38] on the use of the α-cyclodextrin-based GC columns for the separation of glycidol enantiomers. The (*R*), (*S*) elution order is always the same on the α- and γ-cyclodextrin-based columns. Interestingly, a reversal of elution order has been observed on a commercially available Chiraldex™ B-TA column prepared from β-cyclodextrin.[38] This provides a means of avoiding the tailing problem encountered in the enantiopurity determination of highly enriched (*R*)-glycidol. However, the B-TA (trifluoroacetyl-derivatised) Chiraldex™ column is susceptible to hydrolytic degradation, so that regeneration with trifluoroacetic anhydride is frequently required for optimal separation of anhydrous glycidol. The long-chain aliphatic epoxy alcohols such as *trans*-3-ethylglycidol and *trans*-3-propylglycidol require derivatisation with trifluoroacetic anhydride for enantiomeric separation on the A-PH column.

For the determination of the enantiopurity of aromatic epoxy alcohols such as *trans*-3-phenylglycidol, 2-(2,4-difluoro)phenylglycidol and *trans*-3-benzylglycidol, the direct HPLC method is preferred. These compounds, without derivatisation, can be resolved on the commercially available[39] DAICEL™ chiral HPLC columns. This direct analytical method is well suited to the determination of both the enantiomeric purity and the diastereoisomeric purity of the epoxy alcohol products. For example, in the synthesis of *trans*-3-benzylglycidol by the SAE route, the reaction product typically contains 3–4% of *cis*-3-benzylglycidol that originates from the *cis*-4-phenylbut-2-en-1-ol impurity present in the *trans*-4-phenylbut-2-en-1-ol starting material. A Chiralpak® AD column has been used to resolve all four stereoisomers of 3-benzylglycidol in addition to the trace amount of *trans*- and *cis*-allylic alcohols and homoallylic alcohol typically present in the crude reaction product.[25] This chiral HPLC method allows a complete assay of the purity of the desired *trans*-3-benzylglycidol by a single analysis. In this case, *ee* values of 93% and 88% were obtained for *trans*- and *cis*-3-benzylglycidol, respectively. The slightly lower *ee* observed for the *cis*-epoxy alcohol is consistent with the general trends observed for the SAE in this class of *cis*-olefinic substrates.[13]

The chiral HPLC method is also very useful for the determination of the enantiopurity of many synthetically useful glycidol derivatives. We[40] and others[41–43] have published results on the enantiomeric separations of a variety of these derivatives including trityl glycidol and glycidyl arenesulphonates. The direct HPLC method is becoming the standard analytical method for *ee* determination because of the extra synthetic complexity involved in the traditional Mosher ester derivatisation method and the relatively small rotation values from polarimetric measurements. Furthermore, this convenient analytical method can provide the accuracy required for the construction of the melting point phase diagrams of these glycidol derivatives.

18.8 OPTICAL PURIFICATION OF OPTICALLY ACTIVE GLYCIDOL DERIVATIVES

The use of recrystallisation for the optical purification of chiral compounds is common practice in asymmetric synthesis, and it is possible to use this method to upgrade the enantiopurity of the epoxy alcohols obtained from SAE. A single recrystallisation from hexanes was claimed to be sufficient to improve the enantiopurity of *cis*-(2*S*,3*R*)-3-propylglycidol from 81% to 100% *ee*.[44] Another epoxy alcohol that can similarly be purified is *trans*-3-phenylglycidol, which was obtained in >99% *ee* after one recrystallisation.[21] Although the short-chain aliphatic epoxy alcohols cannot be enantiomerically enriched by this methodology, many of their synthetically useful derivatives are crystalline solids and therefore can be purified to higher enantiopurity by multiple recrystallisations.

We have investigated the optical purification of several important epoxy alcohol derivatives.[45] The melting point phase diagrams of these compounds were constructed to understand better their enantiomeric enrichment behaviours (Figure 4). Although the phase diagram of glycidyl 3-nitrobenzenesulphonate (GNBS) exhibits the distinctive V-shape of a conglomerate, optical purification of this compound above 97% *ee* by multiple recrystallisations is exceedingly diffi-

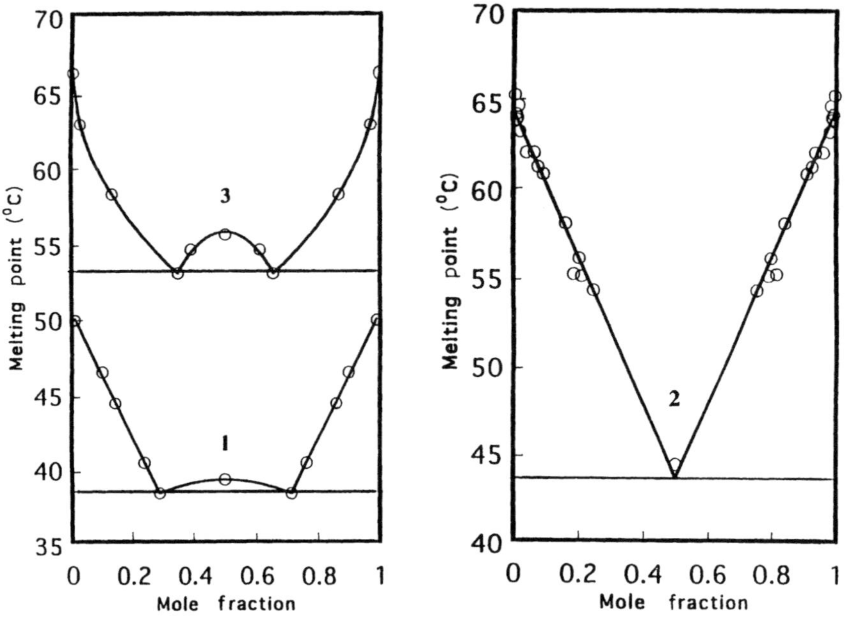

Figure 4 Phase diagrams of glycidyl tosylate (1), glycidyl 3-nitrobenzenesulphonate (2) and *trans*-3-methylglycidyl tosylate (3)

cult. Preparation of enantiopure GNBS is impeded by the existence of 'terminal solid solutions.[46] Recrystallisation of GNBS of low enantiomeric purity is further complicated by the formation of polymorphs, and preferential crystallisation of its seeded solutions did not afford an enantiopure product.[47] It is also difficult to obtain enrichment above 95% *ee* for glycidyl tosylate owing to solid solution formation.

In contrast to these two compounds, *trans*-3-methylglycidyl tosylate was found to be enriched to 99.3% from 89% *ee* by two recrystallisations from absolute ethanol. Although a perfect 100% *ee* cannot be achieved with most of these glycidyl arenesulphonates derived from SAE (since all appear to form solid solutions), many enantiopure pharmaceuticals have been successfully synthesised using the GNBS and glycidyl tosylate of 95% *ee* by recourse to further enrichment in the subsequent synthetic steps.

18.9 SYNTHETIC APPLICATIONS OF OPTICALLY ACTIVE EPOXY ALCOHOLS

Owing to their synthetically useful hydroxy and epoxy functionalities, optically active 2,3-epoxy alcohols prepared by the SAE route have found diverse applications as chiral building blocks in the syntheses of many pharmaceutical and agrochemical products.[48] Chiral glycidol is a particularly useful C_3 synthon that has been used for the syntheses of ß-adrenergic blocking agents[49] and related compounds,[50] alkyl glycerophospholipids,[51–53] acyclonucleotides,[54] monoamine oxidase inhibitors,[55] dropropizine,[56] guaifenesin,[57] and other pharmaceuticals. Applications of optically active 2,3-epoxy alcohols include the syntheses of HIV protease inhibitors,[58–59] AZT,[60] taxol side chain,[61] antifungal agents,[62–63] antibiotics,[64] vitamin D_3[65] and many other biologically active compounds. There are also reports on the use of these optically active epoxy alcohols for the syntheses of novel materials such as ferroelectric liquid crystals[66] and liquid crystalline polymers.[67] Since it is not possible to cover all aspects of these applications here, we have selected only two examples that serve to demonstrate the effectiveness of the Sharpless epoxidation route in the manufacture of chiral compounds.

The optically active aminoalkyl epoxide fragment (**4** in Figure 5) is an important intermediate for the syntheses of a variety of HIV protease inhibitors that are currently under active investigation.[58,68–69] This BOC-epoxide can be prepared very effectively by a multi-step process developed jointly by ARCO Chemical and SIPSY (Avrillé, France). Starting with a Wittig olefination reaction on phenylacetaldehyde (**1**), the desired (2*R*,3*R*)-*trans*-3-benzylglycidol (**3**) was obtained in three steps via the SAE on *trans*-4-phenylbut-2-en-1-ol (**2**).[25] This was followed by a series of three further steps to afford the enantiopure BOC-epoxide.[70] This

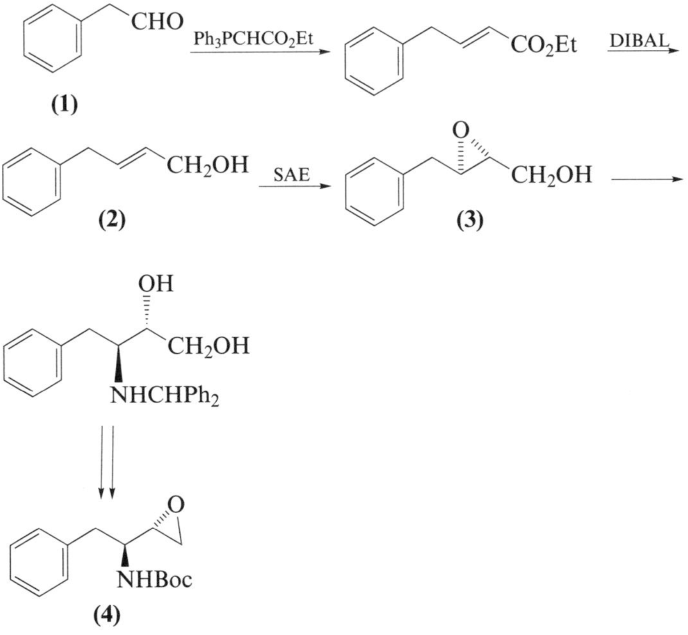

Figure 5 Synthesis of aminoalkyl epoxide intermediate **4**

process has been carried out on pilot-plant scale to produce tens of kilogrammes of the BOC-epoxide for clinical trials of the HIV protease inhibitor.

Glycidol, with its dual functional groups, can readily be derivatised to many pharmaceutically important compounds. Guaifenesin, an expectorant that has been extensively used in anti-tussive formulations in its racemic form, can be synthesised in high yield and high enantiopurity by the direct nucleophilic addition of 2-methoxyphenol to optically active glycidol in the presence of triethylamine catalyst.[57] This route is more effective than other approaches reported in the literature, which include the use of optically active 1-tosyloxypropane-2,3-diol acetonide as the key intermediate,[71] the Sharpless asymmetric dihydroxylation of 2-methoxyphenoxy allyl ether[72] and the lipase-catalysed sequential esterification of racemic guaifenesin.[73] With the availability of enantiopure guaifenesin, several other guaifenesin-related drugs that have been used previously in racemic forms can be readily synthesised with high enantiomeric purity.[74] The list includes moprolol, methocarbamol, mephenoxalone, guafecainol and ranolazine. Furthermore, our simple epoxide-opening reaction provides a convenient means to obtain many enantiopure 3-aryloxypropane-1,2-diols of pharmaceutical and agrochemical importance.[75–77]

In the above examples, the two epoxy alcohols obtained from SAE are not enantiomerically pure: the enantiopurity of *trans*-3-benzylglycidol is 93% *ee* and that of glycidol is 90% *ee*. Both are liquids, and cannot be further enriched enantiomerically by traditional crystallisation methodology. Nevertheless, because of the highly stereoselective nature of the subsequent reactions on the epoxides and the crystallinity of the desired products in their solid-state, the BOC-epoxide and guaifenesin are ultimately isolated in >99% *ee*. Many other biologically active compounds that are synthesised from chiral C_3 synthons or by enzyme catalysis can also be efficiently prepared in enantiopure forms using the optically active glycidol derived from the SAE.

18.10 SYNTHETIC APPLICATIONS OF AQUEOUS OPTICALLY ACTIVE GLYCIDOL

Chiral glycidol made from the SAE can be effectively recovered in high purity by aqueous extraction because of its excellent solubility in water. This aqueous glycidol solution has itself been used for the syntheses of several glycidol derivatives. For example, an aqueous glycidol solution of about 20% was treated with 3-nitrobenzenesulfonyl chloride in the presence of triethylamine to afford a good yield of glycidyl 3-nitrobenzenesulphonate of >96% *ee* after recrystallisation from absolute ethanol.[78] The use of a phase-transfer catalyst was not necessary for this reaction. Other optically active compounds, such as glycidyl tosylate, levodropropizine and guaifenesin can similarly be prepared in good yield and in high enantiomeric purity by this approach. The aqueous glycidol used for these syntheses has an *ee* of 88–90%, which is substantially higher than the 60–65% *ee* of the aqueous (*R*)-glycidol typically obtained from the enzymatic resolution route.[5] The SAE is, therefore, preferred for the synthesis of compounds such as glycidyl tosylate that cannot readily be enriched enantiomerically by recrystallisation.

Derivative synthesis using the aqueous glycidol stream is an improvement over the *in situ* derivatisation method described by Sharpless and co-workers.[21] In the Sharpless method, the residual catalyst and alcohol derived from the hydroperoxide, and also other impurities, prevent a convenient purification of the products by crystallisation. In many cases chromatography is required to reach a purity sufficient for crystallisation. Therefore, aqueous extraction and subsequent derivatisation as described here constitute the preferred route on an industrial scale.

18.11 CONCLUSION

The list of available technologies for producing single enantiomer intermediates continues to grow, and optically active glycidol and derivatives produced by the SAE will continue to play an important role in the synthesis of many new drugs.

The most efficient implementation of the chiral epoxidation technology can sometimes be realised by carrying out several synthetic steps in a reaction scheme and providing an advanced intermediate to the customer. In some cases, isolation of the optically active epoxy alcohol is not required, and an efficient synthesis can proceed with the crude epoxy alcohol in a subsequent step; this can also mitigate some of the safety issues discussed above. In addition, trends in the pharmaceutical industry towards focus on drug discovery and marketing have shifted manufacturing towards third parties for the supply of key intermediates. As the expertise of ARCO Chemical is in epoxidation, the company has licensed the developed epoxidation technology to SIPSY[79], a supplier of key intermediates to the pharmaceutical industry. This technical partnership will continue to pursue the development and implementation of the Sharpless asymmetric epoxidation (SAE) technology.

18.12 ACKNOWLEDGEMENTS

We thank the following for their contributions: Jon R. Valbert, Harry Mazurek, Jian Chen, Angela M. Hughes, Ram Ananth, Carl E. Johnson, Sam Onyekere and Glenn E. Coughenour.

18.13 REFERENCES AND NOTES

1. Baer, E., and Fischer, H. O. L., *J. Biol. Chem.*, **128**, 463 (1939); *Chem. Abstr.*, **33**, 7276, (1939).

2. *US Pat.*, 4 408 063, 1983.
3. Baldwin, J. J., Raab, A. W., Mensler, K., Arison, B. H., and McClure, D. E., *J. Org. Chem.*, **43**, 4876 (1978).
4. McClure, D. E., Arison, B. H., and Baldwin, J. J., *J. Am. Chem. Soc.*, **101**, 3666 (1979).
5. Ladner, W. E., and Whitesides, G. M., *J. Am. Chem. Soc.*, **106**, 7250 (1984).
6. *US Pat.*, 4 732 853, 1988.
7. Wang, Y.-F., Lalonde, J. J., Momongan, M., Bergbreiter, D. E., and Wong, C. H., *J. Am. Chem. Soc.*, **110**, 7200 (1988).
8. Vanttinen, E., and Kanerva, L., *J. Chem. Soc., Perkin Trans.*, *1*, 3459 (1994).
9. Suzuki, T., and Kasai, N., *Bioorg. Med. Chem. Lett.*, **1**, 343 (1991).
10. *US Pat.*, 5 246 843, 1993.
11. *Eur. Pat.*, EP464 905, 1992.
12. Katsuki, T., and Sharpless, K. B., *J. Am. Chem. Soc.*, **102**, 5974 (1980).
13. Johnson, R. A., and Sharpless, K. B., in *Catalytic Asymmetric Synthesis*, (ed. J. Ojima), VCH, New York, 1993, p. 103.
14. Finn, M. G., and Sharpless, K. B., in *Asymmetric Synthesis*, (ed. J. D. Morrison), Academic Press, Orlando, FL, 1985, Vol. 5, 247.
15. Hanson, R. M., and Sharpless, K. B., *J. Org. Chem.*, **51**, 1922 (1985).
16. Rider, T. H., and Hill, A. J., *J. Am. Chem. Soc.*, **52**, 1521 (1930).
17. De Gegerfelt, M. H., *Bull. Soc. Chim. Fr.*, **23**, 160 (1875).
18. Ref. 14, p. 287.
19. Caron, M., and Sharpless, K. B., *J. Org. Chem.*, **50**, 1557 (1985).
20. Kagan, H. B., in *Catalytic Asymmetric Synthesis*, (ed. J. Ojima), VCH, New York, 1993, p. 203.
21. Gao, Y., Hanson, R. M., Klunder, J. M., Ko, S. Y., Masamune, H., and Sharpless, K. B., *J. Am. Chem. Soc.*, **109**, 5765 (1987).
22. See ref. 13, and references cited therein.
23. Corey, E. J., *J. Org. Chem.*, **55**, 1693 (1990).
24. *US Pat.*, 4 764 628, 1988.
25. Shum, W. P., Chen, J., and Cannarsa, M. J., *Chirality*, **6**, 681 (1994).
26. *US Pat.*, 4 935 101 (1990).
27. Kleemann, A., and Wagner, R. M., *Glycidol: Properties, Reactions, Applications*, Verlag Chemie, Heidelberg, 1981, p. 11.
28. *US Pat.*, 5 288 882, 1994.
29. Choudary, B. M., Valli, V. L. K., and Prasad, D., *J. Chem. Soc.*, *Chem. Commun.*, 1186 (1990).
30. Farrall, M. J., Alexis, M., and Trecarten, M., *Nouv. J. Chim.*, **7**, 449 (1983).
31. Dale, J. A., Dull, D. L., and Mosher, H. S., *J. Org. Chem.*, **34**, 2543 (1969).
32. Konig, W. A., Lutz, S., Wenz, G., Gorgen, G., Neumann, C., Gabler, A., and Boland, W., *Angew. Chem.*, *Int. Ed. Engl.*, **28**, 178 (1989).
33. Li, W.-Y., Jin, H. L., and Armstrong, D. W., *J. Chromatogr.*, **509**, 303 (1990).
34. Available from Advanced Separation Technologies, Whippany, NJ, USA.
35. Dougherty, W., Liotta, F., Mondimore, D., and Shum, W., *Tetrahedron Lett.*, **31**, 4389 (1990).
36. Geerlof, A., van Tol, J. B. A., Jongejan, J. A., and Duine, J. A., *J. Chromatogr.*, **648**, 119 (1993).
37. Jin, Z., and Jin, H. L., *Chromatographia*, **38**, 22 (1994).
38. *A Guide to Using Cyclodextrin Bonded Phases for Gas Chromatography*, Advanced Separation Technologies, Whippany, NJ.
39. Available from Chiral Technologies, Exton, PA, USA.
40. Chen, J., and Shum, W., *Tetrahedron Lett.*, **34**, 7663 (1993).
41. Shaw, C. J., and Barton, D. L., *J. Pharm. Biomed. Anal.*, **793** (1991).

42. Kennedy, J. H., and Weigel, L. O., *Chirality*, **4**, 132 (1992).
43. Duchateau, A. L. L., Jacquemin, N. M. J., Straatman, H., and Noorduin, A. J., *J. Chromatogr.*, **637**, 29 (1993).
44. *Eur. Pat.*, EP583 591 1994.
45. Ananth, R., Chen, J., and Shum, W., *Tetrahedron: Asymmetry*, **6**, 317 (1995).
46. For discussions on terminal solid solutions, see Jacques, J., Collet, A., and Wilen, S. H., *Enantiomers, Racemates, and Resolutions*, Wiley, New York, 1981, pp. 126 and 424.
47. *Eur. Pat.*, EP441 471, 1991.
48. For a review on the synthetic applications of optically active glycidol and related 2,3-epoxy alcohols, see Hanson, R. M., *Chem. Rev.*, **91**, 437 (1991).
49. Klunder, J. M., Ko, S. Y., and Sharpless, K. B., *J. Org. Chem.*, **51**, 3710 (1986).
50. Barton, D. L., Press, J. B., Hajos, Z. G., and Sawyers, R. A., *Tetrahedron: Asymmetry*, **3**, 1189 (1992).
51. Burgos, C. E., Ayer, D. E., and Johnson, R. A., *J. Org. Chem.*, **52**, 4973 (1987).
52. Ali, S. and Bittman, R., *J. Org. Chem.*, **53**, 5547 (1988).
53. Thompson, D. H., Svendsen, C. B., Meglio, C. D., and Anderson, V. C., *J. Org. Chem.*, **59**, 2945 (1994).
54. Brodfuehrer, Howell, H. G., Sapino, C., Jr, and Vemishetti, P., *Tetrahedron Lett.*, **35**, 3243 (1994).
55. Gates, K. S., and Silverman, R. B., *J. Am. Chem. Soc.*, **112**, 9364 (1990).
56. Stamicarbon, *Eur. Pat.*, EP 0 349 066, 1990. Shum, W. P., and Chen, J., *Chirality*, **8**, 6 (1996).
57. Chen, J., and Shum, W., *Tetrahedron Lett.*, **36**, 2379 (1995).
58. Parkes, K. E. B., Bushnell, D. J., Crackett, P. H., Dunsdon, S. J., Freeman, A. C., Gunn, M. P. Hopkins, R. A., Lambert, R. W., Martin, J. A., Merrett, J. H., Redshaw, S., Spurden, W. C., and Thomas, G. J., *J. Org. Chem.*, **59**, 3656 (1994).
59. Askin, D., Eng, K. K., Rossen, K., Purick, R. M., Wells, K. M., Volante, R. P., and Reider, P. J., *Tetrahedron Lett.*, **35**, 673 (1994).
60. Jung, M. E., and Gardiner, J. M., *J. Org. Chem.*, **56**, 2614 (1991).
61. Bonini, C., and Righi, G., *J. Chem. Soc., Chem. Commun.*, 2767 (1994).
62. *Eur. Pat.*, EP 0 472 392, 1991.
63. *Eur. Pat.*, EP 0 539 938, 1992.
64. *US Pat.*, 5 352 832, 1994.
65. Okabe, M., and Sun, R.-C., *Tetrahedron Lett.*, **34**, 6533 (1993).
66. Walba, D. M., Vohra, R. T., Clark, N. A., Handschy, M. A., Xue, J., Parmar, D. S., Langerwall, S. T., and Skarp, K., *J. Am. Chem. Soc.*, **108**, 7424 (1986).
67. Taton, D., Le Borgne, A., Spassky, N., Friedrich, C., and Noel, C., *Polym. Adv. Technol.*, **5**, 203 (1994).
68. Kim, E. E., Baker, C. T., Dwyer, M. D., Murcko, M. A., Rao, B. G., Tung, R. D., and Navia, M. A., *J. Am. Chem. Soc.*, **117**, 1181 (1995).
69. Barrish, J. C., Gordon, E., Alam, M., Lin, P.-F., Bisacchi, G. S., Chen, P., Cheng, P. T. W., Fritz, A. W., Greytok, J. A., Hermsmeier, M. A., Humphreys, W. G., Lis, K. A., Marella, M. A., Merchant, Z., Mitt, T., Morrison, R. A., Obermeier, M. T., Pluscec, J., Skoog, M., Slusarchyk, W. A., Spergel, S. H., Stevenson, J. M., Sun, C., Sundeen, J. E., Taunk, P., Tino, J. A., Warrack, B. M., Colonno, R. J., and Zahler, R., *J. Med. Chem.*, **37**, 1758 (1994).
70. Castejon, P., Pasto, M., Moyano, A., Pericas, M. and Riera, A., *Tetrahedron Lett.*, **36**, 3019 (1995).
71. Nelson, W. L., Wennerstrom, J. E., and Sankar, S. R., *J. Org. Chem.*, **42**, 1006 (1977).
72. Wang, Z., Zhang, X., and Sharpless, K. B., *Tetrahedron Lett.*, **34**, 2267 (1993).
73. Theil, F., Weidner, J., Ballschuh, S., Kunath, A., and Schick, H., *J. Org. Chem.*, **59**, 388 (1994).

74. Shum, W., unpublished results.
75. Kleemann, A., Nygren, R., and Wagner, R., *Chem. Ztg.* **104**, 283 (1980).
76. Mazurek, H., in *Proceedings of International Conference on Pharmaceutical Ingredients and Intermediates*, Frankfurt, 1990.
77. Buser, H.-P., Spindler, F., *Tetrahedron: Asymmetry*, **4**, 2451 (1993).
78. *US Pat.*, 5252759,
79. *Chem. Mark. Rep.*, **247**, 24 April, 18 (1995).

19 Asymmetric Synthesis of Sulphoxides: Two Case Studies

P. PITCHEN
Rhône-Poulenc Rorer Inc., Collegeville, PA, USA

19.1 Introduction 381
19.2 Enantioselective Oxidation of Prochiral Sulphides 382
19.3 Synthesis of RP 73163 385
19.4 Synthesis of RP 52891 386
19.5 Conclusion 388
19.6 Acknowledgements 389
19.7 References 389

19.1 INTRODUCTION

The concept of molecular chirality is more than a century old,[1] but it is only in the last 1–2 decades that the asymmetric synthesis of biologically active compounds has really boomed, prompted by the need for more active and selective drugs and rendered possible by the emergence of new analytical and synthetic methods.[2] The object of this chapter is to illustrate this recent trend with two examples: the potassium channel opener RP 52891[3] and the ACAT inhibitor RP 73163,[4] two drugs discovered and studied at the Franco-American pharmaceutical company Rhône-Poulenc Rorer (Figure 1).

Both of these molecules are chiral sulphoxides and, in each case, the scenario was the same: optimisation of lead compounds first led to the preselection of products containing sulphide moieties; it was then discovered that these sulphides were metabolised into the corresponding sulphoxides; the racemic

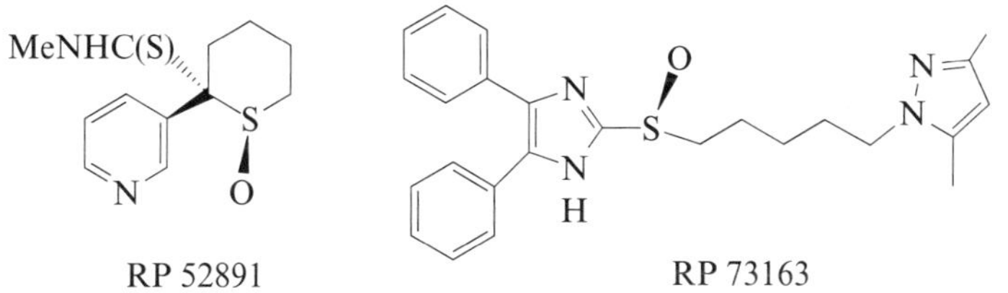

Figure 1 Structure of potassium channel opener RP52891 and ACAT inhibitor RP73163

Chirality in Industry II. Edited by A. N. Collins, G. N. Sheldrake and J. Crosby

sulphoxides were therefore synthesised, thus introducing a new asymmetric centre in the molecules, and resolved on a very small scale. In both cases, the enantiomeric sulphoxides were then subjected to biological testing, and the more active (or less toxic) enantiomer (the eutomer) was selected for further pharmacological and toxicological profiling. The possible development of these candidate compounds depended on the availability of asymmetric syntheses which could be rapidly and safely transferred into existing standard pilot plant equipment for the early manufacture of increasingly large quantities of enantiomerically pure (>99% *ee*) drug substance.

19.2 ENANTIOSELECTIVE OXIDATION OF PROCHIRAL SULPHIDES

The chiral pool strategy[2a] cannot really be applied to the syntheses of RP 52891 and RP 73163, as there are no readily available chiral sulphoxides which could be used directly as starting materials for the stereospecific syntheses of these compounds. On the other hand, there is an increasing number of chemical and biochemical methods for the enantioselective synthesis of chiral sulphoxides.[5] Most of these methods are, however, only suitable for laboratory use. One of the few possible candidates for scale-up is the asymmetric oxidation of prochiral sulphides using organic hydroperoxides in the presence of specific combinations of titanium isopropoxide and diethyl tartrate. This method relies on simple but key modifications to the well-known Sharpless epoxidation of allylic alcohols.

In its stoichiometric or catalytic versions, the Sharpless reaction is based on the use of soluble complexes made from equimolar amounts of diethyl tartrate (DET) and titanium isopropoxide under anhydrous conditions, at –20 °C, in an organic solvent such as dichloromethane or toluene.[6] The reaction has been successfully applied to the preparation of numerous enantiomerically pure epoxyalcohols, some of them to large-scale operations, as discussed in Chapter 18.

The Sharpless reagent, when applied to sulphides, leads to a mixture of the racemic sulphoxide and the corresponding sulphone. As discovered by 1983 by Kagan's group at Orsay,[7] the simple addition of 1 mol equiv. of water to the Sharpless reagent and/or the use of 2 (instead of 1) mol equiv. of DET per mole of titanium isopropoxide generates new soluble titanium complexes which show complementary chemical properties to those of the Sharpless reagent. Indeed, these new species are highly enantioselective for the oxidation of some suitably designed prochiral sulphides, but exhibit no enantioselectivity whatsoever in the epoxidation of allylic alcohols. Similar results were published by Modena and co-workers,[8] who used a larger excess of DET under anhydrous conditions. The first published procedures typically used *tert*-butyl hydroperoxide as oxygen donor and a stoichiometric amount of titanium diisopropoxide. It was then discovered that higher enantioselectivities could be achieved using cumene hydroperoxide, and that it was possible to reduce the amount of chiral complex to 50% (with respect to the sulphide) and even lower in some cases, with limited effect on the enantioselectivity of the reaction.[9] Figure 2 shows the results obtained in the case of the oxidation of methyl *p*-tolyl sulphide. The quoted enantiomeric excesses have been measured in our laboratories by chiral HPLC and might differ slightly from those given in the literature.

The structures of these new reagents have not yet been elucidated, but are clearly different from that of the Sharpless reagent. For instance, the infrared absorption dramatically changes when the DET:Ti ratio of an anhydrous mixture of DET and titanium isopropoxide in dichloromethane is increased from 1:1 (Sharpless reagent) to 2:1. The same phenomenon is observed on addition of 1mol equiv. of water to the Sharpless system.[7b] It must be noted that what is true for one sulphide is not necessarily true for all sulphides, so it is recommended that several combinations [$Ti(OPr^i)_4/DET/H_2O$, 1:2:1, 1:1:1, 1:2:0 or 1:>2:0, in stoichiometric or sub-stoichiometric quantity] are examined when considering the synthesis of a new chiral sulphoxide.

Differences between the sulphoxidation reaction and the Sharpless epoxidation arise in the preparation of the reagent and also from the fact that, unlike allylic alcohols,[10] sulphides cannot form ionic or hydrogen bonds with chiral titanium complexes. The chemoselectivity and stereoselectivity of the reaction are, consequently, mainly governed by steric effects and weak bond interactions. This explains the high substrate dependence of the reaction, a common feature of all biochemical and chemical methods used for the enantioselective oxidation of prochiral sulphides:[5] the more 'unsymmetrical' the prochiral sulphide, the higher is the enantioselectivity. For instance, most aryl methyl sulphides can now be oxidised in more than 90% *ee*, and some of them in more than 99% *ee*. This weak substrate–reagent interaction is also consistent with the fact that the Sharpless epoxidation can be seen as a more 'robust' process than the sulphoxidation reac-

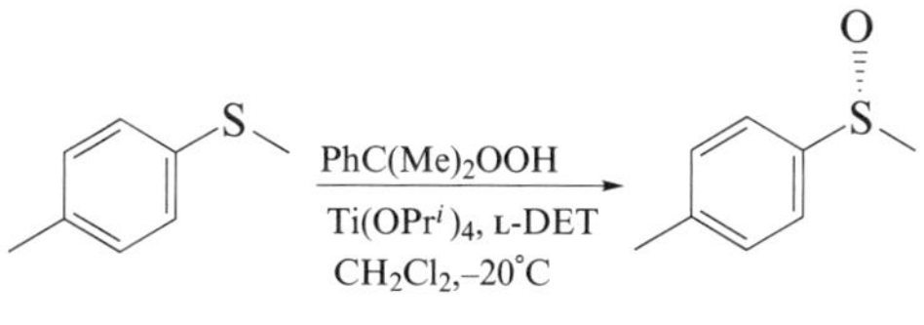

	Mol equiv.			
	$Ti(OPr^i)_4$	DET		*ee* (%)
Sharpless reagent	1	1	Anhydrous	0
Modified Sharpless reagents	1	1	+1 mol equiv. H_2O	96
	1	2	Anhydrous	89
	1	2	+1 mol equiv. H_2O	96

Figure 2 Asymmetric oxidation of methyl *p*-tolyl sulphide by Sharpless' reagent

tion, where minor changes of the reaction conditions (solvent, tartrate, hydroperoxide, etc.) often result in decreased reactivity and enantioselectivity. There is a very good correlation between the absolute configuration of tartrate and sulphoxide when the sulphide bears two substituents of very different size, especially in the case of aryl methyl sulphides: L-DET leads to (*R*)-sulphoxides[5,7b] (Figure 3).

While the Sharpless epoxidation is now used for the bulk manufacture of several epoxy alcohols (see Chapter 18), the above asymmetric sulphoxidation reaction had, to our knowledge, never been used on a large scale prior to this work. We anticipated the suitability of this reaction for our syntheses for the following reasons:

(i) it does not require any special equipment and so could be easily implemented in standard pilot plant;
(ii) the raw materials are the same as those used in the Sharpless reaction and are available in large quantities from many suppliers;
(iii) it is not covered by a patent.

Possible areas for improvement prior to further scale-up and commercialisation included:

(i) the use of dichloromethane as solvent;
(ii) the relatively low throughput (10% w/v was a typical maximum concentration);
(iii) the need to use at least 20–50% of chiral complex with respect to the sulphide;
(iv) the fact that the occasionally required D-DET isomer is approximately 10 times more expensive than the natural L-DET.

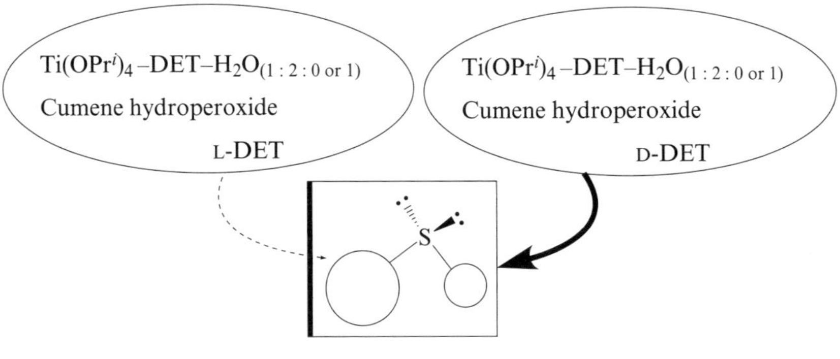

Figure 3 Substrate–reagent orientation with unsymmetrical sulphides

19.3 SYNTHESIS OF RP 73163

RP 73163 is a hypocholesterolaemic agent which acts by inhibiting the enzyme ACAT. The opposite enantiomer of RP 73163 is inactive.

This compound was first synthesised by oxidation of sulphide **1** using *m*-chloroperoxybenzoic acid, followed by resolution with semi-preparative high-performance liquid chromatography. Attempts to resolve the racemic sulphoxide using strong chiral acids failed. Indeed, owing to the presence of the sulphoxide group, the basicity of the imidazole ring is greatly reduced and the pyrazole ring is too remote from the chiral centre. The asymmetric oxidation of sulphide **1** using (+)-8,8-dichlorocamphorylsulphonyloxaziridine[11] in toluene at 40–50 °C afforded the desired enantiomer in 88–89.5% *ee*. However, this reaction could not be considered for the large-scale synthesis of RP 73163, mainly because of the low concentration required (2%) and the cost and availability of the chiral oxaziridine.[12] A first attempt to oxidise the sulphide **1** in dichloromethane at –20 °C using cumene hydroperoxide and the combination Ti(OPr^i)$_4$–DET–H_2O (1:2:1) gave the racemic sulphoxide[13] (Figure 4).

The sulphide **1** is not a good substrate for the titanium-catalysed asymmetric sulphoxidation, as it bears two large substituents, one of them possessing a very flexible five-carbon chain, hence the decision to study the asymmetric oxidation of 4,5-diphenylimidazol-2-yl methyl sulphides, which could be used as intermediates in the synthesis of RP 73163.[12,13] Thus, the sulphide **2** was prepared in a one-pot process from commercially available 4,5-diphenylimidazole-2-thiol, and oxidised using the same reaction conditions as for sulphide **1** (Figure 5). In order

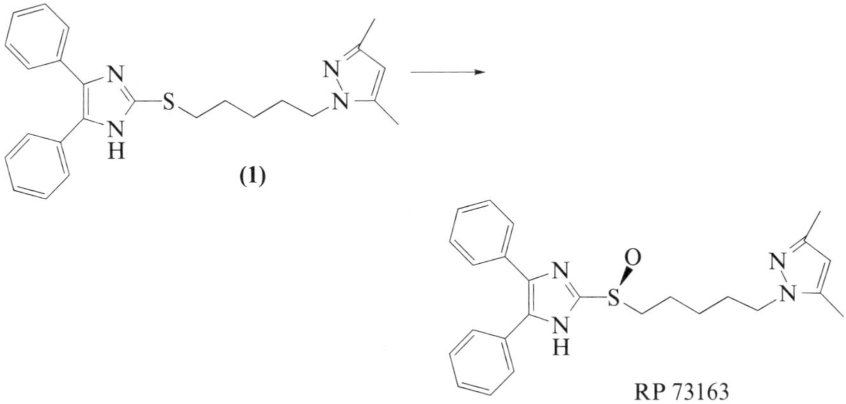

(1)

RP 73163

Ti(OPr^i)$_4$– DET– H_2O (1 : 2 : 1) 0% *ee*

(+)-8,8-Dichlorocamphorylsulphonyloxaziridine: 88–89.5% *ee*

Figure 4 Asymmetric epoxidation of sulphide **1**

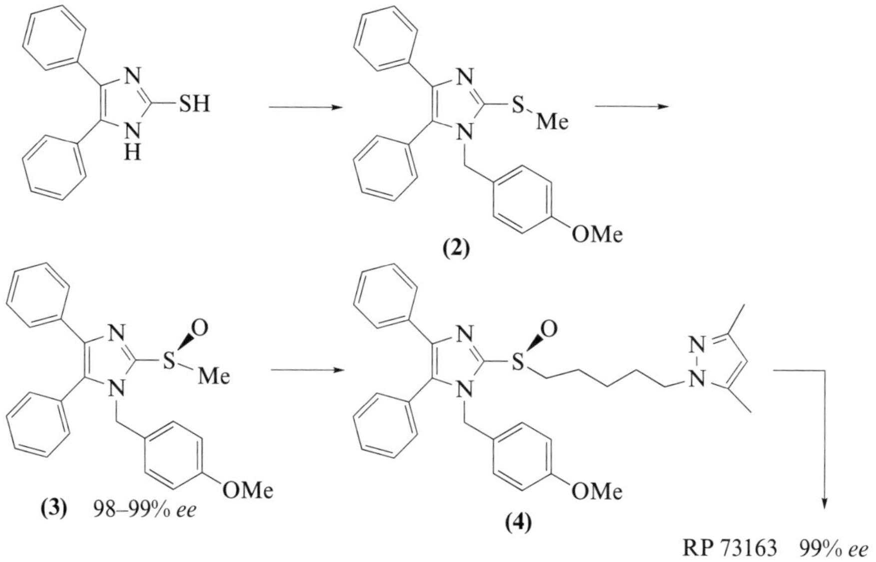

Figure 5 An alternative route to RP 73163

to obtain the desired (*S*) absolute configuration, the unnatural D-DET had to be used.[7] The optimised reaction conditions required the use of cumene hydroperoxide as oxygen donor, 0.5 mol equiv. of titanium isopropoxide (with respect to the sulphide) and 1 equivalent of D-DET in anhydrous dichloromethane at –20 °C. The sulphoxide **3** was found to be surprisingly stable in acidic conditions in which other sulphoxides would racemise.[14] HCl(1 M) could therefore be used to wash out the titanium salts at the end of the reaction, and, after standard work-up, the sulphoxide **3** could be isolated as a crystalline material in 98–99% *ee* and 75% isolated yield.

Alkylation of sulphoxide **3** by 4-chloro-1-iodobutane and introduction of the pyrazole ring were combined in a one-pot process to give directly the intermediate **4**. Removal of the *p*-methoxybenzyl group was achieved without racemisation in refluxing trifluoroacetic acid in the presence of anisole. After a final recrystallisation stage, RP 73163 was obtained in >99% *ee*.[12,13]

19.4 THE SYNTHESIS OF RP 52891

RP 52891 is a potassium channel opener which has possible clinical indications in the treatment of, for instance, hypertension, irritable bladder, coronary artery and peripheral vascular disease and obstructive airway diseases.[3] The opposite enantiomer of RP 52891 is inactive.

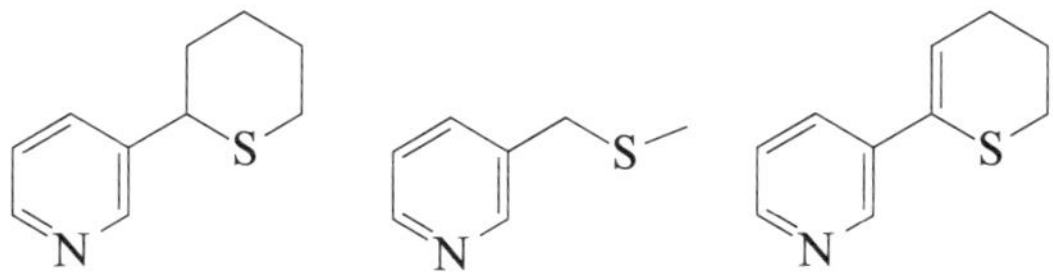

(5) (6) (7)

Figure 6 Possible precursors to RP 52891

The first laboratory samples of RP 52891 were obtained by diastereoisomeric separation of synthetic intermediates and by chiral phase high-performance liquid chromatography.[3] As with RP 73163, several other asymmetric syntheses were then successfully studied in the laboratory and evaluated for future scale-up of RP 52891. Again, the method of choice for the introduction of the chiral sulphoxide moiety was the titanium-catalysed sulphoxidation reaction described above, and the plan was to design a substrate which could lead to the best enantioselectivity in this reaction (Figure 6).

The asymmetric oxidation of the sulphide **5** was obviously not the best strategy in this case, as it is a racemic mixture which could lead to four possible isomeric sulphoxides. As already mentioned, prochiral aryl methyl sulphides usually undergo asymmetric oxidation with good to high enantioselectivities. Thus, the sulphide **6** was subjected to the reaction conditions used with the sulphide **2** in the hope that it would then be possible to cyclise the resulting sulphoxide using an appropriate electrophile, and eventually produce RP 52891. However, the enantioselectivity of this oxidation never exceeded 20% *ee*. This was attributed to facile rotation about the CH_2—S bond. A more rigid substrate was therefore needed where the sp^3 nicotinic CH_2 group would be replaced by an sp^2 centre. To this end, the vinylic sulphide **7** was synthesised in three steps from methyl nicoti-

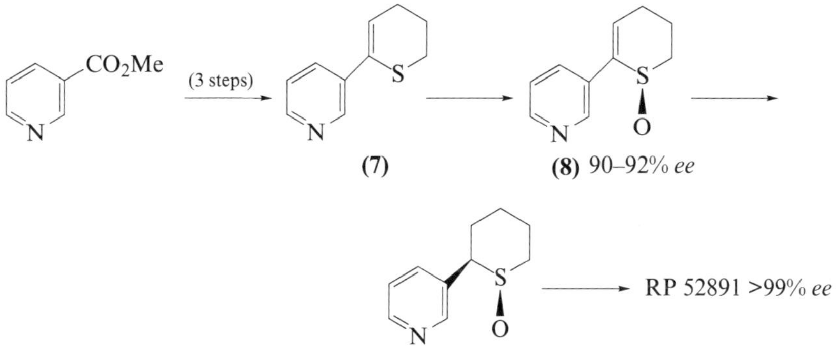

Figure 7 Synthesis of RP 52891 from vinylic sulphide **7**

nate, and oxidised under anhydrous conditions using L-DET as chiral ligand to give the desired (*R*)-sulphoxide (**8**) in 90–92% *ee*[15] (Figure 7).

The electron-deficient double bond of the sulphoxide **8** was reduced with sodium borohydride to afford sulphoxide **9**. In contrast to the intermediates **7** and **8**, the sulphoxide **9** is a highly crystalline material which can be recrystallised to give the desired (*R*, *R*) isomer in excellent enantiomeric, diastereoisomeric and chemical purity. Finally, diastereoselective alkylation with methyl isothiocyanate, followed by recrystallisation, gave RP 52891 in >99% *ee*.[15]

19.5 CONCLUSION

The strategy adopted in the syntheses of RP 52891 and RP 73163 was first to select an asymmetric oxidation method which would be suitable for rapid and safe scale-up, and then to design the rest of the synthesis around this key reaction. Because of the very limited number of asymmetric sulphoxidation methods potentially suitable for scale-up, the choice of the key reaction prevailed over the choice of the synthetic intermediate. The selected reaction was the enantioselective oxidation of prochiral sulphides by cumene hydroperoxide using soluble chiral complexes made from specific combinations of DET and titanium isopropoxide.[16] For better chiral recognition between the substrate and the chiral reagent, the prochiral sulphides had to be specifically designed with the sulphur atom in a very rigid and unsymmetrical environment. This strategy proved successful for both RP 52891 and RP 73163. Using the appropriate tartrate enantiomer, the sulphide intermediates then gave the corresponding sulphoxides in high enantiomeric purity and with the desired absolute configuration, thus allowing the preparations of multikilogram batches of RP 52891 and RP 73163 in >99% *ee* (Figure 8).

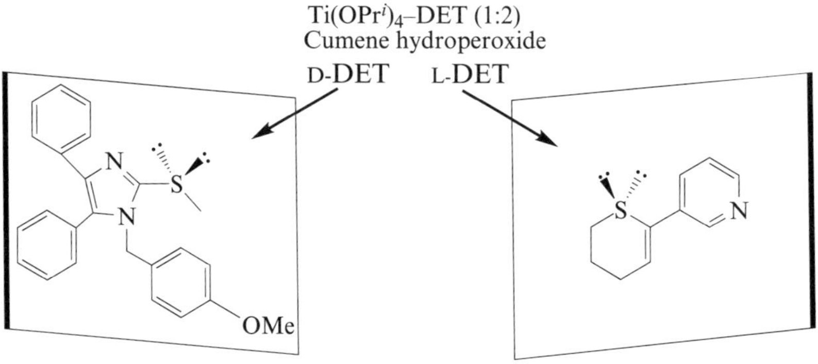

Figure 8 Key asymmetric steps in the synthesis of drug candidates RP 52891 and RP 73163

19.6 ACKNOWLEDGEMENTS

The design and scale-up of the two syntheses presented in this chapter are the result of work done by the Process Chemistry team at the Dagenham Research Centre of Rhône-Poulenc Rorer in the UK. The author thanks all the members of this team for their constant enthusiasm and professionalism. The author also records his gratitude to Professor Henri Kagan for many fruitful discussions.

19.7 REFERENCES

1. See, for instance, Nasipuri, D., in *Stereochemistry of Organic Compounds*, Wiley, New York, 1991, and references cited therein.
2. (a) Crosby, J., in *Chirality in Industry* (ed.) A. N., Collins, G. N., Sheldrake, and J. Crosby, Wiley, Chichester, 1992, pp. 1–66; (b) For a recent article on the subject, see also Cadwell, J.; *Chem. Ind.*, (London) 175, 1995.
3. (a) Aloup, J. C., Farge, D., James, C., and Mondot, S., Cavero, I., *Drugs Future*, **15**, 1097 (1990); (b) Brown, T. J. Chapman, R. F., Cook, D. C., Hart, T. W., McLay, I. M., Jordan, R., Mason, J. S., Palfreyman, M. N., Walsh, R. J. A., Withnall, M. T., Aloup, J. C., Cavero, I., Farge, D., James, C., and Mondot, S. J., *Med. Chem.*, **35**, 3613 (1992).
4. Ashton, M. J., Bridge, A. W., Bush, R. C., Dron, D. I., Harris, N. V., Jones, G. D., Lythgoe, D. J., Ridell, D., and Smith, C., *Bioorg. Med. Chem. Lett.*, **2**, 375 (1992).
5. For a recent review, see Kagan, H. B, in *Catalytic Asymmetric Synthesis* (ed. I Ojima)., VCH, New York, 1993, Ch. 4.3, pp. 203–226
6. (a) Katsuki, T., Sharpless, K. B., and Rossiter, B. E., *J. Am. Chem. Soc.*, **102**, 5974 (1980); (b) Rossiter, B. E., Katsuki, T., and Sharpless, K. B., *J. Am. Chem. Soc.*, **103**, 464 (1981). For a review on the Sharpless epoxidation, see Rossiter, B. E., in *Asymmetric Synthesis* (ed.) J. D., Morrison, Academic Press, New York, 1985, Vol. 5, Ch. 7, pp. 193–246.
7. (a) Pitchen, P., and Kagan, H. B., *Tetrahedron Lett.*, **25**, 1049 (1984); (b) Pitchen, P., Deshmukh, M., Dunach, E., and Kagan, H. B. *J. Am. Chem. Soc.*, **106**, 8188, (1984); (c) Kagan, H. B., Dunach, E., Nemecek, C., Pitchen, P., Samuel, O., and Zhao, S. H., *Pure Appl. Chem.*, **57**, 1911 (1985).
8. Di Furia, F. Modena, G., and Seraglia, R., *Synthesis*, 325 (1984).
9. Zhao, S., Samuel, O., and Kagan, H. B., *Tetrahedron*, **43**, 5135 (1987).
10. (a) Burns, C. J., Matin, C. A., and Sharpless, K. B., *J. Org. Chem.*, **54**, 2826 (1989); (b) Jorgensen, K. A., *Tetrahedron: Asymmetry*, **2**, 515 (1991); (c) Potvin, P. G., and Bianchet, S., *J. Org. Chem.*, **57**, 6629 (1992).
11. Davis, Fig. A., and Sheppard, A. C., *Tetrahedron*, **45**, 5703 (1989).
12. Pitchen, P., *Chem. Ind.*, (London) 636 (1994).
13. Pitchen, P., France, C. J., McFarlane, I. M., Newton, C. G., and Thompson, D. M., *Tetrahedron Lett.*, **35**, 485 (1994).
14. See, for instance, (a) Mislow, K., Simmons, T., Melillo, J. T., Ternay, A. L., Jr, *J. Am. Chem. Soc.*, **86**, 1452 (1964); (b) Mislow, K., Green, M. M., Laue, P., Simmons, T., and Ternay, A. L., Jr, *J. Am. Chem. Soc.*, **87**, 1958 (1965); (c) Casarini, D., Foresti, E., Gasparrini, Fi., Lunazzi, L., Maciantelli, D., Misiti, D., and Villani, C., *J. Org. Chem.*, **58**, 5674 (1993).
15. Pitchen, P., paper submitted to *Proceedings of the Chiral 95 USA Symposium,* Boston, May 15–16, 1995.
16. For the latest improvements to this reaction see: (a) Brunel, J. M., Diter, P., Duetsch, M., Kagan, H. B., *J. Org. Chem.*, **60**, 8086 (1995); (b) Astra Aktiebolag, *Int. Pat. Appl.*, WO 96/02535.

20 Asymmetric Reduction of Prochiral Ketones

J.-C. CAILLE, M. BULLIARD and B. LABOUE
SIPSY, Avrillé, France

20.1 Introduction 391
20.2 The Stoichiometric Approach 392
20.3 The Catalytic Approach 393
20.4 Mechanistic Aspects 394
20.5 Preparation of the Catalyst 395
20.6 Asymmetric Reductions Catalysed by Oxazaborolines 396
20.7 Pharmaceutical Applications 398
20.8 Conclusion 400
20.9 References 400

20.1 INTRODUCTION

Chirality continues to play an important role in the development of new pharmaceutical intermediates and of bulk drugs. Among the numerous techniques that industrial chemists now have at their disposal, asymmetric reduction and, in particular, reduction of unsymmetrical ketones to alcohols, has proved exceptionally useful over the last few years. The reduction is achieved by the overall addition of hydrogen to one face of the carbonyl group, which leads preferentially to the formation of one stereoisomer, as shown in Figure 1.

To achieve the chiral induction necessary for the selective preparation of one enantiomer, industrial chemists have modified the classical reagents with optically active ligands. The aim of this chapter is to survey some of the best practical methods for ketone reduction on an industrial scale, and to highlight in particu-

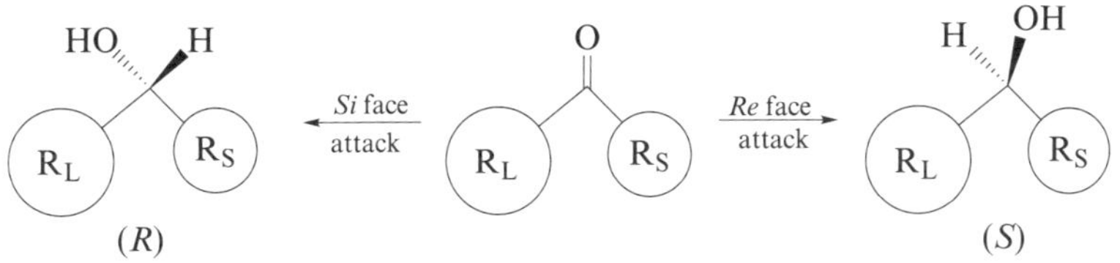

Figure 1 Asymmetric reduction of prochiral ketones

Chirality in Industry II. Edited by A. N. Collins, G. N. Sheldrake and J. Crosby

lar the use of oxazaborolidine reagents. We shall also illustrate the latter with some pharmaceutical targets which are currently prepared by this approach.

20.2 THE STOICHIOMETRIC APPROACH

The most obvious way to form a homochiral secondary alcohol from the corresponding ketone is to use a conventional reducing agent in a stoichiometric amount, associated with an optically pure chiral ligand. In this case the complex formed is able to deliver its hydride preferentially to one face. Among the reagents which illustrate this approach, those prepared from $LiAlH_4$ (LAH) are worthy of note. The Yamaguchi–Mosher reagent and BINAL-H have been most used over the last few years. Although these reagents have proved to be fairly useful for achieving good levels of enantiomeric excess, their manipulation is not straightforward. Both reagents usually require very low temperatures, sometimes below –78 °C to secure a good enantiomeric excess. Weigel[1] at Eli Lilly demonstrated the scope and limitation of the Yamagushi–Mosher reagent prepared from LAH and Chirald® [(2*S*,3*R*)-(+)4-dimethylamino-1,2-diphenyl-3-methylbutan-2-ol] or its enantiomer, *ent*-Chirald®, in the course of the synthesis of an analogue of the antidepressant fluoxetine (Prozac®) (Figure 2). (The merits of the technique for this particular compound are discussed in detail in Chapter 6.)

Another excellent way of producing enantiomerically pure secondary alcohols is to use chiral organoborane reagents. Following the pioneering work of Brown,

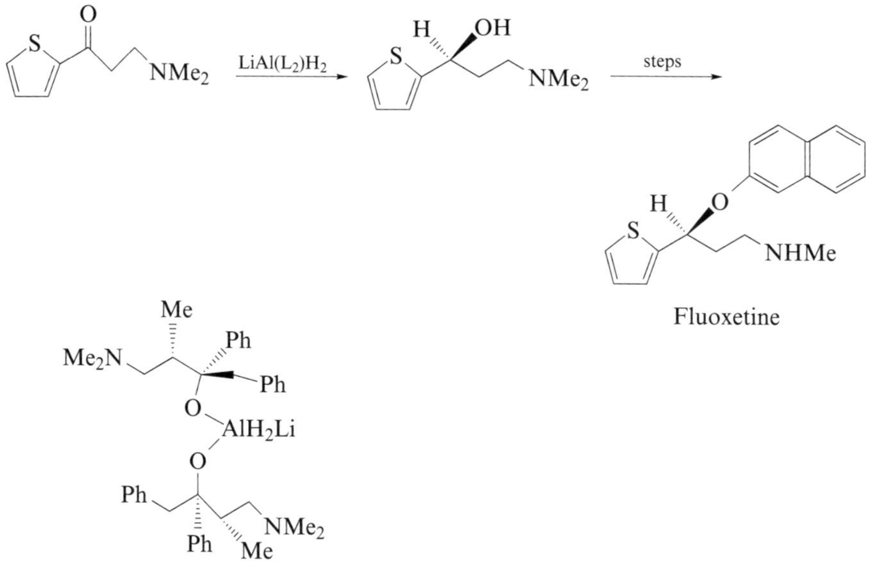

Figure 2 Asymmetric synthesis of fluoxetine

many such compounds have been reported, e.g., *B*-(3-pinyl)-9-borabicyclo-[3.3.1]nonane (Alpine-Borane) and chlorodiisopinocamphenylborane (Ipc_2BCl).[2] The latter is now commercially available from Callery Chemical and is a useful reagent achieving high chemical and optical yields. A good example of its efficacy was reported by Brown and co-workers during the preparation of the antipsychotic BMS 181100 from Bristol-Myers.

20.3 THE CATALYTIC APPROACH

All of the previously mentioned reagents, despite their effectiveness, have an important drawback. They are used in stoichiometric amounts and are, accordingly, often costly. Their use involves separation and purification steps which can become troublesome. Both academic and industrial chemists have searched for catalytic reagents to overcome these limitations and, on a small scale at least, the best technology to date is probably homogeneous catalytic hydrogenation. Numerous articles on asymmetric hydrogenation have appeared over the last few years, and we shall not discuss this area in detail here, because it appears that the method is currently of limited use on a preparative scale. While isolated enzymes and yeasts have been shown to be fairly useful in laboratory preparations, they are of limited use on a large scale owing to the high dilution required and troublesome work-ups.

Apart from these techniques, the most interesting reagents to be described in recent years are the so-called Corey reagents (chemzymes[3]). Capitalising on the excellent work of Itsuno and co-workers,[4] who first described the use of oxazaborolidine as a homochiral ligand, and of Kraatz[5], Corey was the first to report the enantioselective reduction of ketones to chiral secondary alcohols catalysed by oxazaborolidine (Figure 3).

Since then, numerous articles dealing with this versatile reagent have appeared in the literature. Many new catalysts have been designed, usually by replacing the diphenylprolinol moiety by other derivatives of various natural or unnatural amino acids. A large number of new oxazaborolidines have been reported which provide varying enantioselectivities in prochiral ketone reductions (we have counted more than 22 to date), but none has offered an improvement over the proline-derived oxazaborolidines. The only exception is the diphenyloxazaboroli-

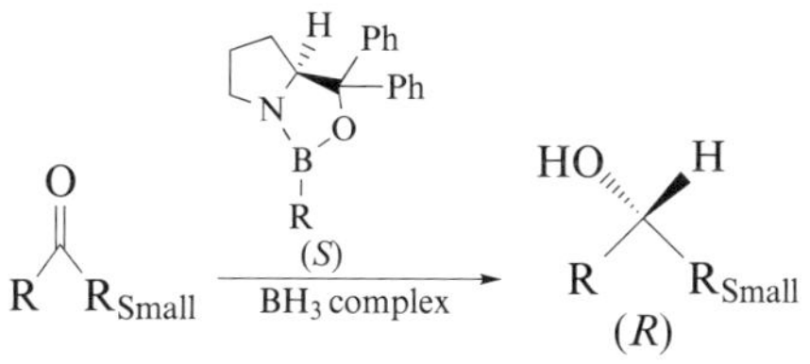

Figure 3 Oxazaborolidine-catalysed reduction of prochiral ketones

dine reported by Quallich at Pfizer, which may be a good alternative since both enantiomers of *erythro*-aminodiphenylethanol are cheap and commercially available.[6]

On the industrial side, the discovery of this technique prompted many groups to investigate its versatility. At Merck, a group of chemists led by Mathre investigated the use of oxazaborolidine to prepare various optically pure pharmaceutical intermediates. In parallel with Corey's work, they studied the mechanistic aspects of this reaction and clearly demonstrated the pivotal role of the borane adduct CBS-B.[7] Their research culminated in the isolation of this stable complex described as a 'free-flowing crystalline solid,' and they were first to report its single-crystal X-ray structure.

20.4 MECHANISTIC ASPECTS

Several groups have attempted to rationalise the enantioselective reduction of prochiral ketones with this type of catalyst, Corey being the first to propose a catalytic scheme. The reduction occurs by coordination of the electron-deficient boron atom, through the lone pair *anti* to its larger substituent, with ketonic oxygen and hydrogen transfer from the NBH_3^- unit to the carbonyl carbon via a six-membered cyclic transition state, following the path shown in Figure 4.

Shortly after his publication, Mathre at Merck in collaboration with Liotta's group at Emory University reconsidered this mechanism. They proposed an alternative 'chair transition state assembly' to explain the origin of enantioselec-

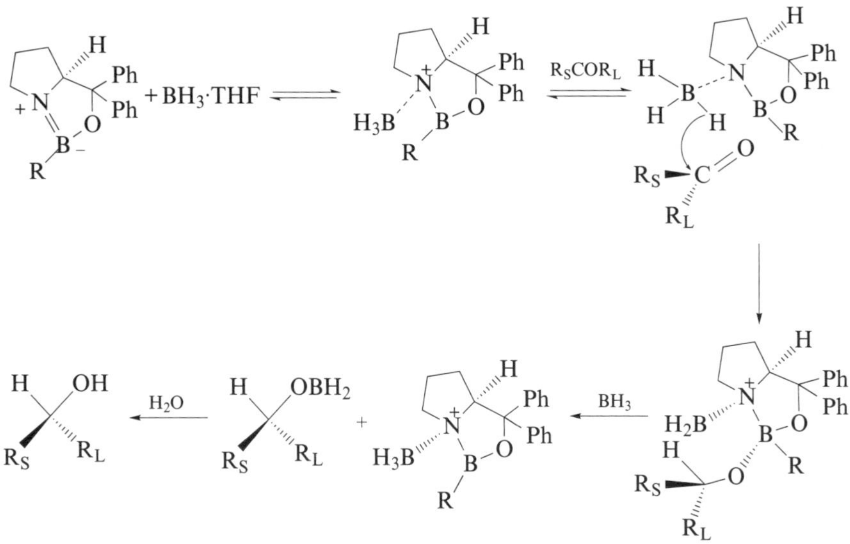

Figure 4 Catalytic asymmetric reduction involving oxazaborolidines

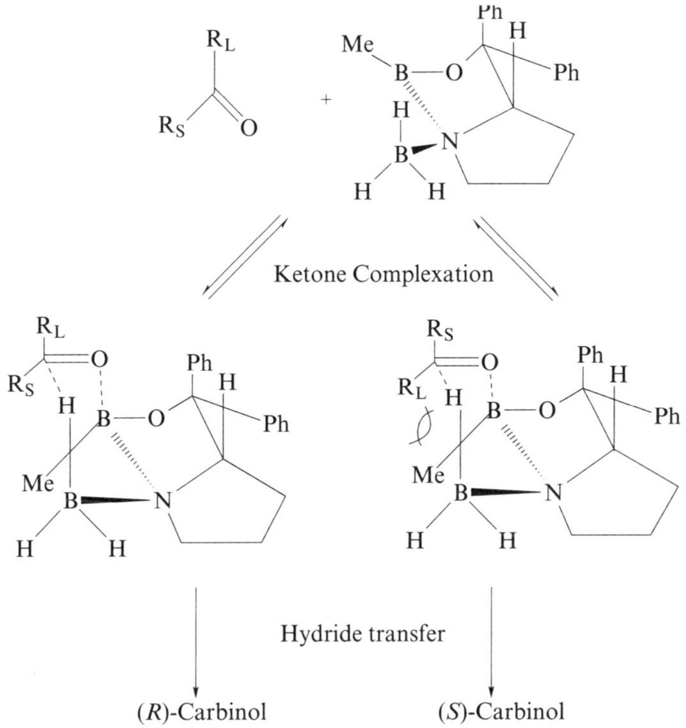

Figure 5 'Alternative' chair transition-state assembly

tivity observed in oxazaborolidine-catalysed reduction of prochiral ketones (see Figure 5). To achieve this goal, they elected to use the reaction surfaces generated using the MNDO Hamiltonian.[8] According to their model, the 1,3-diaxial interaction between the R_L or R_S ketonic substituent (L = large, S = small) and the oxazaborolidine alkyl group differentiates the two diastereoisomeric transition states leading to the major (*R*)- and minor (*S*)-isomers. This model, in our view, accounts much better for the role of the alkyl group borne by the boron atom, which had not been taken into consideration in the previous model.

Other interesting studies of the mechanism and the relationship of structure to enantiomeric excess have been published by Quallich *et al.*[9] in the case of a diphenylaminoethanol-derived catalyst.[9]

20.5 PREPARATION OF THE CATALYST

Two different approaches can be followed for the preparation and use of the catalyst. In the first, it is formed *in situ* by mixing the (*R*)- or (*S*)-diphenylprolinol

(DPP) and BH_3 complex. This route is advantageous because there is no need to use boronic acids (or boroxines) or to remove water to form the catalyst. Another possible way is to use preformed catalysts, some of which are commercially available from suppliers such as Callery Chemical. In this particular case, the instability of the catalyst over time may be a disadvantage. It is worth noting that investigators at Merck circumvented this drawback by making an air-stable complex called CBS-B. This complex is not yet commercially available.

DPP itself is now available on large scale, and can be prepared in several ways. The most straightforward route is based on an earlier publication by Corey: suitably protected (*S*)-proline is treated with phenylmagnesium halide, yielding the diphenylmethanol derivative. An alternative two-step enantioselective synthesis of DPP from proline, based on the addition of proline-*N*-carboxyanhydride to 3 equiv. of phenylmagnesium chloride, has been recently reported by the Merck group.[10] An elegant asymmetric preparation of DPP has been reported by Beak *et al.*[11] The enantioselective deprotonation of BOC-pyrrolidine is achieved with *sec*-butyllithium in the presence of sparteine as a chiral inducer. Subsequent quenching with benzophenone gives *N*-BOC-DPP in 70% yield and 99% *ee* after one recrystallization. It is worth noting that in order to cope with the rather high price of (*R*)-proline, Corey developed an alternative route to this ligand from racemic pyroglutamic acid.

The oxazaborolidine is finally made by refluxing DPP with a suitable alkyl(aryl)boronic acid or, better, with the corresponding boroxine in toluene in the presence of molecular sieves (or by removing water by azeotropic distillation). According to the literature, methyloxazaborolidine (Me-CBS) can be either distilled or recrystallised. The key point is that the catalyst must be free from any trace of water or alkyl(aryl)boronic acid because these impurities decrease enantioselection.

20.6 ASYMMETRIC REDUCTIONS CATALYSED BY OXAZABOROLINES

The best reaction conditions depend on both the nature of the substrate and the catalyst. A wide range of conditions have been studied by Mathre's group at Merck, by Quallich and co-workers at Pfizer, by Stone at Sandoz (Basle) and by our group at SIPSY. Generally, the reaction must be carried out under anhydrous conditions, usually at room temperature. Temperature appears to be the most important parameter of the reaction and, although a clear temperature dependence has been demonstrated by Stone,[12] no good mechanistic explanations have been presented. These will probably follow shortly. The catalyst loading is typically 1–20%, and the DPP can be recovered at the end of the reaction and recycled in some cases. BH_3 complexes of THF, Me_2S and 1,4-thioxane may be used, and diborane gas has also been successfully employed. Catecholborane is a suitable reagent for certain low-temperature reactions. The relationship between enantioselectivity and the nature of the substrate is highly predictable, and the best results are obtained with aryl ketones in which there is a significant size difference between the large and the small substituent, and in which the basicity of the ketone is favourable.

Besides the preparation of aryl alkyl carbinols, many other secondary alcohols have been prepared in good to excellent optical purity using this method, although improvements in the catalytic system still have to be made in order to obtain good results for all types of ketones, especially the problematic dialkyl ketones. Numerous complex target molecules have been prepared with the asymmetric reduction step as the key step. Some selected examples are shown in Figure 6.

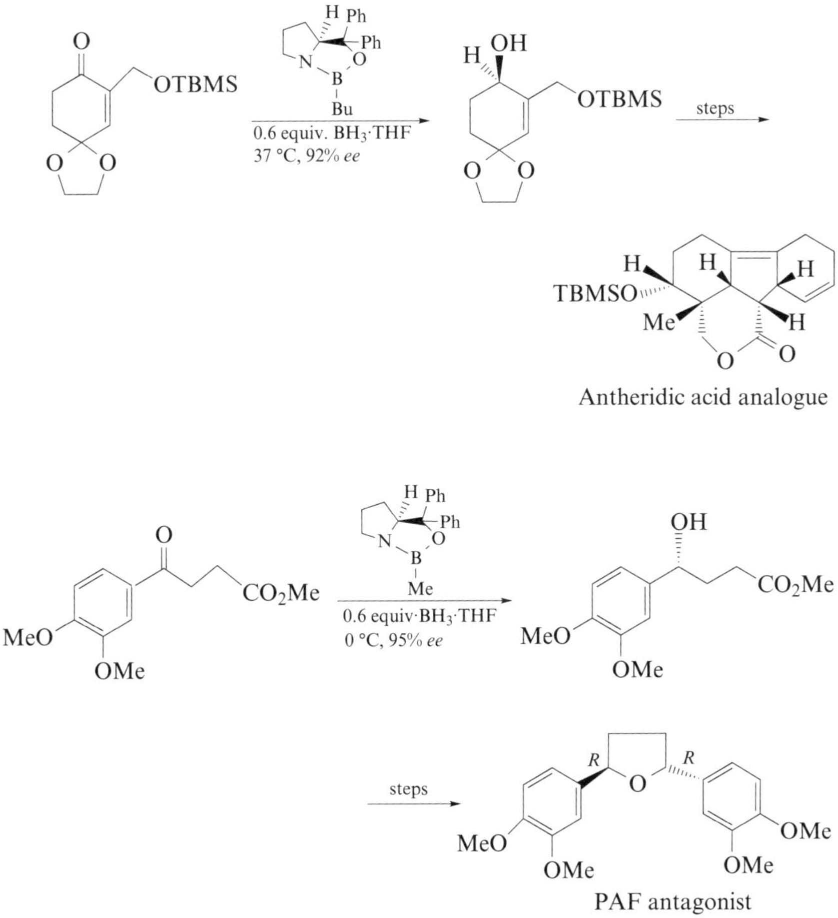

Figure 6 Access to highly complex molecules by oxazaboroline-catalysed reductions

20.7 PHARMACEUTICAL APPLICATIONS

Numerous pharmaceutical intermediates and drug substances bear chiral secondary alcohol groups. We do not intend in this chapter to review all the target molecules that fall into this category, but rather to highlight the most important ones for which an oxazaborolidine-catalysed reduction has been used with success by industrial chemists. In some cases the processes have been commercialised. Corey was the first to become involved in the preparation of pharmaceutical molecules using this technology, and reported several preparations of important drugs. His first success was the synthesis of a pure enantiomer of fluoxetine, the Lilly antidepressant introduced under the brand name Prozac®.[13]

Shortly afterwards, he published the preparation of both enantiomers of isoproterenol, a β-adrenoreceptor agonist and denopamine, and a useful drug for congestive heart failure.[14] From a pharmaceutical point of view, the structure of isoproterenol is remarkable, since the same basic structure is also encountered in various drugs, such as those acting at the adrenoreceptor site. Therapeutic indications are broad, ranging from asthma (β_2-agonist) to obesity and diabetes (β_3-agonist). Halo aryl ketones have been shown to be useful intermediates to these drugs, giving rise, after reduction and basic cyclization, to the corresponding epoxystyrenes, which are important intermediates for the preparation of β_3-agonists (see Figure 7). The regiospecific ring opening is achieved with the appropriate amine to give an advanced intermediate possessing the correct configuration at the carbinol stereocentre. This compound is converted in a sequence of several steps into the active drug.

At SIPSY we have demonstrated the industrial feasibility of this approach in the production of hundreds of kilograms of related molecules.

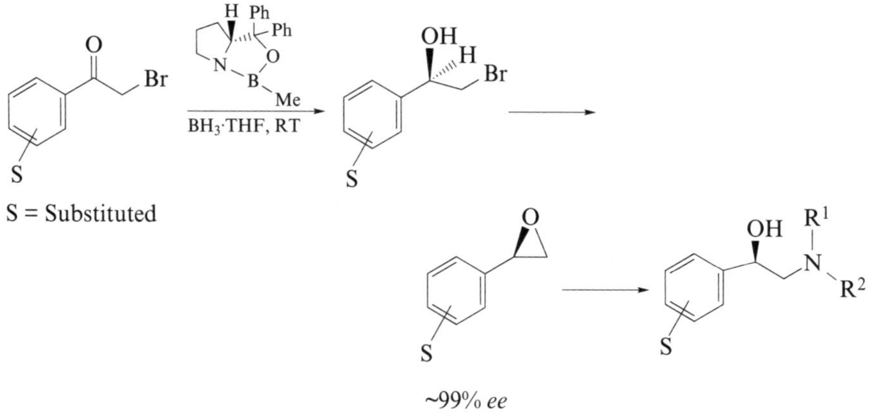

Figure 7 Access to important β_3-agonist intermediates

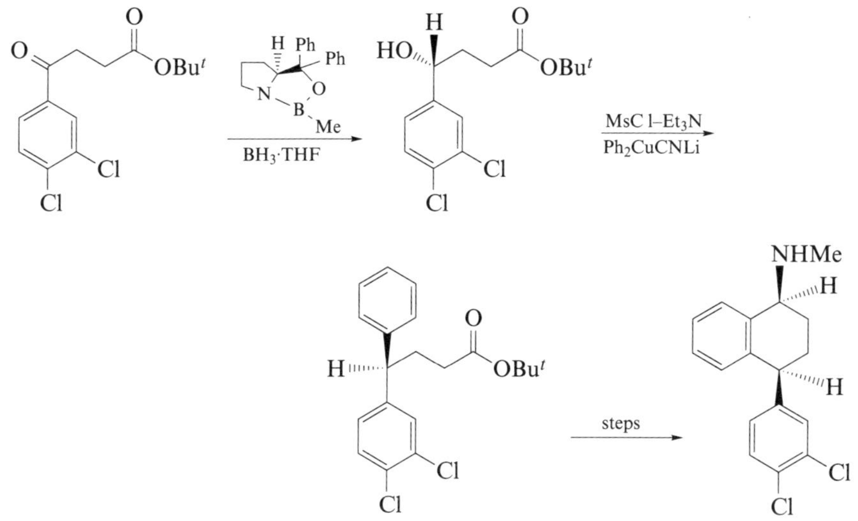

Figure 8 Asymmetric synthesis of sertraline

Along the same lines, chemists at Merck successfully produced substantial quantities of various important pharmaceutical intermediates used in the preparation of drugs currently under development. Among these are MK-0417, a promising soluble carbonic anhydrase inhibitor useful in the treatment of glaucoma,[15] the LTD4 antagonist L-699392[16] and the anti-arrhythmic MK-499.[17] Equally successful has been the group of Quallich at Pfizer,[18] who have reported the preparation of a substituted dihydronaphthalenone, a pivotal intermediate in the preparation of the antidepressant sertraline (Lustral®).

In this synthesis (Figure 8), the newly created chiral centre subsequently determines the absolute stereochemistry of the final product. This illustrates that the use of Corey reduction is a valuable tool to introduce the chirality in a complex molecule owing to its high compability with other reactive centres. Quallich demonstrated that high enantiomeric excesses could also be obtained when other heteroatoms capable of coordinating borane were present in the starting material. Using an excess of borane to chelate/protect the nitrogen functions, a smooth reduction of these compounds could be effected.[19]

Reduction of alkyl aryl ketones is an equally valid route to optically pure secondary arylalkylamines, as researchers from Sandoz showed in an enantioselective synthesis of SDZ-ENA-713[20] (Figure 9).

The future developments of the Corey reduction will undoubtedly include finding ways to reduce effectively alkyl vinyl and dialkyl ketones. The first applications have already been demonstrated, e.g. the key reduction step of a 5-lipoxygenase inhibitor (Figure 10).

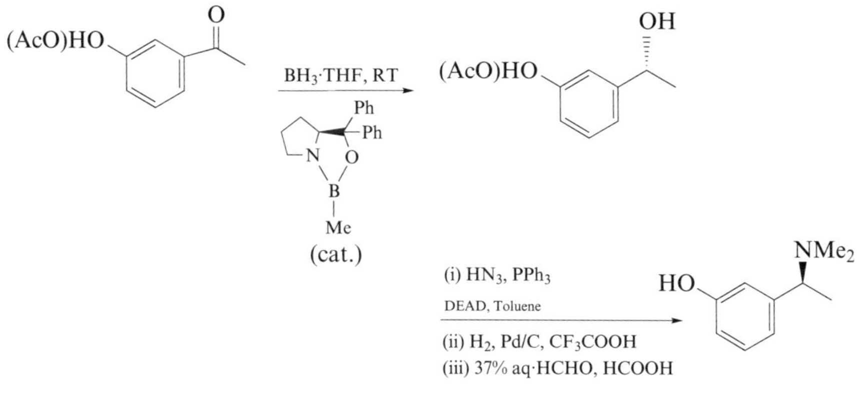

Figure 9 Synthesis of SDZ–ENA–713 intermediate

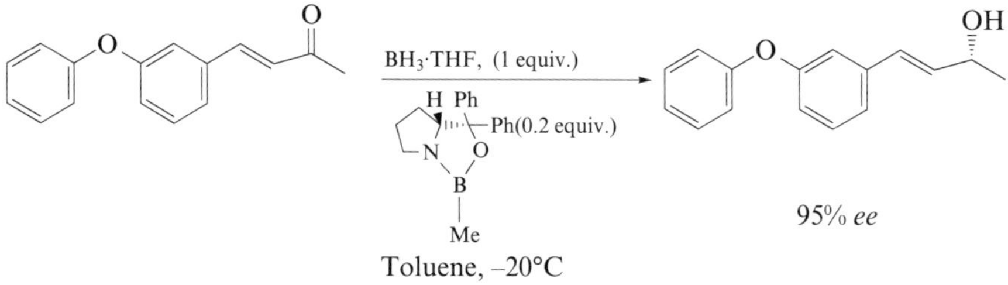

Figure 10 Key step in the synthesis of a 5-lipoxygenase inhibitor intermediate

20.8 CONCLUSION

The representative examples described above are illustrative of the large amount of work going on in the area of oxazaboroline-catalysed reductions. We have demonstrated that asymmetric reduction of prochiral ketones is a powerful tool to produce important pharmaceutical intermediates, and it is also apparent that the discovery of oxazaborolidines has given rise to new insights into enantioselective reactions. Academic and industrial chemists are already looking beyond this type of reagent, and from this basis a new generation of efficient chiral catalysts is being developed.

20.9 REFERENCES

1. Weigel., L. O., *Tetrahedron Lett.*, 7101 (1990).
2. Brown, H. C., *Organic Synthesis via Boranes*, Wiley–Interscience, New York 1975.
3. Corey, E. J., Bakshi, R. K., and Shibata, S., *J. Am. Chem. Soc.*, **109**, 5551 (1987); Corey, E. J., (Harvard University), *US Pat.*, 4 943 635.

4. Itsuno, S., Hirao, A., Nakahama, S., and Yamazaki, N., *J. Chem. Soc., Perkin Trans. 1*, 1673 (1983); Itsuno, S., Nakano, M., Miyazaki, K., Masuda, H., Ito, K., Mhizao, A., and Nakahama, S., *J. Chem. Soc., Perkin Trans. 1*, 2039 (1985).
5. Kraatz, U. (Bayer), *Ger. Pat.*, DE 3 609 152.
6. Quallich, G. J., and Woodall, T. M., *Tetrahedron. Lett.*, **34**, 4145 (1993).
7. Thompson, A. S., Douglas, A. W., Hoogsteen, K., Carroll, J. D., Coreley, E. G., Grabowski, E. J. J., and Mathre, D. J., *J. Org. Chem.*, **58**, 2880 (1993).
8. Jones, D. K., Liotta, D. C., Shinkai, I., and Matthre, D. J., *J. Org. Chem.*, **58**, 799 (1993).
9. Quallich, G. J., Blake, J. F., and Woodall, T. M., *J. Am. Chem. Soc.*, **116**, 8516 (1994).
10. Blacklock, T. J., Jones, T. K., Matthre, D. J., and Xavier, L. C. (Merck), US *Pat.*, 5 039 802.
11. Beak, P., Kerrick, S. T., Wu, S., and Chu, J., *J. Am. Chem. Soc.*, **116**, 3231 (1994).
12. Stone, G. B., *Tetrahedron: Asymmetry*, **5**, 465 (1994).
13. Corey, E. J., and Reichard, G. A., *Tetrahedron Lett.*, **30**, 5207 (1989).
14. Corey, E. J., and Link, J. O., *J. Org. Chem.*, **56**, 442 (1991).
15. Jones, T. K., Mohan, J. J., Xavier, L. C. C., Blacklock, T. J., Mathre, D. J., Sohar, P., Tonner Jones, E. T., Reaner, R. A., Robers, F. E., and Grabardes, E. J. J., *J. Org. Chem.*, **56**, 763 (1991).
16. King, A. O., Corley, E. G., Anderson, R. K., Larsen, R. D., Verhoeven, T. R., and Reider, P. J., *J. Org. Chem.*, **58**, 3731 (1993).
17. Shi, Y. J., Cai, D., Dolling, U. M., Douglas, A. W., Tschaen, D. R., and Veroheven, T. R., *Tetrahedron. Lett.*, **35**, 6409 (1994).
18. Quallich, G. J., and Woodall, T. M., *Tetrahedron*, **48**, 10239 (1992).
19. Quallich, G. J., and Woodall, T. M., *Tetrahedron Lett.*, **34**, 785 (1993).
20. Chen, C.-P., Prasad, K., and Repic, O., *Tetrahedron Lett.*, **32**, 7175 (1991).

Index

(Entries in bold type are references to tables or figures)

abzymes, 5, 9
N-acetyl amino acids, 197–200
D-(–)-(*S*)-acetyl-*β*-mercaptoisobutyric acid, 81
N-acetyl-L-methionine, 176
N-acetyl-phenylalanine, 176
acetylglutamic acid, 152
achiral assay, 13
acyclonucleotides, 375
α-acylase, 176
adrenaline, **93**, 245
β-adrenoreceptor agonist, 398
agrochemicals, 5, 8, 147–8, 209, 278, 283, 310
  market size, 1
Alcalase, 203
alkaline phosphatase, in lamivudine synthesis, 69
L-allylglycine, 200
alminoprofen, **20**
Alpine–Borane, 393
amino acid amides, resolution by aminoacylases, 197–200
amino acids,
  by membrane technologies, 7, 159–73
  liquid crystal applications, 277
  unnatural, 187–90, 194–203, 393
α-amino alcohols, 165
  as chiral proton source, 339–48
D-(+)-*α*-amino-*ε*-caprolactam, **87**, **93**
aminoacylases, 197–99
  combination with esterases, 200
(*S*)-(+)-2-aminobutan-1-ol, **87**, **93**
aminodiphenylethanol, 394
(+)-3-aminomethylpinane, **93**
(1*S*,2*S*)-(+)-*threo*-2-amino-1-phenyl-propane-1,3-diol, 85, **87**, 91, **93**
anisotropic liquid crystals, 264
anthenidic acid, analogue of, **397**
antihypertensive inhibitors, *see* renin inhibitors
antiviral drugs, 41–77
aristeromycin, **185**
(+)-ARSOP, **313**
3-aryloxy-1,2-propanediols, 376
2-arylpropionitriles, by asymmetric hydrocyanation, 312–3
(*R*)-asparagine, 154, 225
aspartame, 3
asymmetric catalysis, recent developments, 3
asymmetric dihydroxylation, Sharpless, 4
asymmetric epoxidation, 5
  Jacobsen, 4, 318
  Sharpless, 3, 6, 363–77, 382 (*see also* glycidol process)
asymmetric hydrocyanation, 309–32
  mechanism, 327–31
  in naproxen synthesis, 25, 331–2
  of norbornene derivatives, 312
asymmetric hydrogenation, *see* catalytic asymmetric hydrogenation
asymmetric ketene-imine Staüdinger reaction, 106
asymmetric oxidation, of sulphides, 381–7
asymmetric protonation, *see* enantioselective protonation
asymmetric reduction, 4, 9, 101–2 (*see also* catalytic asymmetric hydrogenation)
  of ketones, 101–2, 246–7, 253–60, 391–8
  with oxazaborolines, 4, 391–8
asymmetric transformation, *see* second order asymmetric transformations
autocatalysis, 341–2
azabicyclo[2.2.1]hept-5-en-3-one, 185, 188
azlactones, dynamic resolution of, 201–3
AZT, 375

benoxaprofen, **20**
benzonorbornadiene, **313**
(*S*)-benzoyl-*β*-mercaptoisobutyric acid, 83
N-benzoylglycine, 203

D-(+)-*N*-benzyl-α-phenethylamine, 83
(2*R*,3*R*)-*trans*-3-benzylglycidol, 75, 370
cis-3-benzylglycidol, 370–2
beta-blockers, 160, **161**, 245, 375
BINAP, 3, **24**, 31, 313
  in asymmetric hydrocyanation, 312–3
  in asymmetric hydrogenation, 249
  in naproxen synthesis, 25
BINAP-LAH, 102
binaphthyl crown ethers, 167
1,1′-binapthol, resolution of, 165
biocatalysts, whole cell, *see* biotransformation, whole cell
Biopol, 248
bioresolution processes, whole-cell, *see* biotransformation, whole cell
biotechnology industry, development of, 9
biotransformation, 48
  continuous, 214–8
  fed-batch, 217, 302–3
  L-carnitine production, 287–304
  integration into process chemistry, 192, 194–200, 203–4, 298–304
  whole cell, 186–7, 211–4, 246, 253–60
birefringence, in liquid crystals, 266
Blackman kinetics, in L-carnitine production, 300–2
N-BOC-L-thiazolylalanine, (BOC-TAZ), 194–200
bornane-10,2-sultam, 3
(*R*)-(+)-bornylamine, **93**
bovine serum albumin, in enantioselective membranes, 160–61
brefeldin A, 192, **193**
(*S*)-1-bromo-2-methylbutane, 283–4
(*S*)-2-bromopropanoic acid, 208, 220
(–)-brucine, **87**, **93**, 280
(–)-bupivacaine, 2
bupranolol, **165**
(*R*)-1,3-butandiol, 249
2-butanol, resolution with enantioselective membranes, 161
butyrobetaine metabolism, 288–92

c-NECA, 188, **189**
C-3 synthons, 364, 376
(+)-calycanthine, **93**
camphorsulphonic acid, 84, 85, **87**, 95, 102
Candida rugosa, in ibuprofen resolution, 34
captopril, 81, **82**, 85
carbacephalosporin antibiotics, *see* β-lactam, antibiotics
carbinol, **395**
carbocyclic bromovinyldeoxyuridine, 52
carbocyclic nucleosides, 41–77, 184–90
carbohydrate-derived ligands, 315–7
carbonic anhydrase inhibitor, 246, 399
carbovir, 42–47, 185, 188, 192
carnitine dehydrogenase, 289–94
L-carnitine, 8
  metabolism, 288–92
  transport, 292–3
carprofen, **20**
catalytic antibodies, 5
catalytic asymmetric hydrogenation, 25, 249, 339
chemzymes, 393
chiral and non-chiral drugs, distribution, **2**
chiral auxilliaries, 3, 4, 26–7, 32, 102, 105, 111–2
chiral cyanohydrins, *see* cyanohydrins, optically-pure
chiral drugs, regulatory aspects, 11–17
chiral HPLC, 317
chiral intermediates, market size, 2
chiral inversion, *in vivo*, of ibuprofen, (*S*)-(+)-27
chiral nematic phase, 272–3
chiral pool, 3, 107–9, 230, 277, 382
chiral proton source, *see* enantioselective protonation
chiral selectors, in liquid-liquid extraction, 167, 173
chiral smectic C phase, 267–73
chiral solvents, 154
Chirald, 102, 392, **104**
Chiraldex, 372
chirality indices, 94
chloramphenicol, 353, **354**
(*R*)-3-chloro-1-phenylpropanol, 8
(*S*)-2-chloro-3-methylbutanoic acid, 283–4
α-chloroacids and esters, in liquid crystals, 278
chlorodiisopinocamphenylborane, 393
β-chloroisobutyric acid, 85
4-chlorophenylphosphoric acid, as resolving agent, 85, **87**
(*R*)-2-chloropropanoic acid, 220
(*S*)-2-chloropropanoic acid, 2, 151, 177, 207–21
  market requirements, 208–10
chlorphenesin, **365**
cholesteric phase, 266–7, 272

cholesteryl benzoate, as chiral nematic phase, 263
cholesteryl-glutamate, 174
cholesterylnonanoate, 275
α-chymotrypsin, 196
cicloprofen, **20**
cidofovir, **365**
(–)-cinchonidine, 81, 82, **87**, **93**
  in naproxen resolution, 22
classical resolution, *see* resolution, classical
CLECs, *see* crosslinked enzyme crystals
cliprofen, **20**
cloning, 260, 296–7
cofactor regeneration,
  in L-carnitine production, 295
  in membrane-assisted resolutions, 178
Committe for Proprietary Medicinal Products (CPMP), 12, 16
conglomerate, 15, **149**, 153, 373
Corey asymmetric reduction, *see* asymmetric reduction with oxazaborolidines
countercurrent adsorption, 7
(*S*)-CPA, *see* (*S*)-2-chloropropanoic acid
cross-linked enzyme crystals, (CLECs), 5, 34
crown ethers, chiral, 165
crystal packing, in diastereoisomeric salt pair, 89
crystal science techniques, for chiral compounds, 4, 6, 34–6, 119–55
crystal size distribution, 137–40
crystallisation,
  direct, *see* direct crystallisation
  evaporative, 154
  melt, 130
  preferential, 374
crystallisation-induced asymmetric transformations *see* second order asymmetric transformations
cyanohydrins, optically pure, 5, 278, 283
L-2-cyanophenylalanine, 200
cyclic phosphoric acids, as resolving agents, 84, 85, **86**, **87**, **91**
cyclocitrals, 346, 348
cyclodextrin-based GC, 372
cyclodextrins, **163**, 372
γ-cyclogeranates, 343, 346
(*R*)-(–)-1-cyclohexylethylamine, **87**, **93**
cyclooxygenase, 19–21
(+)-(1*R*-*cis*)-4-cyclopentene-1,3-diol monoacetate, 47
cyhalothrin, 147–8
cytidine deaminase, in lamivudine synthesis, 69

γ-damascone, 346, 348
(*S*)-α-damascone, 335, 337
(*S*)-β-damascone, 337
dansyl-phenylalanine, resolution of, 165
Darvon alcohol, **93**
Darvon, 102
p-decyloxybenzylidene *p*'-amino-2-methyl-butlylcinnamate, 268
dehalogenases, in resolution of 2-chloro-propanoic acid, 211–4
(+)-dehydroabietylamine, **87**, **93**
denopamine, 398
'designer' resolving agents, *see* resolving agents, rational design
desupersaturation, 140
dextropropoxyphene, **226**
L-(+)-2,4-diaminobutyric acid, **93**
diastereoisomeric salt formation, *see* resolution, classical
diastereomeric transition states, in asymmetric reduction with oxazaborolines, **395**
(*S*)-2,3-dibromopropanoate, 221
(1*R*,2*R*)-2-dichloroacetimido)-1-[(4-methylsulphonyl)phenyl]-1,3-propanediol, *see* thiamphenicol,
(+)-8,8-dichlorocamphorylsulphonyl oxaziridine, **385**
dichlorprop, **209**
dielectric constant, in liquid crystals, 266
diethyl tartrate, 366
  in enantioselective sulphide oxidation, 383–4
differential scanning calorimetry (DSC), in polymorph identification, 124–5, 151
2-(2,4-difluoro)phenylglycidol, 372
cis-dihydroxylation, of carbovir, 42–47
3,4-dihydroxyphenylalanine, *see* dopa
diisoamylborane, 58–59
D-(–)-diisopropyl tartrate, 366
diltiazem, 4, 176
(*S*)-dimethylcyclopropylcarboxamide, 5
N-(3,5-dinitrobenzoyl)leucine, 165
(+)-DIOP, **313**
(*S*)-diphenyl prolinol, 395
diphenyloxazaborolidine, 393–4
diphenylphosphinite ligands, in asymmetric hydrocyanation, 315–332

(+)-DIPHIN, **313**
direct crystallisation, 148–54
dispersion-free extraction, 170
display devices, 8, 263–87
N-*n*-dodecyl-L-hydroxyproline, in liquid-liquid extraction, 171
dopa, 226
dopamine D_2-receptor antagonists, 227
dopants for nematic mixtures, 275, 280–1
dropropizine, 375
drugs enantiopure, market size, 1
duloxetine, 101–4, **392**
dynamic resolution, *see* resolution, dynamic

(–)-eburnamonine, **93**
electrocarboxylation, route to (*S*)-(+)-ibuprofen, 29
electronic asymmetry, effect of in asymmetric hydrocyanation, 324–6
enantioselective membranes, *see* membranes, enantioselective
enantioselective nitrile hydrolysis, 5
enantioselective oxidation of prochiral sulphides, 382–4
enantioselective polymers, 158–62
enantioselective protonation, 32–3, 83, 335–49
enolates, enantioselective protonation of, 336–9
entrainment resolution, *see* resolution, by entrainment,
enzymatic kinetic resolution, 33–4, 69–70, 167, 175–8, 184–200
  of carbovir, 5, 42
  of ibuprofen esters, 33
  of phenyl-1,3-propanediol, 104
enzymes, "non-nameplate" applications, 5
ephedrine, 84, 85, **87**, **90**, **91**, **93**, **172**, **173**, 339
(*S*)-epichlorohydrin, 363
Epivir, *see* lamivudine,
epoxides, chiral, by enzymatic resolution, 5
2,3-epoxy alcohols, *see* glycidols,
(*S*)-1,2-epoxyalkanes, as building blocks for liquid crystal dopants, 282
(*R*)-1,2-epoxyoctane, 283–4
esterase, 195–6, 210
  combination with aminoacylases, 200
(*S*)-ethyl lactate, 283–4
trans-3-ethylglycidol, 372
(*S*)-1-ethylprolinamide, 232, 234–5
eudismic ratio, 94
eutectic composition, in racemate separation, 6
  of naproxen, 24
  of ibuprofen, 34, 35
eutomer, 382
extremophiles, 6

FDA, 2, 11, 14,15, 27, 83
fed-batch biotransformation, *see* biotransformation, fed-batch
fenoprofen, **20**
fermentation, 252–60
  in CPA production, 213–7
ferrocenylethyl-thiophenol, resolution of, 163
ferroelectric dopants, 277–9, 283
ferroelectric liquid crystals, 375
ferroelectric mixtures, 277
flavours and fragrances, 8
florfenicol, 353, **354**
fluazifop
  racemic switch, 209
  resolution by esterase, 210
flunoxaprofen, **20**
2-fluoro-1-alkanols, as building blocks for liquid crystal dopants, 282
fluoroalkanols, in liquid crystals, 278
(*S*)-2-fluorooctanol, 283–4
(*S*)-fluoxetine (Prozac), 8, 104, 392, 398
(*S*)-(+)-flurbiprofen, **20**, 37, 208
fragrance synthesis, 335–49
fructose, phoshite ligands from, **316**
furcloprofen, **20**

galactose, phoshite ligands from, **316**
gas chromatography, chiral, 233
geraniol, Sharpless epoxidation of, 371
glucophosphinite ligands, in asymmetric hydrocyanation, 318–31
D-(+)-glucosamine, **93**
glucose, phoshite ligands from, **316**
(*S*)-glutamic acid, 154, 230
glycerophospholipids, 375
glycidol process, 3, 363–77
  effect of reactant purity, 368
  recovery of epoxy alcohol, 369–71
  safety aspects, 366, 371
  scale-up, 368–9
(*R*)-glycidol, 3, 364–71
  synthetic applications of, 374–6
glycidyl 3-nitrobenzenesulphonate

(GNBS), 373–4, 376
glycidyl arenesulphonates, by Sharpless epoxidation, 370, 372–3
glycidyl butyrate, resolution of, 176, 364
glycidyl tosylate, 376–7
guafecainol, 376
guafenesin, **365**, 375, 376–7

habit modifiers, 126–9, 154
hepatitis B, 66
herbicides, *see* agrochemicals
Herpes simplex, 42, 185
heterogeneous catalysis, 4
hexaprofen, **20**
hippuric acid, 203
HIV protease inhibitors, 4, 66–67, 185, 192, 370, 375
hollow fibre membranes, 7, 165, 170, 176–7
hollow fibre reactors, 34
homeotropic alignment, in liquid crystals, 267
hydroboration, stereoselective, 58–59, 66, 391–8
hydrocyanation, *see* asymmetric hydrocyanation
hydroformylation, route to (*S*)-(+)-ibuprofen, 29
hydroquinidine, **87**, **93**
4-*endo*-hydroxy-2-oxabicyclo[3.3.0]oct-7-en-3-one, 190–3
(*R*)-3-hydroxy-4-(trimethylammonio)-butanoate, *see* L-carnitine
α-hydroxyacids, as building blocks for liquid crystal dopants, 282
(*R*)-3-hydroxybutyric acid, 248–9, 260
(*R*)-3-hydroxybutyrolactone, 249
D-(–)-*p*-hydroxyphenylglycine,
  as resolving agent, 85
  resolution of, 148
(*R*)-3-hydroxyvalerate, 248
hypocholestemic agents, **193**, 385
(*S*)-(+)-ibuprofen, 2, 6, **20**, 27–37, **172**, **173**, 208, **314**
  by asymmetric hydrocyanation, 331–2
  racemic switch, 22
  resolution of esters by proteases, 177

immobilised enzymes, 5, 187–8
indoprofen, **20**
International Committee on Harmonisation (ICH), 11

(*R*)-(–)-isobornylamine, **93**
isoleucine, resolution of, 168
isoprofen, **20**
isopropyl D-tartrate, as chiral solvent, 154
N-isopropylephedrine, 340, **343**, 349
(*S*)-isopropylidene glycerol, **68**, 363
isoproterenol, 398

Jacobsen asymmetric epoxidation, *see* asymmetric epoxidation

ketone reduction, enantiospecific, *see* asymmetric reduction
(*S*)-ketoprofen, 2, **20**, 37
ketosulphones, asymmetric reduction of, 246–7, 253–60
kinetic analysis, in L-carnitine production, 300–2
kinetic resolution, *see* resolution, kinetic
kinetic resolutions, membrane-assisted, *see* resolution, membrane-assisted

β-lactam, antibiotics, 105, **106**
lactamase, 184–8
γ-lactams, 185–8
lactic acid esters, in liquid crystals, 278
(*R*)-lactic acid,
  as byproduct of dehalogenation, 213–5
  as precursor for (*S*)-2-chloropropanoic acid, 210
Lactobacillus plantarum, in asymmetric reduction, 255–7
lamivudine, **42**, 66–76
  resolution of, 69–70
  synthesis of racemate, 67–69
lasolocid A, 166, **167**
(+)-leucinamide, **93**
tert-leucine, resolution of, 171,
  by oxazolinone resolution, 201
  by aminoacylase based resolution, 199
levobunolol, **365**
levodropropizine, **365**
levopropoxyphene, **226**
lipases, 176, 190–3, 364
  for guaifenesin resolution, 376
  for resolution of ibuprofen esters, 34
  in lamivudine synthesis, 71
  in oxazolinone resolution, 201–3
  Mucor miehei, 71
  Pseudomonas, 48
lipozyme, 202
liquid crystals, chiral, 263–84, 375

commercial requirements, 273–9
market, 280, 284
properties and applications, 263–84
synthesis, 280–4
liquid-liquid extraction, 168–71
liquid-membrane fractionation, 171–3
α-lobeline, **93**
loracarbef, 105–11
loxoprofen, **20**

mandelic acid, 96
as resolving agent, 84
D-mannitol, 363
mecoprop, racemic switch, 209
melting point phase diagram,
ibuprofen, 34
naproxen, 23
epoxy alcohols, **374**
membrane-assisted separations, *see* resolutions, membrane-assisted)
membrane bioreactor, in carbocyclic nucleoside synthesis, **49**
membrane/separation technologies, 6–8, 168–78
membranes, enantioselective, 158–62
membranes,
hollow fibre, *see* hollow fibre membranes,
selective liquid, 162–8
ultrafiltration, 7, 158–62
(–)-menthol, 3, 72, 75
menthyl(–) laurate, 176
mephenoxalone, 376
metastable phase, 122,
isolation of, 140–3
metastable zone width, **143**, 144
measurement of, 134–9
seeding within, 152
metaxalone, **365**
methocarbamol, 376
(*S*)-2-(methoxy)pyrrolidine (MMP), 111
α-methoxy-α-(trifluoromethyl)phenylacetyl chloride, (MTPA chloride), *see* Mosher's ester,
4-methoxymandelonitrile, 332
2-(6-methoxynaphthalene)propionitrile, 326
trans-3-(4-methoxyphenyl)glycidic acid, methyl ester 176–7
methyl α-cyclogeranate, **338**, 343, 348
methyl β-safranate, **338**
methyl (*R*)-3-hydroxybutyrate, 246–8, 260
synthesis, 248–9
α-methyl fructofuranoside ligands, 322–5
methyl oxazaborolidine, 396
(2*S*)-(–)-methyl-1-butanol, 162
methyl-*p*-toyl sulphide, asymmetric oxidation of, 383–4
(*R*)-α-methylbenzylamine 56
(*S*)-α-methylbenzylamine, **87**, **93**
in ibuprofen resolution, 34
as chiral auxilliary, 111
resolution of loracarbef intermediates, 110
(*S*)-2-methylbutanol, 280, 283–4
(*S*)-4-(2-methylbutyl)-biphenyl-4′-carboxylic acid, 283–4
(*S*)-4-(2-methylbutyl)-phenol, 283–4
α-methylDOPA, 152
(+)-*N*-methylephedrine, **87**, **93**, 340
trans-3-methylglycidyl tosylate, 374
N-methylmorpholine-*N*-oxide, 46
threo-4-methylthiophenylserine, 358
mexoprofen, **21**
micellar-enhanced ultrafiltration (MEUF), 158, 174
Michaelis-Menten kinetics, in betaine metabolism, 292
miroprofen, **21**
mirtazapine, **172**, **173**
molecular similarity model, 92
monoamine oxidase inhibitors, 375
moprolol, 376
Mosher's ester, 372
multi-stage reactors, in L-carnitine production, 302–3
mutagenesis, 219–20, 293
(–)-*cis*-myrtanylamine, **93**

(+)-1-(1-naphthyl)ethylamine, **93**
(*S*)-*N*-(1-naphthyl)leucine, 165
L-2-naphthylalanine, 200
(*S*)-(+)-naproxen, 6, 19, 22–27, 177, 208, 310, 314,
by classical resolution, 22
by asymmetric hydrocyanation, 331–2
Zambon process, 26
nematic mixtures, 273–6, 280–1
Neurospora crassa, in asymmetric reduction, 258–60
nickel-DIOP, in asymmetric hydrocyanation, 313–4
nicotine, **93**
nisoxetine, 104

Nonsteroidal Anti-inflammatory Profen Drugs, (NSAIDs), 19–38
mode of action, 19
racemic switch, 22
nopol, **162**
noradrenaline, **93**
norbornadiene, **313**
(*R*)-*endo*-norbornen-2-ol, **178**
norbornene derivatives, asymmetric hydrocyanation of, 312
norbornene, **313**
L-(–)-norephedrine, 75, **87**, **93**, **172**, **169**
resolution by liquid-membrane fractionation, 173–4
NSAIDs *see* Nonsteroidal Anti-inflammatory Profen Drugs
nucleation inhibitors, 142, 154
chiral, 148
nucleation, 134
nucleosides, *see* carbocyclic nucleosides
5′-nucleotidase, in lamivudine synthesis, 69–70
number of transfer units (NTU), in liquid-liquid extraction, 169

(*S*)-2-octanol, 275, 280, **284**
olefin epoxidation, 363
organoborane reagents, chiral, 392–3
L-(+)-ornithine, **93**
oxazaborolidines, 4
in asymmetric reduction of ketones, 391–8
oxazolidinones, 3
oxprenolol, resolution with enantioselective membranes, 160–61
oxynitrilases, 5
oxyprenolol, **161**

paclobutrazol, **153**
penicillamine, **226**
penoxaprop, **209**
pertraction, 170
Pharmaceutical Manufacturers Association (PMA), 11
pharmacokinetic evaluation of enantiomers, 11, 13–14
phase transition temperature (PTT), 132–3, 141
α-phenethylamine, 84, **87**
phenoxypropanoic acids, 208–9
(*S*)-1-phenyl-1,2-ethanediol, 275, 280
trans-4-phenyl-2-buten-1-ol, 75, 373
phenylalanine, 3, 160, 174
phenylalaninol, **93**
phenylethanesulphonic acid, as resolving agent, 148
β-phenylglucophosphite ligands, 318–26
β-phenylglyceric acid, 154
trans-3-phenylglycidol, 370, 372
(*R*)-phenylglycine, 165, 167, **172**
in loracarbef synthesis, 107
D-(–)-phenylglycinol, **87**, **93**
(*S*)-4-(phenylmethyl)-2-oxazolidinone, 3
phosphinite ligands, asymmetric, 26, 31, 314
physostygmine, 245
pilocarpine, 245
(–)-α-pinene, as chiral solvent, 154
(–)-β-pinene, in enantioselective membranes, 160
pirprofen, **21**
plasmid stabilisation, 296–7
poly(α-amino acids), 5, 159–60
poly(L-leucine), 5
poly(methyl-L-glutamate), in enantioselective membranes, 160
poly-L-glutamate, in enantioselective membranes, **159**
poly[(*R*)-3-hydroxybutyrate], 248, 260
polymorphic transformations, 125
polymorphs, 120–25, 373
disappearing, 143
polysulphone UF membranes, 160
porcine pancreatic lipase, 364
potassium channel openers, 386
pranoprofen, **21**
preferential crystallisation, of epoxy alcohols, 374
Principle Component Analysis (PCA), 92
process design, L-carnitine, 298–304
Profens, *see* Nonsteroidal Anti-inflammatory. Profen Drugs,
(*S*)-proline, **82**, 96, 396
as chiral building block, 230
D-proline, 96
proline-*N*-carboxanhydride, 396
(*S*)-prolinol, **93**
as chiral building block, 230
propranolol, **165**, **172**, **173**
N-*n*-propyl-D-glucamine in naproxen resolution, 22
trans-3-propylglycidol, by Sharpless epoxidation, 370, 372
protizinic acid, **21**

protonation, enantioselective,
see enantioselective protonation
pseudoephedrine, **87**, **93**, as resolving agent, 84
Pseudomonas
esterase, in CPA resolution, 210
lipase, in carbocyclic nucleoside synthesis, 48
in lactam resolutions, 186–7
pyrethroid incecticides, 1, 147–8, 310
L-2-pyridylalanine, 200
pyroglutamic acid, 396
pyrrolidines, chiral, 225–40

(+)-quinidine, **87**, **93**
(–)-quinine, **87**, **93**

racemates *vs* single enantiomers, 11, 12, 28–30, 66–7, 226, 382
racemic compounds, 15, 23, 34, 148–9, **150**
racemic mixture, *see* conglomerates
racemic switch, 2, 12, 14
herbicides, 209
ibuprofen, 22
racemisation, 256–60
of enantiomers, *in vivo*, 13
in thiamphenicol synthesis, 359
racemisation-recycle, 148, 153, 196, 358
in situ, 94–96, 359–62
in amino acid synthesis, 199
in duloxetine synthesis, 103–5
L-carnitine, 288, 298–304
raclopride, **227**
ranolazine, 376
recombinant biotransformation strain, in L-carnitine production, 296–7
recovery, of chiral proton source, 339
recycling *see* racemisation-recycle
reduction, enantioselective, *see* asymmetric reduction
refractive index, in liquid crystals, 266
regulatory agencies, policy on racemates, 1
remoxipride, **227**
renin inhibitors, 194
resolution,
classical, 54–9, 75, 81–97, 99–115, 144–8, 186, 228–9, 280, 349, 355–8, 385
enzymatic *see* enzymatic kinetic resolution
'dynamic' of unnatural amino-acids, 201–3
kinetic, enzymatic *see* enzymatic kinetic resolution
membrane-assisted, 48, 157–79
resolutions vs stereoselective synthesis, 27, 99–115
resolutions, by fractional crystallisation of diasteriomeric salts – *see* classical resolution
of Chirald, 102
of Lamivudine, 69–71
of naproxen, 22, 27
of pyrrolidine amines, 228–9
recent developments, 4
efficiency, 4, **86**, **87**
entrainment, 355
of L-carnitine, 288
resolving agents,
criteria, 84
rational design, 81–97
reverse osmosis, 158–62
Ritter reaction, 246
rose ketones, 337, 349
Ru(COD)(BINAP), **249**

salbutamol, **172**, **173**
scale-up issues, 69–70, 75, 100, 102, 187–8, 203–4, 214, 217–8, 226–7, 237–40, 259, 298–303, 362, 384, 388, 398
second order asymmetric transformation, 94–6, **148**, 153, 202, 359–62
selectively permeable membranes, *see* membranes, selectively permeable
seproxetine, 104
sequencing, 260
serine, 160
5-sertraline, lipoxygenase inhibitor, 399–400
Sharpless asymmetric dihydroxylation, *see* asymmetric dihydroxylation
Sharpless asymmetric epoxidation (SAE), *see* asymmetric epoxidation
Sharpless epoxidation, in asymmetric sulphide oxidation, 6, 382–7
E-silyl enol ethers, enantioselective protonation of, 341–3
Simulated Moving Bed technology (SMB), 7
sodium (*R*,*R*)-di-*p*-toluoyltartaric acid, (NaDTTA), 113
sodium erythorbate, in loracarbef synthesis, 107–9
solubility curves, measurement of, 131–3

solvates, 120–25
solvent, choice for crystallisations, 131
(–)-sparteine, **93**
spontaneous polarisation, in liquid crystals, 268, 278, 284
stereoselective assay, 15
stereoselective synthesis, *see* resolutions vs stereoselective synthesis
steroid 5-α-reductase inhibitors, 111–14
stilbene oxide, resolution with enantioselective membranes, **161**
strain isolation, 288
L-(–)-strychnine, **87**, **93**
sulphoxides, chiral, 3, 6, 383, 391–8
sulpiride, **227**
super twisted nematic cell, (STN), in liquid crystals, 275
supercritical fluid technology, 8
superfluorinated materials, (SFM), 275
suprofen, **21**
surface-stabilised ferroelectric liquid crystal (SSFLC) cell, **270**, 271–2, 277, 285
symmetrical countercurrent extraction, 7

Taft σ-values, 84, 320–1, **330**
tartaric acid, 84, **96**, 165, 169, 228–9, 358
tartaric esters, 7, 96, 154, 314, 366, 383–4
tartrate-derived phosphite ligands, 314
Taxol, 375
tazeprofen, **21**
terbutaline, **172**, **173**
terpenes, in liquid crystals, 277–8
tetralin musk perfume component, 348–50
tetriprofen, **21**
thermochromic liquid crystals, 272–3, 279, 283–4
thermogravimetric analysis (TGA), 122–3, 151
thermotropic liquid crystals, 264
thexylborane, 66
thiamphenicol,
derivatives, 354
manufacture, 353–62
market, 354
synthetic approaches, 354–8
(*S*)-3-(2-thienylthio)butyrate, 249]–50
(1*S*,2*S*)-thiomicamine, 358–9
tiaprofenic acid, **21**
tomoxetine, 104
Tonalid, 348
toxological assessment of enantiomers, 13
transition temperature, between crystal forms, 122
trityl glycidol, by Sharpless epoxidation, 370, 372–3
Trusopt, 246–8
tryptophan, **159**
tyrosine, 160

ultrafiltration membranes, *see* membranes, ultrafiltration
undesired enantiomers, FDA guidelines for, 11

L-valine, 168, 282, 283–4
Varicella zoster, 50
viloxazine, **365**
vinylarenes, asymmetric hydrocyanation of, 309–32
vitamin B_T, *see* L-carnitine
vitamin D_3, 375
volumetric productivity, in L-carnitine production, 300

whole cell biotransformation, *see* biotransformation, whole cell

X-ray diffraction (XRD), in crystal lattice determinations, 120–29, 133, 151
ximoprofen, **21**

zoliprofen, **21**